75-

The Rhythm of
The Ages

Portrait of Amadeus William Grabau (1870-1946)

Reproduced from H.W. Shimer: *Proc. Geol. Soc. America for 1946,* Plate 6, with permission.

The Rhythm of The Ages

By
Amadeus W. Grabau, S.D.

With introduction by
Albert V. Carozzi
University of Illinois

ROBERT E. KRIEGER PUBLISHING CO., INC.
HUNTINGTON, NEW YORK 1978

Originally published in Peking China 1940
Reprinted 1978, with new introduction

Printed & Published by
Robert E. Krieger Publishing Company, Inc.
645 New York Avenue
Huntington, New York 11743

© *Copyright 1978 (Introduction) by*
Robert E. Krieger Publishing Company, Inc.

Library of Congress Cataloging in Publication Data

Grabau, Amadeus William, 1870-1946.
   The rhythm of the ages.
   Reprint of the 1940 ed. published by H. Vetch, Peking.
   Bibliography: p.
   1.   Geology, Stratigraphic.   2.   Geodynamics.
I.   Title.
QE651.G66   1978   551.7   78-15615
ISBN 0-88275-694-X

INTRODUCTION

by

Albert V. Carozzi
Professor of Geology
University of Illinois at Urbana - Champaign

Amadeus William Grabau left a multifaceted and monumental contribution to numerous fields of geology written in the United States during the first half of his career and in China during the second. His interests ranged from stratigraphy and systematic paleontology to paleoecology and sedimentary petrology. As a pioneer in most of these fields, and with a unique three-dimensional and dynamic approach to the geological record, he conceived all his works as supporting evidence for his theoretical considerations on the evolution of the Earth's crust and the processes of mountain building.

Grabau's life has been reviewed in biographies by H.W. Shimer (1947) and Marshall Kay (1972).

"The Rhythm of the Ages," published in 1940, was considered by Grabau himself his most important book. He must have intuitively realized that he had attempted to solve some of the most fundamental questions in geology, questions which today remain unsolved and are the object of active research in the framework of plate tectonics and global stratigraphy.

In the Preface, Grabau states that the evolution of the Earth has been dominated by two processes which he embodied into two theories: the "Pulsation Theory" and the "Polar Control Theory." The first one represents the rhythmic movement or pulsation of the sea that is of simultaneous occurrence in all the oceans of the Earth (eustatism), the second is the periodic drifting of the Earth's crust. Grabau assumes a slow pulsatory rise and fall of sea level expressed by worldwide (pangeodic) and essentially synchronous transgressions and regressions, a process he sees as the true basis of subdivision of the geologic record. According to this "Pulsation Theory," each positive pulse beat sends marine waters in slow transgression into all geosynclines as sea level rises, each negative pulse causes the withdrawal of the waters as sea level sinks. Each pulsation period (transgression and regression couplet) is followed by an interpulsation period when, because of depressed sea level, only the lowest marginal depressions of the continents or epi-seas remain flooded, unless sea-level sinking is so low that all shallow water areas disappear and only pelagic and abyssal areas of the sea survive.

Grabau perceives a close relationship between this pattern of transgression and regression and the evolution of life represented by the successive associations of fossil faunas and floras. During the period of transgression, marine littoral faunas spread, develop, and diversify; during the regressive period, a narrowing environment leads to struggles for existence and survival of the fittest. In the epi-seas, the survivors undergo further slow modifications after which they display expansive evolution in transgressive waters. Thus, compared to the earlier ones, a seemingly new fauna appears. If, however, sea level sinks very low, littoral faunas may disappear and only pelagic faunas survive. In the course of time, an entirely distinct marine fauna will appear, with little or no relation to the older fauna.

During inter-pulsation periods, long-continued continental conditions favor the expansive evolution of land faunas and floras preserved in the corresponding continental sediments. The latter may, however, suffer appreciable erosion and be locally absent.

Relying on the work of John Joly (1925), Grabau explains in Chapter I and II the periodic oscillations of sea level by the effect of radioactivity in the sima, namely a periodic increase of heat, followed by loss of heat through conductivity of the cold ocean floor. During the expansion of the ocean floor followed by its contraction, the ocean would rise and fall with the changes of position of its floor and transgression would be followed by regression. According to Grabau, Joly estimated a duration of at least 60 million years for the combination of expansion and contraction to occur. Another 30 million years, at least, would be required for the process to start again. This interval of rest corresponds to Grabau's interpulsation periods.

The worldwide and synchronous transgressions and regressions postulated by Grabau's "Pulsation Theory" represent a catastrophist approach to Earth history. It is in line with the thinking of many geologists of today who, if not catastrophists (Gussow, 1976), at least are becoming convinced that many events in the history of the Earth are episodic in nature and have had remarkable widespread effects (Ager, 1973, 1977). The new global stratigraphy maintains that the answer to intercontinental stratigraphic correlation is represented by synchronous worldwide unconformities (Vail, et al., 1977). Although the synchronous age of most unconformities is far from being established—mostly because of the lack of resolution of recording techniques—the hypothesis is stimulating and renews detailed research in order to establish to what extent it expresses reality. Today, the assumed global unconformities are considered to be caused by eustatic sea level changes that in turn are the direct expression of seafloor spreading processes and related plate movements. This interpretation leads back to deep-seated causes and Grabau reasoned along similar lines. With the available knowledge of the time, he assumed that the worldwide transgressions and regressions postulated by the "Pulsation Theory" were a direct expression of the expansion and contraction of the simatic floor of the ocean.

Grabau stressed the point that the periodic rise and fall of the sea level should be used as the true basis for subdivision of the geologic record. Present-day geologists who believe in eustatic cycles (Vail, et al., 1977) also content that such a geologic clock is the best worldwide time-scale for dating significant events in the history of the Earth.

The fundamental effect of the worldwide transgressions and regressions on the evolution of marine and continental faunas and floras is discussed many times in Grabau's work, namely, the crises in the history of life are related to the breaks in the geologic record that in turn are the consequences of the cycles of expansion and contraction of the simatic ocean floor. The fundamental effect of transgressions and particularly of regressions has been reaffirmed recently by D.V. Ager (1977), and although it is now related to episodes of seafloor spreading and plate movements, both authors have reached similar conclusions.

The second process assumed by Grabau for his synthesis of the Earth is the "Polar Control Theory." At first, the designation seems somewhat misleading because it assumes a displacement of the poles; on the contrary, Grabau considers only a periodic shifting or drifting (in the sense of Wegener and Du Toit) of the Earth's crust or sial under the action of the rotation of the Earth on an axis of constant position and hence under permanency of poles and climatic belts (Chapter I). This drifting brings diverse parts of the sial crust into successively new climatic belts and eliminates the necessity of assuming periods of cooling or of warming all over the Earth (Preface). In other words, the axis of rotation of the Earth and the climatic belts have always been as they are today, only the sial crust drifts. This displacement also allows temporary contacts between continental masses and related exchanges of terrestrial faunas and floras.

"The Rhythm of the Ages" and its splendid illustrations present Grabau's concept of the evolution of the Earth's crust by successive paleogeographies from the Cambrian to the Recent. Its originality lies in the elaborate discussion of the evolution of the pangea during the Paleozoic before its major Mesozoic breaking up. In this manner Grabau traced the work of Wegener and of Du Toit into a more remote past. Of utmost value for a demonstration of the original juxtaposition of North America with Greenland and Northern Europe is Grabau's reconstruction based on distinct trilobite faunas of two more or less parallel geosynclines (the Taconic and the Appalachian) that extend in a continuous way from one continent to the other with the Atlantic Ocean closed and that portion of the pangea reconstructed (Chapter VIII, Fig. 18 and 19). The importance of this spectacular proof of continental drift has been recently emphasized (Marvin, 1973). At present, when the reconstruction of plate boundaries attempts to go back to the same periods, a comparison between modern ideas and Grabau's sketches is an illuminating experience.

Since during the Paleozoic, the pangea did neither break apart nor undergo major dislocations, it is natural that Grabau would be greatly concerned with the processes that led to its original formation. He considered that the Earth consisted of concentric spheres (Chapter I) and that, in particular, it had originally a continuous crust of sial covered by a universal ocean (the great primitive shoreless sea of the Earth). The fundamental question would pertain to the agent that caused the change of the primitive sphere of granitic sial into a southern hemispherical mass of crushed and folded gneisses or pangea generating at the same time a northern hemispherical mass of oceanic waters or panthalassa floored by the sima.

Few scientists have attempted to speculate on how land originally emerged from the ocean as a single continental mass, although such a speculation is logically required before discussing how such a mass could be broken up into individual continental nuclei to be distributed by drifting. Grabau dismisses as causes of elevations and depressions original differences of density in the crust, foundering of masses of sial into sima or the hypothesis that the Pacific Ocean is a scar left when the moon was supposedly torn away from the Earth. He states that only a compression of the original crust into the pangea is acceptable (Chapter I). This concentration of the crust by displacement to one side and exposing of the sima on the other could not have taken place without outside intervention. Thus Grabau assumes an extra-telluric force which must have acted in a poleward direction. Indeed, only such a force could affect the Earth since the latter is revolving on its axis while presenting constantly a new surface for attraction to any force that pulls from any direction other than that of the Earth's axis of rotation. Therefore, Grabau assumes the influence of a stellar body that occupied with respect to the South Pole the same position as the Polar Star versus the North Pole. This celestial body approached the South Pole and remained in such a position long enough to draw the light crust of sial into a large southern landmass that consequently underwent compression, intense folding, and metamorphism. After peneplanation of the first folds, the stellar body is supposed to have also played an important role in the original location of the marginal geosynclines of the pangea. Due to the thinness of the crust, all these processes did not disturb otherwise the position of the Earth in space. The assumed stellar body may have long since vanished into outer space or may have disintegrated altogether.

When "The Rhythm of the Ages" was published in the forties, the intervention of any kind of extra-telluric agent in geological processes was hardly conceivable and hence heavily criticized. Today, more than thirty years later, with the development of space investigations and the increasing knowledge about other planets and satellites, such extraterrestrial influences are seen under a completely different light. In fact, geologists interested in the Precambrian and the original formation of the crust are not at all opposed to large-scale meteoritic actions. For instance, Goodwin (1974) postulates that the Earth underwent giant impacts similar to those that caused the lunar maria, probably at about the same time (4.2 to 3.8 my ago). Similar to their lunar counterparts, the terrestrial mega-impacts were asymmetrical and of cumulative crescent shape. The large-scale impacting initiated deep-seated, long-lived, global thermal convection systems including the mantle plumes that generated in stages the modern continents and ocean basins. According to Goodwin, the crescent-shaped pangea is an outgrowth and reproduction of the mega-impact pattern. Ocean basins are of two main types, either ancestral, that is the negative consequence of the giant impacts, such as the Pacific Ocean, or extentional, that is developed as a consequence of the fragmentation and drifting of sialic cratonic blocks such as the Atlantic Ocean.

Grabau having assumed that the pangea had been shaped by the action of a stellar body during the Early Precambrian proceeds to unravel its evolution through the Paleozoic using the history and succession of glaciations as one of the leading arguments. Polar shifting in a given direction is interpreted as drifting of the sial in the opposite direction. The pangea seems to have rotated in various ways over the South Pole before creeping northwards and undergoing its dislocation into individual continental nuclei. The rotational displacement of the pangea was sometimes clockwise, but more frequently counterclockwise under the effect of the rotation of the Earth, modified perhaps by local changes of specific gravity. Changes in the direction of movement are attributed by Grabau to the temporary existence of pivotal points on which the sial mass was anchored and which were produced by the intrusion in the crust of igneous masses of deep-seated origin (Chapter I). The location of the anchoring intrusions would determine the direction and amount of drifting during a certain time. It would also control the distribution of pressures and frictions that in turn determines the development of geosynclines and therefore their final deformation during orogenies. Hence, for a given geosyncline in the proper position with respect to an anchoring point, compressions and rifting could occur simultaneously.

Before discussing Grabau's concept of the relationship between sial drifting, geosynclines, and mountain building, it is interesting to mention more of his ideas about the anchoring intrusions (1940). He suggests that the location of such intrusions could be found by the fact that they brought to the surface large amounts of basic volcanics as well as heavy metals. Perhaps the famous mining centers of magnetic iron, nickel, and gold could give clues to the location of these intrusions. Conversely, an analysis of the sial movements may lead to further discoveries of anchoring points, hence of valuable ore deposits. These anchoring basic intrusions from the mantle into the crust, as visualized by Grabau, can be compared, when considered in their context, with the present-day ideas of hot spots or plumes that are manifestations of convection in the lower mantle. These points of upwelling when located under the continental lithosphere generate uplifts crested by volcanoes that evolve into triple junctions leading to continental breakup and ocean opening. It is interesting to recall that many failed arms of triple junctions have base metal mineralizations such as syngenetic copper in particular. Grabau had naturally an entirely different interpretation since for him the crust of sial was drifting under the effect of the rotation of the Earth, hence the ultrabasic intrusions had a passive rather than an active role; but regardless of their role, their aspect at the surface of the crust was identical, namely basic volcanism and mineralization.

From his numerous observations, Grabau noticed the coincidence between worldwide interpulsation periods and times of minor and major orogenies. He interpreted to his satisfaction this fundamental coincidence by the combination of the "Pulsation Theory" and the "Polar Control Theory." In order to understand his interpretation, it should be recalled that he visualized geosynclines as structures that belonged entirely to the sial crust and that were located in a marginal or internal position (Chapter III). Geosynclines are long, narrow, and asymmetrical troughs of subsidence with a parallel elevated area of old-land adjacent to them (in the marginal geosynclines it is in a seward position) that supplies the clastics filling the trough. Since the sial is immersed for more than 6/7 of its thickness in the sima that is heavier and resistant, the displacement of the sial results in the deformation of the frontal margin of any segment of moving sial. For a marginal location, this process begins as an uplift of the continental margin and a corresponding subsidence of the belt just landward from it that is being filled by river supplies. The former produces the old-land and the latter the geosyncline. When sinking and accumulation of sediments have proceeded to such an extent that the strength of the crust beneath the geosyncline is exceeded, further pressure leads to an orogenic event folding the sediments into a mountain range juxtaposed to the old-land, destroying or not the geosyncline as a whole. While orogeny proceeds on the side of pressure of any considered portion of the sial, on the opposite side drag produces tension which results in rupturing (rifting) with associated volcanism.

Grabau then turns to the timing between orogenies and the pulsations of the sea-level. Although there is no necessary relationship between sial drifting and the rise and fall of the sea-level, a minor or major orogenic event always occurs during the interpulsation period. Grabau considers this relationship as a fundamental law of the genesis of folded mountain ranges. In other words, when the folding occurs, it takes place while the geosynclines are drained (Chapter I).

Nowhere does Grabau state that orogenies are necessarily worldwide like the unconformities due to the pulsations of sea level, but he says that during every interpulsation period of the Paleozoic, one or more geosynclines in some parts of the Earth experienced minor or major orogenies. At any rate, he sees the need for some kind of articulation between the "Pulsation Theory" that is responsible for global oscillations of the sea-level and the "Polar Control Theory" that regulates the drifting of the sial that in coincidence with interpulsation periods is marked by minor or major orogenic events.

Grabau feels it is important to distinguish between orogenic incidents or minor orogenies occurring during an ordinary interpulsation period and affecting only a part of a geosyncline, and orogenic revolutions or major orogenies occurring near the close of an era and affecting an entire geosyncline and generally terminating its existence (Chapter V and XXIV).

In this volume, orogenic incidents are interpreted by Grabau as an indirect effect of the swelling and contraction of the simatic floor of the ocean. This connection between pulsations and drifting is explained as follows. During the positive phase of a pulsation, the sima not only swells but also stretches, and the rifting that occurs undergoes injection from below. During the negative phase of the pulsation, the sima sinks to a smaller radius, but having become enlarged by the addition of volcanic material in the rifting, it will exert a lateral or tangential pressure. The latter directed from the ocean against the pangea is relieved by local folding in the marginal geosynclines. Whether this process can explain local orogenies in the geosynclines located inside the pangea such as the Appalachian of the Caledonian by the transmission of the compressive forces over thousands of miles from the ocean margin of the time is considered by Grabau himself as an unanswerable question (Chapter V and XXIV).

For major orogenic events such as the Appalachian revolution at the end of the Paleozoic or the Alpine-Himalayan in the Mid-Tertiary, Grabau thinks that a more powerful force is required (Chapter V). He is somewhat inclined, at first, to consider again a cosmic force (later abandoned, 1940), then prefers some direct effect of the drifting of the sial crust that would provide geometrical conditions with respect to anchoring points for stronger compressional effects and would lead to the final closing of geosynclines and to the migration of the belts of deformation.

In conclusion, Grabau found a unifying mechanism relating minor and major orogenic events to the worldwide interpulsation periods. If his reasoning were carried a little further, by considering the major orogenies as worldwide, a coincidence would exist between the latter and the interpulsation periods. Since pulsation-interpulsation couplets are directed by the rhythmic evolution of the sima, then the latter controls all major events at the surface of the Earth, namely worldwide eustatic oscillations and simultaneous orogenies.

Grabau's catastrophist approach leads directly to the question underlying the whole concept of today's global stratigraphy, namely, what is the reason for the coincidence of global unconformities with times of worldwide orogenies? At present, the unifying mechanism is considered to be seafloor spreading and related plate movements, essentially the same deep-seated cause postulated by Grabau in the context of the knowledge of his time.

Urbana, Illinois, March 1978

REFERENCES

1. AGER, D.V., 1973, *The nature of the stratigraphical record,* John Wiley and Sons, New York, 114 p.
2. _____, 1977, Catastrophism reviewed: *Geotimes,* v. 22, No. 2, p. 13.
3. GOODWIN, A.M., 1974, Giant impacts, ocean basins and continental crust (abstract): *Geol. Assoc. of Canada Annual Meeting, Program and Abstracts,* pp. 36-37.
4. GRABAU, A.W., 1940, Present status of polar control theory of Earth development: *Pan-American Geologist,* v. LXXIII, No. 4, May, pp. 241-253.
5. GUSSOW, W.C., 1976, Sequence concepts in petroleum engineering: *Geotimes,* v. 21, No. 9, pp. 16-17.
6. JOLY, J., 1925, *The surface history of the Earth:* The Clarendon Press, Oxford, 192 p.
7. KAY, M., 1972, Biographical essay on Amadeus William Grabau: *Dictionary of Scientific Biography,* v. V, pp. 486-488.
8. MARVIN, U.B., 1973, *Continental drift—The evolution of a concept:* Smithsonian Institution Press, Washington D.C., 239 p. See in particular pp. 110-114.
9. SHIMER, H.W., 1947, Memorial to Amadeus William Grabau: *Proc. Geol. Soc. America for 1946,* pp. 155-166. Pl. 6.
10. VAIL, P.R. *et al.,* 1977, Seismic stratigraphy and global changes of sea level *in* Seismic stratigraphy—application to hydrocarbon exploration (C.E. Payton, editor): Am. Assoc. Petroleum Geologists Memoir 26, pp. 49-212.

THE RHYTHM OF THE AGES

PLATE I

Three Photographs of a 14-inch globe covered by green baize, to represent the sial-sphere (1 mm + in thickness). *a* The undisturbed sial; *b* The same, side-view after pole-ward withdrawal of sial; *c* The same (larger polar view), showing primitive mountain systems. Compare text-fig. 4, p. 11.

THE

RHYTHM OF THE AGES

EARTH HISTORY IN THE LIGHT OF THE
PULSATION AND POLAR CONTROL
THEORIES

BY

AMADEUS W. GRABAU s. d.

FELLOW OF THE GEOLOGICAL SOCIETY OF AMERICA, THE
PALÆONTOLOGICAL SOCIETY, AND THE NEW YORK
ACADEMY OF SCIENCES; CORRESPONDENT
OF THE PHILADELPHIA ACADEMY
OF SCIENCES ETC.
RESEARCH PROFESSOR OF GEOLOGY, CHINA FOUNDATION
FOR THE PROMOTION OF EDUCATION AND CULTURE;
CHIEF PALÆONTOLOGIST, GEOLOGICAL SURVEY OF
CHINA; RESEARCH ASSOCIATE IN GEOLOGY,
ACADEMIA SINICA, ETC.

HENRI VETCH · PEKING
1940

DEDICATED
TO
MY WIFE MARY ANTIN
MY DAUGHTER JOSEPHINE
AND MY GRANDDAUGHTERS
MARGARET AND ELIZABETH ROSS

PREFACE

Among the multitude of phenomena that have influenced the development of our earth, two groups command a dominant position, because they embody the laws which have controlled its evolution and that of its inhabitants throughout the ages.

The first of these is the rhythmic movement or pulsation of the sea, which, unlike tidal phenomena, is of simultaneous occurrence in all the oceans of the earth. This is shown by the record of transgressions and regressions everywhere in the strata of all continents, and I have embodied it in the law of slow pulsatory rise and fall of the sea-level in each geological period, or more briefly: *The Pulsation Theory.* Each positive pulse-beat sends the marine waters in slow transgression into all the geosynclines as the sea-level rises, and each negative pulse-beat causes their withdrawal as the sea-level sinks. Each Pulsation Period is followed by an Interpulsation Period, when, because of the depressed sea-level, only the lowest marginal depressions of the continents, the epi-seas, will remain flooded, unless the sea-level sinks so low that all the littoral or shallow water districts of the earth disappear, while only the pelagic and abyssal districts of the sea survive.

During the periods of transgression the marine littoral fauna spreads and develops in ever increasing luxuriance and diversification. During the regressive periods, a narrowing environment will bring about a constantly increasing struggle for existence, with the elimination of the less adaptable, and the survival of the fittest which must live in an environment of constantly decreasing area, and a condition of existence of progressively increasing rigor. The final survivors in the epi-seas will undergo further slow modification during the continuance of the Interpulsation Period, after which, when the succeeding pulsation again induces the waters to advance, they will undergo expansive evolution. Thus a new, and on the whole, distinct fauna will appear and leave its record in the strata of the new period. Careful observation may reveal its relationship to members of the earlier fauna, for it is from them that the new forms are descended.

If, however, the sea-level sinks so low that no shallow water remains on the continents, and if the land remains out of reach during the next succeeding period or periods of rising sea-level, then the littoral forms will mostly disappear and only pelagic types survive, and in the course of time a wholly distinct marine fauna—such as that of the Mesozoic ammonoids will appear, with little or no relation to the older fauna.

Conversely, long-continued continental conditions during the Interpulsation Periods, especially those including Pulsation Periods where the sea-level does not rise sufficiently to reflood the lands, favor the expansive evolution of the land-fauna and flora, whose record will be preserved in the continental sediments of the Interpulsation Periods.

The second significant concept is that of periodic shifting of the earth's crust or sial-sphere, through the impetus given by the rotation of the earth on an axis of essential constancy of position, and therefore under permanency of poles and climatic zones. This is the law of polar control of the progress of earth development, or more briefly, the *Polar Control Theory*.

This shifting brings diverse parts of the sial-crust into polar position, and in the movement develops geosynclinal depressions on the side of pressure of the moving sial against the resistant sima, in which it is largely sunk. On the opposite side of the moving mass, drag produces tension which may result in rupture and in volcanism.

Translation of the polar site through shifting of the crust is accompanied by the geographic migration of the polar ice-caps at each end of the axis, and of the equatorial and intermediate climatic belts, and eliminates the necessity of assuming universal refrigeration to account for continental glaciation in regions now tropical, or a sufficient rise in temperature all over the earth to account for tropical conditions in regions now within the arctic. In short, under polar control, as here visualized, the axis of the earth and the climatic zones of the surface have always been as they are to-day — only the sial crust shifts and brings different geographic regions into successively new climatic belts. And when a region previously glacial is shifted into a warmer position, it experiences what is commonly called an Interglacial Period —-which may or may not correspond to an Interpulsation Period, since apparently there is no necessary relationship between sial-shifting and the rise and fall of the sea-level.

I am well aware that many of the conclusions herein reached are widely divergent from current concepts, and for the promulgation of these ideas I assume full responsibility, confident that the principles will bear critical examination and test, though changes in detail may be required by further discovery and fuller research. For greater detail the student is referred to my larger volumes, *Palæozoic Formations in the Light of the Pulsation Theory*, four of which have so far appeared, and to the other papers cited in the text and listed at the end of this volume.

I wish here to take the opportunity to acknowledge the assistance I have received in the preparation of this book. First of all credit is due to my former secretary Miss Olga Kravchenko, whose careful comparison of the manuscript and typescript, and reading of part of the proof has saved me much irksome labor. The index was prepared by her in collaboration with Mr. L.P. Chao. The proof of the later chapters was carefully read by my present secretary Miss Trude Schack, and the entire proof was likewise read by Mr. J. Hope-Johnstone and in part by Mr. Henri Vetch.

Important new data, published in the recent scientific literature, and not at present available in Peking, were supplied by Alice Woodland-Grabau of Boston, under a grant from the Richard Alexander Fullerton Penrose Jr. bequest, by the Geological Society of America.

The typing was the work of Mr. L. P. Chao, and the illustrations and maps were all redrawn by Mr. J. Yang from original sketches or other illustrations. The press and map-work was done at the Poplar Island Press under the supervision of my friend Professor Vincenz Hundhausen, and speaks for itself. For all this assistance I tender my thanks.

Peking, December 31, 1939

CONTENTS

ILLUSTRATIONS

TEXT FIGURES

PLATES AND MAPS

TABLES

CHAPTER I

THE CRUST OF THE PRIMITIVE EARTH

1

THE oldest rocks, which form the visible framework of our earth, have all been subjected to great disturbances and profound alteration or metamorphism. We generally speak of them as *crystalline rocks*, and wherever they are in contact with younger rocks that have not been altered, or not to the same degree, the contact is seen to be a discordant one.

There are, of course, many crystalline rocks that do not form a part of this oldest complex. Most igneous rocks of younger age show a crystalline texture. But, even when we exclude these, we must recognize the fact that many younger rocks of "sedimentary" origin have become altered into crystalline schists, marbles or other types, which may be difficult to separate from the older crystalline series, while their age relations may remain wholly indeterminate.

But where we can identify these oldest rocks, we find that not only are they of crystalline character, but that they show evidence of having been intensely crushed; and if they have a schistose or bedded character, these layers show evidence of strong folding and often of extreme distortion such as is shown in our illustration (Fig. 1); and always and everywhere these folded crystalline rocks have been subjected to intense erosion, which has cut off the tops of the folds.

In this manner an erosion surface was produced that was *almost* a smooth erosion plane[1] (a "peneplane"), before the younger rocks were laid upon it. That is the essence

---

1) The spelling *plane* is here adopted to designate a flat surface of erosion, as that has been "planed down"; thus: "erosion plane", "peneplane", etc. *Plain*, on the other hand, is reserved for level surfaces primarily of construction, such as, plain of deposition, river flood-plain, etc.

or real significance of the great unconformity which marks the basal Palæozoic contact with the older rocks in all parts of the earth where it is preserved.

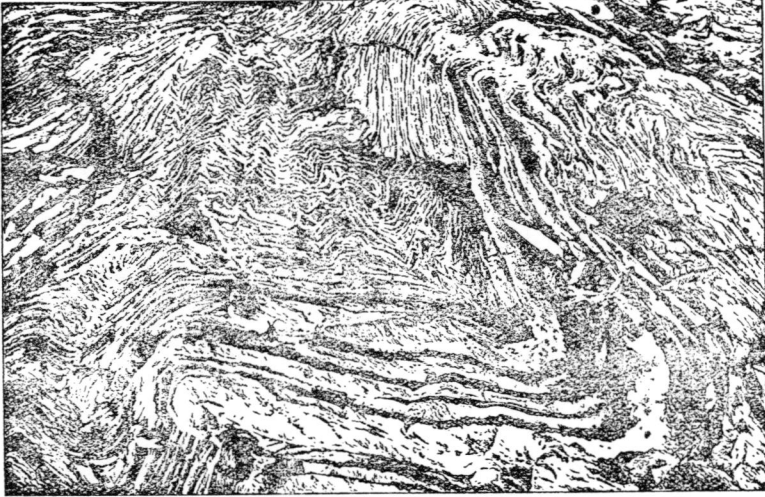

Fig 1. Contorted folds of interbedded crystalline limestone and granular amphibole. Haliburton, Canada. These beds form the typical *Grenville Series* of pre-Palæozoic sediments.

In composition the ancient crystalline rocks range from "acid" to "basic" or even "ultra-basic" types. The former are represented by granites among the igneous rocks, and by the gneisses derived from them by metamorphism, as well as by many schists, etc., while the basic rocks are illustrated by basalts, gabbros, etc., and their metamorphic derivatives.

The acid rocks are generally distinguishable by their prevalent lighter color, while the basic rocks, which are more often found as intrusions into the others, or more or less mingled with them, are of prevailingly dark color, and formed of heavier minerals.

Both types of rock have often been folded together, and crushed and distorted into a complex foliated or banded mass, which is generally spoken of as the "Fundamental Complex".

The lighter, or acid, rocks are characterized by a preponderance of the oxides of silicon and aluminium, as is shown by an average analysis of granite, while the basic rocks, of which we may select basalt as a type, are richer in silicates of magnesium and in iron compounds.

This is shown in the following analyses of average granite (1) and average basalt (2):

TABLE I

COMPOSITION OF AN ACID AND OF A BASIC ROCK [2]

	(1) Average granite Mean sp. gr. 2.67	(2) Average basalt Mean sp. gr. 3.00
SiO_2 (Silicon oxide)	70.47	49.65
Al_2O_3 (Aluminium oxide)	14.90	16.13
MgO (Magnesium oxide)	0.98	6.14
CaO (Calcium oxide)	2.17	9.07
Na_2O (Sodium oxide)	3.31	3.24
K_2O (Potassium oxide)	4.10	1.66
Fe_2O_3 (Ferric oxide)	1.63	5.47
FeO (Ferrous oxide)	1.68	6.45
TiO_2 (Titanium oxide)	0.39	1.41
P_2O_5 (Phosphorous pentoxide)	0.24	0.48
Total	99.87	99.70
Fe (Iron)	2.45	8.86

In the granite, which represents the acid type of rock, the chief constituents are silica and aluminium compounds, and this is expressed by the inclusive name SIAL, applied to them, and derived from the symbols of the two principal elements (Si + Al).

In the basalt, on the other hand, which represents the basic rocks, while both silica and aluminium compounds still dominate, though to a lesser degree than in granites, there has appeared, among others, a comparatively high percentage of magnesium, calcium and iron compounds (ferro-magnesian silicates), which give the rock its darker color and higher specific gravity.

Among other changes, free silicon-oxide, in the form of quartz, is absent, and the feldspars range toward the

2) LEITH and MEAD : *Metamorphic Geology*, p. 66.

basic end of the scale. To rocks of this type, characterized
by silicates of magnesium (Mg) and other substances, the
general name SIMA is now applied (Si + Mag).

2

Most geologists are now in general agreement that
the superficial part of the continents is composed chiefly
of *sial* rocks and their derivatives, and that this "floats"
in a deeper layer of *sima*, in which it is partly submerged, as is an iceberg floating in water. (Fig. 5, p. 17)

It has been estimated that the submerged part of the
sial in the sima represents about 8 times that which projects above it. In other words, taking the average height
of the continents as 820 meters, and the average depth
of the sea as 3,800 meters, the part of the continental
block which projects above the level of the sima, or, as
it has been picturesquely termed by Joly[3], its "freeboard", is
about 4.62 km, while its "draught" is about 35 km,
that being the depth to which it is immersed in the sima.

The upper surface of the sima forms the floor of the
Pacific, Indian and Arctic Oceans, but not that of the
Atlantic, which, according to Gutenberg[4], is believed to be
floored by a thin layer of sial, which covers and floats in
the sima beneath.

Although we have taken granite as the typical acid
or sial rock, we must emphasize the fact that the old rocks
of the earth are no longer of the character of that crystalline rock with the individual mineral crystals intact
(holo-crystalline). Even if we may assume that some such
condition was the original state, the briefest examination
of an area of such ancient rocks will show that they are
now gneisses and schists, derived by metamorphism from
the original sial rock. The foliation of these rocks and
the complex folding and distortion of the foliæ suggest the
changes which have affected them, and the powerful nature
of the force which was instrumental in producing these
changes.

3) JOLY : *Radioactivity and the Surface History of the Earth*, p. 17;
   VAN DER GRACHT : in *Theory of Continental Drift*, p. 13.
4) GUTENBERG : *G. S. A. Bull.*, pp. 158-161.

3

If in imagination we were to restore the original folds, which, as previously noted, have evidently been bevelled and truncated by erosion, and then to smooth out the folds as we might smooth a crumpled newspaper, and also to eliminate the other effects of compression, *i. e.* shearing and foliation, we should probably find that the mass of the sial would become sufficiently extended to cover all the sima of the ocean-floor, and would thus form a continuous shell of sial rock enveloping the entire earth and completely enclosing the sima, though partly immersed in it.

It is commonly held that this original shell of sial had a thickness of nearly 20 miles (approximately 32 kilometers). The sima shell has been estimated to have possessed a thickness of 680 miles (nearly 1,100 kilometers), and since it is apparently still a continous shell, though only with a partly "exposed" upper surface, which forms the floor of the oceans, its present thickness may not be materially different from what it was in earlier days. [5]

We may more easily visualize these relationships if we reduce these figures to a scale which we can more readily comprehend. Taking, then, the diameter of the earth in round numbers as 12,800 km, and the thickness of the original sial crust as 30 km, the relationship is nearly as 427 is to 1. Thus, if we reduce the dimensions of the earth to a globe 427 mm or about 17 inches in diameter, the entire sial would have a thickness of only 1 mm, or the average thickness of match-box wood. Of this nearly 7/8 is immersed in the sima. Or taking the thickness of the immersed part of the sial of the present continental blocks (product of the lateral compression, and consequently of vertical swelling) as 35 km and the part projecting above the ocean floor (above the sima surface) as approximately averaging 4.62 km, we have a total thickness of crust of 39.62 km. This would be in the proportion of 1 to 323. A 323 mm globe would be slightly less than 13 inches in diameter. On such a globe the continental blocks would have a thickness of 1 mm,

---

5) Since the temperature of the sub-crustal material is probably too high to permit crystallization, the lower and thicker portion of the "sima shell" has been called "vitreous sima" (Daly).

of which about 0.86 mm is immersed in, and 0.14 mm projects above, the continuous surface of the sima.

4

Investigations, taking account of the relative specific gravities of these outer shells of the earth (2.75-2.90 for the sial, 3.1-4.75 for the sima), and the specific gravity of the earth as a whole (5.52), have shown a discrepancy, which can only be compensated for by the existence of a vastly heavier core within the earth, for which the sima and the sial form covering shells.

There is probably an intermediate "pallassite zone",[6] between the sima and the inner core, which occupies the depth between 1,200 and 2,900 km. Below, and within this, is the heavy central core of the earth, with a radius of 3,500 km. This core probably consists very largely of such heavy substances as nickel (Ni) and iron (Fe) and similar metallic elements, including gold (Au), and has on that account received the name of the NIFE-SPHERE (Ni + Fe).

We may, then, picture the primitive earth, as formed of- a central heavy core, a "Nife-sphere", including perhaps a deeply buried heart of gold, surrounded by the "Sima-sphere", with perhaps a transitional sphere of material intermediate in composition and density. The "Sima-sphere" is in turn covered by the "Sial-sphere", which may be regarded as a scum of lighter substances, which crystallized out on the surface of the heavier foundation rock. Then, outside of this highest rock-sphere, is the

---

6) Investigations by the seismograph of the rate of transmission of earth-quake waves indicate that this intermediate layer consists of three zones of increasing density, which in descending order are: a) a 500 km zone, between 1,200 and 1,700 km, with a density range from 5.0 to 5.5; b) a 750 km zone between 1,700 and 2,450 km with a density range of 5.5 to 6.0; c) a 450 km zone between 2,450 and 2,900 km ranging in density from 6.0 to 6.5. The temperature increase from the top of a to the bottom of c is given by Daly as from 5.000° C (?) to 10,000° C (?). The "iron" zone at the center, with a radius of 3,470 km, ranges from a density of 10.5 at 2,900 km to 12.5 at 6,370 km. The zone composed of crystallized sima Daly regards as very thin, and as having a density of 3.0; while the remainder, down to 1,200 km, the vitreous sima, has a density ranging up to 4.5 and a temperature range from 1,400° C upward (DALY: The Changing World of the Ice Age, pp. 112-115).

sphere of water, which, if the continents were flattened out to a continuous level surface, and if the amount of water were not essentially different from what it is today, would form a universal ocean, or primitive "Hydro-sphere", the great primitive shoreless sea of the earth, with an average depth of 2.64 km. (Fig. 2).

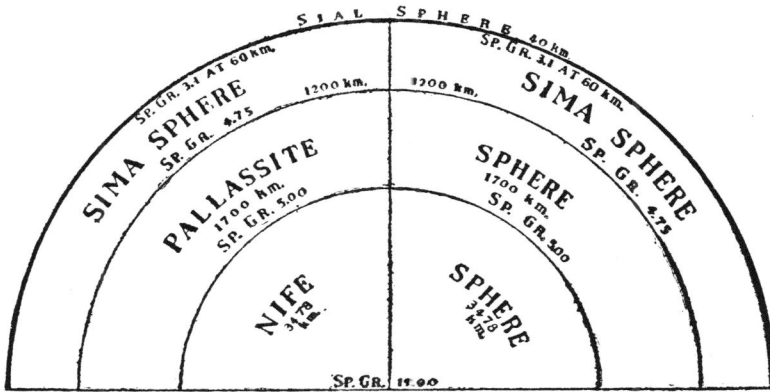

Fig 2. Section of the earth hemisphere drawn approximately to scale, to show the succession of concentric shells.

And last of all, around the shell of water, is the gaseous shell, or "Atmo-sphere", which, however, probably differed from the air of today in containing much of the carbon (as carbon dioxide) which is now locked up in the vegetation and in the coal-beds of the earth.

This primitive earth of many concentric shells turned on an axis which maintained throughout all time essentially the inclination to the plane of its ecliptic that it holds today, such variations as existed being rhythmic. There is no evidence that this relationship has been otherwise since the beginning. We shall make this concept the basis of our discussion.

If the axis of the earth always had an inclination of approximately 66½ degrees to the plane of the ecliptic (23½ degrees from the vertical[7]), then there have always been polar zones more or less glaciated and an equatorial zone of high temperature, and between these, other zones,

---

7) It varies within slight limits, the average obliquity of the ecliptic for 1900 was 23° 27′ 8″.

characterized by a regular poleward decrease in mean temperature. As Prof. G. C. Simpson[8] says, "the evidence of this zonal arrangement . . . . may be expected in every geological period. Also there must always have been an alternation of seasons from summer to winter, and back to summer."

5

We naturally ask, What was the agent that caused the transformation of this primitive spheroid[9] of concentric shells into the earth, as we know it, of separate lands and seas, with part of the sial-sphere removed to expose the surface of the sima at the bottom of the oceans? What brought about the concentration of the sial through compression into the continental masses of today, thus forcing the hydro-sphere to break its continuity and to occupy the newly created ocean-basins? In other words, What caused the lands to emerge from the sea?

Here we enter the realm of speculation, for though we may feel certain that such a separation of land and sea took place in the past by compression of the sial crust, with the resulting production of the deeps for the concentration of the sea-water, the cause we can only infer.

It was formerly thought, that in the original rearrangement of the materials composing the earth, certain regions acquired matter of greater density; that these on solidification shrank to a shorter radius than the neighboring masses, whose greater radius balanced their lighter character, and that thus these several masses were in isostatic equilibrium. Later it was believed that the missing masses of the sial could have dropped down and been absorbed in the sima, and have thus formed the depressions into which the water drained, with the consequent emergence of the land.

Many considerations, however, not the least of which is that the lighter sial rock cannot founder in the heavier sima, any more than an iceberg can normally sink to the bottom of the sea, have led to the abandonment of

---

8) SIMPSON, G. C. : *Discussion of Geological Climates,* 1930, p. 299.
9) The equatorial axis of the earth spheroid is today almost exactly 1/297 longer than the polar axis, but it may have varied slightly in former time. (Geodetic method indicates compression, as 1 : 293 ; astronomical method indicates compression as 1 : 300.)

this theory. Finally, the theory that the Pacific basin is the scar left when the moon was torn from the earth can now be discarded, though it has recently been revived in Germany. There is no real evidence for such an occurrence.

There remains then only the theory of compression of the original crust into isolated land-blocks, or into one hemisphere of land, which subsequently broke up into continents, which, in their turn, floated apart to their present position.

This is the Theory of Continental Drift, first proposed by F. B. Taylor in America, and by Alfred Wegener in Austria.

Wegener has elaborated it in his book *The Origin of Continents and Oceans*, the English translation of which appeared in 1924, and it is commonly known as the "Wegener Hypothesis of Continental Drift".

This is sometimes spoken of as a theory of migration of the poles, but while this is a convenient expression, it must be clearly understood that polar migration is relative. What is really meant is the shifting of the sial crust until a given region is brought over the pole. The poles are stationary in position, so far as the earth is concerned.

Recently the English geologist A. L. Du Toit[10] has again discussed the question and has given the theory its latest expression, while marshalling a wealth of facts in its support.

That there are still many geologists and geophysicists opposed to this interpretation, is, of course, to be expected, for the search for other explanations has not yet come to an end, and until all other theories fail, the theory of continental drift will not come into general recognition.

6

Even more difficult to explain, than the breaking-up of a single mass into fragments, and the drifting apart of these blocks to form the foundations of the present-day continents, is the explanation of the original production of the single mass, or PANGÆA, by the concentration of the former holo-sphere of granitic sial into a hemi-sphere of compressed and crushed gneisses and schists. Creep and

10) DU TOIT: *Our Wandering Continents.*

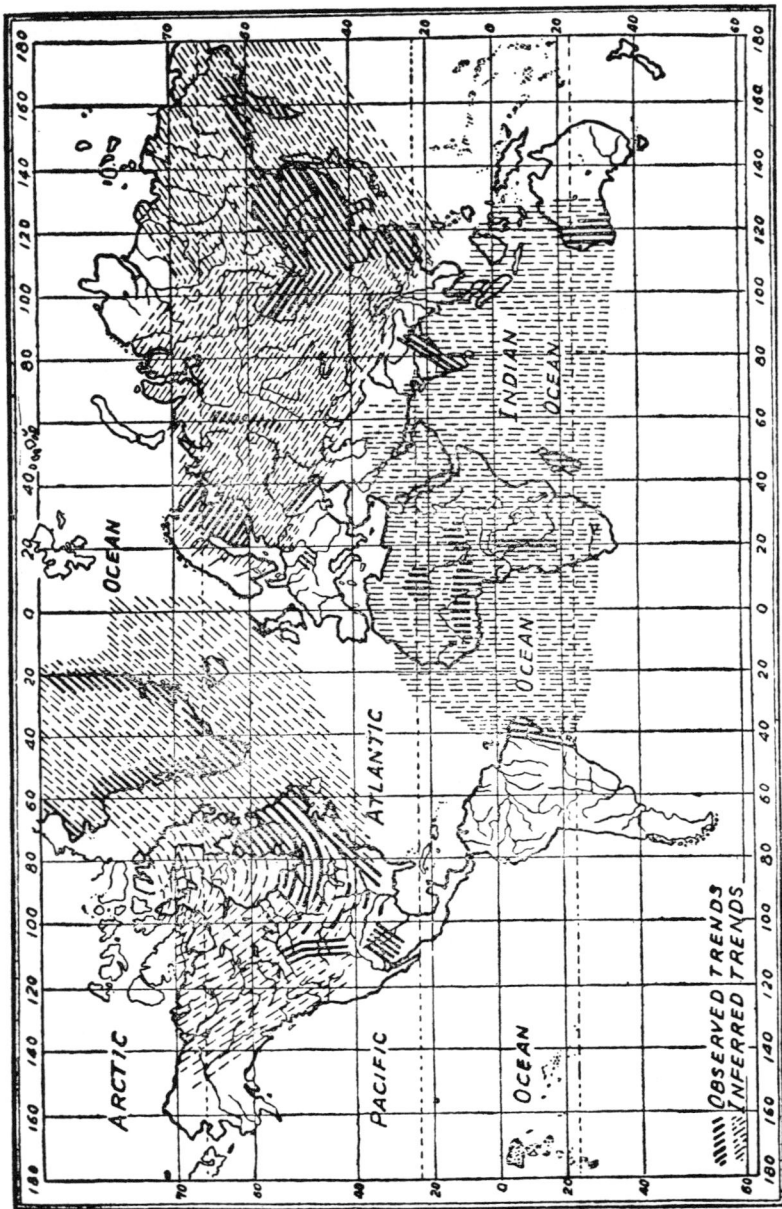

Fig. 3. Trend-lines (observed and inferred) of the pre-Palaeozoic rocks, indicating the directions of the ancient folds after peneplanation. Mercator projection. (After Ruedemann in, *Proc. Nat. Acad. Sciences*, Vol 3.) (KAYSER: *Lehrbuch*, II, p. 244).

the effect of compression, due to shrinking or other causes, have been appealed to, but this is hardly a satisfactory explanation. The earth could no more shrug itself out

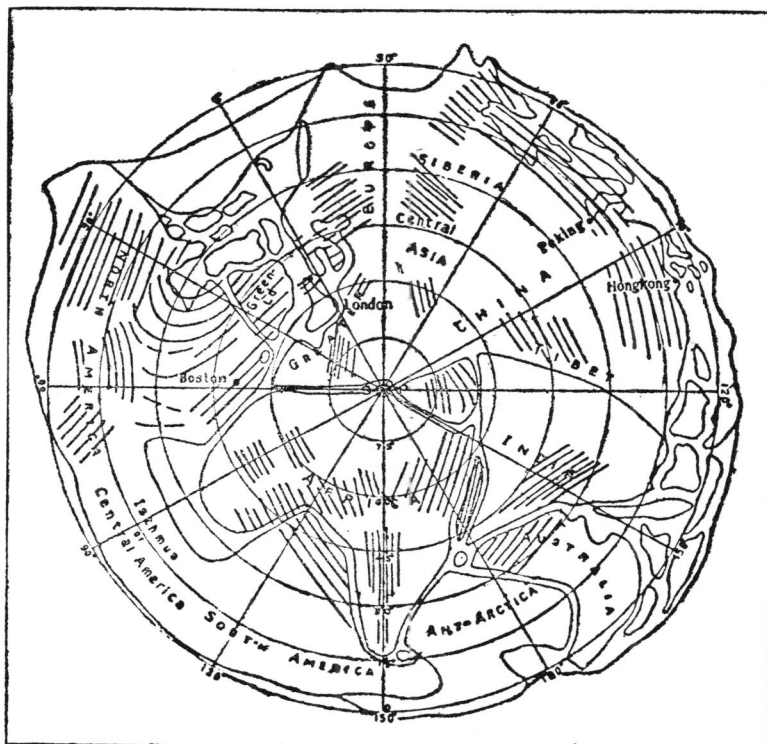

Fig. 4. Archæan trend-lines shown on Fig. 3 (Mercator map), as they appear on the pangæa. (Compare with Plate I, Fig. c.)

of its outer rock-shell unaided, than an animal could shrug itself out of its hide, or a man wriggle out of his skin, or even out of his closely buttoned coat, without assistance either of his own hands or those of others.

But if outside aid was needed, such as some extra-telluric force to withdraw part of the earth's outer shell, it must have been one that acted in a poleward direction. Only such a force can affect the earth, since the earth revolving on its axis, constantly presents a new surface for attraction to any force that pulls from any other direction than the points of emergence of the earth's axis of rotation, whatever position that axis occupied in space.

This being axiomatic, we might invoke the agency of a stellar body, holding the position that Polaris holds with reference to the North Pole, or of a star which occupied at that time the same relation to the South Pole, provided that this star could approach within proper distance of the earth to exert sufficient attraction. Then the sial crust might be withdrawn from the opposite part, over the more resistant sima, and become concentrated into a polar hemisphere of land. In this process the rocks would become intensely folded, and a compressed and metamorphosed sial would result, and this, since the sial-sphere is such an insignificant part of the whole, without essentially disturbing the position which the earth occupies in space.

This may be illustrated by a simple experiment. If we cover a globe with a closely fitting coat of thin cloth to represent the sial, then with suitably arranged devices we might effect withdrawal of the sial from half the sphere into a folded crust in the other half. Then a definite pattern of folds would result from such a concentration, somewhat like the folds of a half-closed globular paper lantern. (See Plate I.) If these folds are compared with the ascertained trend-lines of the oldest rocks of the earth, a certain similarity of pattern may be seen. The folds, or trend-lines, of the ancient Archæan schists can be readily ascertained wherever exposed, since the old folds have been worn own to a more or less level erosion plane.

Fig. 3 is the copy of a map by Dr. R. Ruedemann of the ascertained trend-lines of the old rocks (Mercator projection). In Fig. 4 these trend-lines are plotted on the lands of pangæa, as they naturally fall into association when arranged around a central point in Egypt, assumed to represent the south pole. The resemblance between fig. c. Plate I and text-fig. 4 is at least suggestive.

I admit that, at present, there is no mathematical or astronomical proof that the process outlined was possible; but I also submit that there is today no other explanation that has been offered to account adequately for the deduced phenomena. If another satisfactory explanation is found to account for all the phenomena, I am willing to subscribe to it.

Until then, we may use the "Polar Control Theory" as a working hypothesis, and test it by all observable and deducible facts.

Finally, there remains to be determined the pole around which the sial was concentrated, and here we have more evidence to depend on. Indeed it appears from the history and succession of Palæozoic glaciation, which I regard as polar phenomena, that it could only have been the South Pole around which the land hemisphere was concentrated. It was the South Pole that wandered (relatively) about the African continent throughout Palæozoic time, and finally came to rest in the neighboring Antarctica. Thus, different parts of the continent became subject to polar glaciation in the successive periods.

Polar shifting in a given direction must, of course, be interpreted as the shifting of the sial-cap in the opposite direction. This was a rotary motion, sometimes clockwise, since we are dealing with the southern hemisphere, but more frequently counter-clockwise. The pivotal point on which the sial-cap rotated was probably produced by an intrusion into the sial, at a certain point, of an igneous mass (most probably a dyke, bysmalith or a batholith), which anchored it to the sima at that particular point. Under the influence of the momentum caused by the earth's rotation, and perhaps of a local change in specific gravity, the sial-cap would slowly change its position, until a new point became temporarily stationary over the pole. The location of the anchoring intrusion would determine the direction and amount of change of the polar location ; the pressure produced by the moving mass, and the frictional resistance, would in turn exert a primary control on the development of the geosynclines, and thereby on orogenies. (Chapter XXIV).

7

The original position of the pole in the center of the pangæa probably continued to be maintained throughout the pre-Cambrian periods of peneplanation of the folds, which from time to time may have experienced renewed accentuation.   Indeed, it is most probable that the pre-Cambrian period of erosion was also one of extensive deposition of continental clastics in the synclinal troughs, and that with renewal of folding under the influence of stellar attraction these continental sediments may likewise have experienced deformation.   Even local mountain glaciation may have come into existence in regions beyond the influence of the polar ice, and this may account for the many known occurrences of glacial tillites in the pre-Cambrian rocks of the sial hemisphere.   These glaciations could not all be referred to the polar ice-cap, unless there was actual relative shifting of the polar location in pre-Cambrian time.

It is further suggested, that the marginal geosynclines of the pangæa, which played such an important part in the Palæozoic history of the pangæa, together with the elevated old-lands which flanked them on the seaward side, were in the first place brought into existence by the renewed poleward pull of the stellar visitor, after peneplanation of the old folds had progressed to a marked extent.

Once initiated, the movement might continue, though unequally, under the influence of the pressure induced by shifting sial or expanding sima in the panthalassa, as discussed by Joly[11].   But that pressure could probably not have originated the initial depression of almost completely circumferential character of these geosynclines and the simultaneous development of a circumferential folding of the marginal old-land.

8

While the change in location of the pole with its developing ice-cap might have been accompanied by a change in the maximum elevation of the sea-level by attraction, such was probably counterbalanced by the effect of the centrifugal force which operated in an opposing direction.

Obviously, such elevation of sea-level was quite inadequate to cause the observed periodic submergence and the great marine transgressions, and would not at all account for the regressions, or for the periodicity of the phenomena. Hence, we must conclude that another force was active,

11) JOLY : *Radioactivity and Surface History of the Earth.*

one entirely independent of the forces of attraction and of the centrifugal influence of rotation.

This great source of energy is radioactivity. The rocks of the earth's crust are all radioactive. The radioactive elements generate heat by the energy of their radiations. Periodic increase of heat is followed by loss of heat through conductivity to the cold ocean-floor. Expansion is followed by contraction, and the ocean rises and falls with the changes in its floor; transgression is followed by regression, and a pulsation period is completed. Joly estimates that perhaps thirty million years are required for each of the two processes of such a pulsation, perhaps more. After the cooling and the sinking of the ocean-floor, accompanied by retreat in the geosynclines, another thirty million years or more, are required for the accumulation and conservation of the necessary heat to start the process over again. That interval of rest, I have called the INTERPULSATION PERIOD and it is characterized by low sea-level and drained geosynclines, the sea lingering only in the marginal basins, or epi-seas. Both pulsation and interpulsation periods are realities indicated by the character and interrelations of the rocks, whatever their cause. They mark the pulse-beats of the earth and testify to its perpetually replenished vitality, which is the dominant element of its immortal youth.

CHAPTER II

THE PULSE-BEAT OF THE EARTH

1

WE have endeavoured to follow the history of the earth
since the time that it was 'formless and void', when
it was a sphere of many concentric shells of rock, wholly
surrounded by water and an all-enveloping atmosphere.

The primitive sea was shoreless and therefore without
form or boundary, and it was void, or lifeless, unoc-
cupied by even the germs of organisms, though these it
probably received long before the dry land appeared.
Where they came from is an unsolved problem, but that
they reached the earth from some extra-telluric source
must be assumed, unless we are prepared to believe in
spontaneous generation.

From this primitive state we have followed the history
of the earth through the first great revolution, when the dry
land appeared in obedience to the command, the gravita-
tive urge, of some celestial visitor. This may have been
the newly captured moon, or more probably some star,
which may no longer be luminous, but which at that
early date approached the earth from a south polar
direction, and by its attraction caused the thin crust of
superficial light rock-material to gather together into a
hemisphere of intensely folded and hence much strengthened
sial-rock.

In the space thus stripped of its outer rock shell, the
waters gathered to form the first ocean, no longer a bound-
less sea, but one bounded by the borders of the sial-cap
that formed the land, and floored by the newly uncovered
surface of the sima-sphere.

This land-bordered ocean was now for the first time
able to attack its boundaries, and to attempt to widen its
domain and increase its sphere of activity. Apparently, it
succeeded, aided by the atmosphere, which not only trans-

mitted its motive power to the waves and currents, but carried vapor over the land, where it could fall as rain and snow. This, gathered in streams or transformed into glaciers, began the task of active downward cutting of the lands and of carrying the product of erosion to the sea. Moreover, the disintegrating activities of the air, both moist and dry, helped considerably in rock destruction and the lowering of the lands.

Fig. 5. Section of part of the earth's crust, on the scale of 1 mm = 2 km, showing relation of sial-block to sima.

But what the ocean gained by horizontal expansion, it lost by increasing shallowness, for the rock debris could find lodgement only on the ocean-floor and so diminished its depth. And now too the process of shoaling was aided by the activities of organisms, chiefly at first lowly plants, which began the segregration of lime-salts from the water, and the building-up of "reef" structures on the shallow parts of the ocean-floor and in protected coastal regions.

That this process of lime secretion may have begun before the land appeared is not unlikely; probably it started as soon as the sea-water contained lime-salt and organisms capable of separating it and re-depositing it as hard structures, both within and outside of the living tissue.

2

Before we pursue the study of the geological history of the earth and of the forces that controlled it and gave the impetus to the development of organisms, we must go back and take note of some fundamental characteristics that were established at the time when the earth became a member of our solar system, either by birth and separation, or by capture and adoption.

For untold ages before its geological history began, the earth had probably maintained the relationship to the sun which it holds today; at least there is no evidence that it was ever otherwise.

Certainly for long periods before the land appeared the earth followed its present path round the sun, and rotated upon an axis which maintained the present inclination of approximately 66 1/2 degrees to the plane of the ecliptic, i.e. to the plane of its pathway about the sun. There may have been variations, but these were probably rhythmic; as they are today; fundamentally the motions and the courses are prescribed and fixed, probably for the duration of life of the planet. What seems certain is that since geological history began, since the time of separation of sea and land and for long ages before that, the primary relation of earth and sun, and probably of the moon as well, were the same as now; and that is all that concerns us at present.

But from these relations it follows that the earth was always bi-polar, that it always had an Arctic and an Antarctic region, subject to glaciation, and that this would be so whether these regions were covered by the thin crustal cap or by the ocean. It follows that there always was an equator and a tropical zone, temperate zones of winds and of rain-fall, whether the crust was present in one part of the earth as a hemisphere, or widely scattered, as now, in separate continental blocks. Unless we recognize this we can get only a distorted view of the progress of events.

If the withdrawal of the thin sial-crust from half of the earth, and its concentration into a folded hemispheric mass around the SOUTH POLAR end of the axis of revolution caused any disturbances in the body of the earth, these were adjusted by the time geological history began. All through the early eras of that history, the Archæozoic and the Palæozoic, the earth's surface was divided into a southern PANGÆA and a northern PANTHALASSA, and though the former shifted about in obedience chiefly to gravitational urge, and in doing so displaced the shores of the ocean, it remained intact until after the end of the Palæozoic, which, together with the Archæozoic, comprised by far the greater part of time since the appearance of the lands. (Plate II).

3

We have very little evidence on which to base our deciphering of the pre-Palæozoic history of the earth, for the first of the great eras in which land existed was apparently one of relative crustal rest, after the profound initial disturbance which brought the land into existence. This conclusion is based on the fact, that no matter how intensely folded, crumpled and displaced in segments (faulted) the ancient rocks appear to be, the surface has been worn down to comparative smoothness. So intense has been the erosion of the ancient mountains, as indicated by their truncated folds, that the old wrinkled surface seems to have undergone a sand-papering by a giant workman, until to the grosser vision it has become a perfectly level plane, though to the more microscopic eye of man, perfection is far from having been achieved. We have learned to call such an all-but-perfect erosion plane a *peneplane*. Here and there residual peaks or elevations remain, rising from it as *monadnocks*, old unreduced remnants of the former higher elevation, which dominate the peneplane as Mt. Monadnock dominates the New England upland, or Cretaceous peneplane, on which it forms an erosion remnant.[1]

---

1) The New England peneplane has, however, been raised again, and a new cycle of erosion has begun, with the cutting by the streams of deep trenches or valleys. It is therefore said to be in the $n^{th} + 1$ cycle, in which n has the value of 1 or more, according to the number of antecedent cycles which the region has passed through.

However, the formation of the old peneplane may not be wholly due to erosion. Some of the older and deeper valleys may have been partly filled by material worn from adjacent uplands and preserved because it came to rest beneath the "base level of erosion", that is, below the point to which the peneplane-forming agencies could cut. Some of these valley fillings may be the deposits of ancient glaciers in regions which once lay around the poles of the earth.

The ultimate level which controls the down-cutting of rivers and glaciers is the sea, and from the sea margins even the most perfect of peneplanes will rise gently inland. There will be a sufficient slope to account for the meandering streams, which have remained but are no longer able to cut lower, since they have reached a 'graded' condition throughout their courses.

4

An examination of the ancient peneplane, which everywhere forms the surface of the old, crumpled sial-rock of the early period of the earth, and which the trained eye can recognize even where it has again undergone disturbances, shows that upon it were deposited sands, muds or limestones, which enclose the shells or other hard parts of marine animals, and which could therefore only have been deposited when the old erosion surface had become submerged by the sea. The sea may have accomplished the final sand-papering of the old land-surface, but it did not produce that surface; it occupied it only after long preliminary preparation by the subaerial agencies of erosion.

The sea may locally invade the land when this land sinks beneath sea-level, or the sea-level may rise and so invade simultaneously all the available lowland, whose level it overtops.

The remains of marine animals enclosed in the stratified rocks form a reliable index of the time-period during which these rocks were deposited under the sea. Each geological period had its own organisms by which it can readily be dated in terms of geological chronology.

This dating is not in centuries, but in periods whose length in years, even if it could be ascertained, does not

so much concern us as does their relative position in the
time-scale of the earth's history. And that position is
definitely indicated by the remains of the organisms
living at that period and embedded in the sands and muds
that accumulated in the shallow sea-channels of their
time, or that were built into the solid structures of the
ancient reefs, which, like those of today, were forming in
the past, wherever the environment permitted it.

So reliable are these "index-fossils" of the rocks, that
geologists have learned to place almost absolute confidence
in them; and the experienced student of fossils can usually
date the age of a rock after careful study, sometimes
even by a casual inspection, of the remains.

To be sure, there are some puzzling exceptions that
may at first mislead the palæontologist and sometimes do
mislead him. Barring cases where the fossils of an older
period have been secondarily re-entombed in the sediments
of a newer one—a complication which can usually be
untangled by careful observation of the sections where
the occurrence of a mixed fauna is observed, and sometimes, even by the conditions of the fossils themselves[2]—,
there appear to be cases where an older fauna or group
of organisms, has outlived its proper period. This may
mean that successive generations of the older types have
changed so little that specifically older types may re-appear in
formations younger than those of their normal time period.
But such cases are rare; most of them, on closer inspection and more detailed study of specific characters,
have proved illusory. They are indices of our incomplete
knowledge rather than of the actual longevity of the organisms in question.

5

Renewing then our faith in the reliability of the fossil
marine animals as indices of the geological periods of the
strata which enclose them, we come to the further realization that the marginal portions of the ancient lands, and
sometimes other parts as well, were invaded almost simul-

2) The re-entombed fossils may show signs of abrasion, weathering or iron-
stains, or may be overgrown by colonies which belong to the newer
formation, etc.

taneously by the sea. This did not occur to the same extent everywhere, nor for the same duration in all localities, but essentially it was simultaneous. This implies a simultaneous flooding of these widely distant regions of the earth, either because all the lands of the areas in question sank beneath the sea at essentially the same time, or because the sea-level rose and flooded all the lowlands, and left the record of its invasion in the organic remains entombed in the sands of its temporary floor, *i. e.* the drowned lands.

But the drowning, or the submergence of the low lands, was not of permanent duration. This is shown not only by the character of the rock, but by unmistakable proof that the sea again withdrew after a long period of occupancy; exposure, with its accompanying phenomena — erosion, continental sedimentation, etc., — succeeded the period of submergence. The evidence of such exposure may be more pronounced and more readily recognizable in some sections than in others, but its presence is demonstrable at such widely distant places that there is no question of the essential simultaneity of the phenomena.

But again the new exposure is a temporary one; sometimes the rocks with the older fossils are immediately succeeded by those with fossils of the next newer age; sometimes indeed it is difficult to see a boundary between them. But the very fact that the succeeding fauna is a distinct one, at least specifically, indicates that a time interval has elapsed, long enough for the fauna, which has withdrawn with the retreating sea into the EPI-SEA on the margin of the land, to undergo modification by evolution. Such evolution may become accelerated under the restrictions of the new environment, where the struggle for existence was fiercer.

At other localities the interval is indicated by a disconformity, the presence of erosion features, or even by the almost, if not complete, removal of the previously deposited beds; or it may be shown by sediments which by their character indicate their origin as river, wind, or other continental deposits.

Sometimes the surface of the land is marked only by a period of erosion, which represents the missing interval as well as the next period of sedimentation, because local warpings or other causes have raised the area sufficiently to prevent flooding during the next advance of the sea which is recorded in a neighboring region. Then the new formation, which overlies the first one, contains an assemblage of organisms which clearly indicate that there is a long unrepresented time-period between them.

Of course, in some sections the interval seems to be a longer one than in others. There are several reasons for this. Local warping, or even folding, may have raised a locality, previously invaded by the sea, to such an extent that the returning sea cannot cover it. In other sections, because of the pronounced river activity in bringing sands and muds into the sinking area, the sea may not be able to enter, and so sea-border muds or sands are formed which show no evidence of marine invasion. Such HUANGHO deposits, or deltas of muds, like those formed by the modern Huangho or Yellow River of China, may occupy the entire interval, which is represented elsewhere by submarine sediments. These *huangho* deposits may be subject at intervals to a process of *marining*, that is of flooding during exceptional on-shore winds, by the development of *tsunamis* or seawaves, due to earthquakes or to other causes. Such floodings are indicated by the spreading of marine plankton or floating organisms, or those torn from their anchorage (pseudo-plankton) and swept in by the waves and left stranded on the mud surface, where they are more or less covered by the fine mud of the retreating waters. Examples of these are the graptolite shales of the Palæozoic, and perhaps some muddy sediments on which pteropod shells are spread.

But aside from all the exceptions noted, and others that might occur, the fact remains that the new period is represented in most sections by a series of fossiliferous sediments which indicate the return of the sea after an interval of absence. This is followed upwards in the section by evidence of a new retreat of the sea, and again by its return, this being repeated a number of times according to the completeness of the section. If this occurred in one section only or in a group of nearby sections, it might be regarded as evidence of an oscillatory movement of the land, or as a periodically renewed

sinking of the land in that particular region, followed by
a period of quiescence, after which subsidence recommenced. But, if it can be shown that such a rhythmic
succession is recognizable in most of the sections where
these formations are found in superposition, and this at
widely separated localities, it becomes necessary to interpret
it as a periodic rise of the sea-level, with relatively
stationary land, — in other words, a PULSATION OF THE SEA.

6

In succeeding chapters we shall present evidence that
the Palæozoic pangæa has suffered repeated and regular
floodings of its low-lying portions by the sea, and that the
formations which record the succession and enclose the
remains of their marine faunas, were throughout Palæozoic
time, deposited during these successive inundations.

So regular was the encroachment of the sea upon the
land, and again its retreat, that it is possible to record a
sequence of transgressions and regressions which correspond,
with certain modifications, to the independently established
successions of geological periods, and for the establishment
of which periods in the first place, the changes in the marine
faunas were almost the only reliable criteria.[3]

The causes that induced the changes from one marine
fauna to another, distinct from both that which precedes
and that which succeeds it, have never been adequately
ascertained. These changes were tacitly assumed to be
the result of a general progressive evolution of the organisms,
chiefly in response to a changing environment. But the
nature of the change of this environment, and its causation,
has generally been left among the unsolved riddles of the
universe.

The progress of study of the Palæozoic formations and
faunas has so far revealed the existence of FOURTEEN
PULSATION PERIODS, each separated from the one that
preceded it, and from the one that followed it, by
INTERPULSATION PERIODS.

---

3) The modifications referred to are the partial dismemberment of some
of the older divisions and the reconstructions and recombinations of the
separated parts into new systems. Only in exceptional cases have additions
been made, these being the results of discoveries in new fields, of formations not previously recognized in the countries studied before.

The facts which have been accumulated permit no longer any serious doubt that for the Palæozoic the pulsations and interpulsations were *pangeodic*, that is to say they effected the whole earth at the same time and in the same manner.[4]

In like manner, the withdrawal of the sea from the land was effective in every part of the continental mass at essentially the same time, though not at the same rate of progress, because of difference of slope of the land; and this could only imply a reversal of the process. This periodic, or pulsatory rise and fall of the sea-level, as recorded in the geosynclinal formations of the earth, must be the true basis of subdivision of geological history.

Its ultimate cause must probably be sought in the periodicity of the processes of heating and expansion, followed by cooling and contraction, of the sima-shell of the ocean floor, as already suggested.

Whatever, the cause of transgressions and regressions, the fact is evident that it has acted in a pulsatory manner, and it is most readily visualized as a regular periodic rise of the sea-level with transgressions, followed by an equally regular fall of the sea-level and accompanying retreat of the sea.

It is an expression of the regular pulse-beat of the earth, a slow, powerful rhythm, which continued, apparently without acceleration or retardation, throughout the Palæozoic eras. Each positive pulse-beat sent the waters to flood the lands, to revitalize the organic world, and urge it on to new endeavour; and then each negative beat of the pulse withdrew the waters into their separate marginal basins, there to concentrate the survivors of the living world of the waters and to pass judgement on their fitness; for only the fittest, the most vigorous, were destined to become the ancestors of the next great fauna.

Thus came into existence the new host, selected to occupy the area which was rendered habitable again by the renewed inundation, when the next pulse-beat sent its flood of waters again into the geosynclines and over

---

4) This may seem like a dogmatic assertion, but it can be proved by a comparison of sections. Some of these are given in the subsequent chapters of this book, while for a more exhaustive treatment, reference should be made to my larger work : *Palæozoic Formations in the Light of the Pulsation Theory*, Vols I-IV, 1936-1938, and following in preparation.

the shallow lands.

In this manner was prepared the site for the deposition of the strata in which the generations of organisms of the newer world of life would leave their dead, when, in their turn, they had completed their full measure of individual and of specific existence.

Selection, of course, never ended. On the contrary, it entered on an ever broader field, a wider stage of activity, where more scope to individual development was possible.

7

On the opposite page is given the table of the pulsation and interpulsation periods of the first great life-era of the earth, that is the Palæozoic or period of ancient life, of which no survivors have come down to our day. True, here and there, in deep ocean asylums may linger more or less archæic types, which, though probably no more related to the Palæozoic types than are most modern forms, nevertheless show in their structure a retardation in development, which gives them a closer resemblance to their ancestors of that remote time.

CLASSIFICATION OF PALÆOZOIC SYSTEMS

Palæozoic Pulsation Systems	Systems of Older Classification

14th : Appalachian Interpulsation (Tartarian in part)
Great Terminal Period in North America.

14. Permian   { Upper, Regressive   Supra-Permian S. S.   } Tartarian
      { Lower, Transgressive   Upper (Zechstein)

13th : Post Kungurian (or Lebachian) Interpul. Period   } Permian

13. Uralinskian   { Upper, Regressive   Middle (Artinskian)
       { Lower, Transgressive   Lower (Uralian)

12th : Lochinvarian Interpulsation Period

12. Donbassian   { Upper, Regressive   Upper (Donetzian)   }
      { Lower, Transgressive   Middle (Moscovian)   } Carbonic

11th : Lanarkian (Hintonian) Interpulsation Period

11. Visemurian   { Upper, Regressive   Lower (Namurian)   }
      { Lower, Transgressive   Upper (Viseen)   } Dinantian

10th : Mauch-chunkian Interpulsation Period

10. Fengninian   { Upper, Regressive   Middle (Chiussuan)
      { Lower, Transgressive   Lower (Tournaisian) }

9th : Monteplaisantian Interpulsation P. (Catskillian in part)

9. Devonian   { Upper, Regressive   Upper. . . . . . . }
     { Lower, Transgressive   Middle . . . . . } Devonian

8th : Oriskanian Interpulsation Period

8. Siluronian   { Upper, Regressive   Lower . . . . . }
     { Lower, Transgressive   Upper. . . . . . }

7th : Salinan Interpulsation Period   } Silurian

7. Silurian   { Upper, Regressive   Middle . . . . . .
 (restricted) { Lower, Transgressive   Lower. . . . . . }

6th : Plynlimonian Interpulsation Period

6. Ordovician   { Upper, Regressive   Upper. . . . . .
 (restricted) { Lower, Transgressive   Middle . . . . .

5th : Petrovian Interpulsation Period

5. Skiddavian   { Upper, Regressive   } Lower . . . . .   Ordovician
      { Lower, Transgressive }

4th : Rhobellian Interpulsation Period

4. Cambrovician { Upper, Regressive   } Basal (Tremadocian)
(restr.) (Ozarkian) { Lower, Transgressive { Upper . . . . .

3rd : Arctomian Interpulsation Period

3. Cambrian òr   { Upper, Regressive   } Middle . . . . .
 Acadian   { Lower, Transgressive }   Cambrian

2nd : Pribramian Interpulsation Period

2. Taconian or   { Upper, Regressive   } Lower . . . . .
 Georgian   { Lower, Transgressive }

1st : Pilsenerian Interpulsation Period

1. Sinian or Torridonian. No marine stages known

CHAPTER III

GEOSYNCLINES

1

THE concept of a *geosyncline*, which plays such an important part in our modern view of earth-history, was first formulated in 1859, by that great pioneer in American stratigraphy, James Hall, and was re-defined, in 1873, by James D. Dana, to whose teaching so many American geologists owe their early inspiration.

Hall visualized a great syncline of deposition formed in the surface of the land, and to this concept Dana gave the name *geosyncline*, which it bears today.

In recent years other geologists have considered the subject from various angles, and a number of erroneous definitions of the characteristics and the manner of formation of the geosynclines have been promulgated. Some of these I have discussed elsewhere.[1]

Among all modern discussions we are apt to lose sight of the grand simplicity of the original concept, which alone is entitled to the name of geosyncline, even if some of the disturbing ideas of later date should prove to have a foundation in fact.

I have elsewhere quoted Hall's original description in full[2] and shall here omit certain portions not essential to the understanding of his concept.

"It has been long since shown that the removal of large quantities of sediment from one part of the earth's crust, and its transportation and deposition in another, may not only produce oscillations, but that chemical and dynamical action are the necessary consequences of large accumulations of sedimentary matter over certain areas.

---

1) GRABAU: *Fundamental Concepts in Geology*, pp. 130-133.
2) GRABAU: *Migration of Geosynclines*, p. 209.

When these are spread along a belt of sea-bottom, as originally in the line of the Appalachian chain, the first effect of this great augmentation of matter would be to produce a yielding of the earth's crust beneath, and a gradual subsidence will be the consequence. We have evidence of this subsidence in the great amount of material accumulated; for we cannot suppose that the sea has been originally as deep as the thickness of these accumulations. On the contrary, the evidence from ripple marks, marine plants, and other conditions, proves that the sea, in which these deposits have been successively made, was at all times shallow, or of moderate depth. The accumulation, therefore, could only have been made by a gradual or periodical subsidence of the ocean bed; and we may then inquire, what would be the result of such subsidence upon the accumulated stratified sediments spread over the sea-bottom?

"The line of greatest depression would be along the line of greatest accumulation; and in the direction of the thinning margins of the deposit, the depression would be less. . .

"That this subsidence was periodical, we have the best possible evidence in the unconformability of the Lower Helderberg group upon the Hudson-river group, showing that previous to the deposition of these limestones, there were already foldings and plications. . .

"This successive accumulation, and the consequent depression of the crust along this line, serves only to make more conspicuous the feature which appears to be the great characteristic, that the range of mountains is the great synclinal axis, and the anticlinals within it are due to the same cause which produced the synclinal; and. . . these smaller anticlinals, and their corresponding synclinals, gradually decline towards the margin of the great synclinal axis, or towards the margin of the zone of depression which corresponds to the zone of greatest accumulation." [3]

The name "geosyncline" was applied to these lines of depression by James D. Dana, [4] in 1873, and he extended the application to include all similar earth-troughs of deposition, whether simple or complex. Such geosynclines may

---

3) HALL, J.: *Pal. of New York*, 1859, pp. 69-71.
4) DANA J. D.: *Some Results of Earth Contraction.*

have a length of many thousands of miles with a width of only a few hundred miles.

Unlike Hall, who considered the down-sinking of his earth-troughs of deposition as the response to the loading, Dana held that the formation of a geosyncline was attributable to the activity of compressive forces or primary strains acting upon the land, and that the filling of the depressed area by sediments from the adjoining lands was proceeding *pari passu*, with the subsidence, and was possible because of it. According to Dana, geosynclines are tectonic features, subsidence being the primary factor and sedimentation the consequence. In Hall's view on the other hand, sedimentation is primary and depression the consequence. This is essentially an early expression of the modern doctrine of isostasy.

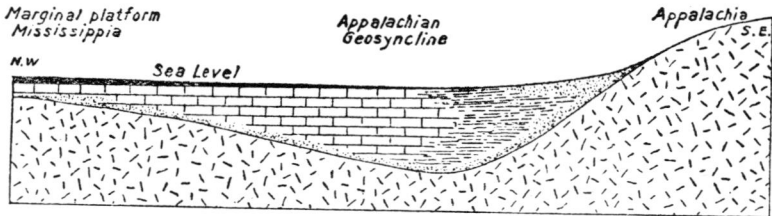

Fig. 6. Cross-section of the Appalachian geosyncline showing change in lithological character of sediment from the old-land outward, and the residual sand of the marginal platform followed directly by limestone (after Grabau).

We may re-define a geosyncline in Hall's and Dana's sense in the following modern terms: (Fig. 6)

a) A geosyncline is a structural feature which belongs wholly to the sial, that is, to the land, being either marginal to the land-border or within it.

b) It is a long, narrow trough of subsidence, parallel to a similar area of elevation, the OLD-LAND, which rises as the trough sinks, and supplies the sediments chiefly by stream erosion.

c) These sediments are deposited in the geosyncline, the coarsest near the old-land margin, and the finer at a distance from the old-land.

d) The length of the geosyncline is usually measurable in thousands of miles, but its width, including the marginal plain or platform, seldom exceeds a few hundred miles.

e)   In cross-section it is asymmetrical, the locus of most rapid subsidence being near to, and parallel with, the axis of the old-land; hence that part of the geosyncline receives the greatest amount of sediment.

f)   From the axis of maximum subsidence, the floor of the geosyncline rises gradually and with marked regularity toward the margins opposite to that of the old-land borders, and it may here expand into an extensive, nearly flat MARGINAL PLATFORM. [5]

g)   The geosyncline is never a symmetrical trough. Since the main mass of the deposits comes to rest in the axial portion of the subsiding trough, near the old-land, this will not only be thicker there, but will consist of coarser material, while the marginal platform will receive only the finer deposits.   The latter may be wholly of a calcareous nature and primarily of organic origin, [6] and will be correspondingly reduced in thickness; or certain portions may thin out altogether on the marginal platform, so that a mild hiatus and disconformity may exist between successive members of a depositional series, some of them being altogether absent. [7]

h)   It must be emphasized, as Hall has originally pointed out, that subsidence proceeds *pari passu* with the deposition of the sediments, and at no time is the geosyncline a depression of profound depth.   The question whether Hall's view, that subsidence is the result of loading, is the correct view of the origin of the geosyncline, or whether the tectonic theory of Dana, which makes sinking the primary factor is to be accepted, will be considered later (chapter IV, sect. 5, and chapter XXIV).

i)   In many geosynclines the surface may never sink beneath the sea-level, and thousands of feet of sedi-

5) It is a mooted point whether this platform partakes of the subsidence of the geosyncline, or whether it is stable like the rest of the country, in which case its submergence is wholly the effect of the rising sea-level.
6) It is a mistake to assume that limestones require deep water for their formation. Most limestones are formed in shallow water, within reach of the waves. They do require water free from sediments and other substances inimical to the existence of lime-secreting organisms. Some limestones are made chiefly, or even wholly, from the sand and dust derived from the destruction of older limestones.
7) If the marginal platform does not partake of the subsidence of the geosyncline but is a stable unit, this temporary withdrawal of the waters, and emergence, may be due to increased rapidity in the subsidence of the geosyncline.

ments may accumulate, each stratum of which at the
time of its formation was a surface deposit exposed to
the air.

2

Of such a nature is the great geosyncline of North India
of today, whose surface forms the Indo-Gangetic Plain.
The material of this plain is wholly derived, as sand and
similar clastic products, from the erosion of the Himalayan
Mountains and their foot-hills.   This is true also of the lower
part of the deposit, which has a thickness of many thousand feet.

In surface area this geosyncline comprises about 300,
000 square miles.   Its width varies from 90 to nearly
300 miles, and its length stretches clear across North
India from the Indus, which enters the Arabian Sea, to
the Ganges, which builds its delta into the Bay of Bengal,
on the east.

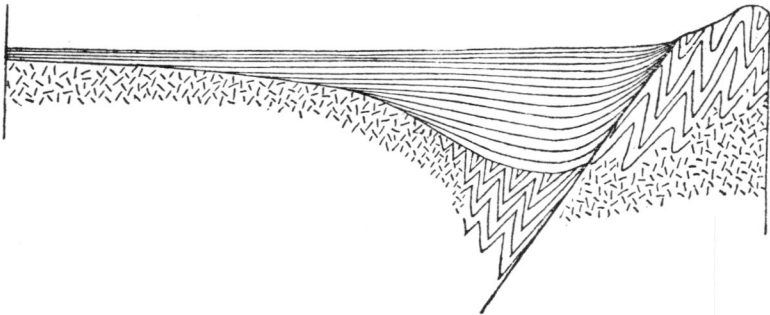

Fig 7.   Geosyncline in fault contact with the old-land.

The head of this plain, near the margin of the old-
land, is 924 ft above sea-level in its highest portion,
but is of much less elevation elsewhere.   A boring at Luck-
now, approximately mid-way between the Indus and
Ganges head-water, and situated 370 ft above the sea-
level, extended nearly 1,000 ft below it; the boring did
not reach the bottom, but the part penetrated had a thickness of 1,336 ft.   This revealed the fact that there is no
marked increase in the coarseness of the beds passed
through, though near the bottom some beds of coarse
material were encountered.

No evidence of the presence of the sea was found in the entire depth penetrated, which, it must be remembered, reached more than 1,000 feet below the sea-level, but all the deposits show that they accumulated on a surface not much different from that characteristic of the plain today.

On the other hand, shells of river and marsh molluscs were occasionally met with at various depths, and remains of land animals, pieces of tortoise-shell, bones of reptiles, shells of fresh-water molluscs, and fragments of wood were found, even below a depth of 400ft.

The relation of the modern Indo-Gangetic geosyncline to its old-land, the Himalayan range, is a faulted one; that is, the line of contact with the old-rocks is a plane of slipping or faulting which is still active, instead of merely a down-bowing, as in other geosynclines. (Fig. 7)

As the old rock-floor of the geosyncline sinks, carrying its load of sand and silt downward with it, the increased grade of the rivers which results, enables them to bring more clastic material to be spread over the surface of the plain, and thus to maintain its level.

The balance is never disturbed for long; always there are the forces of degradation and aggradation, ready to re-establish the equilibrium.

3

We shall have occasion, on a subsequent page, to refer to the origin of the North Indian geosyncline as the result of a migration. In general, this means that the site of the present geosyncline was formerly an elevated old-land, while the geosyncline in which the strata that now form the Himalayan Mountains accumulated, lay along its northern foot. When these strata were folded to form the young Himalayan Mountains, the old-land on the south sank to form the new geosyncline, in relation to which the newly-formed Himalayas played the part of old-land.

Such migration of geosynclines into their old-land has as a consequence, that the new geosyncline is entirely floored by crystalline rocks; in other words, a new cycle of sedimentation has begun.

The first products of erosion of the Himalayas, deposited in the new geosyncline, represented an earlier (Tertiary) period of earth-history. These were sands and clays, similar to those accumulating on the modern plain, but they reached the astonishing thickness of 16,000 feet, or more than three miles, which is a measure of the length of time consumed in the slow subsidence of the first post-Himalayan geosyncline of this region.

None of the strata of this great series show any indication of having been deposited beneath the sea. All are conformable and connected formations of river flood-plain and alluvial sand accumulation. While sands predominate, the strata also include pebble beds, layers of red clay, and calcareous bands. Some include abundant remains of the Tertiary land life of the region. The series is now for the most part consolidated into rock, and it was pressed into a succession of folds, parallel to the northern border of the geosyncline; these folds form the Siwalik Hills.

The cause of the folding was the relative southward pressure or thrust of the Himalayas. Here there was no migration, merely the addition of the faulted foreland to the foot of the Himalayas, an orogenic episode, of local and limited significance; it was an incident, which, when once it had passed, allowed events to proceed much as before. Erosion gradually removed the newly-formed wrinkles; the material eroded from the Siwalik Hills being employed in building the new Indo-Gangetic Plain; while sands and silts from the Himalayas covered the old scars with a protective mantle.

The Siwalik and Indo-Gangetic geosynclines thus represent one type: i. e. the border-fault type of continental geosynclinal deposits. (Fig. 7)

Ancient geosynclines, occupied wholly or in part by continéntal sediments, abound in the geosynclinal records, though their original relation to the old-land that supplied the sediments, whether border-fault contact or down-bowing, is not generally determinable. Most of these sediments have been folded, faulted and eroded, some several times; their original character and extent is only determinable from sections of the remnants which have been preserved. Their origin, as former terrestrial deposits, is shown by their lack of marine fossils, as well as by other characteristics.

Several of these formations will be noted in the course of our survey of the principal pulsation systems. They are one of the features of the interpulsation periods, though not confined to them.

4

A somewhat different type of geosynclinal sediment is illustrated by the great delta or river-plain of the Huangho, or Yellow River, of China. Here the surface is a very level one, and the deposits are chiefly of fine material. The depth of the deposits is not known, but it has been penetrated for nearly 3,000 ft (2,834 ft), and may go to many times that depth. But the significant fact is the prevalence of sands and muds, some of which contain shells of fresh-water or terrestrial molluscs, while marine types are rare or absent.

It is true, there is a fossil oyster-bed of giant oyster-shells, near Taku Bar, at the mouth of the Yellow River estuary, which indicates that the sea stood there for some time. But apart from this the evidence of sea invasion is slight. The only other example on record is a bed 4 ft below the level of the surface, at Tientsin, where shells, like those now found on the beach, have been collected.

On the other hand, shells of the large fresh-water mussel *Lamproteula* found in Chinese rivers today, were obtained from between 40 and 78 ft and between 1,109 and 1,121 ft. Others, which were obtained in the well-section, are the small bivalve *Lutraria*, now living in abundance in estuaries of the rivers in Shantung; the snail *Planorbis*, a flat-coiled gastropod, found living in gardens and among the reeds of rivers and marshes; and finally *Corbicula*, the mud-clam of rivers and mangrove swamps. All these were found alone or in various associations at all depths down to 2,478 ft, below the mouth of the drill-hole which is only 16 ft above tide-level.

These remains clearly indicate that each shell-bearing layer, even the deepest, was at the time of its formation a surface stratum, and that during the accumulation of the sand and mud, the region has subsided more than 2,500 ft.

No one knows how far down is the rocky floor of this geosyncline, which faces the Pacific; it could easily be flooded were subsidence a little more rapid than the additions of the sediments made on top, or by exceptional storm waves, or by *tsunamis*, due to heavy earthquake shocks. That this has happened in the historic past, is shown by the ancient records of tidal waves, which repeatedly swept across the plain.

I have given elsewhere[8] the record of twenty-seven such inundations, during a period of 1,482 years, from A.D. 146 to A.D. 1628, or on the average one disastrous inundation in 55 years. The interval was however more irregular, ranging from one year to 182 years. Twenty-five of the inundations have the moon month recorded, and of these, the record, if correctly interpreted, shows that 18, or 72%, happened during the months of July, August and September, the months of the strongest easterly monsoons in this region. If this was the case, it demonstrates that exceptional storms, during the period of easterly monsoons, may result in inundations of the border of the Great Plain of China.[9]

The destruction of human life varied from none recorded to 20,000 in the eighth moon of A.D. 1166. The destruction of animal life is only incidentally mentioned, but it must have been enormous. The destruction of fish and other animals in the rivers and lakes, through influx of salt water, though unrecorded, must likewise have been very great.

Finally, though the records do not state it, sea-wrack must have been carried far inland by these inundations, as it is carried today by tidal waves due to earthquakes. On withdrawal of the sea-water, masses of seaweed, with attached organisms, are left stranded, and are subsequently covered by thin films of mud. They are thus incorporated as thin, fossiliferous layers of black mud in the strata. Such a partial invasion by the sea, of only short duration, I have termed *marining*, that is, a very short-lived spreading of marine waters over a surface primarily of river flood-plain or delta origin, bringing with it floating organisms, or such as are attached to sea-weed torn from its anchorage and

8) GRABAU: *The Great Huangho Plain, etc.*, pp. 258-261.
9) The storm-driven sea-waters which inundated the low regions of the Atlantic coast of the United States in September 1938, are a recent illustration; the Galveston Flood, of Sept. 1900 was more destructive.

left stranded on the mud-surface of the plain by the retreat of the waters.

That these have not been observed in the deposits of the Huangho plain signifies nothing. It does not imply their absence, as no attention has been paid to them, even on those rare occasions when sections have been opened in these layers. But there can be little doubt that careful search in clay-pits, opened not too far from the coast, will show at least some record of them.

5

It is evident, that such temporary floodings, or marinings, will introduce an alien element into the deposits of the plain by bringing marine organisms, as well as sea-wrack, and leaving them stranded on the surface when the sea withdraws. There they will be buried by the next mud-deposit which is spread over them by the river. Unless their peculiar character as floating organisms is recognized, either as holo-planktonic, or epi-planktonic (attached to a floating substratum such as seaweeds torn from their anchorage, floating logs, etc.), these marine organisms may lead to an erroneous interpretation of the strata enclosing them. They are likely to be regarded as marine beds, whereas they are really super-marine river or delta deposits on the sea-border, temporarily, or only for a short time, invaded by the sea. Since the best known modern example is furnished by the Yellow River or Huang Ho of China, we shall speak hereafter of such deposits, whether recent or ancient, as *huangho* deposits.

In the next chapter some typical ancient examples of huangho deposits modified by marining will be dealt with.

CHAPTER IV

GRAPTOLITE SHALES
AS EXAMPLES OF ANCIENT HUANGHO DEPOSITS

1

THE brief submergence, by the storm waves of the sea, of the huangho deposits, or river-deltas of the *huangho* type, which I have called marining, is well illustrated by the graptolite shales of the Palæozoic, though all deposits of the huangho type may include such records of invasion.

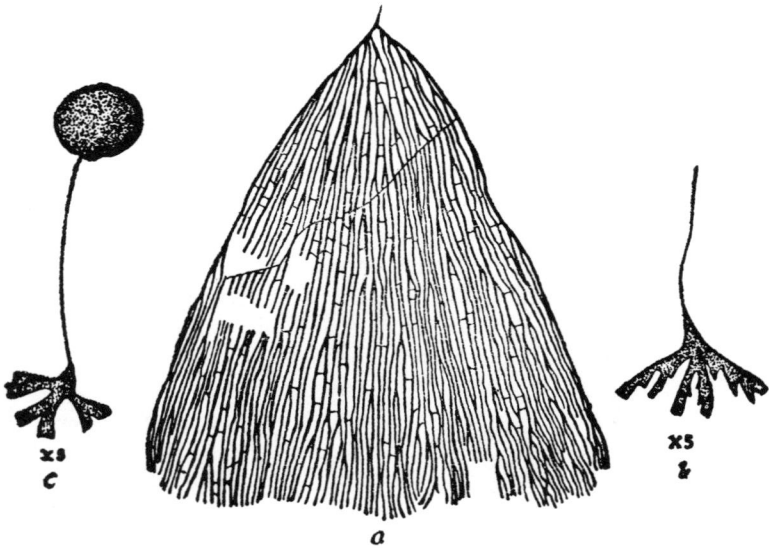

Fig. 8. *Dictyonema flabelliforme* Eichwald ; a. Normal and approximately mature flattened specimen ; b. Young rhabdosome showing several bifurcations, the further growth of the branches and the form of the thecæ; c. Young rhabdosome with nema and primary disk (after Ruedemann).

In post-Palæozoic time after the graptolites had disappeared, such marining may be indicated by planktonic Foraminifera, or even Radiolaria, or by sea-weeds to which

Hydrozoa, Bryozoa or other marine organisms are attached.

The normal graptolite shales are generally thick deposits of mud, usually more or less mingled with fine sand, and barren, as a rule, of organic remains. Only at irregular intervals occur thin layers of fine black mud-rocks, which have their surfaces covered with multitudes of slender thread-like, carbonized bodies, — the graptolites, — lines resembling the spidery characters, sometimes found in beautifully executed Chinese manuscripts drawn by a master calligrapher with a delicately pointed Chinese brush. This is the meaning of the term graptolite : *writings on stone.*

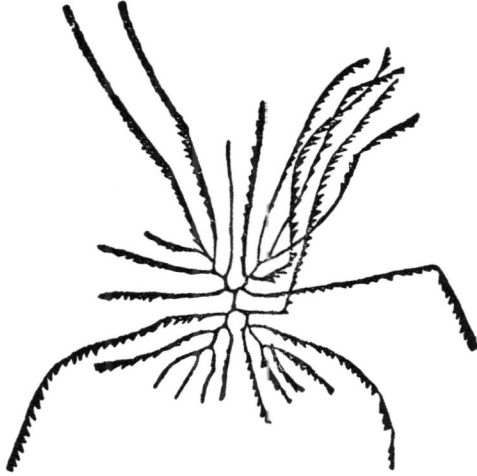

Fig. 9. *Goniograptus thureaui* McCoy, var. *postremus* Ruedemann, an angularly branching graptolite colony. Deepkill shales (Beekmantown) of eastern New York (after Ruedemann).

Graptolites, or Graptozoa, are a class of extinct organisms remotely related to modern hydroids, which abound in the sea today. There they grow attached to sea-weeds or other objects, including, in the case of some species, the submerged rock-ledges of the coast. As in the modern Hydrozoa the graptolites are compound. Many individual "zooids" arise from a common stem, generally in close-set, serried succession, as in the modern genus *Sertularia* among the Hydrozoa. The entire colony is enclosed in a horny or chitinous sheath, with a succession of thecæ, or cups which enclose the individual bud-like polyps.

This chitinous structure, because of its resistance to decay, is preserved, usually in carbonized form, on the surface of the mud-rock on which these organisms are left stranded. (Figs. 8-9).

The oldest graptolites consist of a multitude of branching stems, each a succession of cups, and all generally united by thread-like cross-bars into a solid net. This is *Dictyonema* (Fig. 8), which is also the longest persisting genus among the Graptozoa, since it is found in Cambrovician strata and in those of Devonian age as well. Other graptolites consist of a few branches attached to a floating disk (Fig. 9), and some of these (Ordovician species) show double rows of hydrothecæ, though others (mostly Silurian forms) show only a single row. Fig. 10 shows a restoration of such a floating colony.

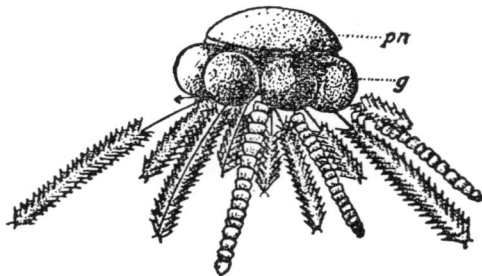

Fig. 10. A graptolite colony (*Diplograptus*), restored (after Ruedemann); *pn.* floating bladder (*pneumatophore*); *g.* reproductive sacs (*gonangia*).

2

During a positive pulsation period, the rise of the sea-level would offer opportunity for repeated marinings of all coasts where huangho sediments were accumulating, at the mouths of rivers and on their extended frontal plain. During each pulsation, while graptolites existed, successive inundations by the sea would spread these organisms on the muds, and so produce normal graptolite shales, where no marine life could have actually existed.

Graptolite shales are, of course, a product of the pulsation period, when the rivers keep the surface elevation of the geosyncline just at flood-level of the sea, but not high enough to prevent excessive floods from sweeping over them and leaving their flotsam spread over the mud-flat.

This signifies that deposition and rise of sea-level are balanced. A falling sea-level, during an interpulsation period, does not permit marining of the river-plain, but instead the rivers, which flow into the sea, have increased slope, and, becoming rejuvenated, entrench their beds. Much of the plain may also be lowered again by erosion, the erosion products being spread farther seawards as normal marine deposits.

The erosion of the old surface also prepares it for an abrupt inundation in the next pulsation period, when the sea rises again above the level of the plain, which has not yet become re-aggraded, *i.e.* built-up to or just above sea-level.

Thus, it follows that the regressive members of a pulsation system make a less impressive showing than the transgressive members of the next series.

This also accounts for the very abrupt superposition of transgressive, purely marine members of a later pulsation, on the mixed, or often non-marine members of a preceding pulsation system.

Many examples of such abrupt invasion of the sea are found in the stratigraphic record, and a number of them will be cited in subsequent chapters.

3

I should here remark, that the history of the graptolite shales outlined above, is not the only one current, for the old doctrine of graptolite deposition, promulgated by Lapworth, which was for so long the orthodox view, still holds the allegiance of a number of his faithful followers.[10]

Briefly, outlined, Lapworth's view, which was also that of Marr, and which counts Ruedemann as its ablest modern exponent, considers that the graptolites were preserved in muds in deep, even abyssal, waters of nearly enclosed basins of the Black Sea type, into which they were carried as floating plankton or epi-plankton. Only the surface waters were able to support life, for, in the

---

10) See for further discussion, GRABAU: *Principles of Stratigraphy*, Chapter 28, p. 1011, and GRABAU and O'CONNELL: *Were the Graptolite Shales, deep or shallow water deposits?* pp. 959-964.

deeper portions, absence of circulation eventually leads to the "poisoning" of the stagnant waters. Therefore, on the floor of the basin no living organisms could exist and act as scavengers, to destroy the plankton which sinks to the bottom. Consequently these bottom layers will preserve all such surface organisms with preservable parts, but will be free from the remains of the normal marine organisms.

Satisfying as this theory might be to the palæontologist, who is concerned chiefly with the organisms themselves, it cannot be acceptable to the stratigrapher, who recognizes that no geosyncline at present, or in the past, could maintain permanent stagnant basins of abyssal depth with their characteristic sediments; for these basins are wholly alien to the geosyncline. The Black Sea, as I have argued elsewhere,[11] is an individual type of basin and has no connection whatever with the geosyncline, and to class it as such leads to confusion and misunderstanding.

Altogether the hypothesis of the stagnant basin origin of the graptolite shales requires too many special conditions to give it a semblance of plausibility.

4

When we turn to the geosyncline in which the chief deposits are of marine types, we find here also abundant evidence to contradict any deep water hypothesis of origin.

Most of the fossiliferous sandstones, shales (lutytes), and even limestones, can unhesitatingly be referred to littoral deposits, using that term in the broader sense in which I have defined it in my *Principles of Stratigraphy*,[12] as including both the shore zone and the zone of the shallow submerged sea-bottom, within reach of direct sunlight.

As in the other types of geosynclines previously discussed, the greatest thickness of the deposit is formed along the axis of subsidence, where these sediments may be largely calcareous; while towards the old-land margin, the limestone strata will merge into sands or pebble-beds, the coarseness of which depends, to a large extent, on the height to which the old-land rises above the sea-level, and

11) GRABAU: *Origin. etc. of Graptolites.*
12) Chapter XV, pp. 641-667.

on the resulting grade of the rivers which supply the clastic sediments.

It must however, be clearly kept in mind, that the successive strata of the geosyncline were deposited each comparatively near the surface of the water, or at a moderate depth below it. Only with progressive subsidence of the axis of the geosyncline, or with the rise of the sea-level, or with both, is space provided for the addition of new layers on the top. If the former predominates, the overlaps will be slight; if the latter is dominant, the attenuated prolongations of the successive strata will be spread widely over the marginal plain.

5

This again raises the question, which is the cause and which the effect? Does the geosyncline subside because of loading; or are new deposits formed because the geosyncline is sinking and the rivers have their grade increased, and so are able to carry more sand and mud into the basin of deposition? In either case isostatic equilibrium would be maintained.

At first thought, it might appear that the former hypothesis gives the most rational explanation of the phenomena, until we reflect, that the actual amount spread over the surface by each river flood, though sufficient to elevate the plain a fraction of an inch or more, can hardly have had sufficient weight to depress the entire geosyncline. If it had, the geosyncline would scarcely be able to withstand the much greater pressure of the weight of the buildings and other structures which we erect upon it. But the evidence shows that the plain is never built very much above the surface at which high storm-waves or other inundations overflow it, except, of course, during interpulsation periods. Repeated marinings of successive surface deposits show that the subsidence keeps pace with the deposition, even though such subsidence is extremely slow. But the accumulation of the sediment proceeds at an equally slow rate.

I hold that there is no evidence that subsidence of the geosynclines is brought about by the weight of the thin layers of sediments spread over the surface, but such sediments are brought there because, by subsidence and

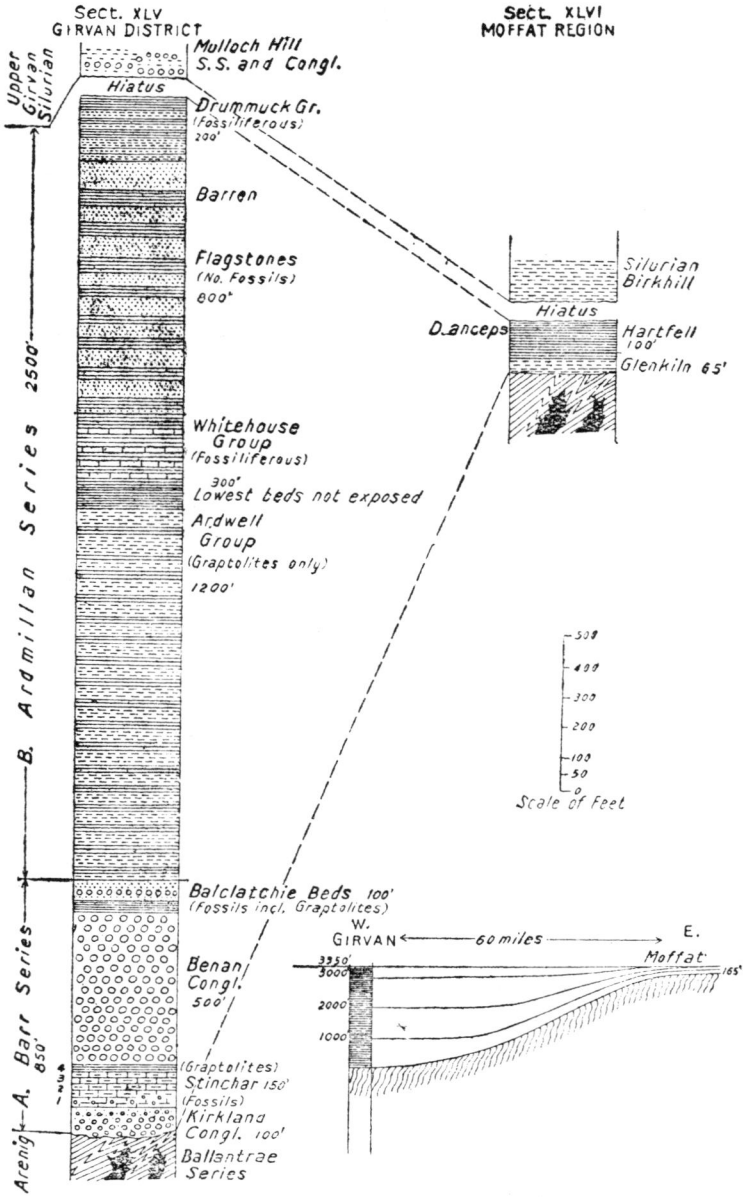

Fig. 11. Comparative sections of the Ordovician (Glenkiln and Hartfell) shales of the Girvan and Moffat districts, of South Scotland. Drawn to uniform scale (after data by Lapworth and others). Also cross-section of the geosyncline.

the consequent lowering of the base-level, the grade of
the river is increased, and it is able to bring more sediment from the inexhaustible source of supply produced by
weathering in the drainage basin. The geosyncline sinks
so slowly that river deposits keep pace with it, — probably
so slowly that no appreciable difference is recognizable in
a generation, or even during many generations. Methods
might be devised for measuring the rate of subsidence,
but these must take into account the periodic change from
upward to downward movement of the datum-plane, that
is the sea-level. The rate of subsidence of the geosyncline,
too, may vary slightly, though this is probably never
actually reversed under normal conditions.

The fact that deposition keeps pace with subsidence
cannot be stressed too strongly, and is especially pertinent
in the case of thick limestone deposits in the geosyncline,
which are of essentially uniform character throughout.
They may not show easily recognizable characters indicating their shallow water origin; but their uniformity
from bottom to top is in itself eloquent of uniform depth
of water during the formation of each layer and consequently
demands a corresponding uniformity in the rate of subsidence,
which is balanced by the rate of deposition. Frequently, such
limestones show actual evidence of intertidal exposure, in
the formation of desiccation fissures (mud-cracks) and
ripple-marks, these testifying to a very shallow zone of sedimentation.

6

While the succession of strata in the geosynclinal axis is
uninterrupted, with thousands of feet of limestone or other
rocks accumulating in unbroken succession during a single
pulsation period, the shallow marginal platform, opposite
the old-land tells a different story. Here, because the
sinking is at a minimum rate, or the platform is characterized in the main by stationary conditions, the amount
of accumulated sediments is correspondingly slight. In
purely continental deposits the sedimentation is mostly confined to the geosyncline, only the final beds overlapping on
to the marginal platform.

In the case of the finer huangho deposits, which
show occasional marining, a thin layer of each of these

deposits may extend on to the marginal platform, while the thicker deposits fall short of it.

A typical example of this relationship and of the differences in thicknesses is seen in the Ordovician series of South Scotland, the region which was taken by Lapworth as best illustrating abyssal deposition of graptolite shales. (Fig. 11, p. 44).

Within the geosyncline in the Girvan district, the Ordovician succession extends through 3,350 ft of gray-wackes, flagstones, silts and fine clays, with not only the complete succession of graptolite zones normal for the series, but also more typical marine strata which contain other fossils. Each zone is separated from the next by a greater or lesser thickness of barren strata.

Sixty miles to the east, in the Moffat region, where the marginal platform was situated, the whole series is represented by 165 ft of fine, black shales. These shales however, contain the complete succession of the graptolite zones, although the thickness is only about 1/20th part of the deposits in the geosyncline farther west. The several interzonal barren beds of the Girvan succession are practically unrepresented in the Moffat region. [13]

When, however, we come to the more purely calcareous phases of the sediments, we find that not only is the representation in the marginal platform reduced in thickness, but minor disconformities may separate the successive members, even to the extent of showing erosion. Some of the members may be entirely wanting locally.

As a further illustration may be cited the marginal platform of Mississippia in Silurian time, as compared with the Appalachian geosyncline.

In the latter the Silurian begins with a thousand feet of almost unfossiliferous sandstones (the Tuscarora, or Medina, formation), which follow upon an even greater thickness of continental interpulsation beds (the Juniata, or Queenston red beds). These separate the Ordovician and Silurian, and correspond to the Welsh Plynlimonian interpulsation series.

Upon the Tuscarora follows approximately another thousand feet of more or less fossiliferous sands, shales and

---

13) See further GRABAU : *Palæozoic Formations, etc.*, Vol. IV, pp. 533-582.

limestones, some of which have the characters of coral reefs.

On the marginal platforms, on the other hand, as shown by the exposures along the Mississippi River in eastern Missouri and in Illinois, the deposits begin with the richly fossiliferous limestones of the Alexandrian series, which represent the product of the first flooding, and probably correspond to the sandstones of the early Tuscarora formation. These vary greatly in thickness and in their individual subdivisions, and have a total range of from 100 to 170 ft. At least four erosion planes are recorded between the several members, and the base rests with a pronounced disconformity and hiatus upon the upper shale member of the Ordovician system.

These beds have sometimes been mistakenly thought to represent a separate oscillation followed again by a retreat, and preceding the main Silurian pulsation. But there is no consistent evidence for such an interpretation, while the fauna is closely related to the succeeding Clinton or Niagara. These beds satisfy all the requirements of strata deposited on the marginal platform of the geosyncline, and as such, reflect the greater sensitiveness of this platform to minor oscillatory movements of the sea-level, which leave the main geosyncline unresponsive.

CHAPTER V

OROGENY, OR THE GENESIS OF MOUNTAIN RANGES, AS REVEALED BY THEIR STRUCTURE[1]

1

MOUNTAINS, once the symbol of permanency, have, under the teaching of the physiographer, revealed to us their evanescent character. The changes seen in them during a human lifetime may be negligible, but those revealed by the study of the mountain structure are profound.

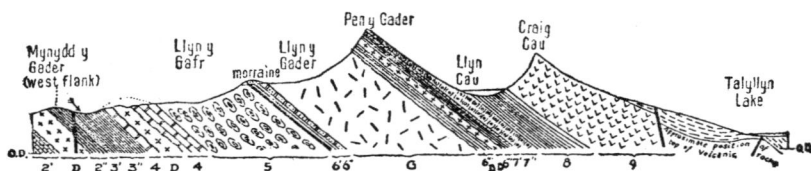

Fig. 12   Section through the Main Peak of Cader Idris, a typical unicline (after Cox). (Horizontal and vertical scale $1\frac{5}{8}$ inches = 1 mile.)

2.   Arenig Series : 2′=Lower Acid or Mynydd-y-Gader Ashes; 2″=Rhyolites;
       3-9.   Llanvirnian Beds and Interpulsation Volcanics:
3=*Bifidus* Beds ; 3′=Bryn-Brith Beds ; 3″=Crogenen Slates ; 4=Cefnhir Ashes ; 5=Lower Basic Group ; 6=Llyn-y-Gader Lutytes ; 6′= Lower Lutytes and iron-ore ; 6″=Ashes ; 6‴=Upper Lutytes ; 7=Upper Basic Group ; 7′=Ashes ; 7″=Spilites ; 8=Llyn Cau Lutytes ; 9=Upper Acid Group ; 10=Bala Series : Talyllyn Lutytes. G=Granophyre ; D=Dolerite.

That mountain chains arise by folding of the strata deposited in a geosyncline, was not understood three quarters of a century ago. That they originated through compressive forces, which squeezed the rocks into a succession of arches, or anticlines, and troughs, or synclines, though now fully recognized, was still debated half a century ago, when up-lift by pressure from below, was more generally appealed to as the cause of the orogeny.

---

1) See also Chapters 24 and 25.

We no longer question the fact that the newly-formed anticlines may be breached, that is to say the top of the arch may be eroded, revealing the inner layers, and that these may suffer selective erosion, the weak layers being more rapidly destroyed, while the hard layers remain as ridges of inclined strata. Such ridges have generally been called "hog-backs" in the south-western United States, but Darwin[2] spoke of them as "uniclines" (Fig. 12), which must not be confused with "monoclines".

Scale will, of course, determine whether these uniclines should be called mountains or hills. Many a formidable mountain chain is only a long uniclinal ridge, made of strata dipping all in one direction, with the hardest forming the crest. Somewhere, perhaps many miles away, is its counterpart, the unicline dipping in the opposite direction if the original anticline was symmetrical, while between them a broad, open "anticlinal valley" is opened by erosion.

All this is common knowledge today, though fifty years ago it was a matter of serious debate.

Then, too, it is known from the evidence furnished by their remains, that the fate of the mountains as elevations may be complete annihilation, until only an erosion level or peneplane remains after the long continued denudation.

Single mountain peaks or "monadnocks" may arise above such peneplanes to form residual peaks, as Mount Monadnock in New Hampshire rises above the New England peneplane. This peneplane, let us emphasize it again, is a surface of erosion cut across strongly folded rocks, and the erosion surface was once continuous, when it stood at the "base level of erosion". That was when the nth cycle[3] of denudation had come to an end, because the country was cut so low that the rivers were no longer effective, and only residual monadnocks remained. Slight uplift, or doming, of the entire surface inaugurated a new cycle of erosion (the $n^{th} + 1$ cycle), which began with the entrenchment of the rejuvenated rivers.

2) DARWIN, C.: *Voyage of the Beagle.*
3) "n" may stand for several older cycles of erosion that the region has suffered, though only the last but one can usually be positively recognized.

The Highlands of Scotland are another such level peneplane which once stood at the base-level of erosion, that
is, near sea-level, but which, like most older peneplanes
of the pre-modern period, has passed into a new cycle of
erosion.   It is now a valley country rather than a. hilly
one; the valleys or glens are the dominant features in an
otherwise monotonous upland level.   To appreciate this
we must climb to the top of the valley-side.

2

Such an old peneplane, instead of being uplifted for
further dissection, may sink to become the floor of a geosyncline, or earth-trough of deposition.   And if it happens
that an old monadnock, or residual erosion peak, was in
the path of the subsiding area, as was the rocky eminence
of Shantung province in the line of the Huangho geosyncline of China, it will sink with the sinking floor, and
become slowly buried by the accumulating geosynclinal
sediments.   (Fig. 13)

Such buried *shantungs*, are sometimes met with in
older geosynclines, or on their marginal platforms.   Barraboo ridge[4] of Wisconsin is an example of a 'shantung,'
which has now been partly re-exhumed, and Caradoc Mountain of Shropshire appears to represent another.

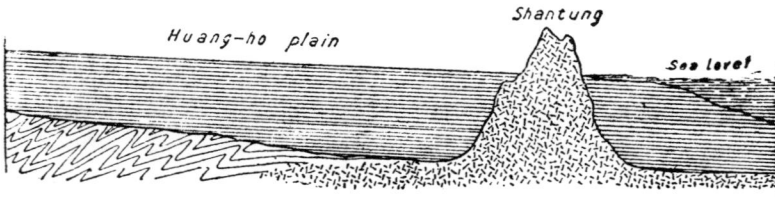

Fig. 13. Diagrammatic section of the Huangho plain of China, with the
residual erosion monadnock or shantung, on the old peneplane, in process of
burial.

Geosynclines are the incubators of mountain chains.
Probably every great folded mountain range (volcanoes
excepted) has been born from a geosyncline, in which
the strata sank deeper and the load above increased, until

4) Such an old buried monadnock which has been re-exhumed has been
called a "barraboo".

finally the revolution occurred and the folded mountain was born.

There is accumulating evidence that in the past all such orogenies, or foldings of geosynclinal strata, into mountain chains have occurred in interpulsation periods, following the sea-retreat. Though these orogenies did not affect all geosynclines alike, it is known that during every interpulsation period of the Palæozoic, one or more geosynclines in some parts of the earth experienced an orogeny.

These orogenies, however, were of a minor type, though often resulting in mountains of great elevation from the human point of view. They were minor, in that they did not involve the whole of the geosyncline in which they occurred. Every orogeny in the geosynclines of the Palæozoic, except the last, resulted merely in the folding of the strata in the axial portion of the geosyncline, the strata nearest the old-land. As a consequence, only a belt of folded mountains, or a folded foreland, was added to the old-land, on its inner or geosynclinal side, without completely destroying the geosyncline itself.

We may summarize this as follows:

Mountains are of interpulsation growth; the folding, when it occurs, takes place while the geosyncline is drained, that is during an interpulsation period. This is a fundamental law of the genesis of folded mountain chains. [5]

Most of the Palæozoic orogenies probably occupied the greater part, if not the whole, of the interpulsation period of emergence, so that when the sea returned to the portion of the geosyncline which was not affected by the orogeny, it found the mountain chain completed and added as a fore-chain to the old-land.

3

The folds are most intense near the old-land, but die away towards the opposite side; the marginal platform and a part of the geosyncline bordering it are unaffected, or develop only low domes and shallow basins. But the folded ranges were still too young to be covered by the transgressing sea during the next pulsation period. While

---

5) This does not include "block mountains" or fault-scarp mountains, though the probability is that if derived by disturbance of recent geosynclinal strata, they too are of interpulsation origin.

the remnant of the geosyncline, in front of these folded ranges, received the deposits of the new inundation, the ranges themselves were subjected to erosion, and no deposits were made in that part of the older geosyncline.

The erosion usually, if not always, occupied the whole of the pulsation period which succeeded the orogeny, and often the following interpulsation period as well. Therefore the post-orogenic pulsation period is not represented by strata in the folded region of the geosyncline.

However, the next pulsation period, that is the second after the orogeny, is generally represented, for when the sea returns for the second time, it finds the folded ranges peneplaned, the anticlines truncated, probably down to the level of the intervening synclines, and on this peneplane the new strata will encroach.

Thus an "unconformity" is produced. The strata of the first pulsation will be steeply inclined and truncated, and on the truncated ends of the folds, the strata of the third pulsation period will be deposited horizontally, while the strata of the second pulsation will be absent.

We may take an example from the Bohemian basin. (See Fig. 14, opposite). Here, the Sinian strata (sect. A), the first of the Palæozoic series, were strongly folded (sect. B), so that they now present a succession of steeply dipping, sometimes almost vertical strata of sandstones and clay-rocks, of great thickness, the Pilsen series. After they were folded — which occupied the interval before the Taconian (Lower Cambrian, old style) period — they were completely truncated by erosion, a very nearly perfect peneplane being produced.

So complete and extensive was the erosion (sect. C), that it could only have been accomplished during a long interval of time. This was the interval represented by the Taconian pulsation period, for marine beds of this age (old Lower Cambrian) are entirely wanting in this region.

There is, however, a thick series of continental beds, forming the Pribram series, which reaches an aggregate thickness of 1,300 ft. This rests unconformably upon the eroded edges of the Pilsen series (sect. D). It is followed concordantly by the *Paradoxides*-bearing Jince beds (sect. E). This Prihram series represents, however, the interpulsa-

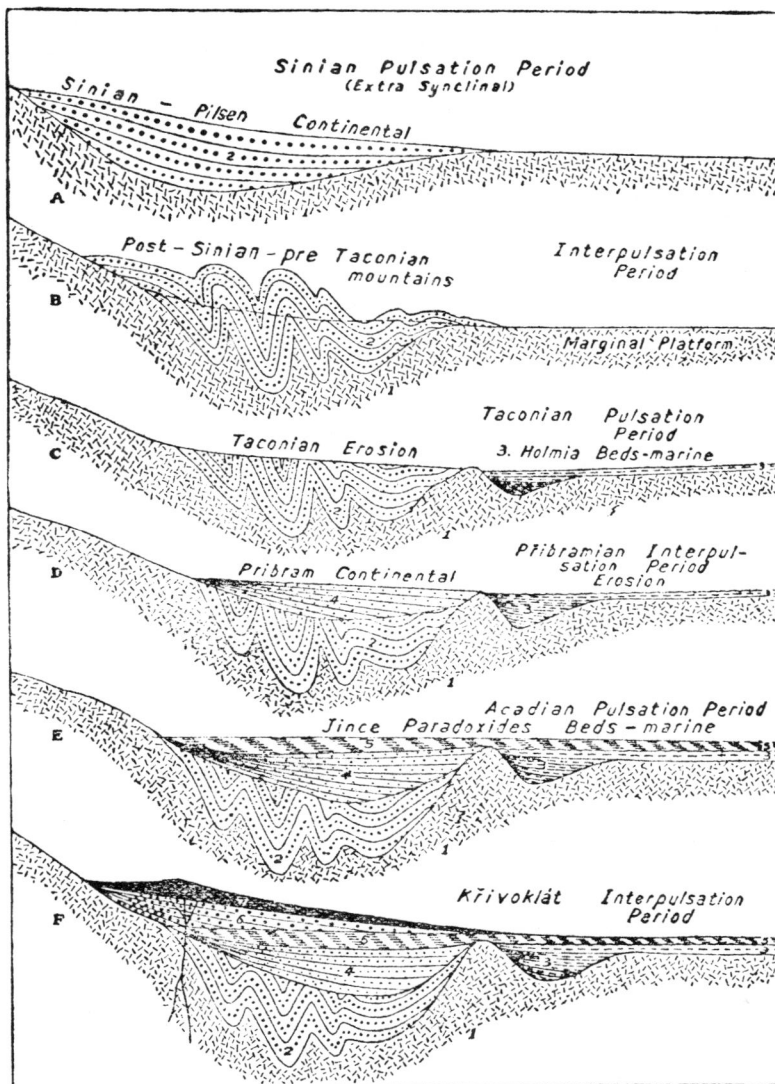

Fig. 14. Five sections in the same region, to show the succession of stratigraphic and tectonic changes in the Bohemian region, during early Palæozoic time. A—SINIAN: Pilsen continental strata (2) deposited in sinking geosyncline; B—The same strata folded and peneplaned; C—Completion of peneplane and deposition of TACONIAN (Lower Cambrian) (3) on continental platform; D—Deposition of PRIBRAMIAN continental interpulsation series (4); E—CAMBRIAN: Jince *Paradoxides* beds (5); F—ARCTOMIAN (Krivoklat) interpulsation period: Brezové Hory conglomerates and sands (6); Krivoklat volcanics (7). (Compiled from various sources).

tion deposits, while the time of the preceding Taconian
pulsation system is here marked by the peneplanation of
the old Sinian folds.

An unconformity of this kind is clearly distinct from
the disconformity which separates the formations of successive pulsation periods, and which is a normal feature
of the interpulsation period. Such a disconformable relation
should never be called an unconformity, even if it involves
a hiatus extending over a greater interval than is represented by one interpulsation period.

In a disconformity, which marks the break between
the deposits of successive pulsation periods, the strata are
parallel, and though there is a great hiatus between the
older and newer series, representing the length of the
interpulsation period, no pulsation period is omitted. Pulsation periods follow each other with only the interpulsation
hiatus between but no unconformity. In section Fig. 14
E, formation 3 and 5, each representing a complete pulsation system, are separated by an erosion hiatus which
corresponds to the non-marine formation 4 in the Pilsen
region. In an unconformity, on the other hand, at least
one pulsation system is absent, in addition to the interpulsation systems. [6]

This suggests a convenient method of naming the
orogenies, after the missing pulsation period. This will be
more fully discussed in chapter XXIV.

                              4

Before the close of the Palæozoic era North America
suffered a major orogeny, the Appalachian revolution. This
terminated the existence of the Appalachian and of the St.
Lawrence geosynclines and of the old-land of Appalachia
as a source of clastic sediment.

The former were changed into the Appalachian and
Notre Dame mountains, respectively, by the folding of the
strata, while the Appalachian old-land sank to become the
new geosyncline, in which Mesozoic and later strata accumu-

---

6) An exception to this seems to be found in the Girvan district of Scotland,
   where beds of Early Bala age (with *Maclurea logani*) rest upon folded and
   eroded Arenig, leaving the Llanvirn and Llandeilan as the sole representative of the time of the erosion interval. (See GRABAU: *Palæozoic
   Formations*, Vol. IV, chapter XVI). This looks like a lag in the orogeny.

lated. This was a complete migration of the geosyncline into the old-land, the first so far as known since the beginning of the geosynclinal periods, and we shall describe it in due course. Nothing like it had occurred before, and nothing like it happened again for many a geological period.

We naturally ask, What is the cause of the periodic folding of the strata of the geosyncline, and why does it fall into an interpulsation period and occur sometimes in one, sometimes in another geosyncline?

It will be serviceable to make a distinction between *orogenic incidents,* which are minor orogenies, and which occur during an ordinary interpulsation period, and affect only a part of a geosyncline, and *orogenic revolutions* or major orogenies, which occur near the close of an era, and which affect the entire geosyncline and cause a migration of the geosyncline into the old-land.

The minor orogenies, the orogenic incidents that produced only a local obstruction in the form of a folded fore-land added to the old-land without greatly affecting the geosyncline, last each only during a single pulsation period. Then the obstruction is removed, and the geosyncline is much the same as before.

Such orogenic disturbances may have a variety of causes. They may be produced by the same agent that was active in the development of the geosyncline which was the birth-place of the orogeny. Or they may be an indirect product of the periodic swelling and contraction of the sima floor of the ocean. During the positive phase of the pulsation the sima not only swells, but it actually stretches, and riftings, into which lava is injected from below, will occur.

During the negative pulsation-phase the sima sinks to a shorter radius, but having become enlarged by the addition of the volcanic matter in the riftings, it will exert a lateral, or tangential pressure from the sea toward the land. This is relieved by local folding of the strata.

This is Joly's hypothesis, and it serves admirably as a cause to explain the local orogenies in the marginal geosynclines. Whether it also could explain the orogenies in the Appalachian and Caledonian geosynclines, which occurred in the middle of the Palæozoic pangæa, thousands of miles from the ocean of the time, by the transmission of the compressive forces over that distance, must remain unanswered for the present.

For the major orogenic movements — the orogenic revolutions —, so far as they are represented by the Appalachian revolution, at the end of Palæozoic time, and the Alpine-Himalayan revolution of mid-Tertiary (post-Oligocene-pre-Miocene) time, it would appear that for these we must probably assume a more powerful force. Such a force may be cosmic, and require the interference of an extra-telluric agent, or the effect may be a direct consequence of the shifting of the sial-cap into a new polar position. We shall return to this subject in chapters XXIV and XXV, where we shall also review some of the minor orogenies and show their close relation to the sial-shiftings of the time.

CHAPTER VI

MASKED DISCONFORMITIES

1

So long as no clear distinction was made between unconformities and disconformities, the significance of the latter was rather submerged, and their occurrences were all too often overlooked. What are called *breaks in sedimentation*, usually characterized by erosion or a non-seqence in deposition, have often been noted, but as a rule not much significance has been attached to them.

It has now become evident, however, that larger disconformities must be expected at regular intervals in the geosynclinal succession of every land, for these major disconformities are the expressions of the retreat of the sea between the pulsation systems.

It may not be easy to prove the existence of the disconformity where no interpulsation sediments were deposited, or in the absence of evidence of volcanism or of erosion. But difficulties do not warrant the abandonment of our search for evidence of the hiatus or of its adequacy, and tacit assumption of its existence or absence, according to the predilections of the proponents or opponents of the Pulsation Theory, is unwarranted.

Often the existence of the hiatus is undetectable because of *masking*. This masking may come about in several ways which it behoves us to examine in detail; for the delimitations of the systems depend, in the first place, on the precise location of their upper and lower limits, and on the contacts with the preceding and succeeding systems, in other words, on the location of the interpulsation hiatus.

It may be stated at the outset, that there are no transition formations from one system to the other, neither in the Palæozoic nor among those of any subsequent period of the earth's history. This statement will of course be challenged, but I maintain that no indubitable proof of the existence of such a formation has yet been adduced. All the so-called transition formations, the *straddle formations* of Schuchert, are such only in appearance.

They are examples of a masked hiatus, or they represent continental interpulsation deposits. In the latter case it remains to be shown that such a deposit begins with the retreat of the sea, during the closing stages of the previous pulsation system, and continues until the sea readvances. Even in that case it is not a transition formation from one pulsation system to the next, but an independent interpulsation system, which separates as well as unites them.

Thus the assumption of transition between the deposits of successive pulsation systems is unwarranted, and is merely an expression of our lack of understanding of the accidental obscuration of systemic boundaries.

It should perhaps be emphasized that the recognition of the age of a system depends in the first place upon its fauna. There is at present less difference of opinion than formerly regarding the typical faunal elements of the standard systems; but we are still far from agreement regarding the dismemberment of some of the older comprehensive systems, and the reconstitution from such severed members, either singly or in combination, of distinct new unit systems.

But that is a phase of stratigraphy familiar to many of us. We have passed through it before. We need only recall the dismemberment of the historic Silurian system into Ordovician and Silurian, which has now been universally accepted, though many of us can remember the fierceness of the controversial battles fought over the proposal.

Perhaps that will make us tolerant of the reluctance, if not actually pronounced opposition, of many geologists to the proposal to again subdivide or split off certain portions of each of the older, now better understood systems, and their recombination into new systems, *if the facts warrant it.*

That sentence should express the crux of the problem; not reluctance to disestablishment of historical precedent, but unmistakable evidence that a rearrangement of our

classification expresses more fully the facts of the case.
But, of course, it is essential that the proponents of dismemberment must show cause why they should be heard,
and show unquestioned evidence from widely separated
areas that such dismemberment is necessary.

On the other hand, we must avoid over-reliance on what
appears to be local negative evidence, against overwhelming
positive evidence from many other regions and parts of the
world. To assert that, whatever may be true elsewhere,
it does not apply to our own region, is provincialism ; and
that has no place in science.

If an assemblage of species of *Orthoceras, Piloceras,
Maclureas*, etc., is positively recognized as characteristic of
the Beekmantown, or Canadian system of North America,
such an identical association cannot be regarded as representing Middle and Upper Cambrian in North Scotland,
however difficult the recognition of the hiatus and disconformity in the latter region may be, and however eminent
the advocates of continuity.

2

And this emphasizes the importance of fossils as the
primary basis of classification. There is no other.

But if a certain species, or an assemblage of species,
indicates Silurian age, and another indicates Ordovician age,
then the finding of both species, or species of both types
together, should not be taken as *prima facie* establishing a
transition from one to the other. If systems are separable on
the basis of pulsation, a normal association of the species
of two systems, is, to put it mildly, extremely unlikely (I
should call it wholly impossible), for it would imply, that
the older species continued unchanged through the long
interpulsation period, granting that there is such an interpulsation period. Such a persistence of type is contrary to all
our understanding of the progress of evolution of organisms.

True, the newer fauna must be regarded as derived
from selected survivors of the older fauna, for no one any
longer considers successive faunas as distinct creations *de
novo* ; but such derivation must have taken place in a
restricted epi-sea, to which the waters withdrew during
the interpulsation period, and under such conditions persis-

tence of specific types without change is unthinkable.

I am aware that these statements may sound dogmatic, since the existence of pulsation and interpulsation periods has not yet been proved to be universal, or contemporaneous, all over the earth. Therefore, we shall state our thesis in this form : If pulsation systems are separated by long interpulsation periods, during which the sea has drained away from all the geosynclines, remaining only in certain local basins, or epi-seas, marginal to the ocean, then the persistence of types unchanged from one system to the other is so extremely unlikely that overwhelming proof must be adduced before it can be accepted.

Then, too, it must be shown why the long-persistent type disappeared completely after the opening of the new period, where, if it occurs at all, it is usually confined to the basal strata. If it can hold its own during the period of adversity without becoming modified, why during the renewal of favorable conditions in the next period should it succumb? Why should it survive only to make a farewell appearance long after its particular era had come to a close?

All this, of course, is based on the correctness of the interpretation of the existence of an interpulsation hiatus.

Let us look at the other side of the picture.

If we find that the initial formations of a higher system follow the terminal formation of a preceding system, without transitional faunas, the two systems being absolutely distinct faunally, we naturally assume an interval of sufficient length to bring about the extermination of the old and the development of the new forms.

What happened during this interval, and how otherwise can it be visualized than as a complete withdrawal of the sea from the area in question? And if the same relation is found in a second and in a third area, in widely separated portions of the earth, even in distinct continents, should we not have cause to think that more than a local phenomenon must be involved?

Let us pause here and recognize the fact that our knowledge is still too incomplete. We do not yet know a sufficiency of facts for all the systems to assert their distinctness, but we do know a sufficient number of widely spaced facts to show that for some of our systems this distinctness is

absolute.   We shall marshall in subsequent pages enough
examples to carry conviction, though not all that are available.
For most of these facts the reader must be referred to the
comprehensive volumes now in course of publication.

And if we have full evidence of pulsation of a universal
character for those of our formations which have been
studied intensively, and some evidence for the others which
still await more detailed investigation, are we not justified
in believing that pulsation is serviceable, at least as a working
hypothesis in the light of which all formations must be re-
examined?

3

But this does not absolve us from the necessity of
explaining the apparent transition of faunas, or the apparent
persistence of older types.   I say *apparent* advisedly, for
I regard such instances as purely apparent, as illusory,
and not as actual facts.   Let us elaborate the problem
and see its solution, as it is explained by the PULSATION
THEORY.

When the sea withdraws from the surface of the
land, where it had rested long enough to leave behind
strata with preservable organic remains, shells of brachiopods and molluscs, trilobite tests, crinoid remains, etc., the
strata in which these have become entombed will be
subjected to weathering on exposure.

Under ordinary conditions a deep regolith may be
formed, consisting of clay from the disintegration of the
lutytes, and this clay may be full of fossils which were
contained in the original shale.   Or, if the original rock
was a limestone, and the fossils which it contained were
more or less silicified, the solution of this limestone will
leave behind extensive surfaces strewn with weathered-out
fossils.

In the course of an interpulsation period, there may
thus accumulate on the surface, the residual product of
the solution of a thick limestone bed, through which these
fossils were once distributed.

Every collector knows that weathered surfaces are
the ideal hunting ground, and the only one from which
there is much hope of obtaining perfect specimens.   I

have seen acres of surface underlaid by Devonian (Hamilton) shales, exposed for a time by the removal of its cover
of glacial till, and this surface was densely strewn with
perfectly weathered-out fossils, many of them separated
valves of brachiopods, — a collecting ground to keep an
enthusiastic palæontologist busy for hours in merely picking
up specimens. What would happen if the returning sea
should rearrange the covering clay-layer, and redistribute the
weathered-out specimens during its earlier stage of occupancy, along with the newcomers which accompanied its
invasion? The answer is obvious. It would form a "transition bed" where the remains of the long dead of a former
dynasty are re-interred by the side of the recently dead
invaders of the newer dynasty. But this would not mean a
transition from one period to the other, but merely the
burial of the dead of two distinct and widely separated periods
in a common graveyard.

4

Travelers have reported extensive surfaces in the Lybian
and other deserts strewn with silicified tests of spatangoid or
cidaroid echinoids, or the massive spines of the latter, to be
had for the labour of picking them up. The enclosing rock
had disintegrated and been removed as dust by the wind.

Johannes Walther[1] speaks of encountering such re-sidual
fossil-strewn fields, extensive "Hamada" surfaces appearing
as if paved with *Nummulites,* oyster-shells, or the tests
of echinoderms, fossils left behind by the weathering of
successive strata, and the removal by deflation of the rock
which had formerly enclosed them and which had weathered into sand and dust; and he laments the necessity of
leaving behind such a rich collecting ground, designed to
gladden the heart of an enthusiastic geologist.

What would happen should the modern sea, with its
wholly different genera and species of the same classes,
encroach gently over such a surface? Would the commingling of Tertiary and modern species in a few feet
of strata of the modern series be regarded as meaning
that there was continuity of deposition in a persistent sea

---

1) *Gesetz der Wüstenbildung,* p. 213.

from Eocene to modern time? The answer can only be *certainly not*.

Of course, a sudden inrush of waters might play havoc with the delicate relics of a former day, though it might not seriously affect the coarser silicified specimens. However, no one assumes that a sudden inrush will take place. A slow creeping up of the sea over a level, or nearly level, surface, strewn with these remains, is the picture suggested by the return of the sea after its long absence, a return due to the slow rising of the sea-level with the beginning of the new pulsation. Not a ruthless conqueror, bent on destruction of all that had been achieved before, but a conservator of the past and its relics, and a contributor to the old record, of the new chapter which it is prepared to write.

And so the faunas of the transgressing Silurian sea, as they become part of the record, will mingle with the relics, still preserved in perfection, of the last deposits of the older Ordovician sea, and the harmonious association of the new with the old will suggest continuity, and to the casual collector it will seem that the Ordovician fauna gracefully died out *sans* struggle, *sans* protest, when the newcomers of the more advanced Silurian type arrived to occupy the territory.

The assertion that these newcomers settled on an old graveyard, where some of the buried dead of the older generation had become loosened in their beds by long exposure, and that the resulting detritus formed a common shroud for old and new alike, would appeal primarily to a philosophical inquirer. Such an inquirer would see an analogy between the mixed fauna of these successive strata and the bones of modern animals buried, accidentally or intentionally, in sands of Oligocene, or Miocene age, which still contain the bones of the saber-tooth tiger, the *Baluchitherium*, or any of the other Miocene or Oligocene animals.

And even the layman, who no longer believes in the existence of giants during the infancy of the human race, would scarcely assert that these were contemporaries, the older awaiting the coming of the newer types before they composed their bones to rest, willingly, or under compulsion, in the common burial ground consecrated to the new and superior race.

Instead of regarding the commingling of fossils of the older pulsation system in the basal beds of the newer formation with the distinct fauna of the newer system, as evidence of continuity, we should hail it as a proof positive of discontinuity and the existence of an otherwise unrecorded time-interval; and our surprise should be at the infrequency with which such facts have so far come to our notice.

CHAPTER VII

A PRE-CAMBRIAN PALÆOZOIC EARTH

1

ONE of the most perplexing problems that the student of earth-history encounters deals with the conditions of the pre-geosynclinal earth, *i. e.* the earth before the orderly development of geosynclines that characterized the Palæozoic era.

With this goes hand in hand the question of the life of pre-Palæozoic time, for, that life existed before the Palæozoic era is shown by its high state of development at the opening of Palæozoic time, especially the high organization of the early trilobites.

No one any longer questions the existence of life long before the time of the earliest Cambrian organisms that we have any record of. But the nature of the organisms, and the reasons for the all but complete absence of a record of their existence, has been one of the major puzzles of our science.

Did the organisms lack all the hard parts that could be preserved in the strata? If so, was it because of a primitive limeless ocean, as Daly has argued, or because of their great activity, as suggested by Brooks and Raymond? These are questions that must be seriously considered.

The problem may be approached from another point of view by asking, Do we know of any pre-Palæozoic strata of marine origin that might contain the remains of such animals? And here, I, for one, hold that the answer is, emphatically, No. If we review all the known pre-Palæozoic formations of the earth we shall find that, with the single exception of the Grenville limestone, not one of them has yet furnished evidence of a marine origin; all the others bear the unmistakable stamp of non-marine formations.

They are either geosynclinal deposits in a series of

inland water-bodies, as was the Sinian limestone; or they
are huangho-plain deposits, typically shown by the Beltian
series, or, finally, these pre-Cambrian strata represent inland
basin deposits, some in arid regions, others in regions with
more or less rainfall. But not one of them is a deposit
in a normal geosyncline flooded by the sea, or a deposit
on the continental margin.

We may briefly review the better known of these
formations.

2

The Torridon Sandstone, typical outcrops of which
are seen in North Scotland (Loch Torridon, Loch Maree,
etc.), rests with a very pronounced unconformity upon the
much disturbed ancient Lewisian (Louisian) gneiss. Though
this gneiss surface on the whole was reduced by erosion
in pre-Torridon time to a subdued topography, nevertheless it retained many minor irregularities; and these were
still in existence when the Torridon sandstone was deposited
upon it. (Fig. 15).

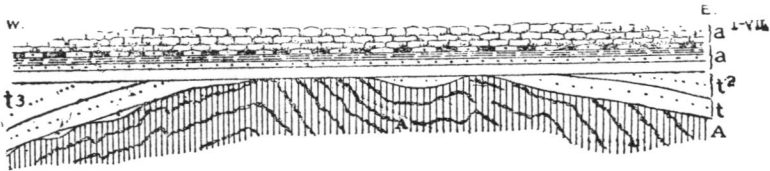

Fig. 15. George-oric, or post-Torridonian Orogeny, St. Lawrence Geosyncline. Diagrammatic section, showing Cambrian succession with basal unconformity crossing post-Torridonian folds. West Highlands of Scotland.

A=Lewisian Gneiss peneplaned and unconformably succeeded by Torridon
sandstones t^{1-3} of Sinian age. A second, but faint, unconformity separates
the Torridon and late Lower Cambrian or Georgian. In the center, the Torridonian has been removed entirely, the Cambrian resting on gneiss.

t^{1-3}-Torridonian: t^1) Diabaig Group; t^2) Applecross Group; t^3) Aulthea
Group.

a-Basal Quartzite, Pipe-rock, Fucoid Beds and Serpulite Grits of the
Cambrian (separately ornamented).

a^{I-VII}-Calcareous series of Cambrian and Skiddavian. (After Peach and
Horne).

One of the surprising features of this younger formation is that it has been so little disturbed by later movements, that it often is still nearly, if not quite, horizontal,
although the underlying gneisses are strongly disturbed.

Nevertheless the Torridon also suffered extensive erosion in pre-Cambrian time, for the *Olenellus*-bearing beds (Taconian or Eriboll (Eireboll) quartzite, *i. e.*, Lower Cambrian of old classification) now rest across their gently bevelled edges, as can be seen in many modern mountains of North Scotland.

The frequent horizontal position of the Torridon today and the slight inclination of the younger *Olenellus* sandstone show of course that there has been a reverse tilting in later time.

Though so little indicated, the hiatus between the Torridon and the Eireboll quartzite is a profound one, for the latter rests on various members of the Torridon or even on the gneiss itself, where the Torridon has been entirely removed by erosion.

The total thickness of the Torridon varies considerably, because of the uneven surface of the old floor upon which it was deposited. However, the maximum does not exceed 20,000 ft, but is generally much less.

In character these rocks bear the clearest evidence that they were formed on land, in inland basins, or possibly in a part of a more extended geosyncline, but they show no indication whatever of the proximity of the sea. In fact the nearest sea-shore was more than 2,000 miles distant at the time of their formation. Often these rocks show conclusive evidence that they were derived from the product of weathering of the old gneiss which forms their foundation. But much of the material gives evidence in its minerals that it was transported from greater distances and derived from rocks no longer exposed in the Scottish Highlands. This is not surprising, when we contemplate the distance to which such material can be carried by streams, and when we realize how small a fragment of the original pangæa is represented by the Scottish Highlands.

Frequently more or less arid conditions prevailed during the deposition of parts of these rocks. These conditions are indicated by the abundance of undecomposed felspar, which forms felspathic sandstones or arkoses of great thickness (Applecross division, about 6,000 ft), and this is also indicated by the presence of wind-facetted pebbles, or *dreikanter*; for these are very characteristic of desert surfaces.

Similar types of rock are represented by the so-called *sparagmite* of Scandinavia, and by some of the newer Algonkian rocks in the region of the Great Lakes of North America, which are derived from the gneisses of the Canadian Shield.

"It is reasonable to infer," say Peach and Horne,[1] "that these isolated relics of old-land surfaces were united in pre-Torridonian time, thus forming a continuous belt from Scandinavia to North America." Not however with their present geographic position.

Cross-bedding is another characteristic of the arkoses, while ripple-marked, sun-cracked and rain-pitted surfaces are found in the shales, flagstones, and old bedded sandstones, of other levels, which were formed during less arid periods.

The activity of rivers is shown by the widely transported pebbles such as those of spherulitic felsite, which find their nearest counterpart in the ancient "Uriconian" rocks of Shropshire. Lenticles and thin beds of impure calcareous rock occur at some levels, but limestone is extremely rare. No evidence of life has been detected, unless the occurrence of phosphatic pellicles and nodules in the shales of the highest division indicate the presence of organisms.

Taken altogether then, these sandstone series indicate deposition in inland basins, or in geosynclines far removed from the sea. The material was washed in by torrents from the surrounding higher ground, and brought by rivers from more distant regions, while standing waterbodies were scarce, or often altogether absent. Chiefly, however, desert conditions prevailed. Probably much wind activity also prevailed, this removing much of the clay and other impurities.

Thic explains the absence of all traces of organisms. Although we realize that the sea was far too distant at that time to allow the marine organisms to enter this region, we might expect the occasional remains of river animals such as characterize the contemporaneous Beltian formation, if the region had been moist enough to develop permanent rivers, and temporary playa-like standing water bodies.

---

1) *Geology of Scotland.*

3

The Belt Formation of western North America is another deposit of this type, but formed perhaps in the
early rimming geosynclines, and under more pluvial conditions, and therefore composed more largely of fine muds,
now split into layers of almost paper-like thinness. These
layers often show on their surfaces the fragmentary remains of what Walcott held were "Merostome-Crustacea",
allied to the eurypterids of later periods. That this is

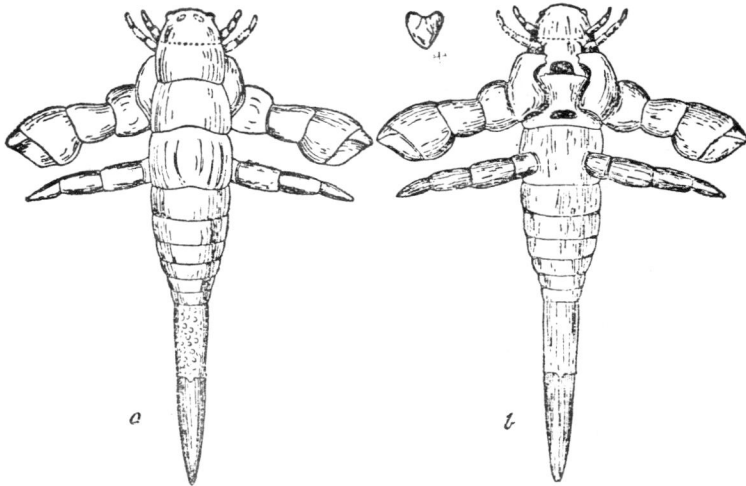

Fig. 16. *Protadelaidea howchini* David and Tillyard, a new type of primitive arthropod of the order Arthrocephala. Restored by David and Tillyard
from fragments abounding in the Teatree Gully quartzites of the Lower
Adelaide series of the Sinian, pre-Cambrian. Samfrau Geosyncline, Adelaide,
Australia. Approximately × 3/8 natural size.

a. dorsal view; b. ventral view. (After David and Tillyard).

true of many, if not most, of the fragments seems probable. They probably represent the broken remains of the
chitinous outer integument of these animals left exposed
on the surface of a playa-lake where they were carried
by river currents. They are found in the Belt formation
of Montana and British Columbia, in strata suggesting
an origin like that of the Huang-Ho plain of China.

These fossils Walcott has named *Beltina danai*. Most
of them show no structure or surface markings. Though
they do not doubt their organic origin, many students
hold that they cannot be Merostome-Crustacea, because they

believe that even pre-Cambrian Crustacea must have a structure or surface-marking, analogous to that of the outer armour or horny integument of the later merostomes.

Perhaps this skepticism is justified, perhaps not. If not Crustacea, they certainly suggest related forms. They are most certainly not marine algæ, as has been suggested, because there was no sea within a thousand miles.

But though the original *Beltina* lacks the crustacean skin markings, one that shows them has been found by the late Professor Stuart Weller in Montana. Not only the surface markings, but the outline and convexity are such that there is little or no dissent among palæontologists that it is an early arthropod. Others found in Alberta, though somewhat less definite, also pass muster. Finally, the finds, reported a few years ago, of thousands of such remains in Australia, show that the geosynclines, when they did appear, preserved some of the teeming life of the rivers of that day. For that these animals were denizens of the rivers, seldom, if ever, even in later periods, entering the salt water, though their remains were often swept into the sea, is, in my opinion, abundantly attested.

From the Lower Adelaide series some 14,000 to 20,000 feet below the *Redlichia* beds (Taconian) a large arthropod *Beaumontia eckersleyi*, related to the Eurypterids, was obtained, and in the Teatree Gully quartzites at a horizon 3,000 feet lower and nearly that distance above the basal schists, the most primitive arthropod yet known was found. This has been named *Protadelaidea howchini*, Fig. 16 a. b.[2], and it has been placed in a new class, Arthrocephala. The material obtained was extremly fragmentary, as might be expected in river laid deposits. Annelid impressions also abound. David, however, believed these animals to be of marine origin, for doubtful impressions suggesting *Lingula* and *Obolella* have also been found in this rock, while Radiolaria and other micro-organisms also seem to be plentiful in some beds.

4

In the fresh-water lakes which at times existed in the geosynclines after these had first formed, extensive

2) DAVID and TILLYARD: *Pre-Cambrian Fossils.*

limestones were deposited, both in East China and Australia
and in western North America. These limestones are largely
the product of structures built by organisms named *Collenia*,
etc., and believed to be some form of lime-secreting fresh-
water algæ.

Of *Collenia* three or more distinct species are known
from the Sinian rocks of China (Cathaysian geosyncline),
several from Australia, and a considerably larger number
from the equivalent beds of western North America
(Palæocordilleran geosyncline). None of these, however,
were marine.

Some minute, more or less sphærical, structures,
which are regarded as shells of Radiolaria, and others
resembling spicules of sponges, have been obtained from
rocks of this early age in France, and reported from the
Adelaide series of Australia, but even if they represent such
primitive organisms of marine type, their occurrence is
such, in both regions mentioned, that they were more probably washed into the basins or blown by winds as dead
shells from a distant sea coast.

There is nothing incongruous in this idea, for we
have at least one authenticated modern example in the
Junagarh limestone of India,[3] and the sea was not so far
removed from the regions of the great peripheral geosynclines of Sinian time, in which these remains were found.

Of course, all the sediments carried by the pre-
Cambrian rivers did not reach the sea. Then, as now,
the great inland rivers washed their load into the enclosed
basins, which were no less numerous than today, probably
more so. There their waters either formed lakes, or
evaporated entirely, according to the climate and degree
of aridity.

That such rivers existed on the ancient pangæa cannot be doubted, and that they supplied sediments to many
a depression and basin, some of tectonic, some of deflational
or other origin, which had come into existence during
the long period of erosion that followed the first great
folding, will hardly be questioned.

It would be folly to assume that, given a great continent with mountains and depressed basins, such inland

3) GRABAU: *Principles of Stratigraphy*, p. 574, with references.

types of river-deposits as we know today were absent. As a matter of fact we know of thousands of feet of such ancient continental deposits in many parts of the world, and only the lingering influence of the old superstition that all deposits of clastics, no matter how thick and no matter of what character, must be marine, has prevented us from giving full recognition to these ancient rocks as inland basin deposits.

5

One of the most interesting series of these ancient sediments is found in what is today Finland, now near the sea, but in those days far inland on the pangæa.

Here, the folded and metamorphosed and subsequently peneplaned Archæan rocks are followed unconformably by an immense series of conglomerates, quartzites, dolomitic limestones, etc., all now more or less metamorphosed. They constitute the CALLAVIAN series. These were also intensely folded and likewise peneplaned, and upon the truncated edges of these earlier sedimentary series lie from 1,500 to 2,000 feet of conglomerates and sandstones, with ripple-marks, cross-bedding and desiccation fissures, exactly as they are found in the deposits of inland basins to day.

And with these occur clay and carbonaceous lutytes, with a bed of anthracite coal 2 meters in thickness, the oldest coal-bed known in the world.

There seems to be no escape from the conclusion, that this coal represents pre-Cambrian vegetation, a luxurient growth in a climate characterized by sufficient moisture, and freedom from excessive cold.

What was the nature of this, the oldest vegetation of which we have any record? We can only conjecture. Of one thing, however, we may be certain, and that is that this coal was not formed by marine algæ, for the sea at this time was more than 2,000 miles distant. The reason why this is so commonly spoken of as evidence of marine vegetation is the old doctrine, referred to above, that these strata could not have been formed anywhere but in the sea. But we are leaving these old ideas behind.

There is, of course, no evidence that land plants existed at that time, though I, for one, would not feel unduly surprised if such were to be discovered. The Potsdam sandstone and the "tree trunks" of the Roxbury conglomerate, described in later chapters are, at least, suggestive. But that there were plants with more or less woody tissues, growing probably in ponds and swamps, is positively suggested by this Finnish coal-bed of pre-Cambrian time, unless the inland fresh water algæ were so highly developed that they had the necessary consistency for the making of this coal.

These JATULIAN deposits, as the entire series which includes the coal is called, also enclose flows as well as intrusions of basic lava (gabbro, diabase), and the whole series is closely folded, so that the strata now stand at high angles, in some cases almost vertical.

These folds were again worn down to a peneplane surface, the third in this region, and after that another thick continental series of clastic strata, also of pre-Cambrian age, was deposited on their truncated edges.

This is known as the JOTNIAN series, and begins with conglomerates formed of pebbles of the underlying rocks. It consists largely, however, of thick masses of quartzites and sandstones with cross-bedding, ripple-marks, and so forth, and of clay-shales as well. The whole, which has a thickness approximating 2,000 meters, is still largely horizontal and little or not at all metamorphosed. This also includes intrusions of diabase and of the famous "Rapakiwi granite". Evidently these later rocks, like the ·Torridon sandstone of Scotland, represent deposits after the series of great disturbances had come to an end, and it is possible that they no longer represent inland-basin deposits, but the first sediments of the geosynclinal period of the Palæozoic earth. But these geosynclines were not yet subject to the ingression of the sea. Over these undisturbed rocks marine waters eventually transgressed, for they are succeeded by fossiliferous rocks of Cambrian age.

Pre-Cambrian rocks, of the Finnish type, though without coal, are found in many places of the old pangæa, and always they have characteristics which clearly enough indicate their origin as continental deposits.

6

In the Lake Huron region and elsewhere in eastern Canada, these old continental formations are represented by many thousands of feet of conglomerates, graywackes, quartzites, shales, cherty limestone and iron-bearing rocks, all very little metamorphosed, but intruded by diabase dykes.

Three distinct divisions are recognized, separated by unconformities, and these have been named in descending order:

3.   The Keweenawan,
2.   The Huronian,
1.   The Keewatin.

Upon these formations lies the fossiliferous Upper Cambrian (Cambrovician).

A very different type of rock, and one that suggests a distinct origin, is the formation known as the Grenville group of eastern Canada. This is a complexly folded and strongly metamorphosed series of ancient limestone, intruded by enormous laccoliths of granitic rock.

In the vicinity of Trembling Mountain and Trembling Lake, Sir William Logan has found 4 distinct bands of this limestone, ranging in thickness from 20 to 1,500 feet, and separated by orthoclase gneiss, and some beds of quartzite. The combined thicknesses of the crystalline limestone is about 3,500 ft.

The beds are intensely folded, the whole series showing in some parts 4 or 5 anticlines in the space of a mile. But besides this there are many minor wrinkles and folds (See Fig. 1, p. 2), the gneiss and limestone being interbedded and folded together.

It was in this Grenville limestone that the far-famed Canadian dawn animal, the *Eozoon canadense*, was found. This consists of globular or irregular masses of white crystalline limestone and green serpentine, sometimes several feet in diameter.

In structure it consists of more or less regularly alternating thin layers of calcite and serpentine, the latter sending off rod-like masses, which penetrate the calcite in a manner suggesting that they fill former canals and tubes, the serpentine having replaced the original organic matter.

They were regarded originally as primitive Protozoa and later as algæ, but the tendency of late has been to consider them the product of crystallization, and thus of inorganic origin. As such these structures have mostly been retired from text-book lore, but they may be slated for a resurrection, for, as long ago remarked by that astute student of metamorphic geology, Dr. Van Hise[4] "it is doubtless true that many of the specimens which have been called Eozoon, are results of the forces of crystallization; but, admitting this, it does not follow that all of the material called Eozoon is of this character".

It seems not impossible that these great limestones may have been an original deposit on the floor of the primary ocean or hydrosphere, when that still covered the whole of the sial to an average depth of 2.64 km.

Of course, some parts of the primary sea were probably much shallower, with greater depths in others, and it was in the shallower portions that limestone deposition may have taken place. The growth of these masses, if they were organic, may often have been interrupted by engulfing into the sial whenever that reached the plastic or molten condition.

There would be nothing incongruous in considering these calcite masses as having been precipitated by primitive bottom organisms of very low organization, which may have occupied the shallow portion of the sea. The astonishing fact is not so much that there is evidence that life existed in the primitive ocean before land emerged, for life of some kind probably began in the sea, as soon as that had come into existence, but rather that any record of it should have survived the intense crushing and metamorphism that have accompanied the emergence of the land, and the transformation of the sial-sphere into a land hemisphere of metamorphic rocks.

That was the end of that particular type of bottom life, but living organisms may well have survived as plankton in the open ocean, and these served as the prototypes from which the higher forms of life in the succeeding eras trace their descent.

7

There is probably no way of correlating the post-Sialic pre-Palæozoic rocks of one region with those of another. For we must realize that intermontane deposits were the

---

4) CHARLES R. VAN HISE: *Correlation Papers: Archæan and Algonkian*, p. 492.

order of the ages in that period of immeasurable length which followed the gradual emergence of dry land from the waters that covered the original earth.

Whenever a period of rest intervened between the epochs of crushing and crunching of the crust and the outpouring and intrusion that characterized the constantly recurring igneous activity, that rest period was employed by rivers in erosion and filling of intermontane basins with clastic material; these in turn were folded and crumpled and metamorphosed during the next period of contraction of the crust.

It is then not to be wondered at that the ancient complex of the earth suggests an almost hopeless confusion, into which an understanding of the sequence of events may be obtained in single localities; but comparative studies in all parts of the world, may, for a long time, perhaps for ever, have to remain merely an approximation.

CHAPTER VIII

THE EARLY SEA-INVASIONS

1

So far as we are able to judge from the facts at our
disposal, the first undoubted ingress of the sea into
the geosynclines of the pangæa occurred in the Taconian,
or "Georgian" era. This was formerly classed as Lower
Cambrian, but since it marks a distinct pulsation period,
it must be treated as such, and given a distinct name.

The name Taconian System [1] was given it by Ebenezer
Emmons, from the Taconian Mountains on the New York-
Massachusetts border, and it has prior claim over all others.

Whether we regard the Taconian earth as a pangæa, or as divided into separate continental blocks, we
can recognize a number of major, and several minor
geosynclines. These are :

1. THE APPALACHIAN GEOSYNCLINE,
2. THE ST. LAWRENCE GEOSYNCLINE,
3. THE CALEDONIAN GEOSYNCLINE,
4. THE PALÆOCORDILLERAN GEOSYNCLINE,
5. THE CATHAYSIAN GEOSYNCLINE, with
   5a. THE HIMALAYAN GEOSYNCLINE, as the radial
       geosyncline,
6. THE PRE-ANDEAN GEOSYNCLINE,
7. THE FRANCISCAN GEOSYNCLINE, and
8. THE SAMFRAU GEOSYNCLINE of South Africa and
   East Australia.

Not all of these were necessarily submerged at the
beginning, but they were probably in existence.

In most cases each geosyncline is characterized by a

---

1) Unfortunately, the name "Taconic revolution" has come into very general
use for an orogeny of very much later date (Sil-oric), and so tends to
create a confusion in nomenclature. Therefore, the term Georgian may
be used as an alternative for the earliest marine pulsation system recorded in the geosynclines, until the name Taconian has dropped out of use
for the Sil-oric orogeny.

distinctive assemblage of extinct organisms, a few of which may be noted.

Trilobites[2] are the dominant types, as well as the oldest of the marine organisms of which we have knowledge. These animals, now entirely extinct, were already of a high degree of organization at the opening of Cambrian-Taconian time. Their abrupt appearance in the oldest known fossiliferous rocks demands explanation — and cannot be dismissed with a mere statement of fact.

This is one of the major problems of the history of the animal world, for in the natural course of events we should expect to find that the oldest rocks, which were formed in the sea, should enclose only the remains of the most primitive animals.

But trilobites can certainly not be considered as primitive animals, though they may be the most primitive of their class.

The integument of these animals is chitinous, or horny, like that of the horse-shoe crab (*Limulus*), common on some sea coasts of today. This animal passes through a larval stage, strikingly suggestive in appearance of a trilobite. The integument of *Limulus* is hardened by impregnation with lime-salts, and so was that of trilobites in general. But even without this such structures are capable of preservation, as is shown by the chitinous integument of graptolites and even more perishable types of animals. So the theory that the pre-Cambrian ocean was without lime in solution, even if true, would not explain the absence of at least traces of these animals in marine strata,

---

2) These animals receive their name from the obvious longitudinally three-lobed character of their bodies, of which the middle part, the axis, is the most conspicuous. The head (*cephalon*) has the axial part most pronounced and flanked on either side by "cheeks" bearing the eyes, which in some of the more specialized types contain a visual surface of many lenses, like those of insects. Generally a "suture" separates the outer, or "free" cheek from the inner, or "fixed" cheek on each side of the "glabella", and these outer cheeks are commonly prolonged into long lateral spines. Generally, the greater part of the body (spoken of as "thorax") consists of a succession of many rings, one in the simplest form, but usually many in more specialized types. These may become dissociated in the dead animal. The terminal tail-piece is called the "pygidium", and is again a single piece, though it generally shows a central axis with rings marked off and a marginal rim. In size it varies greatly from minute to equal in size with the head. Antennæ and thoracic legs have been found in a number of species, and probably existed in all (see Fig. 28, p. 105).

before the Cambrian, PROVIDED SUCH STRATA WERE KNOWN.

If, however, the sial-crust of the earth was withdrawn into a pangæa during pre-Palæozoic time, while only the most primitive of animals and of plants existed in the sea, and this pangæa was not covered by the sea until the beginning of Palæozoic time, then there was no opportunity for the preservation of marine organisms in the rocks formed on the pangæa, because they were all of supermarine origin.

If, then, we find such highly developed marine animals as the trilobites appearing suddenly in the rocks of the earth at the opening of the Palæozoic, we can only consider this as an incontrovertible proof that the sea did not cover the lands prior to Cambrian (Taconian) time, and if we reflect that these animals had attained such a high state of development at the time that the earliest Palæozoic organisms were preserved in the sediments, we must inevitably conclude that pre-Palæozoic time was of immense length, certainly as long as, if not longer than, the Palæozoic, and that during all this time animal and plant life was developing in the sea. Further, it is evident that sea and land were wholly without intercommunication, so far as the life of the former was concerned. The sea was, however, not without influence on the land, for aside from the fact, that it attacked the shores incessantly, it was the source from which the winds derived the water-vapors which they carried over the land. There precipitation in the form of rain and snow furnished the inorganic agencies that shaped the land by erosion of its surface and produced the clastic material from which the mass of stratified rocks of the geosynclines was made.

2

In the Appalachian-St. Lawrence geosynclines the leading trilobite genus was *Olenellus*. (Fig. 17-9). It is found in the southern Appalachian strata from Alabama to New Jersey, in the Taconic slates of the Hudson Valley, above Troy, N. Y., and in the Georgia, or Parker slates of Vermont, where the genus was first discovered.

There the typical *Olenellus thompsoni* (17-9) is associated with *Mesonacis vermontana* (17-2), a near relative, and this species is most characteristic in these deposits on

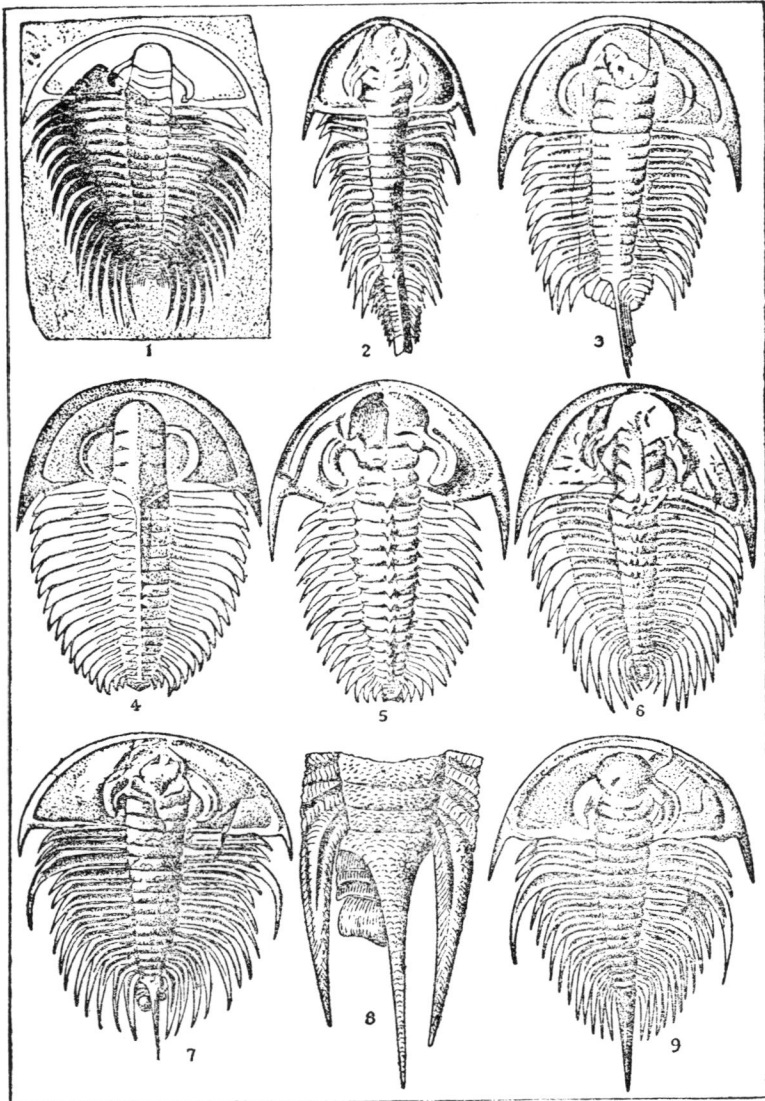

Fig. 17. (1 — 9) Taconian (Georgian) Trilobites (after Walcott).
1. *Nevadia weeksi* Walcott
2. *Mesonacis vermontana* (Hall)
3. *Elliptocephala asaphoides* Emmons
4. *Callavia bröggeri* (Walc.)
5. *Holmia kjerulfi* (Linnarsson)
6. *Wanneria walcottanus* (Wanner)
7-8. *Pædeumias transitans* Walc., 8. *Ibid*: pygidium (greatly enlarged).
9. *Olenellus thompsoni* Hall.

the west coast of Newfoundland (Bonne Bay). Along the
East Greenland coast rocks of this age are found with both
Olenellus and *Mesonacis*, besides many other genera.

So far the occurrence of these beds is such as to
suggest a possible former connection by a continuous coast-
line, but now comes the disconcerting fact that in Ross-
shire, North Scotland, on the other side of the Atlantic,
and more than 2,000 miles to the east, the same Taconian
Olenellus fauna is found in the rocks which overlie the
Torridon sandstone, and these, as we have seen, register
the first marine invasion of that region.

We might perhaps explain this as representing shore-
lines on opposite sides of a Taconian Atlantic, with opportunity for coastal migration between these points. But then
another fact obtrudes itself. When we cross the old Highland barrier into Shropshire, a distance today of only about
350 miles, we find that the oldest fossiliferous rocks preserved there carry the trilobite genus *Holmia* (Fig. 17-5)
and a little higher up the genus *Callavia* (Fig. 17-4).

Now, while these last two belong to the same group
or clan of trilobites, they are distinct from *Olenellus* and
Mesonacis, which are two members that belong to a separate,
though related, clan of early trilobites.

Moreover, the associated fossils are distinct, so that
it is clear that these two regions belong to two separate
geosynclines, the St. Lawrence (extended) on the north,
and the Caledonian on the south, with a barrier, represented by the greater part of the Scottish Highlands, between.

The same genera *Holmia* and *Callavia* characterize
the Taconian beds of southern Sweden and south-eastern
Norway, demonstrating that these regions belong to the
same Caledonian geosyncline. Then the old crystalline
back-bone of Scandinavia reveals itself as a continuation
of the Scottish Highlands barrier, though now no longer
continuous with it, as it was once.

But still another surprise awaits us.

Three thousand miles west across the Atlantic at
Weymouth near Boston, Massachusetts, the Taconian beds
contain the identical species of *Callavia callavei* which is
so typical of the early beds at Comley, in Shropshire.[3]

---

3) This was originally described as *Callavia crosbyi* (Walcott) Grabau, but
is now known to be wholly identical with the British species.

Now, when it is remembered that the Taconic Mountains lie where Massachusetts and New York States meet, and that Troy in the Hudson Valley, where the *Olenellus* fauna is found, is only 150 miles west of Boston, we see that we have here a counterpart in Boston and .Troy, to the Shropshire (England) and Ross-shire (North-Scotland) region, and that Central Massachusetts played the same role of barrier that the Scottish-Scandinavian axis did.

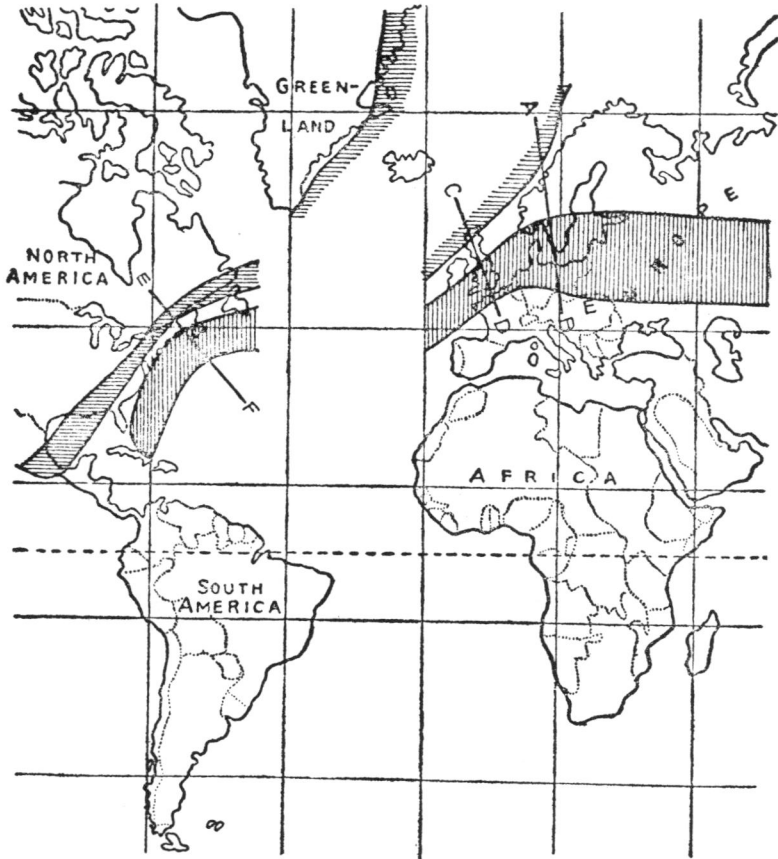

Fig. 18. Map of the continents adjoining the Atlantic, showing the interruptions of the Appalachian - St. Lawrence geosyncline (horizontal lining), and the Caledonian geosyncline (vertical lining) by the Atlantic rift. (*Palæozoic Formations*, Vol. I).

Once more we note that Taconian strata at Manuel's Brook, in south-eastern Newfoundland contain *Callavia bröggeri,* which is a close relative of the British-Boston species and identical with the Scandinavian form.

Then we recall that at Bonne Bay, about 300 miles to the northwest on the other side of Newfoundland, we have these same Taconian beds characterized by the *Olenellus* fauna.

Fig. 19.   The same regions with the Atlantic rift closed as in Palæozoic time, showing continuity of geosynclines.   (*Palæozoic Formations,* Vol. I)

Thus central Newfoundland also plays a part of the same barrier between two geosynclines.

If then we align Troy, Bonne Bay, Ross-shire into one belt, and Boston, Manuel's Brook, Shropshire into another, with central Massachusetts, central Newfoundland and central Scotland and Scandinavia as a barrier between, we have our two geosynclines outlined. (Fig. 19).

But in order to make this barrier effective one of two things is necessary. Either we must build a narrow land-bridge, across the two thousand miles of ocean, from the Scottish Highlands to Newfoundland and Massachusetts, or we must push Europe and America together, until they meet, perhaps at the mid-Atlantic ridge. Then the severed ends of the barrier will join and form one normal continuous barrier between the two geosynclines. (Fig. 19).

Greenland is pushed towards Scandinavia and automatically falls into line. This is seen on comparing Figures 18 and 19, pp. 82, 83, thus showing the early kinship of New England and Old. We are, thus, again led back to the pangæa, for no arrangement of lands and seas such as exists today can satisfy the requirements of the case.

The *Callavia-Holmia* fauna left its trail across central Europe. Traces of it have been found in central Poland and in Esthonia, while some of its minor associates among the brachiopods and gastropods have been traced as far north-east as the Irkutsk region of Siberia.

It was in this region, when the present Arctic coast of Siberia formed a part of the Palæozoic equator, that the epi-sea was located that acted as the feeder to the Caledonian geosyncline; it also served as an asylum, or haven of refuge, to which the survivors of the fauna of the Caledonian geosyncline could retire and continue, after disaster had overtaken the Taconian world of life by the withdrawal of the waters which had made that existence possible. (See Map, Plate III).

3

In the northern portion of the Palæocordilleran geosyncline, in rocks that now form part of the Canadian Rockies, we meet again our old acquaintance the *Callavia* of the Shropshire mud-flats and of the ancient muds of the Boston Basin. Probably because these Canadian regions were so much farther away from the influence of the north pole of the time, *Callavia* was here associated with a number of other trilobites which either did not enter the Caledonian geosyncline or have not yet been discovered. But among its associates in the Canadian Rockies are brachiopods, familiar to the student of European Taconian faunas.

Thus the trail of the Taconian trilobites leads along the coast of Fennoscandia to present day Arctic America, which regions were then in the tropical belt, and terminates in the vicinity, where the pyramid of Robson Peak now stands. This is probably the most striking erosion monument of the Canadian Rockies, rising nearly 13,000 ft above sea-level, and carved almost entirely out of strata which are still horizontal and which range in age from Sinian at the base, to Skiddavian and to Early Ordovician at the top. (Fig. 20).

Fig. 20. — Robson Peak from northwest slope of Mahto Mountain. In the foreground Chetang Cliff, in the center Iyatunga (black rock) Mountain, with the foot of Hunga Glacier at its base, and beyond a portion of Blue Glacier above Berg Lake. In the distance on the right Little Grizzly Peak. (After photograph by Walcott).

It was in the vicinity of this modern land-mark, when its strata were still accumulating in low, more or less monotonous mud-flats of a steadily sinking geosyncline, that *Callavia*, the leader of the Eurasian host, and *Olenellus*, chief of a related clan, whose home was in the distant

Caribbean epi-sea, met, and, as there was plenty of space available, if we may judge from the relative scarcity of individuals, they occupied the territory in unison.

There is some question as to the prior settlement of this region, whether by Eurasian or Caribbean immigrants. This requires further collecting in these strata, when perhaps many interesting new facts will be revealed.

It was apparently in the Caribbean epi-sea that the Appalachian geosyncline, which we have traced to southeastern Greenland and North Scotland, had its feeder. In Alabama and eastern Tennessee the deposits are mainly of clastic origin and measure thousands of feet in thickness, but northward they become thin by overlap of the transgressing series and off-lap of the retreating series, until in North Scotland they are less than 600 ft in thickness, and even less than that in south-eastern Greenland, though there they abound in the characteristic fossils.

It is evident, that the geosyncline came to an end a short distance beyond this, for a broad land-barrier closed it on the north and separated it from the Russo-Scandian epi-sea, which was the home of the *Callavia* fauna, [4] and from which *Callavia* also entered the American Boreal seas.

4

There is another fact of great significance that we are apt to overlook. That is the foreshortening effect of the compression of Europe, during the making of the Alps and the other great east-west mountain chains, which give the country its chief elements of scenic interest.

The same zone of folding on a far grander scale is seen in the Himalayan chain and the many divergent chains of Asia, which form the spreading fan-folds of which the Alps constitute the compressed handle.

The folds came into existence in comparatively recent time, most of them in mid-Tertiary time, when the old pangæa had already broken into at least some of its chief continents.

We shall consider this folding of the Old World and

---

4) In the palæo-geographic map of Taconian time, published in Vol. I, of my "*Palæozoic Formations in the Light of the Pulsation Theory*", a connection between the St. Lawrence geosyncline and the Boreal sea is shown, but this has now been proved not to have existed.

its effects more fully in a later chapter. What chiefly interests us now is that the present northern, not folded regions, Scandinavia, Siberia, etc., were brought closer to Africa and India, this having been estimated by some to amount to as much as 25° or more of latitude.

Before the folding the great east-west rift of the Mediterranean probably did not exist, except as a shallow geosynclinal depression. It probably followed as a rebound, after the great compression, which most probably resulted from the movement of the European lands against the resistant mass of Africa. Leaving these details for the present, we may note that a flattening out of these folds and thrusts removes the southern end of the Scandinavian mass some 20 or more degrees further north, or approximately to latitude 30° south (in Palæozoic time), or some 60° from the Sinian south pole.

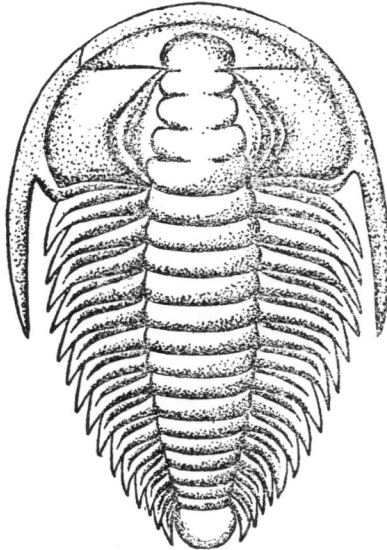

Fig. 21. *Redlichia chinensis* Walcott. The characteristic trilobite of the Cathaysian geosyncline of S. China. (From *Stratigraphy of China*).

The Cathaysian geosyncline of China was distinct from the others, and had a distinctive fauna in which the trilobite *Redlichia* (Fig. 21) played the leading part. This likewise entered the branch which radiated towards the pole of that time, and so we find that the oldest marine rocks of

the Himalayas are characterized by the *Redlichia* fauna.
This fauna also occurs in the Kimberley district of North
West Australia.

5

We must here pause to note that the conditions which
characterized Taconian and probably Cambrian (Acadian)
time as well, on the pangæa, indicate that the south pole
had shifted or migrated, some 20° to the west [5], so as to
come to lie in about 10° E. and 20° N. of today.

This, of course, implies a new set of meridians, though
the general designation is still nearly the same. But the
rear-west has become the smallest of the land quarters.
It is probable that the ice of the south polar ice-cap of
that time drained into the Caledonian geosyncline, and the
fine silt of the ground moraine was deposited as the Cambrian
slates of that region. Even the limestone of Nahant and
Weymouth may be calcareous muds ground from some
ancient calcareous rock among the Archæan of the African
basement, though none are now exposed, because of the covering by the desert sands. The Taconian (Lower Cambrian)
limestone-lenses of eastern Massachusetts are certainly not
of organic origin, though some beds enclose shells of hyoli-
thids, and rarely other fossils, but instead of being com-
minuted, they are as complete as in other kinds of clastic
mud deposits. [6]

5) It will be convenient to divide the pangæa into quadrants by the east-
west meridians of 90° and the north-south meridians of 0° and 180°. If
the reader imagines himself standing on the south pole and facing in the
direction of the 0° meridian (which is always drawn through Greenwich,
near London, as today), and his back to the 180° meridian, then the two
quadrants on his right and left respectively, will be the fore-east (*Antero-
dextral*) and fore-west (*Anterosinistral*) quadrants, and those behind him the
rear-east (*Posterodextral*) and rear-west (*Posterosinistral*) quadrants. If each
of the larger sections of the pangæa falls largely within one of these
quadrants, they may be called by that name. Thus, in the original
pangæa, with the south pole in the center, the fore-east is also the Eurasian,
the fore-west the North American, while the rear-east is the Austral-
Asian and the rear-west the South American. With a change in the
location of the south pole a new set of meridians comes into existence,
but the 0° meridian will always pass through the London district.

6) The Boston region was perhaps 10° distant from the margin of the Cambrian ice-cap, or 30° from the pole of that time, which would mean Lat.
S. 60°. It is possible, however, that in Lower Cambrian (Taconian) time
the pole had not yet moved to its new location, this being occupied only
by Cambrian or Acadian time. When the pole was in Egypt, Boston was
approximately in 42° S. Lat.

It is not because life was scarce in the Early Cambrian seas that we find so few remains of it in the longest studied sections, but because these sections happened to be in regions of unfavorable environment in early Palæozoic time.   A sea influenced by the cold waters which drained from the polar ice allowed only those forms of life to exist which were hardy enough to endure the prevailing low temperatures.   Such was the geographical position of all the sections of eastern North America and of some European regions as well, and, in what is now eastern North America, life became more difficult as the Early Palæozoic eras progressed, and only the most resistant species survived in the end.

In the western regions, however, in the Palæocordilleran geosyncline, which has since been transformed into the Rocky Mountains, conditions were much more favorable, because this region was closer to the equator of the time, and the wealth of organisms preserved is correspondingly greater. Though mountain glaciers creep today over their burial ground, during the life time of the organisms this was a region of tropical seas. (Compare maps: plates III and IV).

CHAPTER IX

CAMBRIAN HISTORY AND CAMBRIAN LIFE

1

CAMBRIAN history, both that of Lower Cambrian (*sens. lat.*), or Taconian, and of Middle Cambrian (Acadian or Cambrian (*sens. strict.*), must be read in the light of the geography of the earth in Cambrian time, for in no other setting can the sections of the formations and their faunas be correctly interpreted.

Formerly it was generally assumed, that the distribution of the continents and ocean basins in the Cambrian periods had been essentially as it is today, and, on this assumption, geologists have endeavoured to interpret the early history, each of his own country, with all too frequent disregard of geographical indications furnished by the formations of another region. We have seen some of the difficulties that are encountered in the Taconian history of an unassembled earth, and that the apparently irreconcilable facts fall into harmonious groups when the continents are assembled into a pangæa.

An entirely distinct series of problems confronts the palæogeographer who deals with a Palæozoic pangæa instead of with dismembered fragments or individual land blocks.

Not the least of these important readjustments is in regard to the location of the pole of the period. For it is the pole, in the Palæozoic pangæa the south pole, that controls the physical conditions which influence the development of the life of the period, by virtue of the influence it has on the distribution of climates, and all that that implies in the creation of a habitat for the organic world.

It is obvious that even open waters in the vicinity of the polar ice-caps are less capable of supporting life than those situated in a region of milder climate, unless the water is deep enough to be out of reach of the ice.

This is well illustrated by the contrast, not onlyin

the thickness of formations, but also in the abundance and variety of organic remains, between the Cambrian beds of the Boston region and those of the Palæocordilleran geosyncline in western North America. In the former the thickness of the series is a few hundred feet, with a sparse fauna of few species, though sometimes many individuals. In the latter the thicknesses measure several thousands of feet, with a fauna which, though as yet only partly recovered, is unrivaled by that of any other known Cambrian region of the earth.

Our reconstruction places the Boston region within 30° of the Cambrian south pole, or within 10° of the front of the polar ice-sheet. That the waters of the Caledonian geosyncline of this region received the drainage from the ice-front during the seasons of melting, the Antarctic summers, seems evident, and also that the waters must have been frozen in winter, though as yet the ice-front was too far away to have a destructive effect on the marine life[1]. It, however, limited the number of species that made their home in these waters.

The Boston region is unique among the areas of the Caledonian geosyncline, because it lay nearer the ice margin of the time than any other locality of that shallow marine water-way. If we correctly interpreted the relationships it was situated in 60° S. Latitude. The ice, of course, was still confined to Africa, but it was slowly advancing on the Boston region, because relative movement of the pole was in that direction, and it reached that region by the opening of Early Ordovician (Cambrovician, or Skiddavian) time.

True, Christiania (Oslo), and St. Petersburg (Leningrad), all lie today on or near the parallel of 60° N. and

---

1) Both in the Arctic and Antarctic marine life, both vertebrate and invertebrate, abounds in the ocean, at a few fathoms below the surface and downward, but the pools along rocky shores are barren, because of the destructive effect of the ice. On the Palæozoic pangæa, however, the Antarctic polar regions were always far removed from the ocean, and the only water-ways which the ice encountered were the geosynclines. These were always shallow, frequently entirely drained, and the sites of sedimentation. Such water as there was in the continental neighborhood of the polar ice was frozen during the winter, and that within the ice-cap area all the year round. This, of course, would mean that the normal organisms of the geosynclinal waters were driven to seek shelter in milder regions away from the polar-ice.

Stockholm is only a fraction of a degree south of that, but these localities are on, or near, waters open to the North Atlantic, whose climate is controlled and greatly ameliorated by the Gulf Stream.

No such waters favored the Boston region in Cambrian time; the climate was a continental one and therefore more nearly comparable to that of the Siberian region, *i. e.* that of Tobolsk or Yeniseisk. On the American continent, the modern boundary line between Alberta and Mackenzie provinces in Canada would perhaps be climatically comparable.

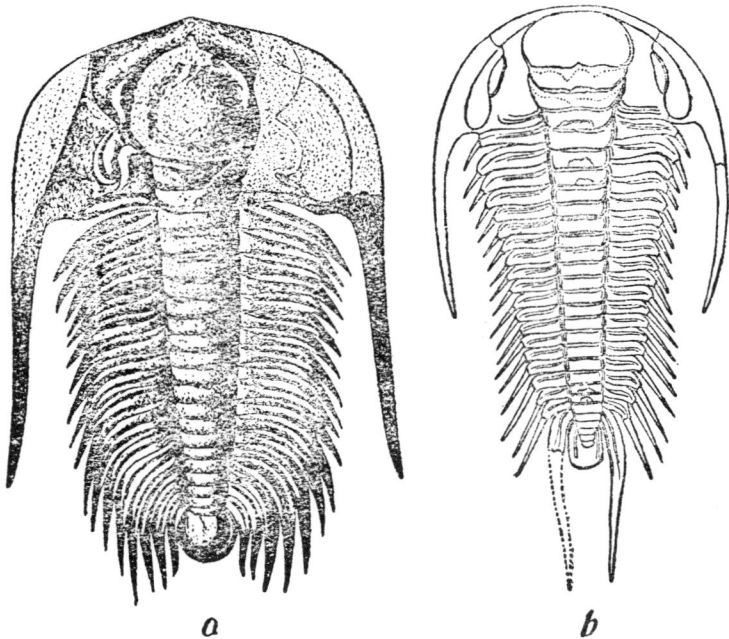

a   *b*

Fig. 22. Cambrian (Acadian) trilobites of the Caledonian geosyncline. a. *Paradoxides harlani* (Green), Eastern Massachusetts; b. *P. bohemicus* Barr., Bohemia. (Grabau: *Text-book of Geology*).

The leading trilobite of the Boston (Braintree and Quincy) region is *Paradoxides* (Fig. 22), which may reach a length of a foot or more. Only 2, or at most 3, species are known from eastern Massachusetts, and even those are not abundant. If we add to this a *Ptychoparia,* an *Agraulos* and an *Agnostus* we have practically listed the complete trilobite fauna. In addition to this, 2 species of *Hyolithes* and a brachiopod are known from single, or at most a few, specimens. The entire known fauna thus comprises altogether 5 trilobites and 4 other forms.

Passing down the geosyncline of that time we reach Spain, where beds of this age are widely distributed between the 1st and 6th meridians of west longitude of today. A section, showing some 200 meters of marls, shales, etc., with fossils at interval, and overlain by about 300 meters of barren lutyte and sandstone is known from the province of Aragon, while other deposits are known from south central Spain and from the provinces of Leon and Asturia in north-west Spain of today. In this region the Acadian beds contain a much richer fauna, which comprises altogether 32 species of trilobites, of which 7 belong to the genus *Paradoxides*. Besides this, there are 6 species of fossils belonging to other phyla.

A few degrees further north (p. d.), at Montagne Noir and elsewhere in south-eastern France, a much thicker series of beds is found with a fairly rich fauna of 19 trilobites and 3 non-trilobite species.

Most abundant, however, are the remains of this fauna in that part of the old Caledonian geosyncline which is the Bohemian region of today. Here 38 species of trilobites, each usually with many individuals, occur, and of these 16 species belong to the genus *Paradoxides*, this being probably the largest number of species of this trilobite obtained from any one region. In addition to this the known fauna contains 27 other species (8 brachiopods, 4 pteropods, 11 gastropods, 3 cystids and 1 graptolite) making 65 the total number of species known today.

Bohemia is also the best known region from which these faunas have been obtained and one that is unique for the light it throws on the early Palæozoic history of Europe. It is, moreover, classical ground for the palæontologist, because of the foundation-laying researches of Joachim Barrande. This region on our chart lay in about latitude 45° South, or comparatively removed from the influence of the south polar ice-cap.

In Bohemia (Czechoslovakia) the Taconian (Lower Cambrian) era was preceded by a folding of the older continental strata of Sinian age, the Pilsen beds, into a

marginal mountain chain which extended in a general
way east and west, as present compass directions read.

This folding was followed by a period of erosion, or
peneplanation, until the folded strata had their arches
entirely cut away.   This, of course, occupied a long period
of time, and it is not too much to assume that it represents the whole of the Taconian or Lower Cambrian
pulsation period which is elsewhere represented by fossiliferous sediments.

On the top of the truncated folds was subsequently
deposited a series of continental sands, the Pribram series. [2]
These were geosynclinal strata, as their lithic character
and bedding indicate, but they were entirely non-marine.
This can only mean that they were deposited during a
period of falling sea-level, in other words, they are
interpulsation deposits.

We may, then, interpret the progress of events in
the Bohemian region as follows : (see Fig 14, p. 53).

The southern margin of the old Caledonian geosyncline, which extended in a general east-west direction,
was in Sinian time receiving river-borne sands and muds
from the highlands of present day southern Europe, until
a vast thickness of continental sediments had accumulated.
Most of these were river-plain deposits, and were formed
during the constant sinking of the geosyncline.   There
may have been occasional standing water-bodies, but these
had no marine connection, since at this time the sea was
far distant.

Meanwhile, rising followed by falling sea-level was
no doubt in progress in the distant ocean, but this sea
was kept out of the geosynclines on the pangæa by effective barriers.

Then, if we may judge from the analogy with the
progress of events in subsequent periods, while the sea-

2) These Pribram beds, which are concordantly succeeded by the *Paradoxides*-
   bearing strata, are almost always referred to as representing the Lower
   Cambrian. This would mean that not only the folding of the Pilsen
   beds, but the erosion of the folds, *i. e.* their peneplanation, all took place
   in pre-Cambrian (pre-Taconian) time. There is no real evidence for this,
   and moreover, the Pribram beds, by their continental and clastic character, indicate that the sea was retreating, not transgressing as it did
   during the Taconian period. It is most logical therefore to refer these
   beds to the interpulsation period, and consider that the peneplanation
   occupied the Taconian period.

level of the distant ocean was depressed during the interpulsation period, the strata of the Pilsen series (Sinian)
were elevated by intense folding into an east-west mountain chain, this being added along the then north-western
border of the old-land.

This mountain folding probably consumed the time
of the interpulsation period, which preceded the final
successful attempt of the sea to break across the barrier
and invade the geosynclines of the pangæa. As these
Pilsener Mountains, now represented only by their stumps,
are the best dated of the mountains formed during this
interval, the interpulsation period may be named after
them, *i. e.*, the first or the PILSENER INTERPULSATION
PERIOD.

3

Then followed the Taconian pulsation period, characterized by the first invasion of the sea.

The portion of the Caledonian geosyncline which lay
northwest of the Pilsener Mountains, and which is represented by the whole region that is today divided into
Poland, Austria, Germany, the southern Baltic region, and
the southern Scandinavian coast, then forming a foreland
of the Bohemian region, was submerged by the Taconian
sea, and received the sediments which the rivers brought
from the Pilsen Mountains.

In these sediments *Holmia* and *Callavia* left their
cast-off integumental armor, or their dead bodies, while
the other organisms of the Taconian fauna also contributed
their shells and other remains.

But the Bohemia of today received none of these,
because Bohemia was the Pilsen Mountain country, which
supplied sediments but did not receive any. So generous
was the supply, that by the time the Taconian pulsation
period had ended, and the sea began to retreat, the greater
part of the Pilsener Mountain-chain had been reduced to
a peneplane.

Thus, erosion of the Pilsener Mountains of Bohemia
coincided with the deposition of the Taconian strata, with
their marine (*Holmia*, *Callavia*) fauna, just as the erosion
of the much more recent Siwalik Hills of India coincided
with the deposition of the early sediments in the North

Indian geosyncline of that day. Then, just as the truncated Siwalik folds were covered by horizontal sands and
pebble-beds from the Himalayan old-land behind them, to
form the modern plain, so the truncated Pilsener folds
were covered by horizontal sand and pebble deposits from
the old-land between them and the south pole of the time.

And again, as in the modern example, these sediments
were purely non-marine, because the sea had retreated.
Therefore they should be classed as representing the interpulsation deposits of the second or Pribram interpulsation
period, and not of Taconian or Lower Cambrian age. For,
it must be clearly recognized, that geosynclinal subsidence,
whatever its cause, is independent of pulsation. The
geosyncline may sink continuously, even though the sea
retreats, during the negative stage of the pulsation period,
and the geosyncline may keep on sinking, when the sea-
level rises again, during the positive stage of the next
pulsation period.

In the first case, that of a subsiding sea-level in an
interpulsation period, and of independent sinking of the
geosyncline, continental sediments brought by rivers will
be spread over the sinking surface, and many an interpulsation period is marked by such continental sediments,
which are generally erroneously referred to one or
the other of the contiguous pulsation periods. In the
second case, that of a rising sea-level, the geosyncline will
be flooded, or at any rate, marining of huangho deposits
will take place.

Generally, the difference in texture indicates the state
of the sea, for with rising sea-level, the river-grade is
diminished, and the only sediments, which the river can
carry are fine muds and silts. But with falling sea-level,
coarser sediments are brought forward for deposition, while
the previously deposited finer sediments of the geosyncline,
not reached by the new sediments, will become subject
to erosion.

There may, of course, be variation in the rate of
sinking of the geosyncline, and in exceptional cases it may
pause altogether. But the reverse movement,[3] that is eleva-

---

3) Local doming, or up-arching, of barriers is here considered as a minor
   manifestation of an orogeny.

tion, takes place only when an orogeny is effected, though apparent elevation or relative rise, due to sinking sea-level, occurs probably during every interpulsation period.

The region now included in the Baltic provinces (Esthonia), apparently received no sediments during the second marine invasion, the Acadian, but remained unsubmerged, though it had been the site of deposition of fine clays and green glauconitic sands during the Taconian invasion. It represented the marginal plain of the geosyncline, and during the Taconian period it experienced oscillation and frequent emergence and the formation of coastal sand-dunes, — conditions quite analogous to those existing on a modern flat coast. During the succeeding Pribramian interpulsation period, few recognizable changes took place here, except perhaps erosion of the previously deposited beds. The same thing may be said for the succeeding Acadian pulsation period. The Baltic region was apparently not flooded during the Acadian transgression, or, if it was, the sediments which were formed were removed again during the following interpulsation period. So the rocks of the Cambrovician pulsation period rest disconformably upon the Taconian sediments.

During the Acadian pulsation period, the Bohemian region for the first time became the site of active marine deposition, and of the appearance of an abundant trilobite fauna. This was the period when *Paradoxides* reigned supreme in the waters of the Caledonian geosyncline. Bohemia was one of their strong-holds, for here they developed in greatest variety, some 20 species being known from this region alone, including Poland, or 31, when the marginal platform of southern Sweden is included.

It is probable that the center of origin of this trilobite fauna was in the Irkutsk region of Siberia, *i. e.* the Siberian epi-sea, which, it must be remembered, was then on the equatorial margin of the surrounding ocean. Though *Paradoxides* itself has not been recorded from the Siberian region, *Centropleura*, a member of one of the dynastic periods of the *Paradoxides* supremacy, has been found in Bennett Island (now in 76°30′ north latitude, 149° E.), one of the New Siberian Islands. A connection with the Bohemian-Scandinavian basin is possibly indicated by the presence in these beds, both in Scandinavia and in far away Bennett Island of the characteristic trilobite *Centro-pleura loveni*.

4

A related species, *Centropleura vermontana,* together with a number of others that belong to its faunal as-semblage, is found in the St. Albans slates of the state of Vermont in eastern North America.

The St. Albans consists of about 200 ft of dark, micaceous shales, which rest with a disconformity on the *Olenellus*-bearing slates of the Taconian, and begin with a basal conglomerate. The latter, it will be recalled, formed a part of the continuous deposits of the Appalachian-St. Lawrence geosyncline, the fauna of which was entirely distinct from that of the Caledonian geosyncline.

The St. Albans slate fauna, on the other hand, shows its relation to the Caledonian fauna, by way of Bennett Island, and it thus becomes apparent that a new arrange-ment of the geosynclines characterized the Acadian period. A direct arm of the Siberian epi-sea, extended between Fenno-Scandia and Appalachia on the one hand and Green-land on the other, along the present St. Lawrence Valley to Little Metis,[4] Quebec, where records of the presence of the Caledonian-Acadian fauna are found (*Acrothele sagittalis*), and on into the St. Albans Vermont region. The slates are preserved in the St. Albans vicinity. Every-where else they were removed again by erosion during the next interpulsation period, and their site covered by the transgressing Cambrovician sea.

It appears, however, that the St. Lawrence arm of the Siberian sea did not extend much farther, for a low transverse elevation, hardly more than a swell, or limen, the Albany Axis, had arisen 150 miles, or more, west of Boston (p. d.), in response to the same general warping effect on the land, which caused the extension of the arm of the Siberian sea into the St. Lawrence region. (Plate IV).

In other words, the old Appalachian geosyncline, which in Taconian time extended the entire eastern length of the North American continent, as far as Greenland and

---

4) On the modern St. Lawrence River, 175 miles below (north-east of) Quebec City.

North Scotland, had now become dismembered into a Southern or Appalachian geosyncline proper, which extended from Alabama, to the Hudson River (Stissing Mountain), and a Northern, or the St. Lawrence geosyncline, which was now united to the Siberian epi-sea, and extended south to the region of the present Mohawk River. Dividing the two geosynclines, was the Albany Axis.

5

The sequence of events in the southern Appalachian is far less dramatic. Vast masses of unfossiliferous sands and shales begin the sedimentary series, the first fossils appearing some ten thousand feet or more above the base. These lower sands may well represent continental deposits, formed during the Sinian period, or during the Pribram interpulsation period, or during both. But they indicate a constant and steady sinking of the geosyncline, though only towards the last did the beds actually become calcareous and enclose the typical *Olenellus* fauna.

The series is terminated by a thousand feet of Red Sandstones (Rome formation), which mark the retreat of the sea in the geosyncline and the return in the main of continental sedimentation with all its accompanying characteristics, though still with occasional fossil-bearing beds, showing that the sea was not wholly excluded.

After the interpulsation period, the record of which is rather obscure, the Cambrian (Conasauga) sea followed the pathway of its predecessor, transgressing over the more or less eroded beds of the previous pulsation system, or the continental beds formed during the interpulsation period. The organisms of the Cambrian (southern Acadian) sea, in the Appalachian geosyncline, were wholly distinct from those of the St. Lawrence geosyncline, this distinctness being maintained in the succeeding Cambrovician period, when the southern Appalachian waters were the home of the Ozarkian fauna.

In general the Cambrian (mid-Cambrian, old style), of the southern Appalachians is referred to as the Conasauga formation, and four divisions are recognized in some sections.

These, in descending order, are:

4.	Nolichucky shale [5]	400-750 + ft.
3.	Maryville limestone	150-750 ft.
2.	Rogersville limestone	70-250 ft.
1.	Rutledge limestone	200-500 ft.

Since the transgressing Conasauga sea frequently encountered weathered debris of the Rome formation of the Taconian, this material was reworked as the basal beds of the new transgression. Thus it may frequently have happened that weathered out fragments of the older fauna were secondarily included in the transgressing deposits of the newer period, and so a deceptive commingling of Taconian (Georgian) and Cambrian (Acadian) fossils may characterize the basal Conasauga beds; this would suggest a transition between these formations, a relationship which is, however, entirely non-existent.

There were, of course, no *Paradoxides* in the Cambrian (Conasauga) beds of this region, for they belonged to entirely distinct marine provinces. Instead the leading trilobite genera were *Norwoodia*, *Olenoides* (Fig. 23-a), and *Crepicephalus*. (Fig. 23-b.)

These transgressions extended up the Appalachian geosyncline to the vicinity of Albany, where the Albany Axis separated the Appalachian from the St. Lawrence geosyncline, on the north. This axis, it must be remembered, was not effective in the Taconian period, for *Olenellus* is found both north and south of it.

6

The Mississippian marginal platform of North America was flooded by the slowly advancing Cambrian waters of the Conasauga sea, these eventually extending beyond the line of the Taconian transgression, so that often the Conasauga strata overlap the Taconian beds, and rest directly upon the Old Crystalline rocks.

---

5) Resser refers the Nolichucky to the Dresbach of the upper Mississippi region, which is probably correct. But he also refers it along with the Dresbach to the Upper Cambrian. In this I cannot follow him. (See Vol. III, *Palæozoic Formations*, etc, pp. 150 *et seq.*). The characteristic fragmentary fauna in these formations indicates long exposure, deep weathering and the rough reworking of the weathered material.

This is seen in the northern Mississippi Valley region (Minesota, etc.), and where the base is exposed in eroded domes as in the Arbuckle Mountains and the Ozark uplift. Here the Conasauga Cambrian beds are represented by a few hundred feet of fossiliferous limestone.

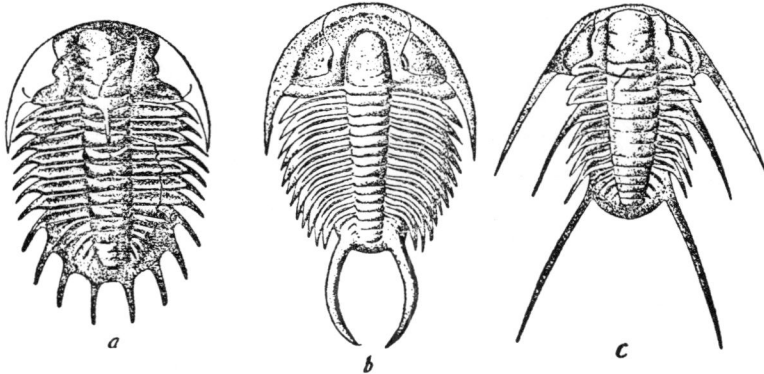

Fig. 23. Cambrian (Albertan) trilobites of the Palæocordilleran geosyncline, etc.
    a. *Olenoides curticei* (Nevada).
    b. *Crepicephalus texanus* (Southern Appalachians).
    c. *Albertella* (Canadian Rockies).

Of exceptional interest are the deposits of the Cambrian (Acadian or Albertan) sea, in the Palæocordilleran geosyncline, where we have already noted the great thickness and undisturbed character of the preceding Taconian (Georgian) deposits. The best known section is that of Mount Bosworth, in British Columbia, where the Taconian (1,453 ft) is followed by 4,580 ft of Cambrian or Albertan [6], and this by 3,590 ft of Cambrovician, none of the strata being much disturbed or altered. Throughout this geosyncline the lower division of the Cambrian (Albertan) series is charaoterized by the trilobites *Albertella* (Fig. 23-c) and *Neolenus*.

But, it is the Mount Stephen section that has gained worldwide fame, because of the wonderful preservation of the fossils of one of its formations, the Burgess shales. Here the most delicate sponges still show their siliceous

6) A name to designate the exceptional deposits of the Palæocordilleran geosyncline, equivalent in time to Acadian or to Conasauga, *i. e.* the old Middle Cambrian.

structure.   Holothurians, and worms have their soft parts
clearly outlined.   Crustacea of many genera and species,
with their delicate gills and antennæ clearly visible, have
their jointed legs outspread, and even their internal organs
are visible through the transparent outer integument.
(Figs. 25-28).

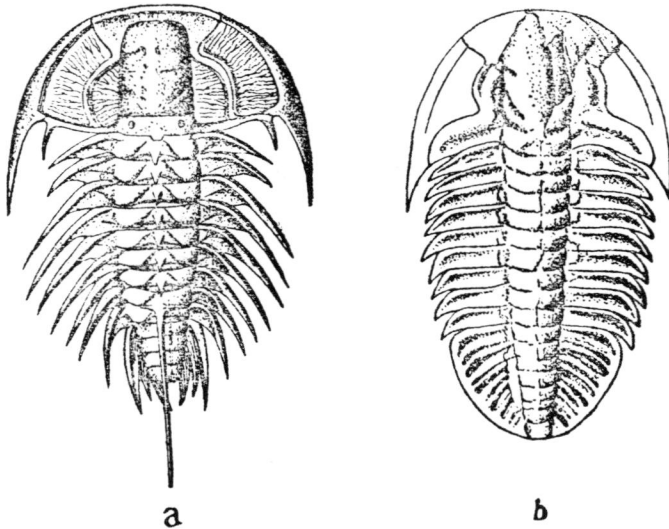

Fig 24.   Cambrian trilobites: a. *Zacanthoides typicalis*, Chisholm
shale, Pioche, Nevada; b. *Bathyuriscus rotundatus* (Rominger), Ogygopsis shale, Nevada. (After Walcott).

The main fossil-bearing bed is only 1 ft. 4 inches
thick, and it has furnished 17 species of algæ without
hard parts, which are seldom preserved elsewhere; 26
species of sponges; 4 of holothurians, a medusa, or jellyfish, most delicate and evanescent of organisms, composed
as it is of 90 % of water; 19 species of worms, and 29
distinct species of crustaceans.   All these were obtained from
that single bed, a foot and a third in thickness, and exposed
on the mountain side.

Of course, the treasures of these beds have hardly
been touched, as the greater part of the bed passes into
the mountain, and could only be mined by tunnelling.
But think of the vast store-house of life of the Cambrian
sea which was here buried and preserved for hundreds of
millions of years in unprecedented perfection.   Here, then,

we have an excellent illustration of the abundance and
variety of life in the sub-tropical region of the Cambrian
earth, which contrasts so strongly with the eastern region,
then partly under the chilling influence of the polar ice-
sheet, which centered in the Sahara region of today.

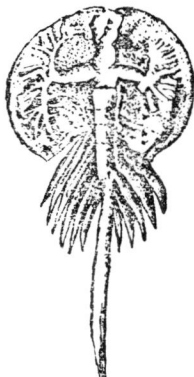

Fig. 25. *Burgessia bella* Walcott ($\times$ 2 1/2). Dorsal view of a specimen
showing the internal structure through the thin carapace. (After Walcott).

These fossil-bearing beds are some distance below the
middle of the Cambrian (Albertan) series, and they are
overlain by 2,700 ft or more of limestones (Elgin formation) before we reach the evidence of sea-retreat.

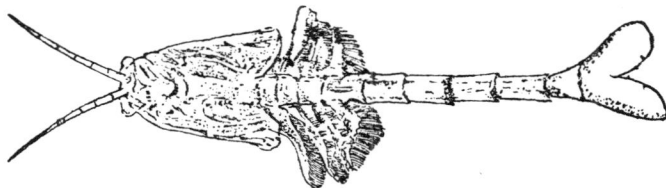

Fig. 26. *Waptia fieldensis* Walcott ($\times$ 1 1/3). Dorsal view of a specimen
flattened on the shale. (After Walcott).

Even greater thicknesses are found in Esmeralda county,
southwestern Nevada, and in California. Here the Cambrian
(old Middle Cambrian) series has a thickness of from 5,000
to 6,000 ft, and consists almost entirely of limestone
(Eldorado limestone), while the Taconian series also has
a great thickness, and consists on the whole of more clastic
formations.

From the same center of origin, the Caribbean epi-
sea, the waters extended for some distance into South
America along the pre-Andean geosyncline, which lay to
the east of the present Andes Mountains. The sections
are, however, still very imperfectly known.

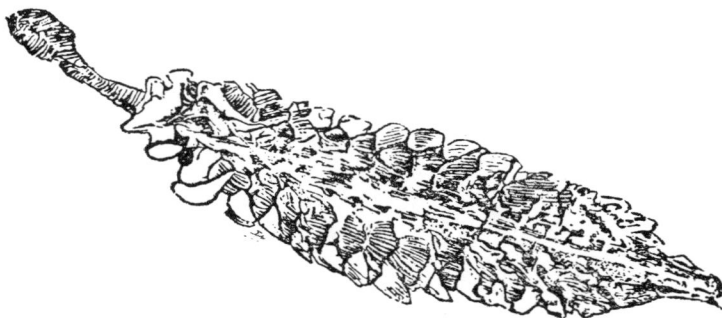

Fig. 27. *Opabinia regalis* Walcott (slightly enlarged). Dorsal view of a
male specimen flattened on the shale. (After Walcott).

When we compare the Taconian and Cambrian (Conasauga or Albertan) sections of the south-western United
States (p. d.), that is the deepest portion of the Palæocordilleran geosyncline, with the development of these
beds in the best known European sections, the striking
difference in thickness is at once apparent, and this explains why in the early history of the science when these
thinner deposits were the only ones known, all these beds
were classed together as a single formation, the Cambrian,
so named from the Cambrian Mountains of North Wales.

The explanation of these varying thicknesses is, of
course, seen in the different rates of subsidence, and to
a lesser degree in the amount of interpulsation erosion,
while the retarding influence of the polar ice-cap of the
period must not be left out of consideration.

7

The succeeding beds, which were also formerly classed
as Upper Cambrian, are now included in the Cambrovician,
and between them and the Albertan of the West-American
sections lies the Arctomys formation, the representative
of the interpulsation period.

The Arctomys formation includes a series of continental strata, with ripple-marks, desiccation fissures, and salt-pseudomorphs, signalizing the withdrawal of the sea

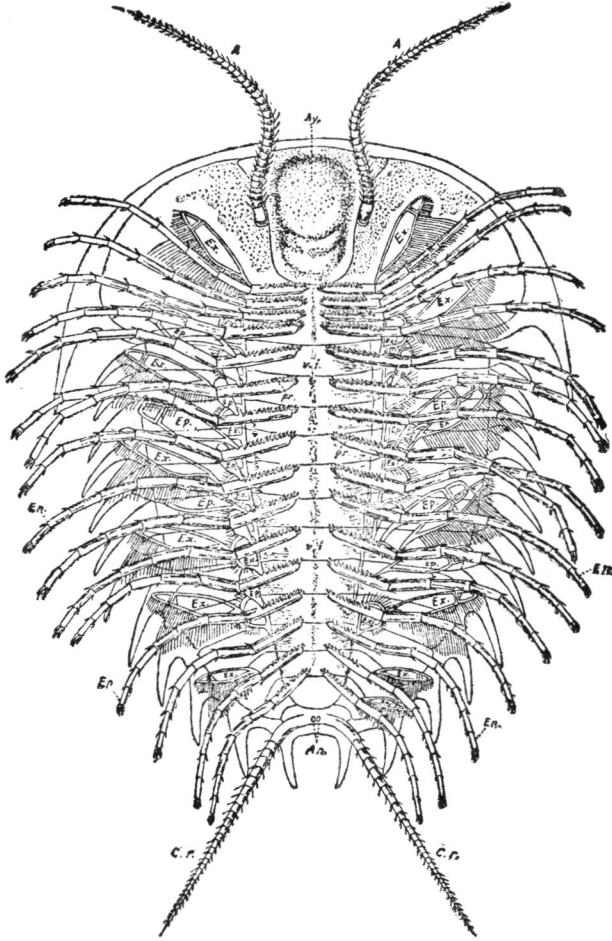

Fig. 28.   *Neolenus serratus* (Rominger). Restoration of ventral surface with legs, antennæ, etc. about × 1 2/3. (After Walcott). Cambrian Burgess shale of Stephen formation near Field B.C. *Legend:* hy. hypostoma; *A.* antennæ; *An.* anal aperture; *Cr.* caudal rami; *En.* endopodite; *Ep.* epipodite; *Ex.* exopodite; *exi.* exite; *Pr.* protopodite; *vi.* ventral integument.

and the complete emergence of the geosynlinal floor with the substitution of semi-arid climate for the former moist

conditions. Only a comparatively small thickness (200 to
1,200 ft) of the Arctomys formation is known in the
various sections, yet this is the most suggestive deposit
of the interpulsation period, which generally is represented by erosion only. On that account, and since no
other name seems available, the interpulsation period between the Acadian and the Cambrovician may be termed
the ARCTOMYAN INTERPULSATION PERIOD.

The duration of the interval is, of course, not measured
by the thickness of the deposits. It is emphasized in some
sections by the fact that the whole of the Elgin formation (2,700 ft or more) has been removed by erosion,
before the deposition of the Arctomys formation, which
rests directly on the Stephen, or its equivalent, the Murchison formation.

Only in Bohemia is a counterpart known to the
Arctomys formation. Here, on emergence the interpulsation period was characterized by the deposition of about
200 meters of coarse conglomerate, the Brezové-Hory
formation, composed of pebbles washed from the old-
land into the basin, and this was followed by the outpouring of great floods of acid lavas, — an older series of
porphyrites, felsophyres and keratophyres, followed by later
eruption of quartz porphyries and their tuffs, until many
hundreds of meters of these old lavas had accumulated.

Then followed a prolonged erosion period, during
which in most places all but 100 to 200 meters of these
lavas were worn away again, this lasting to the end of
the interpulsation period, when the sea returned at the
beginning of Cambrovician time. This Cambrovician
sea reached the Bohemian region only towards the last,
for only beds of Tremadoc age, the upper part of the
Cambrovician system, were deposited, the earlier part, the
beds formerly referred to as Upper Cambrian, being absent.
These, however, were formed in Poland and in the Scandinavian part of the basin.

It is true that the sea lingered only long enough in
the Bohemian part of the geosyncline to permit the deposition of less than 200 ft of fossiliferous Cambrovician
sediments, while in the Polish and Baltic regions, on the
marginal platform, scarcely more than one fifth of that
amount of sediment remains to represent this period.

Even less than this was probably formed in the Scandinavian region, for throughout the northern portion of that marginal platform, submergence was probably wholly due to rising sea-level, and so was very slow, and all the sediments which reached it, were of the finer Cambrovician muds.

This means, of course, that the old-land of the Scandinavian back-bone, was itself a part of a low-level peneplane, too low and too flat to be able to contribute any appreciable amount of sediment to the rivers which ran over it, and into the shallow sea which covered Baltica and southern Scandinavia.

The essential marshy and lagoonal conditions which prevailed over this region in Cambrovician time are revealed by the bituminous muds, almost without any sizable animal remains, which characterize the deposits of this period on the mud-flats. Concentration of highly bituminous lime-carbonate gave rise to thin layers or lenses of foetid limestone (Stinkkalk), but with only here and there a trilobite head of small form (*Peltura, Parabolina,* etc.), indicating that life was not wholly absent. These calcareous beds became subject to solution during the following period of sea retreat, when the mudflats again emerged (Rhobellian interpulsation period), and subsequently, when the sea returned in the Skiddavian period, it found these hollows ready to receive the first deposits of green glauconitic sands.

In one respect the Bohemian region was unique. All the interpulsation periods, the Krivoklat between the Acadian and Cambrovician, the Rhobellian (Komarov) and the Petrovian are characterized by interpulsation volcanics. After this, volcanism shifted to the Welsh side of the Caledonian geosyncline, where it continued through the remainder of the Ordovician period.

CHAPTER X

THE RHOBELLIAN INTERPULSATION PERIOD

1

THE ancient volcano of Rhobell-Fawr, on the south-eastern flanks of the Harlech Dome, in Merionethshire, North Wales, aptly lends its name to the fourth interpulsation period, which separates the Cambrovician and Skiddavian pulsation periods.

The volcano has been largely destroyed by erosion, but from the remnants left Dr. A. K. Wells has been able to restore it and picture the stages in destruction, which it has passed through. (Fig. 29).

When eruption began the region had already suffered some slight deformation, and an enormous amount of erosion, for the space on which the volcano was built up had been reduced to an almost perfect peneplane.

But it is the rocks on which this erosion plane was cut that are significant, for they belong to the Cambrovician pulsation system. In other words, they are rocks of Upper Cambrian and Tremadoc age.

Here, then, is evidence of prolonged erosion after the deposition of the Tremadoc, the last formation of the Cambrovician pulsation period (in the restricted sense in which this term is now used), and before the commencement of volcanic activity.

As every student of physical geography knows, the time interval occupied in the production of a peneplane by subaërial erosion is of great length, and the measure of its duration is the perfection of the old erosion surface.

Due regard must of course be paid in the first place to the petrographic character of the rock, and to the amount of deformation, while account must next be taken of the slowing up of the process as the erosion plane approaches more and more the base-level of erosion, which, as in the case in question, is the sea-level.

In other words, the rate diminishes in proportion to the degree of perfection which the peneplane attains.

After the peneplane was formed the volcano was built up by the eruption of andesite lavas, and when eruption ceased erosion began again, together with further slight deformation and intrusion of some diorite-porphyry dykes, so that when the second peneplane was completed the

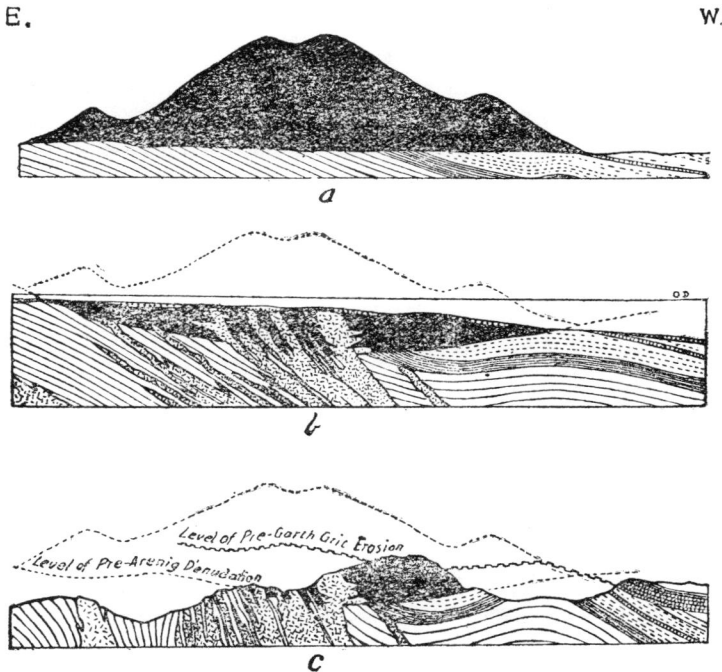

Fig. 29.   Stages in the development and destruction of the volcano of Rhobell Fawr. (After A. K. Wells).

a.   The volcanic pile resting on denuded (peneplaned) Cambrovician (Upper Cambrian and Tremadocian).

b.   In Arenig ("Garth") grit time, showing the local base of the Skiddavian resting in hollows on the truncated base of the eroded volcano, and on the Cambrovician outside of the volcanic area. This shows that it was formed after the second peneplanation, which cut away most of the volcano.

c.   Present day conditions, showing Ffestiniog, Dolgelly and Tremadoc beds with the relic of the volcanic pile, penetrated by intrusions of diorite-porphyry. The "Basement Group" of the Skiddavian System rests upon the volcanic rocks in the center and on the Tremadoc beds in the east.

basal portion of the old volcano was preserved because it was depressed in a broad, though shallow trough (See Fig. 29). Not until then came the renewed transgression which characterized the fifth, or Skiddavian, pulsation period.[1]

At first, only river-borne sands were spread across the peneplane and came to rest sometimes on the old volcanic remnants, and at other times, by overlap, on the Cambrovician rocks on which the lavas themselves rest. These sands now form the "Garth grits", or sandstones.

The approach of the sea was heralded by the fact that in the more distant portions of these covering sands early Skiddavian graptolites are found, while succeeding beds contain other characteristic fossils of the Arenig series, *i. e.* the transgressive deposits of the Skiddavian pulsation period.

This example is not unique, except in the perfection in which it preserved the record of the two erosion periods, with the volcanic stage between.

In Pembrokeshire, South Wales, there is another volcanic series, which belongs to the same interpulsation period. It agrees in a remarkable manner in respect to petrographic character of the volcanic rock, and even in its color, with the lavas of Rhobell-Fawr. This is the Trefgarn andesite series, the remnant of a flow which was contemporary with the eruption of Rhobell-Fawr, 90 miles to the north-east. Like the latter, it was preceded by erosion of the Cambrovician rocks.

It is true that the succeeding early Arenig beds began

---

1) This term is preferred to the name Canadian for the "Lower Ordovician" series of the older classification, because the term Canadian covers parts of two pulsation periods with an interval of erosion. It is true that the Lower Canadian, or Beekmantown, limestone of America represents the calcareous facies of the transgressive member of the pulsation system, whereas the Arenig, or Lower Skiddavian, is of the nature of a "huangho deposit" with graptolite layers separated by sandy beds, while other fossils are rare or in some cases absent. But the Beekmantown is followed by an erosion surface,—no retreatal members being positively known, for those that existed have been reworked during the Petrovian interpulsation period, largely by wind, and are now a part of the St. Peter sandstone, a pure eolian interpulsation deposit. The succeeding Chazy limestone, which was part of the original Canadian series, is now recognized as the basal transgressive member of the Ordovician. The higher Skiddavian, the Llanvirnian beds are also graptolite shales, but represent the retreatal phases of the Skiddavian.

with basal deposits formed of the partly reworked ashes of the eruption which had not yet hardened to rock when the first incursion of the Skiddavian sea brought its graptolites (of the *Extensus* zone), and left them stranded on the ashy surface at the turn of the tide.   As a result there seems to be a gradation from the ashes to the Arenig sands, but this is purely deceptive.

In Bohemia, too, this interpulsation period was characterized by an intense volcanicity, for though the Komarov volcanic series is often thought to be the product of a submarine eruption in Skiddavian time, such an interpretation is unwarranted, because it takes no account of the fact that these lavas rest on an erosion surface cut on the underlying Tremadoc, of which only about 55 meters (160 ft) are preserved; and of the further fact that they are followed by the Arenig beds, with the graptolites of Zones 4 and 5, the same bed which covers the andesites of Rhobell-Fawr.

As we have seen, these volcanic rocks are separated by a prolonged period of erosion, both above and below, from the enclosing rocks, and in spite of the deceptive continuity, we are justified in postulating the same length of time between the Tremadocian (Krusna Hora) beds of Bohemia and the Skiddavian beds with graptolites, which indicate Zones 4 and 5 of the entire graptolite time scale, or the second and third of the Skiddavian divisions.

2

But we are not reduced to this one measure of the length of the time of the Rhobellian interpulsation period.

In various parts of the Lleyn peninsula, a hiatus, ranging from about 2,500 ft to 3,000 ft or more, is recognized by British geologists, between the Cambrovician, or older system (Upper Cambrian and Tremadoc), and the overlying Lower Ordovican (Skiddavian).   The higher strata of the former division are absent, through nondeposition, because of the withdrawal of the sea, and in part, because of erosion during the Rhobellian interpulsation period.

The Lower Ordovician (Skiddavian) is partly absent because of overlap of the strata during the new transgression of the Skiddavian sea. In other parts of North Wales the missing interval is even greater, reaching some 5,000 ft, and showing that this is not a local feature due to retreat because of local elevation, but is caused by the general withdrawal of the sea and exposure for a long period of time to erosion, followed later by the readvance of the sea.

In this sea, meanwhile, a new population of marine organisms had arisen, which had developed from the survivors of the older host in the distant epi-seas, on the margin of the great ocean.

It was one of the many examples that occur in the geological record of natural selection on a vast scale, induced by the hardships of the constantly narrowing environment, the extinction of the less adapted and the survival of those best able to develop new characteristics under the stress of adversity.

The destruction was of such magnitude, that when the survivors were again able to spread and multiply to occupy the new territory as it was conquered by the transgressing sea, scarcely a single species of the old inhabitants remained, though many of the new forms bore within their organization the evidence of derivation from ancestors that had lived in the older period.

Several other examples from more distant regions will emphasize the universality of the interpulsation period, which must be regarded as a world phenomenon, explainable only by the rhythmic pulsation of the sea, a sinking sea-level enduring for a very long time, during the negative stage of the pulsation period and the interpulsation period, and an equal period of the positive stage of the pulsation, during which the sea transgressed and which endured for an essentially similar length of time.

The rock formations of the southern Appalachian geosyncline bear similar testimony to the withdrawal of the sea at the end of Cambrovician, or as it has been named by the American geologist, E. O. Ulrich, Ozarkian time. [2] This, however, is much less clearly indicated than

2) The establishment of the Ozarkian System as an independent system is the monumental achievement of Dr. Ulrich, and that he has triumphantly maintained it in spite of all opposition is a tribute to Ulrich's scientific acumen. As one of his former critics I take pleasure in acknowledging my conversion to the "Ozarkian doctrine."

the second, the Petrovian interpulsation period, but the same type of clearly washed continental sands, separates the successive systems.

The withdrawal must be regarded as one of the major interpulsation events of the Palæozoic, and therefore its extent and duration was probably commensurate with the European examples.

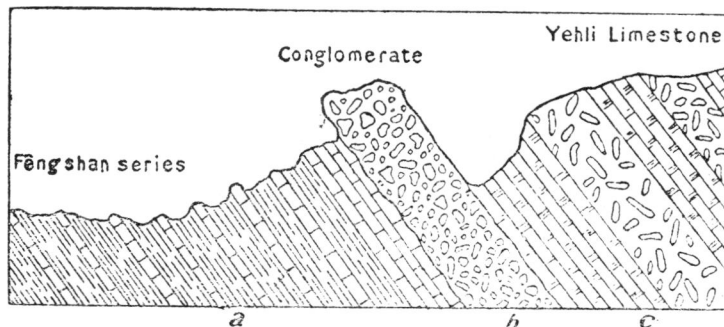

Fig. 30. Section of the Cambro-Ordovician contact and disconformity east of Yehli, Chihli. a) Fengshan limestone (Upper Cambrian, Cambrovician); b) Basal conglomerate, 1 m; c) Yehli, limestone (Lower Skiddavian).

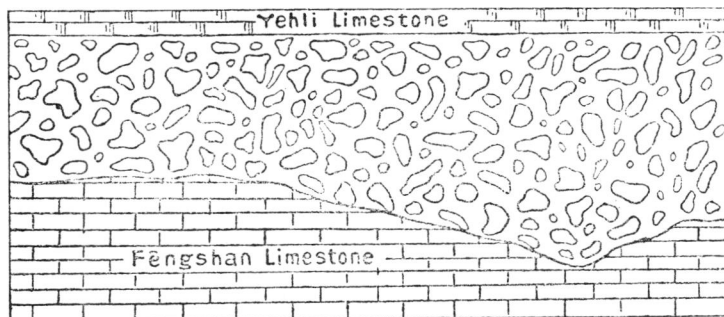

Fig. 31.   Details of disconformable contact of basal Skiddavian conglomerate on Upper Cambrian (Fengshan, Cambrovician) limestone, east of Yehli, Chihli. (Grabau: *Stratigraphy of China*, Pt. I, p. 68).

As a final expression of the universality of the interval, we may note the pronounced hiatus and evidence of prolonged exposure at the end of Cambrovician time in the Cathaysian geosyncline of East China.   Before the advent of the sea, in which the Early Skiddavian beds, the Yehli limestones, were deposited, the Upper Fengshan series suffered channelling.   In these ancient channels a bed of limestone conglomerate is now found with pebbles well waterworn, and ranging up to an inch or more in major diameter. The outer surface of each pebble is marked by a zone of deep brown oxidation, recalling the desert-varnish of pebbles exposed for a long period to the desert sun. (Figs. 30 and 31).

When the pebble is freshly broken, this oxydized zone is seen to form an outer layer about 1 mm in depth and of very uniform character. On the other hand, where the pebble was chipped or broken before it was embedded in the matrix of sand, it is clearly seen, that the oxydized zone is absent from the margin of the broken end, or where the pebble was only slightly chipped, only the outer dark rind is lacking, while the deeper lighter colored portion of the oxydized zone is still recognizable.

The obvious conclusion to be drawn from this is, that the oxydized zone developed while the pebbles were still unconsolidated, while in fact they represented surface pebbles, subject to the influence which produced the discoloration.

Judged by analogy with similar pebbles, which develop the desert varnish at the present time, we must consider this as evidence of long exposure to a desert climate, or at least to a periodically dry climate, and the influence of heating by the sun, which was the primary cause of the formation of the varnish.[3] This occurred before the readvance over this region of the Skiddavian sea, in which the fossiliferous Yehli limestones, which succeed, were deposited. As the map (Plate V) shows, the Peking region at that time was close to the equator, and during the interpulsation period the region was probably not reached by moisture-bearing winds.

Folding of the Cambrovician and older rocks has also occurred in some sections during the Rhobellian interpulsation period, but the truncation of the folds by erosion continued through the Skiddavian period, so that the resulting unconformity lies between the Cambrovician and Ordovician rocks.

An example of this is seen in the Caradoc country, of Shropshire, England, where the Hoar-Edge-grit, of Lower Bala-Caradoc age, rests on the steeply folded and sharply truncated Shineton shales, which include Tremadoc-Upper Cambrian, and older formations.

─────────────

3) GRABAU: *Principles of Stratigraphy*, 1913, p. 57.

CHAPTER XI

THE PETROVIAN INTERPULSATION DESERT PERIOD

1

To the student who is not particularly interested in the succession of extinct faunas and their distribution the retreatal phenomena and the succeeding interpulsation periods are of greater interest than the transgressive phases of the pulsation periods. The latter present chiefly the structural phenomena of progressive overlap, and, while there may be considerable diversity in the sedimentation, this is primarily a matter of detail.

The interpulsation periods, however, offer a greater variety of physical phenomena, such as is illustrated by the volcanism of the Rhobellian interpulsation and the wide-spread phenomena of special continental sedimentation of the type illustrated by the St. Peter sandstone, which gives the name to the fifth, or Petrovian, interpulsation period. The latter separates the Skiddavian and Ordovician pulsation periods and is typically developed in North America.

The St. Peter sandstone may be described as forming a zone of epithelial lining to the huge stratigraphic jaw which encloses the Petrovian hiatus. If these jaws were opened, in other words, if the strata composing them were absolutely horizontal, the depth and width would be a thousand miles, and its gape would expand from a narrow gullet in Oklahoma to a mile or more in Arkansas and Missouri and finally to a maximum capacity of a mile and half in Minnesota and Wisconsin. (Fig. 32).

The arenaceous lining, formed by the St. Peter sandstone, would vary from a minimum of 15 ft, or less, to a maximum of several hundred feet, distributed irregularly between the upper and lower jaws, like the warty lining of the toothless jaw of some giant amphibian of the past.

2

At the end of the Skiddavian or Canadian transgression, the Caribbean sea filled the whole of the Appalachian geosyncline, as far north as the Albany axis, which was still a barrier between the Appalachian and St. Lawrence geosynclines, the latter being filled by an arm from the Siberian or East Boreal epi-sea.

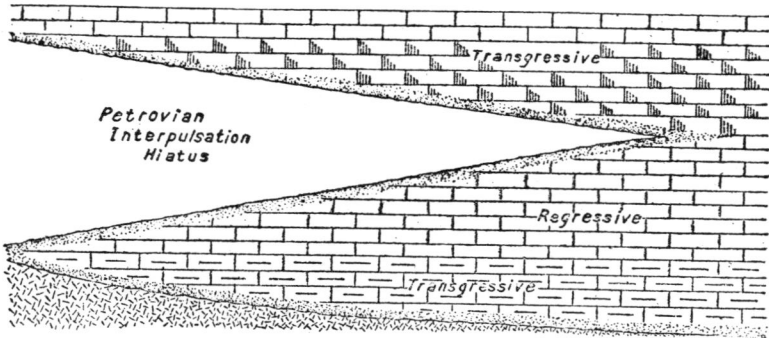

Fig. 32. Diagrammatic north-south section from Minnesota to the Arbuckle Mts. in Oklahoma, illustrating the Petrovian interpulsation hiatus in the central United States. The older transgressive beds are followed by the regressive Beekmantown, with off-lapping strata, and a covering of eolian sand (St. Peter). After the hiatus the Ordovician (Chazy, Black River, Trenton) beds readvance with progressive overlap and reworked basal St. Peter Sand.

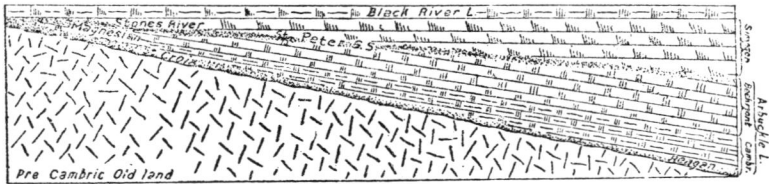

Fig. 33. The strata shown in Fig. 32 in contact (hiatus closed), showing the change in thickness; also the basal (Regan-St. Croix) and the interpulsation (St. Peter) sandstones.

But the Caribbean had also transgressed over the marginal platform of Mississippia, submerging all the area now included in the states of Arkansas, Missouri, Iowa, Illinois, Indiana, Michigan, Ohio, Wisconsin, and Minnesota, and a part of the area further to the north-west.

In the Lake Superior region it encountered the old sandstones of the Upper Cambrian (Cambrovician), the Potsdamian or St. Croix sandstones, but over the rest of the area normal marine deposits, chiefly calcareous, accumulated during the slow advance of the sea. These limestones overlap against the old sandstone (Roubidoux or New Richmond), which recorded the earlier retreat during the Rhobellian interpulsation period. That sandstone represents an earlier, less complete, manifestation of phenomena which were developed on a much grander scale during the Petrovian interpulsation period.

The limestones which mark the maximum extent of the Cambrovician sea are only 40 ft thick in Wisconsin, and 60 ft in Minnesota, where they constitute the Shakopee dolomite of the marginal platform.

Nearer to the margin of the geosyncline, the thickness of the transgressive series is measurable by a few hundreds of feet, while within the geosyncline itself, its thickness was from 1,000 to several thousand feet.

All this was formed during the transgressive stage of the Canadian (Skiddavian) pulsation period. When regression commenced the retreating shore laid bare the limestone surface, or covered it by an outwash of sands derived from the older sandstones which were exposed on the inner border of the platform. These older sands originated in the first place from the disintegration of the old crystalline granites and gneisses of the Laurentian basement rocks, *i. e.* the Archæan, or ancient sial rocks which were exposed in Laurentia.

At first it was no doubt a very impure residual sand, but by St. Peter time much of the felspar, which was originally mixed with the quartz grains, had become decomposed to clay, and had been removed by currents of water or by wind, and so a sand richer in its proportion of quartz and freer from foreign admixture, remained to be washed outward by the retreating sea.

There can be no doubt that the retreat of the sea was so slow as to be imperceptible during the years, had there been anyone to observe it. Ebb and flow of the tide were daily phenomena, though different from today, in conformity with the different distribution of land and sea. But year by year, century by century, the tidal advance became less, as the level in the outer ocean gradually fell.

How many centuries were occupied in this retreat, we cannot even guess.[1] All that we have to guide us is the observation that during the retreat of 200 miles from central Wisconsin to northern Illinois there were deposited at least 100 ft of dolomite in the latter place. After this region had also emerged by the continued withdrawal of the sea, this limestone became subject to erosion, and in some sections the entire mass, including the previously deposited transgressive portion (a total of 240 ft in Iowa), was again removed. This could only happen, however, after the sea-level had fallen considerably, with a corresponding retreat of the shore-line, though there may have been local elevations of the emergent area as well; and underground solution of the limestone may have been in progress.

The irregular erosion surface of the limestone, where preserved, is locally covered by a red clay-shale, suggestive of laterite or residual oxydized soil, left after the solution of the limestone. Over this is spread the St. Peter sandstone.

In the Ozark Mountains of southern Missouri and Arkansas more than 200 miles farther to the south, an increase of over a thousand feet in thickness is seen in the lower limestone and dolomites between the sands of the Rhobellian emergence (Roubidoux) and the St. Peter sandstone, and here again half of this, or 500 to 600 ft, may be credited to the retreatal series.

Similar conditions are seen in Arkansas, where, however, a greater thickness of the lower limestone (Jefferson City) represents the transgressive series of the Canadian, above the Rhobellian (Roubidoux) interval. Finally, the thickness of the retreatal phase represented by limestone in the Arbuckle Mountains, in Oklahoma, is more than 10 times this amount.

Of great significance is the character of the St. Peter sandstone itself. Throughout the area between Minnesota on the north and Oklahoma on the south, the rock is an almost pure assemblage of rounded quartz grains. In 51 samples the lowest silica content was 98.01 %, the highest 99.72 %. The amount of Aluminium oxide (Al_2O_3)

---

1) If we follow Joly, we may believe that about 300,000 centuries, or 30 millions of years elapsed between the beginning and the end of the retreat.

was generally less than 0.5 % (range 0.11 % to 0.64 %), and that of iron oxide (Fe_2O_3) even less.

When we remember that the source of the quartz is the old granite and gneiss, which has nearly 70.5 % of silica, though of this only 39 % can be credited to the free quartz, it is evident that in a process of purification, the felspar and other minerals must first be removed. The first step in this process is their decomposition into clay which may be removed by wind. This means that the pure quartz sand is the end stage of a long process of sifting and sorting by whatever agents were active in the region during the interpulsation period.

The difference between the St. Peter and an ordinary sandstone may be deduced from the following summary of the mineral composition of an average granite, the parent rock, and an average sandstone, the normal residue of the disintegration product. [2]

TABLE   II

MINERAL COMPOSITION OF AVERAGE CRYSTALLINE ROCKS

Mineral Group	Average Granite	Average Sandstone[3]	St. Peter Sandstone (range)
Felspars	52.45	16.39[4]	0.64 — 0.11[5]
Ferro-magnesian Minerals and Mica .	14.33	1.38	Trace[6]
Quartz	31.39	69.76	98.01 — 99.72
Accessory Minerals . .	2.03	1.70	} Trace — 0
Lime Minerals . . .	0.00	10.77	
	100.20	100.00	

2) LEITH and MEAD : *Metamorphic Geology*, pp. 74 and 76.
3) Detailed mineral analysis of an average sandstone gives the following percentages out of a total of 100%:

Quartz	69.76%	Gypsum	0.12%
Kaolin	7.98	Orthoclase Felspar	8.41
Chlorite	1.15	Magnetite	0.58
Limonite	0.80	Rutile	0.12
Dolomite	3.44	Ilmenite	0.25
Calcite	7.21	Apatite	0.18

4) Includes 7.98% kaolin or pure clay.
5) As Al_2O_3 or kaolin.
6) As Fe_2O_3.

This analysis at once shows the enormous reduction in the felspars, from 52.45 %, or more than half the mass of the rock, to 8.41 %, or to less the 1/6 the amount present in the parent rock.

The presence of nearly 8 % of kaolin (included in the condensed analysis with the felspar) shows the fate of the remainder of the felspar, though this 8 % is only a part of the amount of kaolin (which is the chief product of the decomposition of the felspar) that has remained behind after the re-assortment.

The ferro-magnesian minerals, including the micas, have all been destroyed, and the product of their decay removed, but a considerable amount of calcareous material has been added as a new constituent. The iron and ilmenite have been reduced to minor amounts.

These analyses further show by comparison that the St. Peter sandstone has proceeded much farther in the same direction, so that practically only the quartz remains behind. All the other minerals have been decomposed and removed, either by solution or in the form of dust by the wind. It should be emphasized, that the second analysis is of an average sandstone, such as is everywhere found as the result of deposition in the sea, or by rivers. Therefore we conclude that the St. Peter (3rd column) is an exceptional type, and must have been produced under exceptional conditions. That these were not local is shown by its distribution over a million square miles of country.

3

Obviously, an agent of exceptional character, one not commonly active in the formation of sandstones, is indicated. Only the wind, acting under desert conditions, is adequate to produce these results, which are comparable to those found in the larger deserts of the modern world.

Those who regard the purity of the sand as mainly due to the work of water overlook the fact that the removal and sorting by water, either during the retreat, or during the transgressive stage, would require that the product of weathering, especially the clay, should be carried to some other part of the geosyncline and there deposited

as clayey sediment. But the fact that is especially significant is the absence over the entire area of such sediments
in association with the St. Peter sandstone.

The sand, as shown by analysis, is everywhere an
almost pure quartz sand, which rests upon the limestones.
On the other hand, if a certain amount of sorting was
performed by the retreating sea, and if after the region was
fully exposed, wind activity came into play, sorting the
grains according to size and blowing away the product
of felspar decay as loess, we should have, on the return
of the sea in Ordovician time, an absolutely pure quartz
sand resting upon the old, more or less wind-eroded
limestone surface of the Canadian rocks, or locally on the
residual laterite left by solution of the limestone. Moreover this sand would consist of grains of approximately
uniform size and exceptional roundness, while it would
be covered by the similarly pure Ordovician limestone
with clay rocks in only negligible amounts.

4

We may put the quartz sand itself to a test to see
whether it conforms to this analysis and furnishes any
other evidence of wind activity. Quartz grains are rounded
by attrition in water, but the process is a constantly
slowing one, and will cease long before the limit of
sphericity is reached. This is due to the fact that a
film of water is held between the adjoining grains by
capillary attraction, and this acts as a cushion to absorb
the shock, preventing further attrition.

We know, however, from observation, that the grains
of the St. Peter sandstone are not only rounded to a
marvellous degree, but are mostly small: (1 mm to 0.5
or even 0.25 mm in diameter), while less than 13 %
by weight are larger than 1 mm, and less than 4 % larger
than 2 mm. At the same time only a very small amount
of the material in a large number of samples is smaller
than 0.0625 mm, while almost 90 % is retained on the
screen with mesh-openings of 0.125 mm. Moreover, and
this is significant, all the quartz grains show a marvellous
perfection of rounding, which is even seen in the smallest
grains. The following table will enable the reader to
compare different sands in respect to size of grain.

TABLE III

COMPARATIVE TABLE OF SIZE OF GRAIN IN DIFFERENT SANDS[7]

Size of Grain	River Sands	Beach Sands	Dune Sands	St. Peter Sand-stone	Fraction of mm
Above 2.0 mm	—	—	—	4.0%	over 2
2.0 mm to 1.0 mm	0.8%	0.1%	—	8.0%	2 - 1
1.0 mm to 0.5 mm	24.9%	14.5%	0.1%	—	1 - 1/2
0.5 mm to 0.25 mm	37.8%	81.8%	43.30%	—	1/2 - 1/4
0.25 mm to 0.175 mm	27.4%	3.6%	55.80%	89.0%	1/4 - 1/8
0.175 mm to 0.0875 mm	3.2%	—	0.80%		1/8 - 1/16
0.0875 mm to 004375 mm	4.0%	—	0.10%	1.0%	1/16 - 1/32
0.04375 mm to 0.021875 mm	1.2%	—	—	—	1/32 - 1/64
0.021875 mm to 0.0109375 mm	0.3%	—	—	—	1/64-1/128

Experiments made at Columbia University by Dr. J. J. Galloway,[8] have shown, that in water flowing at 4 miles per hour, rounding of quartz grains below 0.05 mm is extremely slow, only a few being rounded in 200 hours, and this he regards as the lower effective limit in water. The lower effective limit for wind-blown sand is taken at 0.03 mm. Again he concludes, that: "if a non-calcareous sand has more than 50% of its grains well rounded, it is of eolian abrasion."

Taking all the physical and chemical characters of the St. Peter sandstone into consideration, as well as its relationship to the enclosing strata, and its distribution, there seems to be scant room for doubt that our interpretation of it, as a record of a great desert during the Skiddavian-Ordovician interpulsation interval, is the correct one, for it meets all the requirements of such a deposit.

5

If we look for a modern parallel, the Takla Makan desert of Central Asia comes to mind, as presenting many

---

7) TWENHOFEL : *Treatise on Sedimentation*, 1932 edition, p. 219-223.
8) *Ibid*: p. 223.

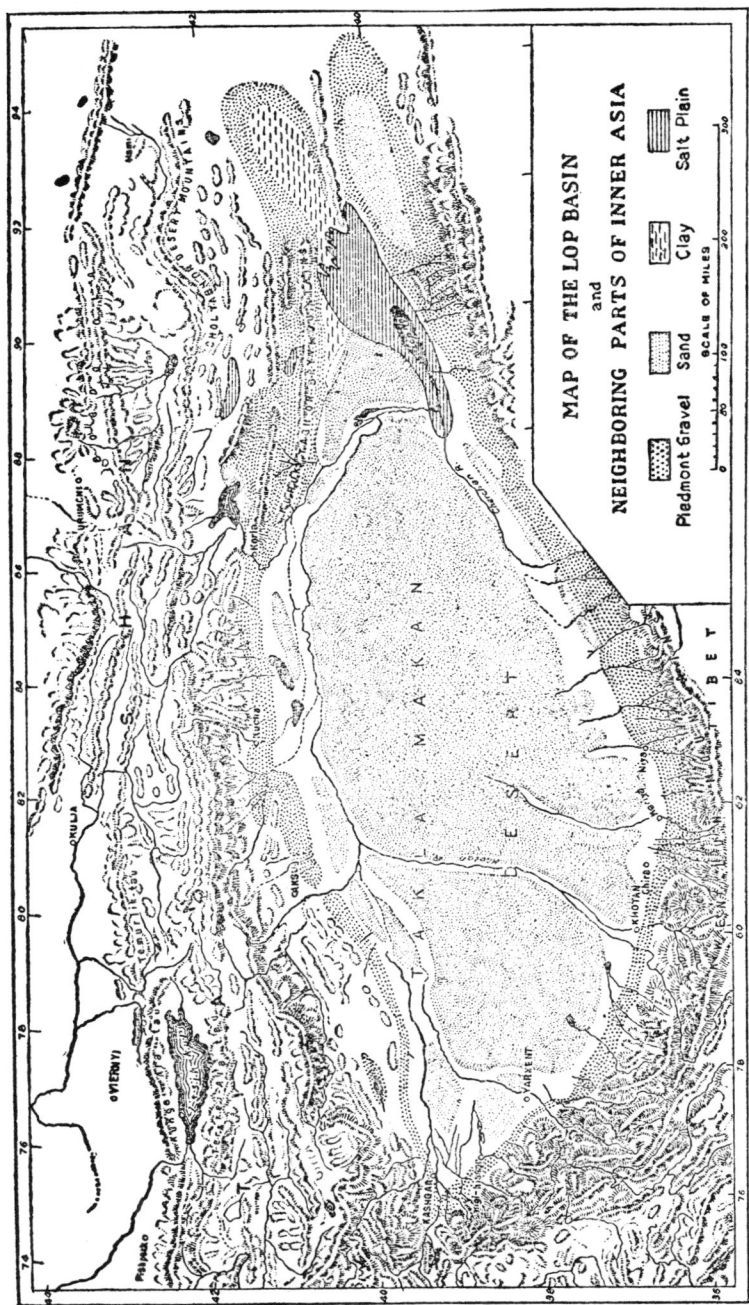

Fig. 34. Map of the Takla Makan desert basin and the salt lake Lop Nor, between the Kwen Lun (Kuen Lun) and Tianshan (Tienshan) Ranges of Central Asia. The numerous mountain streams dwindle away in the desert which is covered by dunes of pure rounded quartz sand. The salt plain of Lop is of connate origin. (Redrawn after Huntington)

analogies, though lacking the broad open character of the St. Peter desert. This has been fully described by Huntington. [9] The Takla Makan is 1,400 miles in east-west extent, averaging 18° of longitude at latitude 40°, and in its widest part it is not much more than 400 miles in width, decreasing to almost half that in the Lop Nor region (Fig. 34).

It differs, of course, radically from the St. Peter desert in the fact that it is surrounded by high mountains on the north and south, these closing in to form a complete barrier on the west, at Kashgar. That the St. Peter desert was of exceptional character, in that it did not lie within a mountain-enclosed basin, as do so many deserts today, I readily concede. But that it was a desert of shifting sands without vegetation, and to a large extent one of dwindling rivers, is evident from the character of the deposits.

We are apt to forget, in making comparisons, that we are dealing with a period of time when land vegetation was of a relatively low order, and this no doubt contributed largely to the barrenness of the desert, which, under the influence of arresting vegetation, might have been less stark and forbidding.

Bearing these circumstances in mind, and noting the character of the rock and its extent, we cannot escape the conclusion that the St. Peter desert was a prominent geographical feature of the early Palæozoic of America, and though not wholly unique, was a phenomenon of such extent and significance that it may well be taken as the outstanding feature of this interpulsation period, and justifies the application to it of the term the PETROVIAN INTER-PULSATION PERIOD.

9) HUNTINGTON, E. : *The Pulse of Asia*; 1907, p. 92.

CHAPTER XII

THE PETROVIAN INTERVAL IN OTHER REGIONS

1

VERY different from the conditions which characterized the Petrovian interval in the southern Appalachian geosyncline and its marginal Mississippian platform, and which permitted the wide-spread development of the St. Peter desert, were those of the St. Lawrence geosyncline, the strata of which now form the Notre Dame Mountains of Quebec, which are the northern continuation of the Appalachian folds.

A short distance north of the international boundary, an exceptionally interesting line of cliffs faces Lake Champlain. They are part of a slice of Skiddavian (Canadian) and Ordovician rocks. These rocks after they were folded, were forced to slide across rocks of similar age but of less disturbed character which formed the foreland of the mountains on the west. They are in turn over-ridden by other rock slices, each beginning as a rule with rocks of more ancient date than those across which they have been thrust, until the geological column in these mountains seems to stand on its head. (Fig. 35).

It is, however, with the rocks of the Philipsburg slice, as the lowest thrust block is called, that we are especially concerned, for these represent something more than 200 ft of lower limestone of Canadian age (Beekmantown, with perhaps some underlying Cambrovician beds), followed by an upper shale, limestone and chert series, 2,500 ft or more in thickness, and of Ordovician age. These lie on the border of the former Taodini (Taudeni) ice-sheet (Fig. 40).

Between the Canadian and Ordovician beds of this region lies the famous Philipsburg breccia, or "Mystic Limestone Conglomerate." This consists of 250 to 300 ft, of black limestone conglomerate formed of the ruins of

massive beds like those on which it rests. The fragments
or "pebbles" range in size from a diameter of 1 inch to
blocks containing 50 or 60 cubic feet. They are closely
packed and cemented by a calcareo-magnesian paste, apparently the triturated dolomitic limestone.

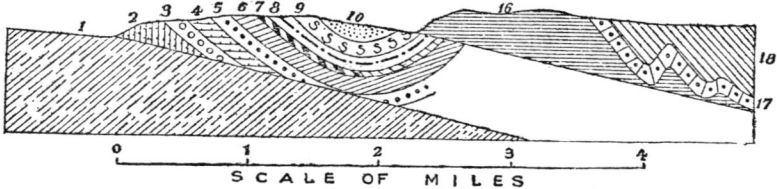

Fig. 35. Section of the Philipsburg-Quebec region, showing Philipsburg
and Rosenberg overthrust slices. (After T. H. Clark). Scale in miles.
Foreland :

1	Iberville Sh.	6	Hastings Creek Ls.
Philipsburg Slice :		5	Morgan Corner Ls.
14	Stanbridge Sl.	4	Wallace Creek Ls.
13	Mystic Ls. Cg.	3	Strites Pond Ls.
12	Basswood Creek Sl	2	Rock River Dol.
11	Corey Ls.	Rosenberg Slice :	
10	St. Armand Ls.	18	Georgia Sl.
9	Solomons Corner Ls.	17	Highgate Ls.
8	Luke Hill Ls.	16	Milton Dol.
7	Naylor Ledge Ls.	15	Mallett Dol.

There are also some lighter colored dolomitic blocks,
1 ft in diameter, intimately mingled with the blocks of
darker limestone. These limestones contain the same species
of fossils that are found in the massive bedded Beekmantown limestone which underlies it, while some of the
intercalated, bedded strata contain fossils of Chazy, that is
Early Ordovician, age.

These breccias have all the earmarks of a collapsed
cavern roof, though we may perhaps not wholly exclude
the possibility of their representing ancient rock streams or
rock glaciers, similar to the "rock rivers" of the Falkland
Islands, the "stone glaciers" of Alaska, or the "rock flows"
of Bear Island in the North Atlantic.[1] All of these modern
examples occur in regions of rigorous climates, much moisture, and of considerable elevation of the rocks, from which
these flows proceed.

On the whole, the theory of the origin of the
Philipsburg breccias through collapse of cavern roofs, is

1) GRABAU : *Principles of Stratigraphy*, 3rd ed., p. 542.

more in harmony with the fact that the fragments belong to the type of rock which still overlies the breccia and contains the same organic remains, thus showing that the cavern was dissolved out in the upper layers of these rocks.

The collapse of the roof may have accompanied the return of the sea to this region in Chazy time, which would explain the fossils of that period in the mud layers intercalated between the brecciated rock. Rocks of Chazy and Trenton age succeed the breccia, showing that the sea-return signified the beginning of the new, *i. e.* the Ordovician, pulsation period.

It thus appears that while the St. Peter sands were shifting about on the desert surface of the Mississippian marginal platform during the Petrovian interpulsation period, the exposed Canadian limestones of the St. Lawrence geosyncline were subjected to underground solution and the formation of caverns, which collapsed with the readvance of the Ordovician sea.

This underground solution, in regions just emerged from the sea, indicates the great extent of the retreat and the corresponding subsidence of the sea-level, and emphasizes moreover the great length of time occupied by the Petrovian interpulsation period.

2

Elsewhere, the interval appears usually to have been one of exposure with off-lapping strata, and renewed transgression with overlap of the members of the older series by the succeeding transgressing formations. This usually results in a disconformity, marked primarily by erosion, and though careful search would probably, in most instances, disclose the physical evidence of erosion and off-lap, on the whole this evidence is not easy to recognize.

Where distinctive faunas characterize both the off-lapping beds of the retreatal series, the Skiddavian or Canadian, as well as the overlapping beds of the transgressive series, the Ordovician, the retreat and readvance can readily be determined, despite the fact that the stratigraphic jaws are tightly closed on the hiatus.

Such structure is typically seen in the Baltic provinces of Esthonia, where the Petrovian interpulsation hiatus is well marked. (Fig. 36). There, the faunal zones have been lettered, as well as named, and since the extent of each has been carefully traced in the field, it can easily be summarized.[3]

W.   E.

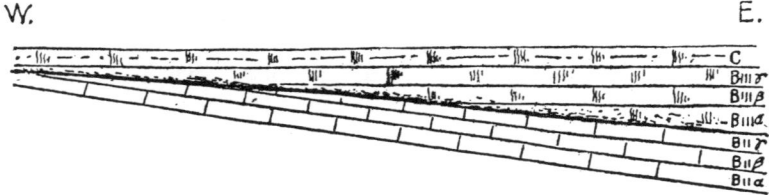

Fig. 36. Diagrammatic east-west section, showing the Petrovian interpulsation hiatus and the thinning of the formation westward by off-lap of the Skiddavian (B-II) and overlap of the Ordovician (B-III and C). Baltic Coast of Esthonia, etc. (Compare Fig. 33).

From west to east the sections reveal the following succession in the lower series.

At Packerort it begins with the zone of *Megalaspis planilimbata,* or zone B-II-α.

Some miles farther to the east the second zone B-II-β, (the zone of *Asaphus bröggeri*) appears above it, and thence continues with increasing thickness eastward, always overlying the earlier zone.

Some miles still farther east, the third, or zone of *Megalaspis gibba* and *Asaphus lapidurus* (zone B-II-γ), appears, and continues as a cover of the older divisions.

The hiatus is still marked as an erosion surface at Volchow, 300 miles east of the first section, where the *Asaphus lepidurus* beds are followed disconformably by beds with *Asaphus expansus* (zone B-III-α), the first of the post-hiatus transgressive series.

Those of zone B-III-β with *Asaphus raniceps,* follow and extend for 150 miles eastward to Narva, overlapping the earlier zone.

Finally, zone B-III-γ, with *Megalaspis longicauda, Pseudasaphus globifrons, Onchometopus stacyi,* and other typical fossils restricted to it, overlap for another 150 miles, as far as Packerort. (See Fig. 36). Thus at Packerort the lowest zone of the earlier retreatal, or Skiddavian, series

(B-II-α), is followed directly by the highest zone of the transgressive Ordovician series (B-III-γ).

None of the species characteristic of the Lower (B-II) group (Skiddavian) that have been fully identified pass across the intervening hiatus into the higher division (B-III) (Ordovician). Of the species characteristic of B-II, many occur in two or even all three of the subdivisions of the Skiddavian, and the same is true for the species of the Ordovician (B-III), while they are unknown in the Skiddavian (B-II). Many Ordovician species pass through all three, or at least occur in two of the divisions of B-III.

This distinctness of the lower from the upper faunas is to be expected, since the Petrovian interpulsation interval, which separates them, is one of long duration and marks a complete drainage of the geosynclines and a retreat of the fauna to a restricted area, the marginal epi-seas in which in the course of time they would undergo a complete change. The marvel is indeed not that the species do not pass across, but that so many of the genera do.

3

An almost identical illustration of the relationship of the retreatal Canadian and the transgressive Mohawkian (Early Ordovician), is seen in the St. Lawrence geosyncline. Here, too, the parting sandstone is absent, or only locally and sparingly developed.

W.   E.

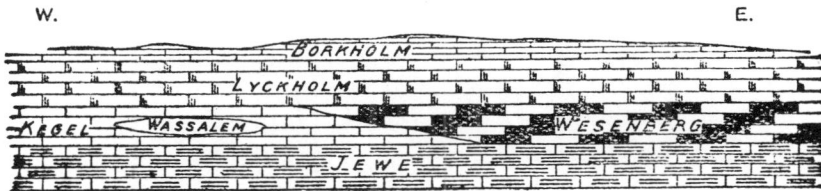

Fig. 37. Section to show the relationships of the Kegel, Wassalem and Wesenberg formations to one another and to the underlying Jewe, and the overlying Lyckholm. (Drawn from data published by Raymond).

This type of disconformity is in reality wide-spread. The hiatus is generally unmarked by readily recognizable deposits, and indeed is often masked, so that only the studies of the faunas will disclose the relation.

Illustrations of it are found on the northern shore of Scotland, in North China, including Manchuria, and in South China as well.

It is recognized in the graptolite shales of Scania (southern Sweden), by careful plotting of the zones, and is seen between the graptolite shales of the Deepkill and those of the Normanskill divisions, in the Hudson Valley in New York. I am confident that it will be found in all parts of the world where these rocks are exposed, when careful attention is given to their relationship, for the pulsation was a universal one. The withdrawal drained all the geosynclines, and not until long after were they occupied again by the new transgressing series with the new fauna.

4

But, while the Ordovician hiatus so often wears an unobtrusive and deceptive expression with tightly closed, unrevealing lips, so that only an expert in terrestrial physiognomy could hope to recognize it, now and again, as in the type region, it reveals itself more clearly, and where it is marked by volcanism, where it has taken a mouthful of lava, it is as readily recognizable as it is where its lower jaw is dislocated and the upper one closes upon it unconformably.

Volcanic rocks occupy the hiatus in the region of Mount Arenig, Mount Snowdon and Cader-Idris, and though in many cases the flows of lava began during the retreatal period, their main extrusion falls in the interpulsation hiatus. In none of these cited is the evidence so sharply cut and clearly presented as it is in the case of Rhobell-Fawr, the volcano of the preceding interpulsation period, but by the geologist it will be readily recognized.

In Bohemia volcanics likewise occupy the Petrovian interpulsation period. But the greatest known development of lavas that occupy this interval is seen in the Lake District of North England, where the Great Borrowdale volcanic group separates the Skiddavian below from the Ordovician or Bala above.

Volcanism here began during the period of retreat of the sea (Llanvirnian), while the occasional marinings still washed the graptolites of the *Bifidus* zone over the ash-strewn mud-flats of the river plains. The *Murchisoni* zone, the highest zone of the Llanvirnian series, does not occur in the Lake District of Cumberland, the shore by that time being already too far removed for the marinings to reach this region.

Eruptions began with the explosive ejections of ashes and lapilli together with larger blocks, and this was interrupted by an occasional flow of andesitic lava. These older pyroclastic rocks are sharply succeeded by the great flows of rhyolite lava which make up the bulk of the Borrowdale volcanic series, and which accumulated to a thickness exceeding 2,000 to 5,000 ft.

Not until Middle Bala time, when eruptive violence was exhausted, did the sea return to this region, and the *Corona* beds, the shales with the brachiopod *Trematis corona*, were deposited upon the old lava surface, the volcanism by this time being wholly spent.

These shales were followed by other fossiliferous beds, and finally by pure shell limestone, known as the Keisley limestone. This probably marks the maximum transgression of the sea at this period, for though the waters were not over deep, they were pure, so that even corals could flourish along with the brachiopods, while trilobites in numbers and in great variety crawled and swam among the fixed types. No less than 43 species of trilobites have been described from this limestone. To these we must add 40 species of brachiopods, besides cephalopods, bi-valved molluscs and gastropods. The total number of species that has been recovered from the Keisley limestone exceeds 100, though the limestone is only 40 ft in thickness. And of course the number of individuals is uncountable.

Regression and the deposition of more shales followed, then erosion, which in the region of Coniston Lake cut down to the limestone, so that the next hiatus, the Plynlimonian, follows directly upon the transgressive part of the older series without the intervention of the normal regressive strata of that series.

5

Such are a few of the facts that have been revealed by the study of these formations in widely distant parts

of the earth.   All tell the same story of retreat and off-
lap, of interpulsation deposits, either of desert, volcanic,
or other continental type, or of erosion of the exposed
rocks.   Not until the interpulsation period had come to
an end, and the sea-level slowly rose again in obedience
to the new pulse-beat, did the sea creep up again, to fill
the geosynclines, and to cover the erosion surface and the
lava flows and the desert and other continental deposits
which were formed during its absence.   But by then a
long period of time had passed, during which the organisms
had undergone profound changes in their distant retreat,
so that the returning sea brought a new host specifically
distinct from its predecessor, though generically as well as
genetically bound to it by ties of ancestral relationship.

To ignore the interpulsation periods, the periods of
falling sea-level, when the base-level of erosion is lowered
everywhere, is to neglect half of the history of the evolution of our earth.   It is to neglect the great periods during
which the rivers are rejuvenated and begin once more
to incise their courses, to diversify and then again to
smooth out the topography of the land and prepare it for
the next great transgression, or to build extensive continental fans, which separate the fossiliferous formations.
It neglects the periods of the world's great deserts, and
those during which volcanism is chiefly active.   And finally, it ignores the periods of adversity in the organic world
when the luxurient life of the geosynclines becomes
restricted to constantly narrowing limits, accompanied
by, and inducing, the elimination of the unadaptable,
with a corresponding increase in the fierceness of the
struggle for existence among the survivors.   And it
ignores the main fact, that those that survive become
the founders of a new dynasty, which, on the return
of the waters to their old geosynclinal beds, will develop,
for weal or for woe, into the new fauna that for some
millions of generations will occupy the area vacated by
its predecessors.   And having contributed their share to
the history of the growing earth, they, in turn, will
disappear as the seas once more retreat, and a new period
of trial and struggle begins.   But like their predecessors,
they too will leave their preservable remains in the rocks
of the earth, there to record as on a tombstone the fact

of their existence and the part which they played in the history of an ever changing and, in spite of the many set-backs, of a constanly advancing world.

CHAPTER XIII

PHASES OF APPALACHIAN DEVELOPMENT

1

NOWHERE is the progress of the history of the earth during Palæozoic time more clearly preserved, and nowhere is it more readily deciphered than in North America, for nowhere else was the record written in such completeness, and nowhere else has it been so little disturbed by violent mountain making revolutions, and by the crush and tear and destruction that accompanies these. Europe has been a country of repeated minor and of some great revolutions; Asia has had them on a grand scale. Africa and Australia were seldom free from polar glaciation during the Palæozoic, and Antarctica of today is buried in polar ice, and will not reveal its earlier history.

True, America also has had its revolutions, but they were on such a broad scale, and accompanied by so little actual destruction, and so little of that hopeless confusion of detail that characterized such disturbances elsewhere, that it has become the standard by which all the other records are to be judged, and their shortcomings corrected.

In North America, Palæozoic rocks extend in almost undisturbed condition from the mouth of the St. Lawrence, in 50° N. Lat., to the banks of the Tennessee in Lat. 35°; from the Catskills, in Long. 75°, to the Black Hills of Dakota, close to Long. 105°, and south on this meridian to the Mexican border. The area is roughly measured at over two million square miles, which is approximately half the area of Europe between the North Sea and the Urals, and extending from the Arctic shores of Norway to the Smiling Isles of Greece.

Over this vast area, forming the eastern half of the United States, and extending north into Canada, the Palæozoic strata are so little disturbed that they still appear for the most part in horizontal sheets, yet are sufficiently arched locally to have enabled erosion to uncover them, often to the base. Thus are formed innumerable sections, where superposition of formation on formation can be studied, from the base of the series to its top, where this passes beneath the covering of younger strata. Even the broad anticlinal folds of the Alleghanies on the south-east, and the even gentler synclines of the Canadian Rockies on the northwest, show so little actual disturbance of the strata, and exhibit such satisfying sections that they can be read as readily as the crumpled pages of a book, even without the need of much smoothing of the wrinkles.

But let us not deceive ourselves. Though America presents us with the most complete record found anywhere, it is not perfect, and if we rely wholly on American sections we get an incomplete view of Palæozoic history. Comparsion with sections elsewhere will enable us to complete and amplify the picture, but its grand lineaments, as well as the major details, are found only in the American succession.

This is why American sections must always be the standard of comparison, and why they alone will give the sequence of Palæozoic periods in all their completeness, together with the evidence of the lapses in the record when the mighty pen of the recorder was seemingly at rest. But the record must be read by an eye adjustable both to microscopic and to telescopic vision, and possessed of a faculty of seeing the minute fraction in its relation to the whole. Such a faculty can be cultivated by anyone who studies geology as the science of the whole earth in the light of local detail.

2

When the Ordovician south polar ice-cap centered in the western Sahara, and the ice-front lay along the whole of the present Atlantic coast of North America from New-foundland to the Caribbean, the entire American portion of the Caledonian geosyncline was buried beneath its mass. (Fig. 40). It was then that some of the old conglomerates of the New England region were formed as glacial moraines, conglomerates which have long puzzled the geologists and whose true character as indices of early Palæozoic glaciation has only recently been recognized. Among these by far the most outstanding and best known is the "Roxbury Pudding Stone", which has been immortalized by Oliver Wendell Holmes in his delightful verses on the "Dorchester Giants", who threw lumps of this pudding with its stony plums, all over the Boston region, where they lie today.

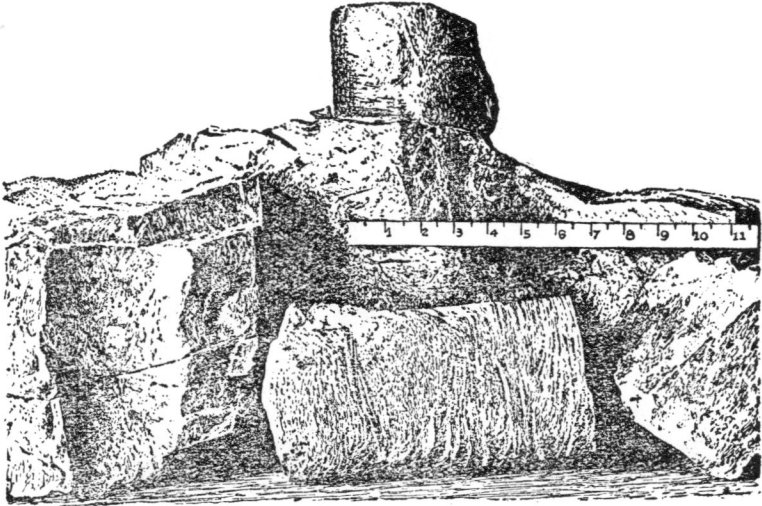

Fig. 38. Fossil "Trees" of the Roxbury conglomerate of Ordovician age. Scale in inches. (Redrawn from photograph by Burr and Burke).

It was once thought that this rock, now recognized as an old solidified glacial moraine, and the *varved* slates, which represent the annual deposits of fine mud from the dwindling glacier in its waning stages, belonged to the Carboniferous or Permian period of New England history, because the upper sandy beds of this conglomerate have furnished some rather remarkable remains, which suggest "tree trunks" of presumably Carboniferous age. (Fig. 38). But similar, though much longer and much thicker "tree trunks", so large that they are locally called fossil telegraph poles (Fig. 39), are known from the Potsdam sandstone, which lies at the base of the Lower Ordovician (Cambrovician) in the Thousand Island Region of today, and which was a part of the same general lowland, on which at that, or somewhat later time, grew the Roxbury trees. This lowland was invaded by the ice-sheet from the south pole, when that occupied the present desert region of north-west Africa.

Fig. 39. Fossil trees ("telegraph pole concretions") in the Potsdam sandstone at Rideau, Ontario, in the Thousand Island Region. (Size shown by figures of men). From photograph, Geol. Surv. Canada.

3

There are many other conglomerates and varved or seasonally banded clays which preserve the record of this former ice invasion. The localities, marked on the accompanying map (Fig. 40), show those discovered so far. Among the more interesting of these are the limestone conglomerates, which extend along the St. Lawrence River from Quebec City to the end of Gaspé Peninsula, a distance of 360 miles. They contain boulders, or erratics, which range in weight up to twenty five tons and in rock character from the dominant limestone blocks with Cambrian fossils to blocks of sandstone, basalts, and amygdaloid volcanic rock. Both above and below these old

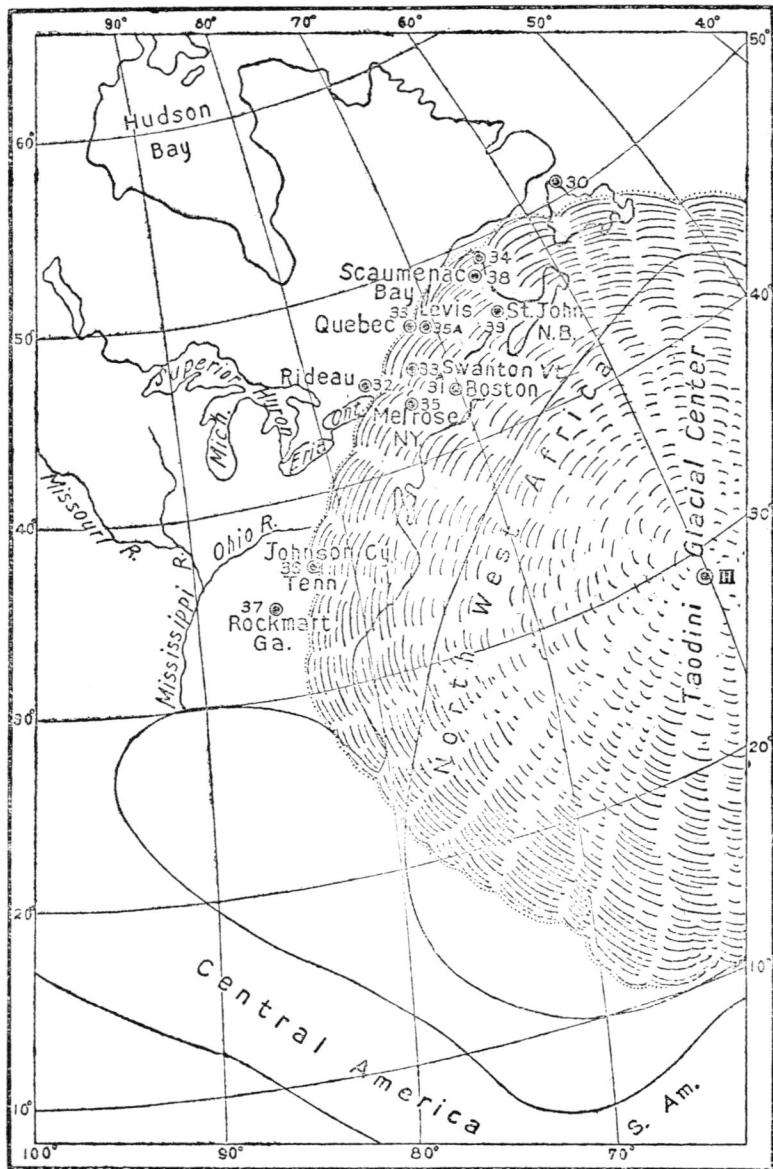

Fig. 40. The Taodini Glacial Center, or the south polar glaciation in Ordovician time. This was responsible for the Roxbury tillites and the other glacial beds of Eastern North America.

glacial moraines are beds with marine fossils, showing that
the work of bringing these conglomerates together was—
as Coleman assures us in his delightful book [1] — "done by
an ice-sheet of the continental type on low ground."
These rocks are a part of the deposits in the old St.
Lawrence-Appalachian geosyncline, showing that this was
subjected to glacial invasion at the time. This geosyncline was generally occupied by the sea in which the
so-called Trenton limestone was deposited. In the Quebec
region the Trenton limestone, which was forming in this
geosyncline, shows evidence of the influx of much cooled
water in its dwarfed fauna, a fauna of diminutive representatives of the normal species. In these fossiliferous
limestone beds has also been found a block of granite,
weighing 8 tons, an erratic probably brought by floating
ice and dropped where the calcareous muds of the Trenton limestone were accumulating.

The Trenton [2] limestone is the deposit in the transgressing sea which followed the Petrovian interpulsation
period. Sometimes its thin-bedded basal portion rests directly on the old erosion surface of the Beekmantown dolomite
of Skiddavian (Canadian) age, but more often an earlier
Ordovician deposit intervenes. In all but the eastern sections, moreover, a variable thickness of St. Peter sandstone,
—the interpulsation deposit—separates the Canadian and
Ordovician beds.

In the lower St. Lawrence valley and the region of
the present Lake Champlain, which was covered by the
Ordovician sea invasion before it reached the Mohawk
Valley, the typical Trenton limestone is preceded by a more
massive rock, the Chazy limestone, with the remarkable
flat-spired and left-handed coiling gastropod *Maclurea*. This
fauna was apparently brought in by the sea which invaded
the region from the Appalachian geosyncline. It originated
in the Caribbean epi-sea, and its traces are found all along
the line of its migration in equivalent formation (Stones
River, etc.) in Tennessee, New Jersey and Pennsylvania.

It continues north to the Mingan Islands in the present
St. Lawrence Gulf and the Lake Champlain region from

1) COLEMAN A. P. : *Ice Ages, Recent and Ancient.*
2) Named from the Trenton Falls, on a tributary of the Mohawk River,
south of the Adirondacks.

which near the village of Chazy the formation derives its
name. It is generally followed by another limestone, the
Black River,[3] in which cephalopods and corals occur, which
had a wide distribution in all the water-bodies fed by the
Siberian epi-sea, where this fauna apparently had its center
of distribution. It has been found in strata of equivalent
age in North China on the one hand (Cathaysian geosyncline) where it is known as *Actinoceras* or Machiakou
limestone, and throughout the St. Lawrence geosyncline as
far as Quebec on the other, and it penetrated even into
the Caledonian geosyncline.

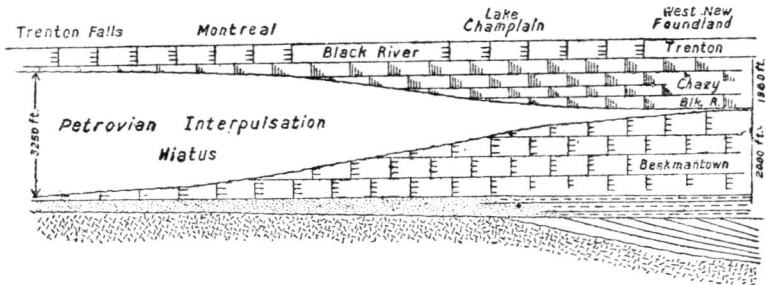

Fig. 41. Relationship of the Canadian and Ordovician beds in the lower
St. Lawrence Valley, showing the off-lapping and overlapping strata and the
augmentation of the Petrovian interpulsation hiatus (Compare Figs. 32, 33
and 36).

Where the transgressive series is fully represented in
the St. Lawrence geosyncline, as it is in western Newfoundland, both Chazyan and Black River divisions are
present. (Fig. 41). The former, 1,380 feet thick, rests
with a disconformity upon 2,000 feet or more of Beekmantown (locally known as St. George series), which in
turn rests on 1,700 feet of Cambrovician graptolite shates,
sandstone and some limestone (Green Point series). In lhe
Lake Camplain region 900 ft of Chazy rests with a disconformity and hiatus (Petrovian interpulsation hiatus) on
1,500 ft of Beekmantown (Canadian, or Skiddavian), which
in turn rests disconformably on Cambrovician that overlies the crystallines and begins with a basal sandstone
(Potsdam). The Chazy is followed by from 30 to 60 ft
of Black River, above which follows Trenton limestone.

By the time we reach Montreal, the formations have

3) Type locality Black River, west of the Adirondacks.

thinned significantly. The Chazy is represented only by its upper 200 ft or less, followed by Black River (75 ft) and Trenton (400-500 ft). The Chazy rests with a marked disconformity on the erosion surface of the Beekmantown, which is here 200-250 ft thick, and in turn overlies from 200-300 ft of Potsdam sandstone (Cambrovician), under which lie the crystalline rocks of the Canadian shield. Finally, in the Mohawk Valley, the Chazy may be entirely overlapped and the Black River, or Lower Trenton, rests directly on an erosion surface of Beekmantown.

This indicates a constantly widening hiatus as we ascend the St. Lawrence River, due to the progressive wedging out of successively lower members of the Canadian (Beekmantown) series and of successively higher members of the Ordovician series. This implies that the Canadian sea retreated regularly away from the Montreal region, and that the Ordovician sea steadily readvanced toward it. The hiatus thus enclosed is the Petrovian interpulsation hiatus, but it is not occupied by the sandstone which characterizes it in the interior. Here only erosion prevailed and the surface of the Beekmantown everywhere bears evidence to this effect.

4

In those days, Bear Island and Spitzbergen, now the outlying lands of the Arctic, were enjoying a tropical climate, lying within 15° to 20° of the equator. Bear Island was within the geosyncline, as was also East Greenland of today, but both lay nearer the margin, and both were reached by the transgressing sea in Black River

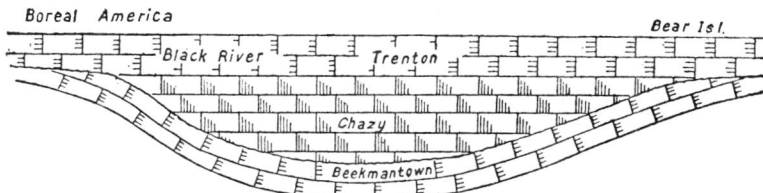

Fig. 42. Diagrammatic cross-section of the St. Lawrence geosyncline in Ordovician time, in the region between Greenland and Bear Island, before dismemberment. The Lower Ordovician (Chazy) overlaps on the Canadian (Beekmantown) and is in turn overlapped by Black River and Trenton.

time. On Bear Island more than 240 meters of richly
fossiliferous limestone with the Black River type of fauna
rest disconformably on the eroded surface of the Beekmantown equivalent, *i. e.* the Canadian dolomite with *Archæoscyphia*, *Pyloceras* and *Protocycloceras*. The fact that
typical Chazy beds are absent shows that that faunal
phase did not extend so far down the St. Lawrence
geosyncline, or if it did, was overlapped by the Black
River group. These beds were the first deposits of the
transgressing Ordovician sea on Bear Island, overlapping the
Chazy, which, if present, was restricted to the center of
the geosyncline. (Fig. 42).

The Black River formation of Bear Island, like that
of north-west Newfoundland, represents, of course, much
more than the Black River of the region about Watertown,
New York.   This Watertown record marks only a temporary
ingress of the sea from the distant Siberian epi-sea, a short
sojourn of the delegates from the Bear Island fauna, only
long enough to record its visit in 50 or 75 feet to fossiliferous limestones.   The influence of the cold from the
nearby polar ice-sheet caused the fauna to retire to its
tropical home in Siberia and leave the colder waters of
the St. Lawrence-Quebec region to the Trenton fauna.
This consisted of delegates from two other faunal centers,
both marginal to the pangæa, and therefore both from
centers of temperate climates.   These were the animals of
the Galena, from the Alaskan epi-sea (then in Lat. 15°),
and those from the Chicamauga-Moccassin limestones, of
Caribbean epi-sea origin (temporarily in Lat. 30°-45°). Many
of these organisms that entered the Trenton sea of the St.
Lawrence geosyncline survived only in a stunted condition
in the iceberg infested waters.

But, even they had to give up after a while, when
the shoaling of the geosyncline allowed only fine muds to
be deposited, which were subject to repeated exposure,
followed by storm-wave flooding, or "marining", and the
spreading of planktonic graptolites over the surface. The
record of this is now seen in the Utica graptolite shales,
which contain hardly any other fossil than graptolites,
except the pelagic trilobite *Triarthrus*. Mud deposits of
this type had existed throughout the Ordovician, and
throughout the earlier Skiddavian, as well as in the still

earlier Cambrovician pulsation period, in the region of the present Hudson River. These muds were the huangho deposits of the rivers that flowed from Appalachia, fed probably by the glaciers which for ages had occupied, through all this time, their West African center, and covered the east coast of North America. They blotted out the Caledonian geosyncline, and repeatedly overtopped the dividing ridges of the Appalachian old-land. Whenever they sent their muddy streams into the Appalachian geosyncline, they built huge delta plains of varved mud deposits. These muds besides their regular variation in texture, a seasonal effect, exhibited time and again the effects of momentary submergence, or "marining", by the waters of the geosyncline, which on retreat left behind its stranded flotsam, the graptolite colonies.

Thus the "Hudson River" slates, with their graptolites, were formed during Skiddavian, as well as Ordovician time, while further away in all directions, where the polar influence was not felt, richly fossiliferous limestones were accumulating. Such was the fossiliferous Galena limestone of Wisconsin and the central states, the fauna of which was derived in part from the Alaskan or west Boreal epi-sea. This follows upon the fossiliferous Stones River beds of the geosyncline, whose fauna was derived from the Caribbean epi-sea. In these north-western and central regions of the present United States, the strata represent deposits formed away from the polar influence of the West African (Taodini) ice-sheet, which held New England and New York in bondage, and they are correspondingly rich in marine organic remains. Here, the richly fossiliferous Cincinnati beds were forming, while in the regions dominated by the glaciers only sandy shales with graptolites or with a sparse fauna of brachiopods accumulated. As already noted, the "Hudson River" shales comprise Lorraine, Utica and Normanskill, as well as the Canadian Deepkill shales, all of which are huangho deposits, formed in the Hudson River Valley at that time, within easy reach of the glacial waters.

Then, as the sea again retreated in all directions towards the distant epi-seas, and the geosynclines were abandoned, and as these lands moved away from the pole and its glaciers, which migrated relatively, in the opposite

direction to the new Silurian polar station in Central
Africa[4], the strata of the old Appalachian geosyncline
experienced their first collapse. Unable to resist the pressure
which the moving African colossus exerted on the weakened
geosyncline, as it slid in what is now a north-westerly
direction, they were pressed together to form the long
Taconic ranges of mountains, which extended from Newfoundland to New Jersey.

Probably the Cambrian and Ordovician beds, the
latter chiefly of glacial origin in the Caledonian extension
of the Boston region, and perhaps elsewhere in eastern
North America, received their initial deformation at this
time by the same irresistable pressure.

5

In the southern Appalachians the Ordovician formations have distinct names, partly because they carry the
richer and at times quite distinct fauna of the Caribbean
epi-seas, and partly because of lithological differences from
those formed in the northern or St. Lawrence geosyncline.
The beds corresponding to the Chazy limestone in age,
are known as STONES RIVER, and they have a thickness
of about 1,000 ft. They contain the large *Maclurea magna*
and seven other species found in the Chazy of Lake
Champlain, but aside from this the fauna differs markedly.
Since the identical species indicate more or less free
connection at the time, the prevailing differences must be
due primarily to differences in temperature, the Appalachian
geosyncline of the Pennsylvania-Tennessee region being
less influenced by the glacial waters.

In Pennsylvania, the Stones River is followed by the
CHAMBERSBURG limestone (up to 600 ft), and this by the
MARTINSBURG formation, the latter representing the richly
fossiliferous Cincinnati group (Eden, Maysville, and Richmond beds) of the Cincinnati section.

---

4) It must be kept in mind that a south-eastward (p. d.) movement of the
   pole as between the Ordovician position at Taodini and the Silurian
   location in Nyassa Land, a distance of nearly 50° on a Great Circle, is
   to be regarded as a north-westward movement of Africa to the same
   amount, with enormous pressure against the east coast of North America,
   which, in spite of the sliding of North America westward, and against
   the resistance of the Pacific sima, — was amply sufficient to crush the
   strata of the Appalachian geosyncline into the Taconic Mountains.

Although known by many local names, the strata which succeed the St. Peter sandstone form essentially a uniform series from the Appalachians to Canada and from the Black Hills to Texas. Of course, they vary in thickness because of overlap of the successive beds, and because of the vagaries of the St. Peter sand-dune topography, which the waters of the transgressing sea attempted to overcome. Moreover, all of them suffered post-Ordovician erosion at the end of the period.

Throughout the American Arctic region of today, beds of Trenton or Galena age appear to be the first to be deposited, often resting directly upon the crystalline basement rock. This may indicate that the Trenton and perhaps Richmond faunas had their center of distribution in this now northern area, when it was a part of the equatorial belt of the pangæa, while the earlier Ordovician beds (Chazy, etc.) had their centers in the Caribbean, and, the Black River fauna was at home in the Siberian epi-sea. From there they spread to China on the one hand, and to New York State and Canada on the other.

A point of some significance is seen in the fact that the Galena-Richmond, of the Alaskan epi-sea, had an extension into north-west Greenland of today, where at Cape Calhoun, in West Greenland, a rich fauna of over 150 species of cephalopods, gastropods, trilobites, brachiopods and corals characterize these rocks, which in age apparently range from Black River to Richmond. Typical are *Vaginoceras*, *Protocycloceras*, *Gonioceras*, *Receptaculites* and *Columnaria*, forms which are now found in rocks of this age at opposite ends of the earth.

CHAPTER XIV

VOLCANISM AND POLAR SHIFT

1

THE Taconic folding at the end of the Ordovician period
had, apparently, no effect on the Ordovician strata of
the European continuation of this geosyncline, nor so far
as we may judge, on the formations of the Caledonian
geosyncline of Europe. In all of these regions Silurian
strata follow upon the Ordovician concordantly, though
always separated by a disconformity and hiatus. This may
be masked, or there may be an interpulsation deposit. The
Ordovician, as well as the Skiddavian and Cambrovician
deposits of the Caledonian geosyncline are predominantly
clastic; often an entire series consists of huangho deposits
with graptolites as their only fossils, though in the parts
which were more distant from the old-land the extension
of these huangho beds may continue below the sea-
level of that time as normal, though always shallow water,
marine deposits.

Limestones on the whole are rare in the older sediments of this geosyncline. They are found more often in
the portions distant from the polar center. When they
do occur they are of slight thickness, and probably formed
by the drifting together of dead shells often dominantly
of the same species, in local protected areas. Only in the
distant Irkutsk region, nearer the center of faunal dispersion,
are these strata formed by lime-secreting organisms growing
in situ. Here the Lower Cambrian (Taconian) has a
thickness of over 4,000 meters (nearly 14,000 feet), while
the "middle" Cambrian (Acadian), which is nearly all
limestone and dolomite, is 700 meters thick.

2

There is, however, a dominant type of material that

characterizes the older Palæozoic sediments of this geosyncline, especially in its European development, and that is volcanic lavas and volcanic clastics, both ashes and lapilli, with sometimes coarse volcanic breccias.

In no other geosyncline of these older Palæozoic periods was volcanism so active, and we are led to enquire into the possible cause for the concentration of igneous activity in this region. To have a basis for judgement, we will first of all briefly review the principal volcanic occurrences.

In Wales volcanic activity began as early as the Lower Cambrian or Taconian, and igneous beds are found in the Cambrovician ("Upper Cambrian" and Tremadoc) of the Malvern Hills in the Midland district. But it was at the end of the Cambrovician pulsation period, and especially during the succeeding interpulsation period, that the first great volcano of which we have reliable records, was built.

This was Rhobell-Fawr, from which the Rhobellian interpulsation period derives its name. (See Chapter X). The next volcanic eruption was locally preceded by very gentle folding of the older rocks, but this was not sufficiently pronounced to be regarded as an orogeny. Later there follows a great series of volcanic agglomerates and beds of andesite-ash, that precede the Ordovician series in the Mount Arenig region, and these are followed after an interval by the main volcanic rocks and ashes, over a thousand feet thick, that built up Mt. Arenig and Cader Idris; these volcanics make up most of the Skiddavian series.

The extrusion of these volcanics occupied the Petrovian interpulsation period, as did so many other volcanic beds of North Wales. Again in South Wales, we have the volcanics of the Pembroke district, and those of Cærmarthenshire, which also lie at the contact of the Arenig and the Ordovician (*sens strict*). The volcanic activity often continued into the base of the Ordovician, while older volcanic ashes were often secondarily redeposited as huangho beds, subject to more or less marining, as shown by the enclosed early Ordovician graptolites.

In western Ireland too, large parts of the Ordovician beds were frequently formed of volcanic tuffs which are more or less rearranged into huangho beds by streams. They enclose graptolites at intervals, where the low river delta of volcanic material was momentarily flooded by the sea, on the margin of which these deltas were forming.

The magnificent scale of the Petrovian volcanicity is shown in Wales by the enormous lava masses of Mt. Snowdon, which aggregate nearly 5,000 feet of lavas and ashes, all of post-Skiddavian age. In the Lake District of North England the same interval is marked by the even greater Borrowdale volcanic series, which Marr has estimated to range from 10,000 to 20,000 feet in thickness. More recent estimates, however, [1] taking into account repetition by faulting, have reduced this thickness to between 4,000 to 5,000 feet. This is a succession of andesite lava-flows, alternating with tuffs and capped by a rhyolite flow.

In the southern uplands of Scotland (Moffat and Girvan districts) volcanism is less pronounced, but some intercalated flows have been recorded. But it is in this region that we meet the evidence of disturbance of the rocks, referred to the Ballantræ series, followed after an erosion interval by Lower Bala beds, which rest on igneous and altered rocks of the older series. A similar unconformity is seen in Shropshire, England, where Lower Bala beds rest on tilted and eroded Shineton shales, as has been noted in a previous chapter. These unconformities indicate that the forces of compression were not wholly inactive in this part of the geosyncline, though the folding preceded that of the American series by the length of time of a whole pulsation period.

In Bohemia, intense volcanic activity followed the deposition of the Brezové-Hory conglomerates of the Arctomyan interpulsation period, which succeed the Jince *Paradoxides* beds. These volcanics represent an outpouring of lavas, [2] to the extent of 325 to 650 + feet. These eruptions are terminated by volcanic tuffs, representing explosive ejections towards the end.

After extended erosion, sands and mud-rocks follow, which contain a marine fauna of brachiopods, and trilobites and shales with *Dictyograptus (Dictyonema) flabelliformis*, indicating the Cambrovician (Ozarkian) pulsation period, and more especially the Tremadoc horizon.

During this period of sedimentation volcanic pheno-

---

1) For details see *Palæozoic Formations*, Vol. IV, p. 479.
2) These comprise porphyrites, felsophyres, and keratophyres, followed by quartz porphyrites.

mena were in abeyance in Bohemia, but with the retreat
of the Cambrovician sea an interpulsation period of intense
volcanic activity began. This was the Rhobellian interpulsation period, when the Komarov volcanic series was
extruded. This was also the period of maximum extrusion of diabasic rocks and tuffs that the region has witnessed.

The return of the sea in Skiddavian time again inaugurated a period of normal marine and of huangho
deposition, comprising 165 to 330 feet of transgressive
Arenig, and 200 to 265 feet of retreatal Llanvirnian beds,
both with intercalated diabase volcanic flows. Thus, the
volcanic activity continued throughout the Skiddavian period.

Once again, with the retreat of the sea and its absence
during the Petrovian interpulsation period, extensive volcanic eruptions of porphyrites with volcanic ashes and
breccias were formed. This was the last of the older
volcanic outbreaks, for during the whole of the Ordovician
period purely marine sediments accumulated in Bohemia.
During the succeeding Plynlimonian interpulsation period
only continental sediments of huangho type, but without marining, were formed. The whole of the succeeding Silurian period was one of quiescence and normal
marine sedimentation, but in the next, the Salina, interpulsation period volcanicity began again with eruption of
lavas and tuffs. (Fig. 43).

Volcanism in the New England extension of the Caledonian geosyncline is less satisfactorily indicated, chiefly
because of the lack of faunas to date the deposits. To
be sure, there is the white (Etcheminian) limestone interbedded between basaltic or diabasic flows, in the midst of
the Lower Cambrian or Taconian strata of Nahant in
Massachusetts, but if these diabase sheets are subsequent
intrusions, instead of contemporaneous flows, their exact
date cannot be determined with certainty.

On Cape Breton, at Dugald Brook in the intermediate
portion of this geosyncline, the Lower Cambrian is followed disconformably by 100 feet of felsitic conglomerate, a
purely continental interpulsation deposit. The fact that
the pebbles of this rock are felsite, a compact fine-grained
igneous rock or ancient lava, shows that somewhere in
the vicinity volcanic rocks were exposed to erosion. Of
course, these may belong to a very much older geological

period, for the ancient crystalline rocks with their accompanying old volcanics formed the old-land of the geosyncline.

But, whatever their age, it is certain that volcanism was active in this region immediately after the opening of Middle Cambrian time, for the conglomerate referred to is followed by 25 feet of shale (Coldbrookian), with a few early marine brachiopods, *Hyolithes* and some ostracods; and this is in turn covered by 185 ft of felsite — an unquestioned volcanic flow, followed by about 60 feet of trap-rock[3] with some shaly beds. Only after this is sedimentation of fossiliferous beds resumed, when the basal beds of the early Middle Cambrian[4] (Acadian) were deposited.

It is thus seen that by the end of Cambrian time volcanism became active in the Caledonian geosyncline. It was most pronounced of course during the interpulsation periods, when the geosynclines were drained and when presumably the heat generating impulses, which during pulsation periods were sub-oceanic, became sub-continental.

4

But this does not account for the remarkable localization of the igneous activity in the Caledonian geosyncline.

If, however, we note that throughout the older Palæozoic from the Sinian to the Ordovician, the pole was, apparently, moving from its originally central station in Egypt, to the western Sahara, and if we keep in mind that this was in reality a movement in the opposite direction by the pangæa, the pole being stationary in position, some light is thrown upon this problem.

Let us realize that it is the old colossus of crystalline rocks, Africa the most stable of all the units of the old pangæa, — which is really moving north-eastward, then eastward (p.d.),[5] shifting its original polar station in Egypt to the Ordovician center at Taodini in the western Sahara. This is a journey of 40° on a great circle in two stages,

---

3) "Trap" is a general name for a dark, fine-grained, basic, volcanic rock, usually an intrusive sheet or dike, but it may also be used for a surface flow as is probably the case here.

4) These are the Dugaldian beds, 507 ft thick, followed by the Hanfordian series, or *Protolenus* beds and then by the *Paradoxides* beds.

5) p.d. is used to indicate present day directions. For the time involved the direction of movement was always south, *i.e.* towards the south pole.

one of 25° N.E. (p.d.), the other of 13° E. (p.d.), or a total of about 2,400 miles, and over the distance it dragged the "western" continents with it.   Now the weakest spot in this part of the pangæa was the Caledonian geosyncline, which had the old Fennoscandian, Greenlandian and Laurentian old-land masses backing it on its west, and the Africo-Asiatic old mass on its east.   The Appalachian-St. Lawrence geosyncline had hardly begun to function as a depression in the American block, and did not assume importance until the movement of the African mass neared its maximum extent in early Skiddavian time.   It seems inevitable that the strain of the eastward-moving African land colossus should be most felt between the two great, relatively stable continental masses, and that therefore the connecting weaker part of the sial would be stretched. How much of this stretching force is also to be attributed to the straining away from the pole of the rotating pangæa, a straining which would be most active in the Laurentio-Fenno-Scandian rigid mass, is a question which deserves some attention.

But that the strain was there is certain, whether due to the eastward pull of the moving African mass, or the straining in the opposite direction under polar centrifuge, or both working in opposite directions.   The result was inevitable; the dividing region of the Caledonian geosyncline was stretched, with a relief of pressure on the underlying sima-sphere, and this relief of pressure resulted in the production of lavas, which could force their way upward through potential cracks in the weakened crust.

Thus might be explained the remarkable localization of the volcanic activity in one geosyncline, while the others were almost wholly free from such manifestations at this period. And incidentally we might note that this correspondence to the requirements of a theoretical inter-relation, becomes an additional factor in the establishment of the essential validity of our belief in a Palæozoic pangæa.

6

Volcanic activity, however, was apparently not wholly confined to the Caledonian geosyncline,[6] for a bed of altered

---

6) NELSON: *Volcanic Ash in Ordovician* ; 1922, pp. 605-616.

volcanic ash, or bentonite, has been recorded from the Ordovician of Rock Ridge County, Virginia. "The coarseness
of the ash fragments and the fact that the layer is about
10 feet thick show that it is near the location of this
ancient volcano." Dr. Nelson would locate the vent under
the Cumberland Mountain, near the point where Kentucky,
West Virginia and Virginia meet, approximately in Long.
87°30' W., and Lat. 37°30' N., of today, *i. e.* some 700
miles south-west from Boston on the modern map.

The distribution of this ancient volcanic ash is described by Dr. Nelson to have been over an area extending
from south of Lake Erie on the north, to the Gulf of
Mexico on the south, and from Western Virginia and
North Carolina on the east, to Missouri and Arkansas on
the west, an area about 800 miles long by 450 miles wide.

The formation occurs 8 ft below the Lowville of the
Black River group and is overlain by the Hermitage beds
of the Trenton. It occurs in the Lowville and is sometimes directly succeeded by Silurian beds. At High Falls,
Ky., the maximum thickness averages 5 ft, although 10 ft
is said to occur at one point.

It is, of course, not absolutely certain that the location
suggested for the volcano is the true one. If it could be
located from 150 to 200 miles farther to the south-east
it may well have been in the line of continuation of the
Caledonian geosyncline. If it lay in the Appalachian
geosyncline it represents an isolated example of volcanicity
in that region of otherwise quiescent sedimentation. Farther north, however, in the St. Lawrence geosyncline, the
same tension was felt, and here intense volcanicity was
active in the region occupied today by the Shickshock
Mts., on Gaspé Peninsula.[7]

When, at the end of the Ordovician, the mass of
Africa commenced its great westward trek, which brought
the pole to rest some 3,000 miles to the south-east in
Silurian time, the former tensional strain was transformed
into a compressive one, from which the Appalachian and
St. Lawrence geosynclines were the most obvious sufferers,
for there the strata of the geosynclines were compressed
into the Taconic mountains, which played so important a

---

7) PARKS: *Geology of Gaspé* ; 1931 ; pp. 785-800.

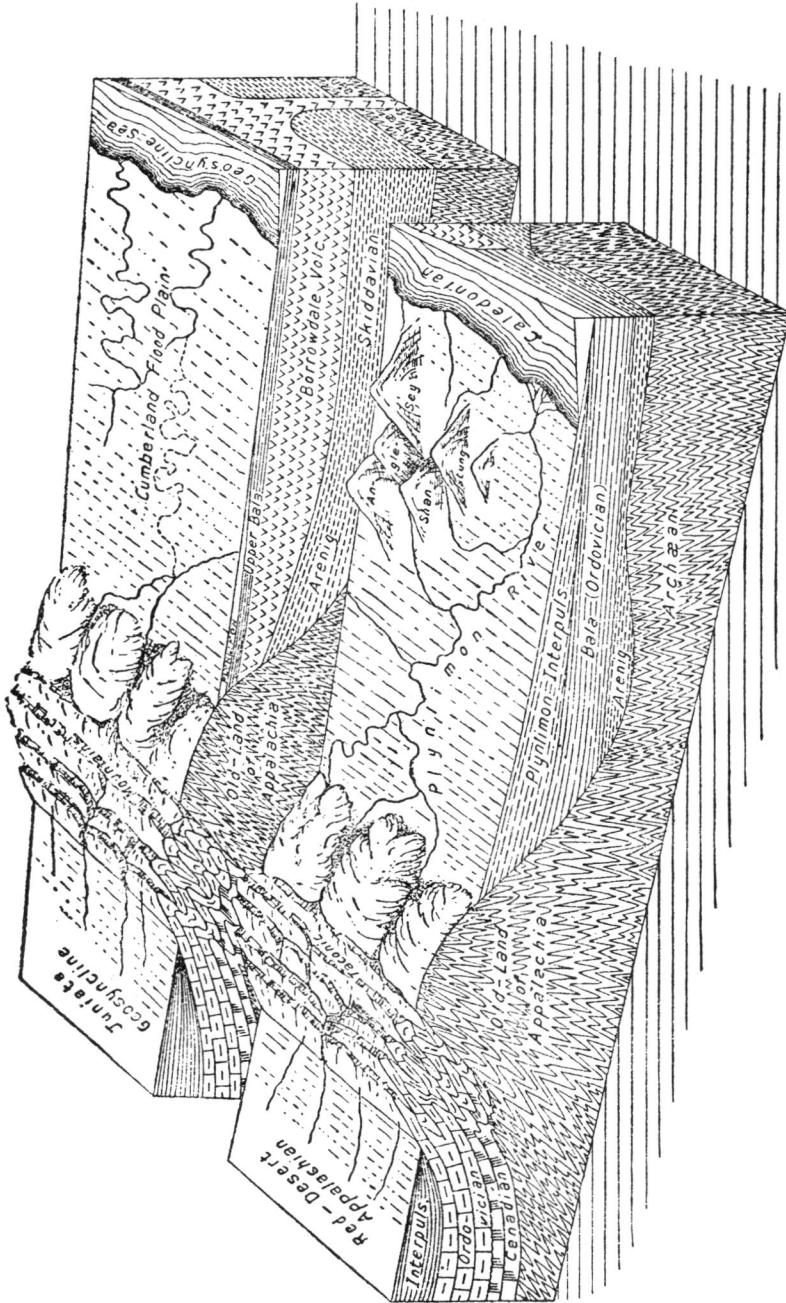

Fig. 43. Block diagram to show the former relationships of the older Palaeozoic beds on opposite sides of the Appalachian old land. In the Caledonian geosyncline (right) the Arenig (Skiddavian) and Bala are shown separated by the Borrowdale volcanics. Then follow the continental Plynlimonian river-plain deposits representing the Ordovician-Silurian interpulsation period. In the Appalachian geosyncline (left) the Canadian and Ordovician strata have been folded to form the Taconic mountains, the erosion of which furnish the material for the interpulsation river-plain deposits, that now form the Juniata red beds.

part in the history of the succeeding period. There can be no question that the strata of the New England extension of the Caledonian geosyncline received their first compression at this time, but the subsequent compression, during the Appalachian orogenic period, complicated the structures to such an extent that, in the absence of fossils to determine the age of the strata, the dating of the orogenies becomes an uncertain undertaking.

No such effect was produced in the European part of the geosyncline at this time, though compressive folding occurred at the end of Skiddavian time in some of the sections, as noted in an earlier chapter. Perhaps this meant a temporary relaxation and a reversal of the tensional strain responsible for the volcanic outbreaks, or perhaps it was due to a rotary movement of the sial crust, the Eurasian mass moving westward against the geosyncline, as Africa moved east. If so, the movements must have been consecutive, for tension and compression could hardly occur in the same geosyncline simultaneously. But at the end of the Ordovician there was no folding comparable to that in America, for the Silurian beds lie concordantly, though disconformably, upon the more or less eroded surface of the Ordovician, except when the two are separated by the continental deposits of the Plynlimonian interpulsation period. (Fig. 43).

CHAPTER XV

AN EARLY PALÆOZOIC 'HUANGHO' IN WESTERN EUROPE

1

WE have named the Ordovician-Silurian interpulsation period, the *Plynlimonian*, from the region of Mount Plynlimon in Cardiganshire, near the Montgomeryshire boundary, in South Wales. The events of the interpulsation interval were of a far less spectacular kind than those we have followed during the same interval in America. There we saw mountain ranges arising because the geosynclinal strata were compressed under the irresistable westward push of a giant, an essentially homogeneous and undifferentiated land block, the colossus of Africa, which has today much the bulk and outline that it had in early Palæozoic time.

No new mountain ranges were formed at this time in Europe, where the strata remained for the time unaffected by the pressure which crumpled eastern North America. Volcanicity had subsided for the moment, and the region was entering a period of quiet, characterized mainly by the building of continental river deposits. Both volcanic and tectonic forces gathered energy for a while; steady subsidence and sedimentation, either below or above sealevel, marked the passage of another era, and the organic world multiplied and became differentiated, unaware of the slow march towards the horrors of another revolution.

The rejuvenation of the rivers, caused by the fall of the sea-level in the interpulsation period, had a pronounced effect, which was expressed in the increase of sediments which they carried. And, since at this time the waters were rapidly draining from the Caledonian geosyncline and its marginal platform (Shropshire, etc.), continental deposition, *i.e.* the formation of subaërial deposits, was inevitable. Likewise inevitable was the merging of the last Bala marine sediments with the succeeding terrestrial deposits without break, for emergence of the old marine series and increasing river sedimentation followed as a direct consequence of the falling sea-level.

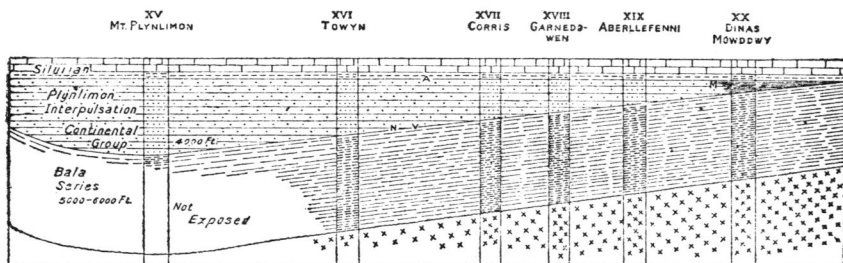

Fig. 44. Diagrammatic section of part of the Caledonian geosyncline in the Central Wales region, based on the sections in Fig. 43. The Plynlimon beds are represented as continental huangho deposits derived from the old-land on the west, and overlap eastward on the Bala, with the Narrow Vein (N.V.) between. The highest beds (M) are reworked by the transgressing early Silurian sea (*Dalmonites mucronatus* stage) followed by the Abercorris or *Persculptus* stage, which covers the whole continental series (A), and by higher Silurian. The base of the Bala rests on the volcanics.

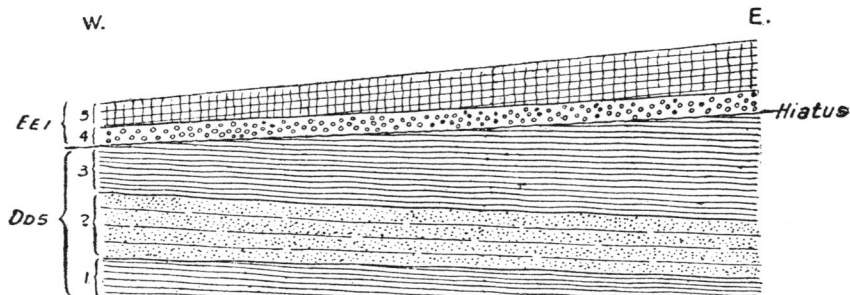

Fig. 45. Section on the north side of Slivence (Slivenec) Valley, near Beroun, Bohemia, showing the Silurian disconformably on Bala. (After Marr). Length of section about 15 yards.

D-d-5=Upper Bala :
    1. Olive green shale ; 2. Grits ; 3. Olive green shales 2 ft thick at the west end, and 5 ft at the east end.
E-e-1=Silurian :
    4. Conglomerate bed, 6-9 inches ; 5. "Wafer" graptolite-shales.

2

Mount Plynlimon, 2,468 feet high, is a conical erosion hill, a remnant of an anticline which lay within the embrace of the Silurian contact-line. This, as seen on a modern geological map, here formed a southward loop.

Its basal strata rest upon the highest of the Bala beds, which contain the last of the graptolite zones of the Ordovician, that of the *Dicellograptus anceps*. Its mass includes a series 4,000 ft thick of river-muds, flags, sands, and conglomerates, with quartz-pebbles the size of a hen's egg, or larger.

These beds are all without fossils, but contain mud-pebbles which show a curious, contorted internal structure, visible on weathered surfaces, which seems to be best explained as due to claygalls, that is, masses of mud, cracked and curled up on drying, and carried or blown among the sands, as happens on many a modern river-plain, or playa-surface.

The Plynlimon type of deposits can be traced for many miles north and east, growing rapidly thinner. At 10 miles distance from the type region the beds have decreased to half the thickness which they show at Mount Plynlimon, and, while chiefly fine-grained, still contain their large quartz-pebbles. Five miles further they have decreased to a thickness of about 500 ft, which shows that they are approaching the margin of the alluvial fan.

Southward they pass under the Silurian cover, and when the Ordovician beds appear again, where the entire series was folded in the Cærmarthenshire anticline, and the pre-Silurian beds are exposed by erosion, these parting continental beds have thinned out entirely.

Elsewhere, as in Bohemia, we find equivalent strata, though these may represent independent accumulations. Not all continental beds that lie between the Ordovician and Silurian are, however, referable to the interpulsation deposits, for some may be purely retreatal Bala beds, or river sediments formed during the retreat of the Bala sea. In that case, the first of the subjacent marine beds should not be the uppermost, but one which lies below the youngest of the marine zones of the Bala, *i. e.* below the zone of *Dicellograptus anceps*. That zone would be represented by unfossiliferous deposits, while the uppermost fossiliferous marine bed present should grade into the overlying non-marine beds.

If, however, there is a marked erosion hiatus between the last of the marine and the first of the non-marine beds, or if the last marine bed is that of *Dicellograptus anceps*, the non-marine beds following must be regarded as the interpulsation deposits. This reference holds in any case, when the non-marine deposits are of great thickness, as they are in the Plynlimon region.

It must also be borne in mind that the returning Silurian sea would rework the upper beds of the interpulsation deposits, and not only leave Silurian fossils in them, but would give the appearance of gradation downwards, from fossiliferous Silurian to unfossiliferous Plynlimonian beds, and thus obscure the original contact. As an example of this may be cited the relationship which is seen in the Dinas Mowddwy region on the margin of the Harlech Dome.[1]

The lowest Silurian beds of Wales are characterized by the trilobite *Dalmanites mucronatus*, and the brachiopod *Meristella crassa*, though these usually form separate zones and may represent sequential horizons.

3

Of course, the original areal extent of the interpulsation deposits was much greater than their present distribution indicates. Especially is this true for the area west of Mount Plynlimon, the flood plain territory of the Plynlimon River.[2]

If it had been a river comparable to the modern Huangho or Yellow River of North China, whose plain extends 435 miles back from the sea, its alluvial plain would have extended entirely across Wales and the western sea-border of Ireland, which was then a part of the old-land of Appalachia, the main part of which, it is believed, joined Ireland and Scotland to Greenland, while beyond that lay the catchment basin of the Plynlimon River.

If it may be assumed that the Plynlimon plain equalled in extent the Huangho Plain of today, we may consider that its preserved thickness of 4,000 feet is an indication of the depth of the modern river deposits, though only about half that depth has so far been penetrated by the borings in Tientsin.

---

1) GRABAU : *Palæozoic Formations*, Vol. IV, p. 226.
2) It must be remembered that at that period, the old-land of Appalachia lay adjacent to the present Ireland and beyond it.

Certainly the Plynlimon was one of the major rivers of the interpulsation period, and perhaps its best authenticated example. It is for this reason that the interpulsation period is named the PLYNLIMONIAN.

4

It is estimated, that the modern Huangho brings annually to the delta plain from 400 to 650 million cubic yards of alluvial material. If the Plynlimon carried only a fraction of this it would still have been able to build a significant deposit during the time at its disposal.

That it was the same river which brought the Bala muds to the Caledonian geosyncline, when the latter was flooded and could supply the graptolites preserved in these shales, seems very probable.

There is reason for believing that it was essentially this same river, or its similar ancestor, which supplied the thousands of feet of Arenig strata to another portion of the geosyncline during the Skiddavian pulsation period, when the older graptolites floated in the waters and were spread over the river-plain at every temporary transgression of the sea.

Permanent submergence by the rising sea-level was prevented by the excess of load carried by the river, which was sufficient to keep the surface of the plain above the sea-level, except during the intervals of diminished flow, with resultant diminished load, when the sea would encroach on the plain.

The interplay of positive and negative forces, induced by a sinking geosyncline, a rising sea-level, and a river abundantly supplying sediment, must have required a delicate equilibrium to maintain its balance, and any slight disturbance would result in a temporary modification of the sedimentation.

Thus layers of shell-bearing muds, or those containing trilobites, may become interspersed with those carrying only graptolites, while others may be entirely barren. And when the sea retreats, the continued sinking of the geosyncline would make possible the accumulation of thousands of feet of barren muds, as illustrated in the deposits of the Plynlimon interpulsation period.

A shifting of stream mouth, from the Skiddavian of the Lake region of North England to the Bala and Plynlimon country of Wales, a distance of 200 miles, was less than the one that was made by the Huangho 85 years ago (1853), from the south to the north of the Shantung Peninsula, a distance of approximately 250 miles, while the preceding one occurred 362 years before that, or in 1491, the year before America was discovered. This was a similar shifting, but in the opposite direction. Such changes have recurred seven times in the historic period, the river each time changing the direction of its course.

5

After four or five thousand feet of Skiddavian graptolite beds had been deposited by the Palæozoic 'Huangho' (Plynlimon River), in the Lake District of North England, there came the great outpouring of the Borrowdale volcanics, to an extent of many thousands of feet, which has been described on a preceding page.

This apparently induced the shifting of the mouth of the streams to South Wales, about 200 miles south, on the opposite side of the Anglesey head-land, the early Palæozoic shantung.[3]

In Pembrokeshire, South Wales, the Skiddavian is from 1,500 to 2,000 ft thick, in Western Wales about 1,500 ft and 600 to 800 ft in North Wales.

In the Lake District of North England, on the other hand, the thickness is many thousands of feet, though accurate measurements are not available because of intense folding at a later time. In the Shelve district of Shropshire, between the north and south region, the thickness is still 3,500 ft, but the fauna is a more normal marine one, showing that the geosyncline was here less influenced by the river sediments.

With the shifting of the river mouth from the northern to the southern region, consequent on the eruption of the Borrowdale volcanics, a great change took place.

In North Wales there were also interpulsation volcanics (Mount Arenig, etc.), but they were not very thick, generally from 400 to 500 ft up to 1,000 ft. Then followed

3) The term shantung has been defined on p. 50.

the marined huangho beds, the Ordovician (Bala) graptolite shales, with a thickness of 5,000 to 6,000 ft.

Above these lie the typical Plynlimonian barren beds, netting another 4,000 ft of huangho deposits, after which follow the similar huangho deposits of the Valentian, or lower half of the Silurian (3,650 ft), and an unknown thickness of the Salopian. Thus, during the Ordovician, Plynlimonian and Silurian time, the Plynlimon River, the Palæozoic counterpart of the modern Huangho of East China, deposited some 15,000 ft of strata in Wales, with the fossils, chiefly graptolites, indicating successive marinings. Only occasionally was the submergence long enough to permit other animals to be buried there.

During all this time, the Palæozoic huangho did not shift its outlet, and hence the region of the old mouth, north of the shantung of Anglesey, received very little sediment. Where in Skiddavian time had accumulated many thousands of feet of huangho deposits with graptolites, followed by the enormous outpouring of the Borrowdale volcanics in the Petrovian interpulsation period, there were deposited in Bala time 180 ft of normal fossiliferous limestones, shales and conglomerates, the *Dicellograptus anceps* beds, followed by 16 ft of Plynlimonian interpulsation muds, and this by true Ashgill shales, etc., of basal Silurian age, but only 66 feet thick.

The succeeding Stockdale shales of later Silurian age are only from 200 to 450 ft thick, but they are followed by 12,000 feet of later Silurian flags and grits, probably formed by a Silurian successor of the Plynlimon River.

We have then in the modern huangho deposits of China an almost ideal illustration of the history of British early Palæozoic deposits by the Plynlimon River, from the old-land of Ireland and Scotland, at that time joined by a continuous land mass to Greenland (See Map, Plate V). We have at least the record of one change of outlet from north of the headland of Anglesey shantung to south of it, comparable to the historic changes in the mouth of the modern Huangho of China. Since that time, some 80 ft of huangho deposits were formed in the Tientsin area, beneath which lies a bed with the large shells of the river mussel *Lamproteula*. This is of unknown thickness and is probably separated by a hiatus from the latest deposits above it, which, if they began 85 years ago, would have grown at the rate of about a foot a year.   This is probably an imperfect estimate of the amount of sediment brought by the Huangho, since much of the river material was used to extend the plain forward.   Beneath the *Lamproteula* beds is an older deposit of the same type, with layers containing *Planorbis* and other shells of still living land and fresh-water species.   The series has been penetrated to a depth of 2,834 ft, without reaching the bed-rock beneath.

The determination of the total depth and the nature of the underlying rock will need either a deeper boring, or an investigation by the seismographic methods, used so successfully elsewhere.

CHAPTER XVI

THE ADVENT OF THE SILURIAN SEA

1

IN the typical Silurian country, Shropshire, England, the Plynlimon interpulsation deposits are absent. The contact with the Ordovician is not only marked by erosion, so that commonly a part of the later Ordovician beds are absent, but also by an overlap of the several divisions of the Silurian, as a result of which the earlier members of the Silurian are likewise missing.

An instructive series of sections is found in the type region, the country once occupied by the ancient tribe of aboriginal Britons, the Silures, ruled by their king Caradoc, their most notable chief.

The valley, which lies between Mount Caradoc on the west, and "Wenlock Edge" or escarpment on the south-east (Fig. 46), and the escarpment itself are classical ground. Here can be seen the typical Silurian rocks in their relation to the underlying formations, and the combination of these sections results in the generalized view shown in Fig. 47, which represents the relationships of the formations which existed towards the end of Silurian time.

At the north-eastern end of the region lies Wrekin Hill, and between it and the villages of Shineton and Much Wenlock, the older exposed beds are strongly folded. This folding involved the Cambrovician (Tremadocian) and older beds, which are here known as the Shineton shales.

The folds are strongly eroded, and on their truncated edges lie the overlapping Silurian beds, which are still horizontal, and whose subdivisions are known in descending order, as follows:

SILURIAN BEDS:
5. Ludlow shales;
4. Wenlock limestone;
3. Wenlock shales;

2.   Purple shales ;
1.   *Pentamerus* bed.

HIATUS AND UNCONFORMITY
CAMBROVICIAN BEDS

Shineton shales (folded and eroded).

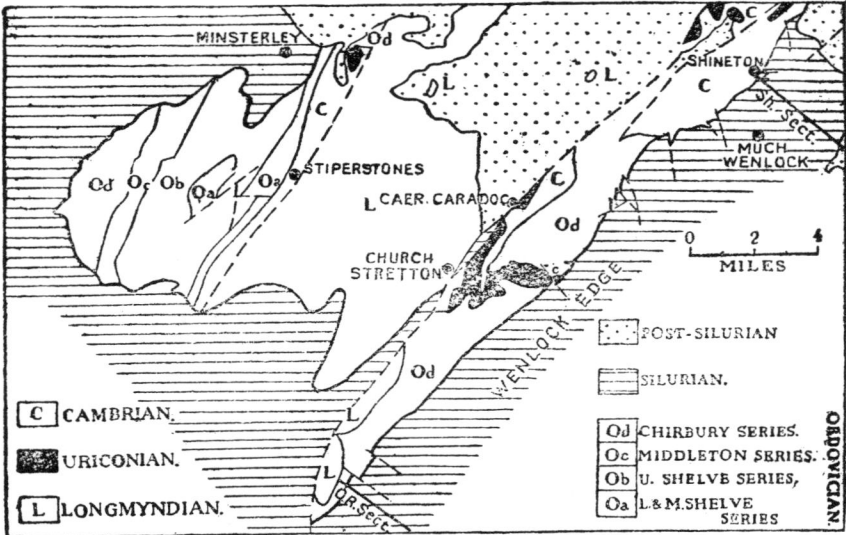

Fig. 46. Outline geological map of the Shelve and Caradoc regions of southern Shropshire. (Redrawn with slight modifications from Whittard). The pre-Silurian, Palæozoics, etc. are in white, with divisions marked by letters as indicated.   The locations of the Shineton section (Sh. Sect.) on the north-east and the Onny River section on the south-west (O. R. Sect.) are given, as is also the general position of Wenlock Edge.

The highest beds are not represented in Fig. 47, though they are present in the region.   The section shows that two entire pulsation systems, the Skiddavian and Ordovician are missing in this district, besides the basal Silurian.

The Skiddavian is accounted for by the orogeny, that is, it is represented by the time during which the Shineton (Cambrovician) folded beds are eroded, these having been folded during the preceding Rhobellian interpulsation period.   Normally, they should be, and probably were, succeeded by the Ordovician beds, and these are still present, except their highest members, in their proper position in the south-western end of the section, where only the uppermost Bala beds were removed by erosion during the next interpulsation period.

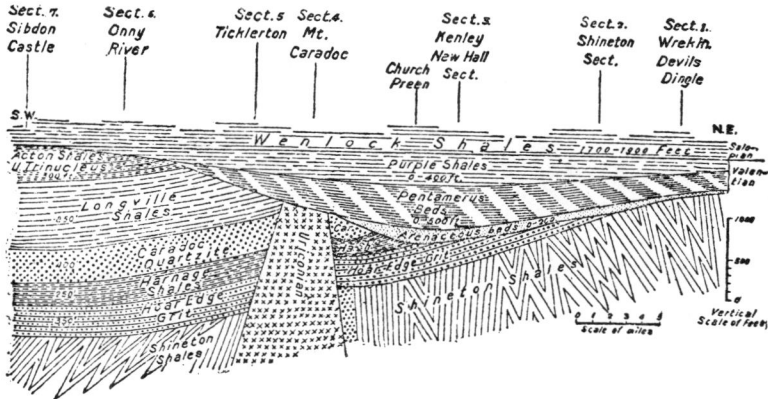

Fig. 47. North-east—south-west section through Shineton, showing the unconformable overlap of the Silurian strata of the typical region of Shropshire on the strongly folded Tremadocian (Shineton or Cambrovician). The section lies about 2 miles north of Much Wenlock and an equal distance south of the Wrekin. This illustrates the Skidd-oric orogeny. The Ordovician (Bala) is overlapped. (Redrawn with additions from Stubblefield and Bulman after Whittard).
VALENTIAN: Comprising from below upward: Basal Sandstone (0-200 ft); *Pentamerus* Beds (0-500 ft); Purple Shales (0-720 ft).
SALOPIAN: Wenlock Shales.

But in the Shineton region these Ordovician beds were either completely eroded during the post-Ordovician (Plynlimonian) interpulsation period, or else not deposited because of original doming of this region, so that it was land during most of Ordovician time.

In the south-western region the Ordovician beds are followed directly, but disconformably, by the Wenlock shale; the lower two members, present in the north, are overlapped.

2

The Shropshire Silurian beds, though not in the center of the geosyncline, were nevertheless within its actively subsiding portion, for the total thickness of these sediments is well over 3,000 ft, and all are richly fossiliferous. They all indicate water of moderate depth within the zone of wave activity, for evidence of such is frequently seen. For example the *Pentamerus* beds contain an abundance of the coarse brachiopod shells, *Pentamerus oblongus*, which attain a length of 3 inches, and evidently were shallow water organisms.

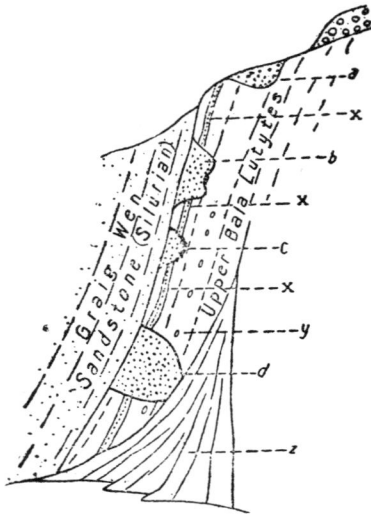

a = Decalcified sandstone full of
    fossils at the base.
b = Sandstone filling of pocket.
c = Sand-filled pocket, with con-
    centration of phosphatic
    nodules.
d = Very deep pocket filled with
    sandstone.
x = Bed of silt-stone in the Bala;
    cut repeatedly by the ero-
    sion hollows.
y = Phosphate nodules in Bala
    lutyte.
z = Scree of recent origin.

Fig. 48. Diagrammatic view of the north-west wall of Graig-Wen Quarry, near Meifod, showing the disconformable contact of the basal Silurian Graig-Wen Sandstone (main sandstone of Quarry), with the eroded surface of the upper Bala beds. (After W. B. R. King).
a-d : pockets eroded into the surface of the Bala lutytes and filled by Silurian *Crassa* sandstone, often with weathered-out fossils and phosphatic nodules from the Bala beds. All the beds are tilted nearly vertically by later folding.

This is further indicated by the fact that in the Wrekin district occurs a conglomerate which is composed almost exclusively of crystalline (Uriconian) pebbles and shells of *Pentamerus oblongus* set among quartz grains, which are cemented by calcite. Moreover, as Prof. Whittard remarks : "The rock has many other features of a compacted shell beach."

The fact that the whole of the *Pentamerus* series is overlapped by the Purple shales, and these by the Wenlock shales, in the space of less than 35 miles, can only mean that these were shallow waters, spreading over a previously exposed land surface as the geosyncline subsided or the water-level rose.

The Wenlock limestone undoubtedly represents the period of greatest transgression, for it is formed of a series of coral mounds or reefs of moderate thickness scattered originally at irregular intervals over the sea-floor, close to the surface, so that the space between the mounds could be filled by the coral sand and debris. This material was worn from the mounds themselves by the waves of the Silurian sea and mingled with the shells of animals that lived between the mounds. One is reminded of the living reefs along the outer coast of Hawaii, which the visitor may inspect through a glass panel in the flat bottom of his boat, as he is slowly rowed over the living coral-garden. This is so near the surface that the living polyps are sometimes almost brushed by the bottom of the boat.

The water over the Silurian reefs was equally shallow, and often the coral reefs were exposed at low tide, just as the modern coral polyps of the Great Barrier Reef of Australia are exposed today, especially during exceptionally pronounced ebb-tides. The water was free from mud, because the mud supplying shores were distant. This is again paralleled by the Australian Barrier Reef. This condition, however, continued only for a time, long enough to allow something over 100 feet of Wenlock limestone to accumulate. Then the retreating sea permitted the current-borne muds to spread again over the site of the reefs, and of these Lower and Upper Ludlow muds which accumulated, some 500-800 feet remain. There were probably more of these, for when the sea had drained away, and the long Salina interpulsation period succeeded, erosion removed again an unknown amount of these Ludlow muds.

3

The name SALINAN is given to this interpulsation period, because of the remarkable series of desert salt deposited during its continuation in America. These we shall describe presently.

In western Europe, it is barely recognizable as an erosion interval and a hiatus, and is so little marked that the first of the succeeding series of beds, which indicate the return of the sea in Siluronian time, is not separable by a readily recognizable erosion hiatus from the preceding Ludlow beds. There is no variation in the thickness of the beds between the Aymestry limestone and the base of the Tilestone (Downton, and Ledbury beds) to indicate erosion, but there is indirect evidence of a land interval.

The evidence of the return of the Siluronian sea is widely marked by the occurrence, at the base of the Tilestone series, of the Ludlow Bone Bed. This is a thin bed composed of fragments of the dermal armour of the early fish, and of eurypterid tests.

This bone bed ranges from a quarter of an inch to three or more inches in thickness, and is believed to cover an area of not less than 1,000 square miles. It may be regarded as a product of the destruction of the fish by the influx of the sea-water in the estuaries, or river mouths, in which these normally fresh-water fish lived. In Lanarkshire, Scotland, beds of essentially this horizon have been widely known for the number and variety of Merostomes which they contain. These are fresh-water animals[1] related to scorpions on the one hand, and to the modern horse-shoe crab *Limulus* on the other. Besides three species of *Eurypterus*, and two of *Pterygotus*, several other related genera are represented. Finally, these beds have furnished the oldest known true land-scorpion (Fig. 49) and a myriopod, while another scorpion has been found in the bed of sandstone that separates the Silurian and Siluronian on the island of Gotland, Sweden.

Everywhere the Silurian is separated by a hiatus, which may be apparent or masked, but despite its masked character the time represented by the hiatus is a long one, for in America it was marked by a period of unprecedented aridity, with the formation of desert salts of great thickness.

As shown on our map (Plate VI), both the type regions of the British Silurian and the American Niagaran lay in what had by this time become the same geosyncline, formed by the union of the Caledonian and Appalachian geosynclines after the Taconic orogeny, when the former American continuation of the Caledonian, and the European continuation of the St. Lawrence-Appala-

---

1) This is, however, not the opinion of the majority of palæontologists.

chian geosynclines were obliterated. The new compound geosyncline and its marginal plain now extended from Latitude 15° South, to Latitude 65° South. Great Britain lay on the north-west side in Latitude 65°, and had a back-country of little elevation in Greenland of that period, while the American type region lay on the "inside" of the geosyncline, 20° to 30° of Longitude from Britain, and in the "rain shadow" of the recently formed Taconic mountains. These stretched along the southern[2] border of the geosyncline in an unbroken chain of high parallel ridges, eminently capable of levying an excessive toll of moisture from the winds that attempted to cross these frontiers.

Fig. 49. *Palæophonus caledonicus,* a fossil scorpion. Continental interpulsation beds, base of Upper Ludlow, Lanarkshire. (After Geikie).

4

Let us now consider the Silurian of the Appalachian-St. Lawrence geosyncline of North America.

The geosynclinal formations involved in the Silurian sequence of North America vary from east to west, or, translated into terms of the Silurian map, from south to north, in accordance with the climatic belts of the time, and the nearness of the old-lands, which supplied the clastic sediments.

A glance at the map of pangæa (Plate VI) shows that the St. Lawrence-Appalachian geosyncline extended from Latitudes 15° to 55° South, while the region of the marginal platform lay nearer to the equatorial belt of that time, or between 20° and 35° South Latitude.

This was especially true of the region which is now Wisconsin and Michigan, and here the most characteristic

2) To-day the eastern border.

coral-reefs are preserved. These Wisconsin reefs are in the form of dome-like mounds, which, at the time of my visit in 1900, were partly exposed in the older quarries of Cedarburg, Grafton and Milwaukee.[3]

In one of the most extensive of these, the exposed portion of the reef from north to south was about 300 feet and the vertical depth 30 to 40 ft, but the actual diameter of the mounds is much more than that.

The reef rock is massive, no bedding being visible, and it consists in the main of the ancient *Stromatoporoids,* and of *Favosites,* and other corals in the position of growth. The first is by far the most abundant of the reef-forming organisms. As a rule these are visible only on the weathered surface, on account of the general uniform and semi-crystalline character of the rock.

On the margins of the reef, however, bedding-planes become visible, for here the flanking lime-sand, the product of contemporaneous wave-work, rests against the reef-mound, often interfingering with extensions of the reef, and dipping away in all directions from the central mound. The inclination of these beds is often very steep.

I have observed dips ranging from 18° to 28°, while Chamberlain, at an earlier date, found exposures that showed some of the beds of the lime-sand, close to the reef-mound, with dips ranging from a maximum of 54° to 34°, 31° and 30°, on the flanks of one of these reefs, and dips of 40° to 20° on the flanks of another.

All these appear to be original dips, for the mounds show no evidence of differential subsequent uplift, nor the beds of later sagging. Moreover, the individual beds (layers) of lime-sand increase in thickness as we approach the reef-mound, and finally merge with the structureless reefs. (Fig. 50).

Fig. 50. Diagrammatic section of a Silurian coral reef, showing proportional size and general massive character of the consolidated reef of *Stromatoporas* and *Favosites,* etc., and the steeply dipping marginal layers of clastic lime-sand, with interfingering reef organisms.

3) GRABAU : *Palæozoic Coral Reefs,* 1903 ; pp. 337-352.

On the flanks of the major reefs numerous smaller mounds or monticules are commonly found, which constitute subordinate reefs, and these may even occur some distance from the main reef. They are formed by the same lime-secreting organisms, *Stromatopora* and *Favosites*, while *Halysites* and smaller cup-corals also abound.

Besides the actual reef-forming corals and corallines, these mounds formed the center around which the animal life of the period congregated. Vast numbers of crinoids, chiefly preserved by stem-joints, bryozoa, and brachiopods existed here, and added their remains to the reef-mounds, or formed bedded strata on their flanks.

Away from the mounds these bedded lime-sands form horizontal strata, remarkably free from organic remains, though now and then stranded cephalopod-shells became embedded in them. For the most part, however, the bedded strata between reef-mounds consist of lime-sand originating from the shells of dead brachiopods which the waves had pounded to pieces, and the ground-up coral-sand washed from the reef itself.

It is not unlikely that much of the coral sand originated through the activities of crustaceous animals, which ground up the smaller corals to extract the organic matter for nutriment, as happens on modern coral reefs. These crustaceans were mostly trilobites; but cephalopods, of which shells of huge species are sometimes found in these bedded strata, may have been among the agents active in the destruction of the smaller corals, bryozoans and brachiopods, grinding up the lime-structures into sand.

Within the city limits of Milwaukee, 20 miles south of Cedarburg, Wisconsin, a group of three large coral mounds was formerly located, these having been uncovered by quarrying operations in the past. These mounds lie a few miles distant from one another, forming a triangle, and within this area the great limestone quarries of later date were opened. In these are found only even-bedded limestone layers consisting of coral-sand and crinoid debris, in which completely preserved shells or other remains are rare.

Similar reef-mounds, growing at intervals, have been recognized in Cedarburg and at other localities, and here too the space between the quarries is occupied by the bedded limestone. Though fossils as a rule are uncommon in these lime-sands, they may be quite abundant locally in thin strata between the thicker beds, indicating a spread of smaller organisms over the surface of the sandy plains between the reefs.

The following sequence of strata has been observed in one of these quarries in the reef areas:

3.	Porous, vermicular dolomite, brittle, with stratification poorly developed. Full of cavities, where small corals have been dissolved out.	20 ft
2.	Hard, white granular lime-rock.	8 ft
1.	Soft, friable, brown lime-sand.	10 ft
	Total	38 ft

Often the bedded lime-sandstone (calcarenyte) has been changed to dolomite, in the crystallization of which the original structure of the component grains was further altered. This also occurred in the reef portion, and as a result the organisms of the reefs have their characters frequently obscured.

So far as their structure is concerned and the dip of the flanking and marginally interfingering beds of lime-sand and organic beds, these reefs closely parallel those of modern shallow seas, as for example those of the Bermudas, and of the Great Barrier Reef of Australia. But, of course, the organic components are entirely different.

The Silurian reefs occur both in the Waukesha and Racine limestones, which have a combined thickness of probably over 400 ft. They rest on the Mayville limestone and are followed by the Guelph.

5

The reefs of Wisconsin are traceable south to the Chicago area and northward into the Northern Peninsula of Michigan and the Manitoulin Islands, which continue the line of outcrop north of Lake Huron. They are traceable into the Northern Peninsula, where in the St. Ignace well-sections it is seen that the limestone is about 600 ft thick, and essentially uniform throughout. All of these regions lay, in Silurian time, between 15° and 25° South Latitude.

In the Hudson Bay region individual reefs are repre-sented, reaching a thickness of 75 ft or over. These are now in 50°-55° N. Lat., 80°-90° W. Long., and the limestones extend to nearly 75° N. In Silurian time all these regions lay between 15° and 30° S. of the equator.

The New Siberian Islands are to-day in 75° N. Latitude, and well within the Arctic Ocean, but in Silurian time they too were essentially in the tropical belt (approx-imately in Latitude 5°-10° South), and their position near the East Boreal epi-sea indicates that they were essentially in the region of more or less continuous submergence, which supplied both the Caledonian and St. Lawrence geosynclines with their coral faunas.

At the Alaskan end of the Palæocordilleran geosyn-cline, in essentially the same parallel of latitude, lime-stones of Silurian age, aggregating 2,000 ft in thickness, also carry this coral fauna, though on the outer coast of the Alaska peninsula evidence of the presence of glaciers has been found at this period. This is probably to be regarded as the work of mountain glaciers, as the region was too close to the equator for continental glaciers to form.

The formations on Grand Manitoulin Island show the following succession :

SILURIAN PULSATION SYSTEM

Manitoulin Group

    Guelph Dolomite

    Racine Reef

    Wakeshaw Reef

Cataract or Alexandrian Group

    Mayville Limestone

    Cabot Head Beds

These beds of Grand Manitoulin Island and the neigh-boring coast of Upper Michigan (Point Detour), were the source of the many well-preserved silicified corals, which were described by Rominger in Volume III of the *Michigan State Survey*, one of the earliest authentic works dealing with these interesting fossils.

450 miles east of the ancient Wisconsin reefs are

the outcrops of the Silurian rocks of Niagara Falls,[4] and
60 miles further east are those of Rochester, described
by Hartnagel.[5] This was a region situated from 30°
to 35° south of the equator of the time, and therefore
still within the border of the sub-tropical zone. The climate
may have been that of the Florida coast to-day, or the
Bermudas, in the western Atlantic, or the coast of Queens-
land, Australia, including much of the Great Barrier Reef.

That the region was less favorable for coral growth
was due to its proximity to the mud-bearing currents
from the old-land, which at that time lay on the south.
This made conditions inimical to polyp life, and the great
reefs of the Cedarburg type dwindled to lenses of small
dimensions.

One of these is exposed in the sections of the Niagara
Gorge in the Upper Clinton limestone. It has a maximum
thickness of 8 feet, and a diameter of perhaps 30 ft. Others
are found in these limestones near the edge of the escarpment above Lewiston, while in some of the small streams
that have bared these rocks further east, they are fairly
abundant, but small, and so close together that it is
possible to step from one to the other. Bryozoa are
generally more abundant in these lenses than corals, and
they have been designated "bryozoan reefs". They were
favorable areas for attachment of crinoids, brachiopods,
and other sessile animals, and served as feeding grounds
for trilobites and other predatory types. As a result the
remains of such animals contribute largely to the composition of the miniature reef-mounds.

In these regions, which were closer to the axis of
the Silurian geosyncline, limestone deposition was more
frequently interrupted by deposits of mud or even sand,
and the great succession of almost pure limestone, characteristic of the deposits in the purer waters of the Wisconsin region, are unknown.

The succession at Niagara emphasizes this, and though
this is generally taken as the type section for North
America, it is no more typical than many others which

4) GRABAU : *Geology and Palæontology of Niagara Falls*; Bull. 45; N.Y.
State Museum, 1901.
5) HARTNAGEL: *Geologic Map of the Rochester and Ontario Beach Quadrangles*;
Bull. 114; N.Y. State Museum, 1907.

show constant variation. This is seen in the variable
thickness of the beds in the Rochester section as compared
with that of Niagara (Fig. 51), showing that on the whole
the shale increases eastward at the expense of the limestone.

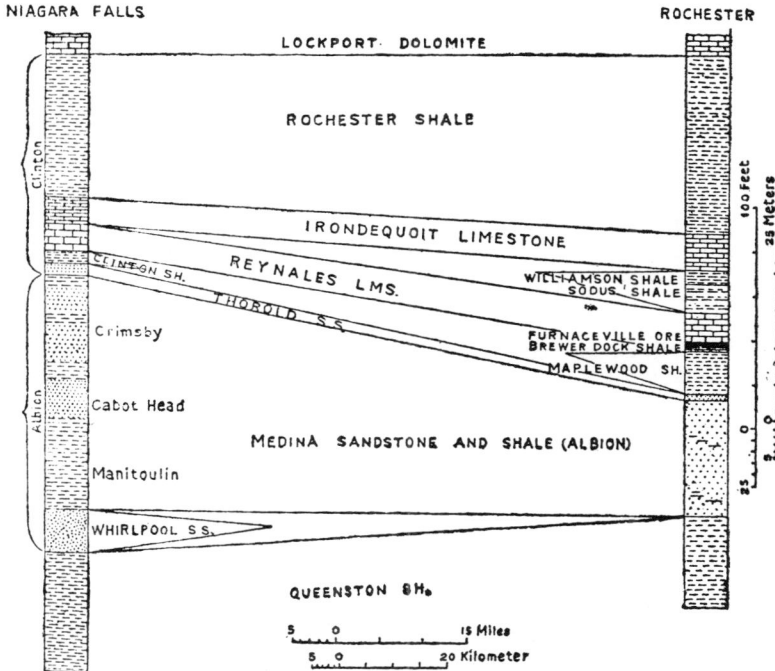

Fig. 51. Columnar sections, showing the stratigraphic succession, thickness and character of formations and approximate correlation of the strata
of the Niagara and Genesee River (Rochester) regions (*16th Intern. Geol.
Congress, Guide Book 4*). The Queenston represents the interpulsation deposits,
the others represent the Silurian pulsation system.

A line drawn from the Niagara, or Rochester region,
parallel to the trend of the Appalachians, brings us to the
exposures in Central Tennessee, which, though less satisfactory, may be considered as in a general way a counterpart of the Western New York section, being, at the time
of deposition of the Silurian strata, essentially in the same
latitudes.

As we proceed eastward toward the axis of the geosyncline, limestones diminish and shales and sandstones
increase, until, in the section exposed in the eroded

Appalachian ridges, nearest to the old-land, only shales (Clinton, or Rockwood) and sandstones (Clinch, Tuscarora, etc.) remain, preceeded by the Juniata Red shales and sandstones, the American equivalent of the Plynlimonian interpulsation deposit.

In the region of the old Taconian mountain folds (Sil-oric orogeny) no Silurian strata are present, for the old folded mountain range was added to the Appalachian old-land, as a folded foreland which had to be eroded before deposits could again be made upon their truncated edges. This took time, and probably the whole of the Silurian pulsation period was occupied by this erosion, so that Silurian strata could only be deposited in the narrowed geosyncline, beyond the newly folded ranges and over the marginal platform.

Here, however, also a few low domes had been raised (blister-like) by the compression, and they had to be worn down, so that their inner beds became exposed, before Silurian strata were deposited upon them. As a result the Silurian beds, when they finally covered these eroded monticules or blisters, came to rest with successive overlap of their higher beds, on various members of the underlying Ordovician series, with a mild unconformity.

At the end of the Niagaran coral-reef period, when several hundred feet of fossiliferous limestone had accumulated on the marginal platform, and great thicknesses of shales and sandstones had been formed in the geosyncline, the negative pulsation set in.

The sea withdrew and laid bare the previously deposited beds, or left residual pools in the deeper, more or less down-warped, areas. These began to concentrate under the influence of drying winds. For, with the addition of the young Taconic (Sil-oric) mountain chain to the border of the old-land, and the lowering of the water-level, and its final withdrawal from the geosynclines, a condition of aridity once more spread over the land, and the winds which blew across the emerged plain, became drying winds.

This implies evaporation and the higher concentration of waters in the region of incomplete circulation. Increased density and salinity resulted, and this was reflected in the type of rock formed during the Guelph period, and in the character of its organisms.

The Guelph is the last of the limestone deposits of the Silurian, and in its highly magnesian character and in the peculiarity of its fauna, it may be regarded as forming the closing chapter of marine Silurian deposition, preparing the ground for the desert salts which succeeded, after the retreat of the Silurian sea was completed.

CHAPTER XVII

THE SALINAN INTERPULSATION PERIOD AND ITS SALT DEPOSITS IN NORTH AMERICA

1

THE Silurian was a period of erosion of the folded strata, and the marine Silurian sediments were confined to the geosynclinal region beyond the folded belt, and to the marginal platform.

There are no Silurian beds in the folded belt and the designation of any strata as such rests on a misconception. It is difficult to understand why any stratigrapher, finding a sandstone overlying *folded and truncated* Ordovician rocks, should refer this sandstone to the Tuscarora, the lowest of the Silurian strata of the Appalachian geosyncline.

Can anyone really hold that Ordovician strata can be so intensely folded as those of the Taconic belt and the folds truncated by a smooth plane of erosion, which has removed hundreds of feet from the folded ridges (See Figs. 52-53), and yet occupy so little time that it is practically unrepresented in the stratigraphical record?

The reason why such sandstone, as the beds in the Delaware Water Gap section and others, had been referred to the Tuscarora or basal Medina, is because of the presence in them of the relief casts of the trails known as *Arthrophycus harlani*, or *A. alleghaniensis.* Such trails were evidently made by some animal with a bilobed tail, or perhaps bilobed gill-covers, by the aid of which it moved over the wet sand when it found itself stranded on such a surface, as has been shown by J. B. Woodworth for similar, though much larger, trails known as *Climatichnites.* Such a caudal character is found in some eurypterids, which are animals that existed throughout the early Palæozoic, inhabiting the rivers of the successive periods. Eurypterids have been found in shales in the Shawangunk conglomerate, in which this trail occurs, but they have not yet been found in the true Tuscarora and Medina sandstones, in which the trail also occurs.[1]

Fig. 52. Unconformity between the Shawangunk conglomerate (Siluronian) and the Hudson River shales (Ordovician) in the railroad cut west of Otisville, N.Y. (Drawn from photograph). The folding represents the Taconic or Sil-oric orogeny (Plynlimonian, or Juniatan, interpulsation period), and the erosion of the folds occurred during the Silurian, which is unrepresented.

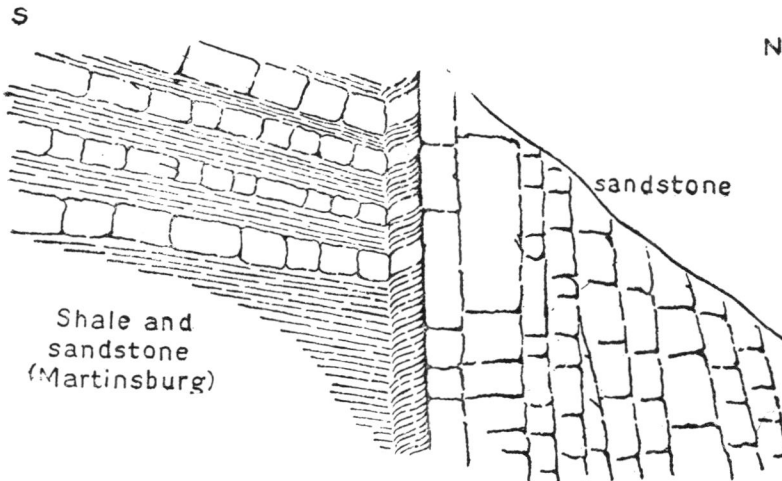

Fig. 53. Unconformable contact between the Martinsburg shale (Ordovician) and the so-called Tuscarora (probably Shawangunk) sandstone, on the Schuylkill River, near Port Clinton, Pennsylvania. (After G. W. Stose).

1) Eurypterids have, however, recently been reported from the Upper Richmond, which is of uppermost Ordovician age.

It may probably be considered as almost beyond doubt that these animals were brought by rivers in flood on to the flats of the Tuscarora and Medina shores, and that some day their remains will be found in these rocks.

But because these trails also occur in the Shawangunk conglomerate, which was formed after the long period of erosion of the Taconic folds, their presence there is no excuse for calling that formation basal Silurian, or Tuscarora, when it is obvious from the structural relations that this cannot be its age.

The fact of the matter is that some palæontologists are too prone to overstress the organic side, or the indication of the presence of remains which on *a priori* grounds, are thought to be characteristics of a definite horizon, whereas the one in question is a facies fossil, associated with certain types of sediments. Also the structural relations of the higher of the two beds which contain the *Arthrophycus*, *i. e.*, the Shawangunk conglomerate, leave no room whatsoever for doubt that it belongs to a post-Niagaran age, *i. e.* is of Salinan if not younger age. There ought to be no need any longer for an argument on this point.

And if in some of these beds a doubtful fossil is found that suggests a Clinton pelecypod, as cited by Schuchert, does that mean that these beds are to be referred to the Clinton, when they lie with a violent unconformity on Ordovician beds? Not at all. Dozens of secondarily included fossils from an older formation, which was most likely exposed in many places to weathering, would have no weight whatever, except to emphasize the post-Niagaran age of the strata.

It seems that a little attention to tectonics is badly needed in some of the age determinations of the strata, but that, and sane judgment, seem to be put to rout by the presence of the most fragmentary piece of fossil, which could have been supplied over and over again by the older strata that were undergoing extensive weathering.

Let it be said once for all that the beds of Shawangunk conglomerate, which lie with an angular unconformity upon the folded and peneplaned Ordovician sandstones and shales, cannot be older than Salinan in age. They represent the first river-laid sediments of the Salinan interpulsation period, when the Silurian sea had again retreated, and the old folds in the Ordovician strata had at last been peneplaned, — peneplaned during a period of time so protracted that the whole Silurian, in the restricted sense here used, or the Niagaran of the older usage, was barely long enough for its completion. [2]

2

Let us, however, add in fairness that not all the beds that have been identified as Shawangunk belong to this horizon. Some are truly basal Silurian, *i. e.* Tuscarora, but these lie with a disconformity, not an unconformity, upon the Ordovician rocks.

Discussing the salt problem of the Salina, we shall see that it could only have accumulated in desert basins, far from the sea. It was another interpulsation period and as such, a period of aridity, though no new mountains were formed to intercept the wind, as happened at the end of Ordovician time. The aridity then produced a series of red rocks, which were deposited during the Plynlimonian interpulsation period in the Appalachian region at the foot of the newly formed Taconic mountain ranges, and were the first product of the erosion of these mountains during the interpulsation period. In the Niagara Falls region they are known as Queenston Red beds, and have all the characters of a fossil loess; in Pennsylvania and elsewhere they are known as Juniata sandstone.

The Salinan interpulsation period was one of worldwide emergence, though no other country was situated, in surroundings especially favorable for desert disintegration, as was America.

Attention should be called to the fact that the Salinan eurypterids are related to, if not identical with, many of the species found in the Shawangunk grit.

These are found in black shale layers intercalated in the Pittsford shale series, which overlies the Guelph and underlies the Vernon shale, and represents the last outpost of the river deposits and faunas, before the aridity and the Vernon loess overwhelmed this region.

The Bloomsburg Red shale, which follows upon the

---

2) See further, Chapter XXV.

Shawangunk conglomerate, and the Vernon loess variant of the Bloomsburg, which follows on the Pittsford, present similar problems to those already faced in the case of the Juniata and Queenston.

There, however, we had the evidence of the recently formed mountain barriers along the old-land front, and the explanation was a simple one. In the case of the Salinan Red beds, the reason for the localization of aridity is not so apparent, unless we may assume that the relative height of the old-land was increased sufficiently by the falling sea-level to render it again an effective climatic barrier. The aridity may also have been increased by greater retreat of the sea, owing to more marked subsidence of the sea-level.

That this barrier was not sufficient at once to produce the necessary aridity is shown by the extensive pebble and sand delta-plain, built by torrential rivers outward from Appalachia. Eventually, however, the old-land had risen to such an elevation, through the continued fall of the sea-level, that it became effective again as a barrier to the moisture-laden winds and the red Salinan (Bloomsburg and Vernon) loess was formed.

3

Much has been written in the attempt to evaluate the significance of the Salinan deposits in terms of Palæozoic geography, and to explain the origin of the salt-beds between the marine Niagaran below and the Monroan or Cayugan marine bed above.

Most commonly the salt has been held to be of sea-margin origin, the conditions of deposition being assumed to be paralleled by those of the modern Karabugas Bay of the Caspian Sea.[3]

From the very first, this has been an uncritical comparison, for an analysis of the conditions characteristic of the Karabugas shows that they in no way conform to those which existed in Salinan time, or for that matter during any other known Palæozoic period of salt depositon.

The sequence of events which led to the cutting

---

3) Between 40° 31′ to 42° N. Lat., and 53° to 54° 45′ E. Long.

off of the head of the Gulf of Suez [4], the *Sinus Heroöpoliticus*, or western arm of the Red Sea, and its conversion into the Bitter Lakes of the Isthmus of Suez, gives a far more complete illustration of the origin of sea-salt deposits than does the Karabugas, and is indeed the only good modern example.

The complete separation is believed to have been effected by about 600 B. C., but long previous to this, the bar formed a temporary barrier, and during its effectiveness the waters within the cut-off evaporated under the influence of the desert winds, until the concentration was sufficient for the commencement of salt deposition. Long before this began, however, all except the most euryhaline animals, *i. e.* those capable of living in concentrated brines, had been killed off, and their remains buried in the mud at the bottom of the cut-off lagoon.

These beds, which underlie the lowest evaporation deposits, or 'evaporates', are therefore highly fossiliferous, the remains being of the contemporary species living in the Red Sea, just across the bar, and buried in the sands which constantly accumulated on its bottom.

That is the first fact, so commonly ignored, in the uncritical reference to sea-border lagoons of all natural salt deposits.

Ochsenius explained the supposed absence of organic remains in the Karabugas deposits by assuming that the fish that accidentally entered the Karabugas Gulf would, on finding the water not to their taste, turn round and swim out again.

He never saw the Karabugas, or he would never have made such a statement. He would have known that escape from this natural fish trap is impossible, and the doom that awaits the multitude of organisms, carried in a constant procession into this natural death-trap, is inevitable.

Andrussow tells us that: "the quantity of dead fish that lie upon these shores in March can be measured by the fact that the [migratory] gulls [which] at this season of the year [congregate there], feed only on the eyes of the fish, and do not even take the trouble to turn over the fish to get at the other eye."

---

4) This region lies in 30° N. Lat.

As for the molluscs, crustaceans, and other organisms, that lived on the bottom of the lagoon, when it became closed and evaporation began, what chance had they for escape? Even before the inlet across the bar was wholly closed they would be unable to leave. The fact of the matter is that destruction of the animal life in the lagoon, once concentration has begun, is inevitable, and just as inevitable is the formation of a richly fossiliferous bed at the bottom of the lagoon, before salt begins to separate out.

Now it is just this fossiliferous layer which is absent beneath the Salinan salts in all cases where the actual contact between the underlying formation and the salt bed is known. Usually the salt is underlain by marls, or directly by red beds which have all the characteristics of an ancient loess, or by Niagaran limestone of an earlier pulsation period and bearing the evidence of erosion prior to the deposition of the Salinan beds.

The first salt to separate out in the evaporation of impounded sea-water is sulphate of lime, either in the form of gypsum or of anhydrite. Then follows the sodium chloride, and this continues until the outer sea breaks in again, and the salinity is reduced by the influx of normal sea-water.

Ordinarily, this would re-dissolve the salt that has been deposited, except for the fact that, because of the lower level of the lagoon, the inrush at first will be great, and much muddy sediment will be carried into the gulf, and this will settle to form a protective layer over the salt. With the comparative freshening of the lagoon-water the animals of the outer sea will again enter the compound and continue to exist there as long as the inlet is open and normal salinity is maintained. Therefore the covering layer of clay will likewise be fossiliferous, especially if after a second closing of the inlet evaporation begins again.

Then a second layer of gypsum will form, following the second fossiliferous clay bed, and this will be followed by a second layer of rock-salt. And this will be repeated as often as the bar is breached and the level of the lagoon restored with accompanying return to normal salinity. Such an event might occur annually with the advent of the on-shore monsoons, but it is more likely that the breach might be closed for a number of years.

While the breach is open no salt can be deposited, though the salinity may be considerably higher than that of the feeding sea, as is the case in the Karabugas lagoon. Here, as a result of the lower lagoon level due to evaporation, there is a constant influx of Caspian waters and, as already stated, a continuous destruction of life. If, however, the inflow of water is sufficient to maintain a normal salinity in the lagoon, it can also tolerate a normal marine fauna, until the next closing of the bar.

Thus, a succession of similar deposits will be formed, each begining with a clay-layer, in which are preserved the remains of the contemporary animals of the outer or feeder-sea of the lagoon, followed by a layer of gypsum and then by a layer of rock-salt, and the order of succession will be repeated an indefinite number of times, depending on the continuance of the relative conditions.

When the Suez Canal was cut in 1861 to 1863, it revealed in the center of the great Bitter Lake, which it traversed, a salt bed 13 km long, 6 km broad, and averaging 8 meters in thickness.

In the center of the lake it reached an estimated thickness of 20 meters. The salt deposit consisted of parallel layers of rock-salt, varying in thickness, and separated from one another by layers of earthy matter and small gypsum crystals. A mass 2.46 meters thick, consisted of 42 layers of salt and gypsum varying from 3 to 18 cm in thickness and separated by earthy layers only a few millimeters in thickness.

It took 42 separate overflows, each representing at least a year (more probably dozens of years), to produce this thickness of the series, and the bar was closed 42 separate times while this amount was being deposited, remaining closed a variable number of years, if we may judge by the variable thickness of the individual salt layers. And this represents less than 1/8th, the maximum thickness of that salt deposit.

Even if the overflow occurred annually, it would have taken more than 300 years to form the entire mass, and since it is almost certain that it took many years, after each closing of the bar, to concentrate the water to the point of salt deposition, the time consumed for the entire process is indefinitely prolonged.

LEROY RETSOF LIVONIA

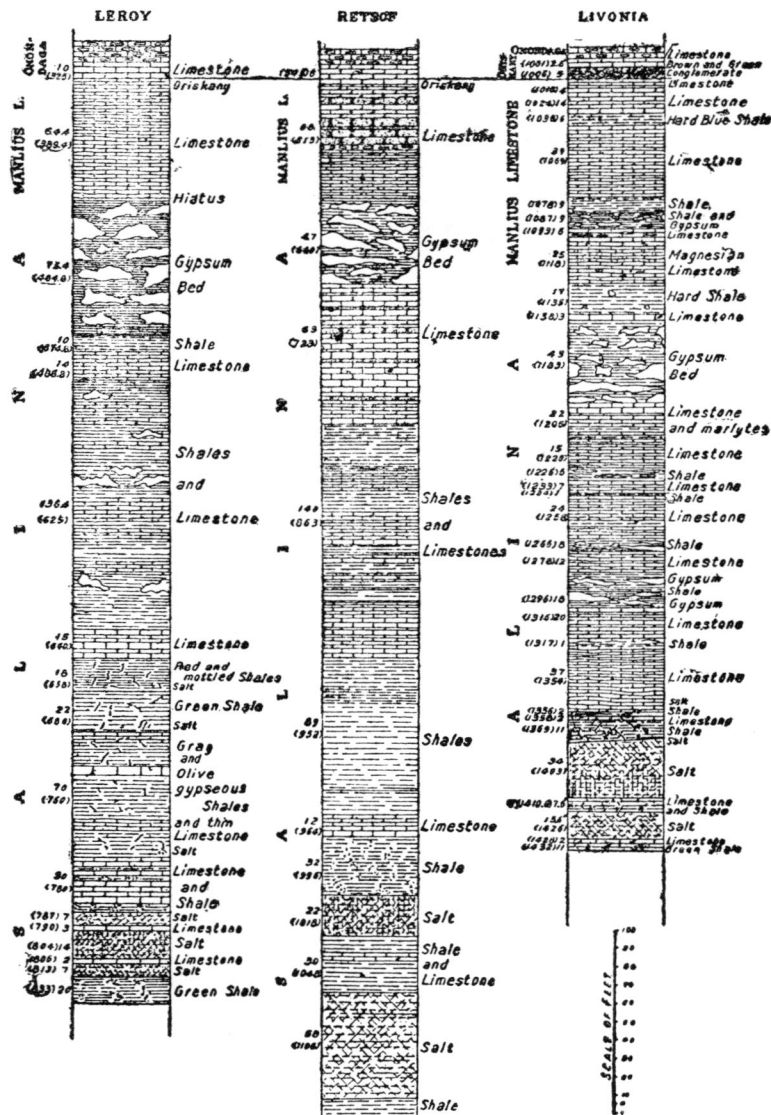

Fig. 54. Sections of bore-holes in the Salt district of New York State, showing the salt beds and the overlying strata of the Salinan with their gypsum (alteration products), followed by the Manlius limestone of the Siluronian, with the Oriskany interpulsation hiatus separating it from the Devonian (Onondaga). (Compiled from well-records).

One fact should be especially emphasized, namely that the numerous clay-layers in the deposits were, as a rule, richly fossiliferous, containing the shells of marine genera and species of molluscs still living in the Red Sea.

Now it is precisely this fact and the constant alternation of the salt with these fossiliferous beds and with the gypsum beds, that is fatal to an interpretation of the Salinan salts by an appeal to the bar theory.

While in one or two sections the salt is believed, though not positively known, to rest on the fossiliferous Niagaran limestone, which by an excessive stretch of the imagination might be compared with the normal base of the series (though in reality these limestones are of much greater age, and separated from the salt deposits by a disconformity), the almost universal rule is that the salt beds of the salt district are underlain by a great thickness of unfossiliferous sediments, *and there are no fossils in the parting beds.*

Moreover, these parting beds are most often granular dolomitic rocks of clastic origin, or massive beds, separated out as evaporates, and the gypsum more often overlies the salt bed than underlies it, and is known to be, in many cases at least, an alteration product from limestone, though this is sometimes questioned.

I have elsewhere [5] discussed at length the various modifications of the Bar Theory which have been proposed to explain the deficiencies of the salt deposits in not conforming to its requirements, and I need only note here that none of these satisfy the conditions demanded to explain the Salinan salt series.

The wells in the salt district of New York (Fig. 54), Ontario and Michigan (Fig. 55), show great thicknesses of salt, with hardly a trace of foreign admixture; the well sunk at Goderich shows six solid rock-salt beds, of which the lowest and thinnest is 6 ft, while the highest and thickest is 30.9 ft in thickness, the intermediate beds varying from 30.5 ft to 25.4 ft. They are separated by layers of marl, dolomite and some anhydrite, and there are no fossils.

The lowest bed is separated from the Guelph dolomite

---

5) GRABAU: *Principles of Stratigraphy*; also: *Principles of Salt Deposition.*

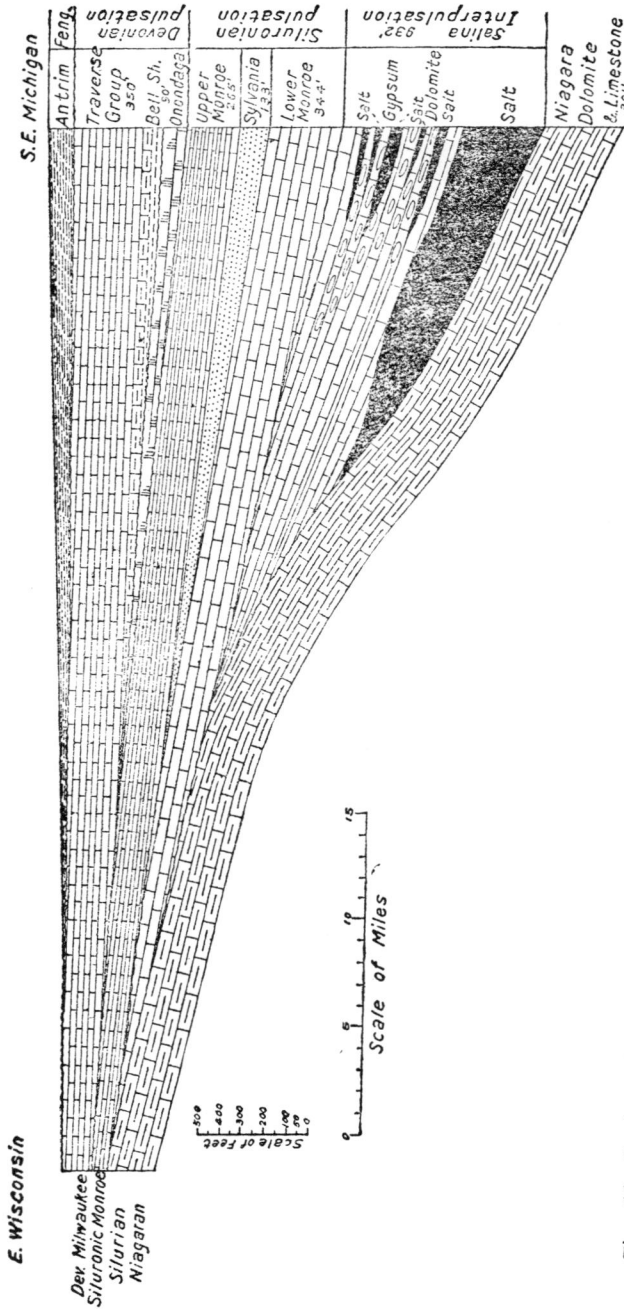

Fig. 55. Sections showing the Silurian and Devonian formations of S. E. Michigan, and their thinning away by overlap and erosion in the eastern Wisconsin region. The latter region was still within the marginal platform. In the former, the continental salt-bearing Salinan interpulsation beds are developed in full thickness, followed by the Monroan (Siluronian) marine strata. In the Wisconsin region the Upper Monroan beds lie disconformably upon the Niagaran, and are followed disconformably by the later Devonian Milwaukee dolomites, the earlier ones, as well as the Oriskany and Helderbergian, being absent by overlap. (Compiled from various sources.)

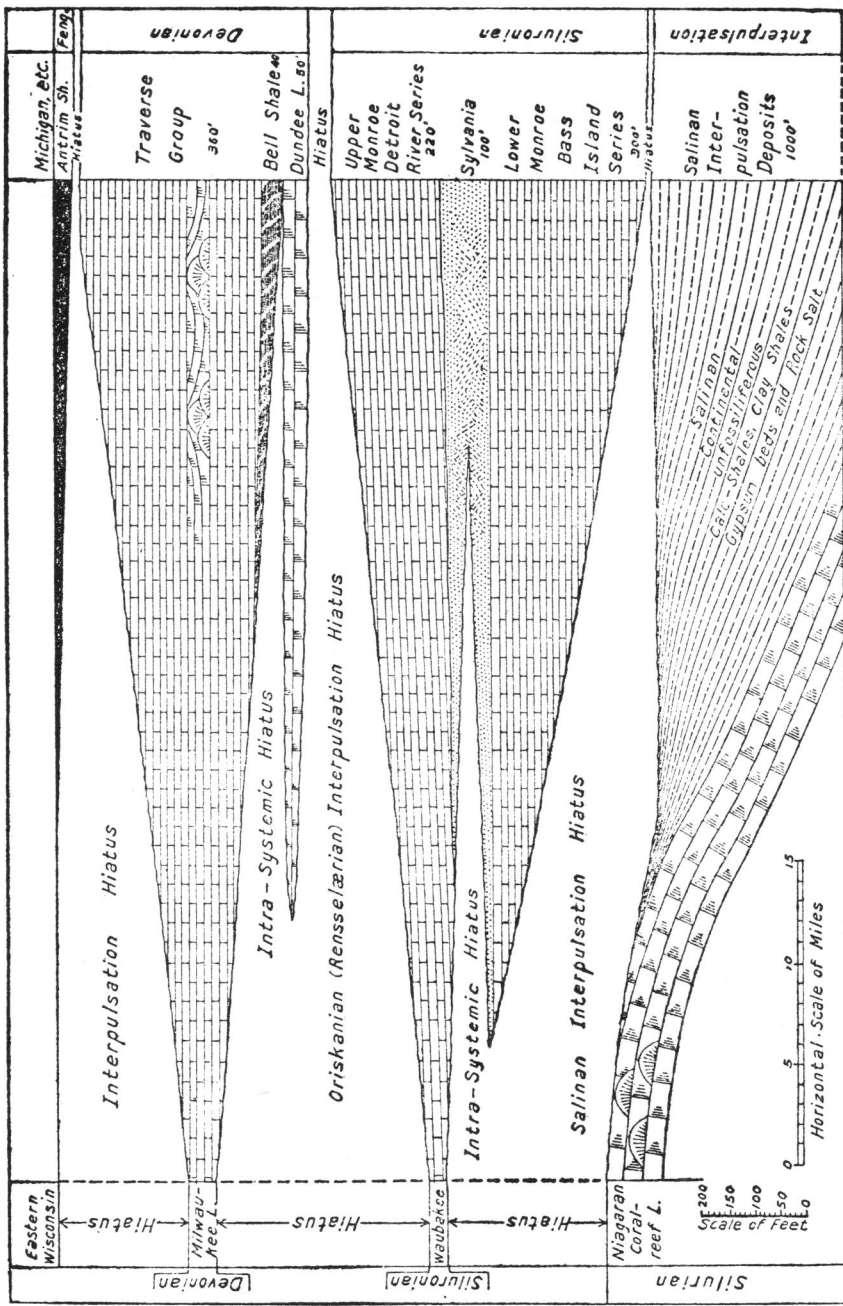

Fig. 56. The same section, but with the strata horizontal, showing increase in the interpulsation gaps.

of the Silurian pulsation system by 132 ft of unfossiliferous marls, with dolomite and anhydrite. The rock-salt is 99.687 % pure NaCl.

In the well section in Onaway, Michigan, salt beds ranging from 15 ft to 225 ft in thickness were penetrated. The salt series ranged through 1,175 ft of beds above the Niagaran limestone; 12 beds of rock-salt were found, the lowest and the thickest 225 ft, and the thinnest 15 ft thick.

Altogether the rock-salt aggregates 840 ft, which is more than 2/3 of the entire succession. The intervening layers have a thickness of 18 to 30 ft, and in one case of 100 ft. They are shales and anhydrites, but in a normal association.

The lowest salt, 225 feet thick, rests directly on the Niagaran limestone, but the details of the contact are not recorded.

Usually, the anhydrite overlies the salt, whereas normally it should underlie it. Moreover, the anhydrite, or gypsum, often proves to be an alteration product of the limestone. Fossils are unknown.

In another well section (Royal Oak), 9 salt beds, aggregating 609 ft, occur in a series with a total thickness of 1,010 feet, the separating layers being shale or limestone. Here, too, the lowest salt bed, 180 feet thick, is represented as lying directly on Niagaran limestone. The separating layers here, are dolomite, often mixed with salt and some anhydrite, but rarely shales.

It would be folly to compare these salt deposits with those of the Bitter Lakes, the most typical lagoon deposits known.[6] There is not a single significant characteristic which the ancient and modern salt deposits have in common, and not even by the most liberal stretching of the imagination can they be shown to resemble each other. That geologists have persisted in the attempt to parallel them, when they do not even have the slightest features in common, and that hard-headed businessmen should have wasted money in trying for potash and other mother-liquor salts in these deposits, when a brief review of their characteristics should have convinced everyone of the hope-

---

6) The Karabugas, which is not even depositing salt, may be disregarded.

lessness of such search, simply shows the folly of assuming that all salt deposits were built on the same pattern, because of the assumed uniformity of geological history.

5

There is a good modern parallel to the Salinan salt deposits, in the salt beds of Lop-Nor, in the Takla-Makan desert of Chinese Turkestan. This region and its deposits, which have been so graphically described by Huntington,[7] deserve consideration in some detail. (See Map, Fig. 34, p. 123.)

The basin is a depressed block which lies between the Kwen Lun mountain range on the south and the Tien-Shan ranges on the north, these meeting on the west in a complex of many virgations. Some of these mountains rise to heights of from 15,000 to 25,000 feet, and they are lofty throughout, except at the north-east, where a comparatively low, maturely dissected mountain region separates the Tarim basin from the desert of Gobi. The area of the basin is equal to that portion of the United States east of Lake Michigan, and north of Tennessee, or approximately three times that of Great Britain and Ireland. The drainage of the basin, where it does not die away in the sands of the desert floor, is toward the salt lake Lop-Nor in its eastern end. The principal river, the Tarim, gathers in the streams from the western and northern slopes, while a smaller stream, the Cherchen, carries a part of the drainage from the southern slope, the remainder withering away in the desert. The climate of the basin is determined by its position in the middle of the temperate zone and in the center of the largest of the continents, as well as by the fact that it is surrounded by lofty mountain ranges, which shut out the moisture-bearing winds. The precipitation is thus very low, and the extremes of temperature very great. Some rain falls in the basin during June, July and August, and a little snow during the winter months. The total annual rain-fall, in the center of the basin, does not amount to more than an inch or two. In the mountains, however, at an elevation of 10,000 feet or more, rain is plentiful and vegetation abounds.

---

7) HUNTINGTON, E.: *The Pulse of Asic.*

The floor of the basin, which lies between 3,000 and 5,000 feet above the sea, appears essentially level, and is mostly a barren desert with little or no vegetation except along the streams. It is covered with sand dunes of great height. This desert plain is 900 miles long by 300 miles wide, and is surrounded by a relatively narrow border of vegetation, developed where the streams from the mountains enter upon the plain and for the most part disappear by the evaporation of the water, or by its sinking into the thick deposits of sand and gravel. Outside of the belt of vegetation is a broad piedmont belt from 5 to 40 miles in width and sloping inwards toward the plain. It is composed of the gravel brought by the streams from the mountains.

In the eastern end of the basin lies the great salt plain of Lop, in which is situated the contracted modern salt lake of Lop-Nor or Kara-Koshun. This lies at an altitude of 2,600 feet above the sea. In places the piedmont gravel deposits border the plain, in others there are vast stretches of desert sands and clayey loess which encroach upon its margins. The salt plain has a length of over 200 miles, with a width of 50 miles or more in its broadest portion. Large areas consist of very pure solid white salt with its surface "resembling the choppiest sort of sea, with white-caps a foot or two high, frozen solid." (Huntington). The salt is so hard that Huntington found that when he tried to drive his iron tent pegs into what appeared a soft spot, most of them bent double. He explains the roughness of the surface as follows:

"During the long-continued process of drying up, the ancient lake of Lop deposited an unknown thickness of almost pure rock-salt. When the salt finally became dry it split into pentagons from five to twelve feet in diameter, the process being similar to that which gives rise to cracks when a mud-puddle dries up. . . . The wind, or some other agency, apparently deposited dust in the cracks; when rain or snow fell, the moisture brought up new salt from below, and thus the cracks were solidly filled. When next the plain became dry, the pentagons appeared again. This time the amount of material was larger, and the pentagons buckled up on the edges and became saucer-shaped. By countless repetitions of this process, or of something analogous to it, the entire lake-bed became a mass of pentagons with ragged, blistered edges."

6

Where does this salt come from? The answer is, from the rocks. It is *connate salt*, that is old sea-water, imprisoned in the pore-spaces of ancient marine rocks, and liberated and concentrated on disintegration of these rocks.

The pore-spaces of rock vary enormously. That of samples of Niagaran limestone is 0.77 % from one locality, and 6.4 % from another in Wisconsin. Wisconsin sand-stones range from 20.7 % to 28.2 %. Sandstones of rounded quartz grains from 30 % to 40 %, and clay loams from 40 % to 50 %. (Fig. 57).

Taking 30 % as a fair average and the salinity of normal sea-water at 3.5 % (35.0 per mille = 3.5 %), we arrive by a simple calculation at the conclusion that the average amount of salt included in marine strata is 1 % of the mass

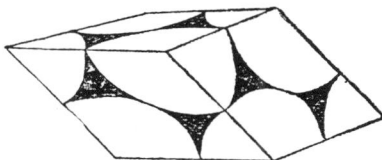

Fig. 57. Diagram to show relative pore-space (black), in rock of rounded sand grains of uniform size.

of the rock. Of this 77.76 % is NaCl, which would amount to nearly 0.365 % of the volume of rock.

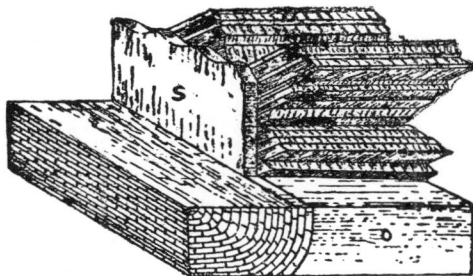

Fig. 58. Section at Kingston, New York, showing the strongly folded (Taconic folding) and peneplaned Ordovician graptolite shales (O), unconformably overlain by the Siluronian (S) Cobleskill and Rondont water lime — the equivalent of the Bertie. The peneplanation represents the Silurian period, which is otherwise unrepresented. The beds are again disturbed by the Appalachian folding.

Thus, a marine formation 1,000 ft in thickness would enclose a sufficient amount of salt to produce, after concentration, a bed 3.65 ft thick over the same area. Concentrated into an area 1/10th of the original extent of the parent rock, a salt bed 36.5 ft thick would be produced.

This would be practically pure salt, for the gypsum derived from the connate water would separate out first, and long before deposition begins in the central basin where the salt is deposited, gypsum is crystallized out and is left in beds or in scattered crystals in the surrounding sandstones or clay strata.

Nor would there be any fossils except terrestrial organisms accidentally mired in the salt lakes; or older fossils, weathered out of the mother-rock that supplied the salt and clastic sediments.

The thickness of the upper salt bed of the Lop-Nor salt plain is, according to Sven Hedin, not over 2 meters, (about 6 feet) or equal to some of the thinnest beds of the Salinan. The salt plain has a length of 200 miles, and a width of 60 miles. Assuming, on the basis of other considerations, that this amount of salt has accumulated in the Holocene or modern period, we have a rough measure of the time required to form the successive salt beds of the Salinan salt series and therefore of the length of the Salinan interpulsation period.

The interpulsation period is thus seen to be fully equal in length to the pulsation period, and the reason we have heretofore neglected it entirely was because of the inherited belief, or the tacit assumption, that the salt deposits were only an accidental separation of marine salts in local lagoons during Silurian time, that the Niagaran sea took the salt series in its stride, as a minor incident, — that, in other words, no great time interval was involved in the separation of the salt.

But we are learning that deposits of this type give us a truer picture of Palæozoic history, and a new and more accurate conception of the length of the time interval, so modestly marked by deposits to which we stratigraphers have heretofore paid only cursory attention, obsessed as we were by the magnitude and importance of the fossiliferous strata.

7

In the eastern portion of the Salinan salt basin the rock, which immediately underlies the salt, is often a bright red loess-like deposit, the Vernon shale of New York and

the Bloomsburg shale of Pennsylvania. Similar deposits
are found in the neighborhood of the salt basin of Lop-
Nor, shown as clay beds on the map, and they represent
the accumulation of the finer siliceous product of disintegration.   Their color was not necessarily red at the time of
deposition, but ochery, due to admixture of small amounts
of iron-hydroxide.   These beds are mostly absent from the
salt series in the western area, which was distant from
the rim of the Salina salt basin.   In the limestone region
between the Niagara and Mississippi rivers, the separating
beds are mostly lime-sands or dolomite-sands and marls, with
gypsum or anhydrite layers more or less discontinuous.
Outside of the salt basin, as in Central Pennsylvania, the
red loess (Bloomsburg) is the only representative of the
Salinan period.

8

The total absence of deposits of Salinan type in the interpulsation period of the European succession, is a significant
feature, indicating a great climatic difference between the
emerging geological regions within the same geosyncline.

In general, the succession and faunas of the older
series present only minor differences, such as may be due
to local environment and the amount of sandy or muddy
sediments brought in by the tributary rivers.   But while
the different parts of the geosyncline were probably in constant interrelation, the typical European deposits were more
influenced by the nearness of the polar ice of the period,
than were those of America.   Also the former were more
constantly populated by organisms from the Barents epi-sea,
and the latter from the Caribbean epi-sea.   Nevertheless
the corals of Gotland and Shropshire correspond closely to
those of the Wisconsin reefs.

But the interpulsation period, which in America was
represented by the Salinan deposits with its salts, etc., is
in Europe developed mainly as an erosion plane, or represented only by other types of continental deposits.

There is even a slight arching, — an echo of the greater
disturbance in the Silurian (Wenlock) and older beds, which
is well seen on the Island of Gotland in south-east Sweden.
In addition to this, there is a local deposit of an interpulsation sandstone, the GANSVICK SANDSTONE, which shows

æolian characteristics (See Fig. 59) and marks the interval of retreat. At this level was found one of the oldest land-scorpions, previously referred to.

Fig. 59. Section on the Baltic coast, showing the Gansvick æolian sandstone disconformably followed by the Siluronian oölites and limestones. The Gansvick takes the place of the Salina interpulsation deposits of America and the overlying beds, that of the Monroan, etc.

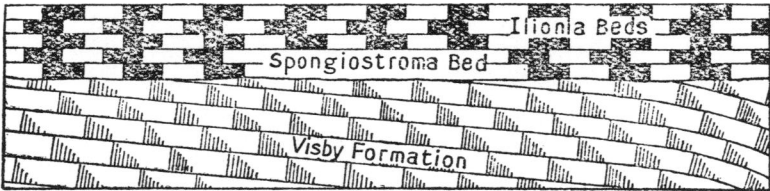

Fig. 60. Faint unconformity between the Silurian Visby strata (Lower Gotlandian), which were gently arched and eroded during the Salinan interpulsation period of exposure, and the Siluronian (Upper Gotlandian), in the Lindström quarry, Gotland.

When the readvancing Upper Silurian or SILURONIAN sea spread these sands across the erosion surface, which had been cut on the various Lower Gotlandian strata during the period of the Salinan (Gansvickian) interpulsation emergence, the succeeding calcareous beds of the transgressing Siluronian sea, the Upper Gotlandian beds, progressively overlapped against these sands as on a basal bed. It may be imagined what a source of puzzlement this section had become, so long as geologists still continued to regard it as a consecutive series of deposits, for where the intervening sandstone is absent, some of the basal beds of the higher Gotlandian (Siluronian) rest upon a variety of beds of the lower Gotlandian (Silurian) formations. Thus before the intervening hiatus was recognized there seemed to be no order in the succession of the strata of this small area, no two of the local sections showing similar formations in superposition.

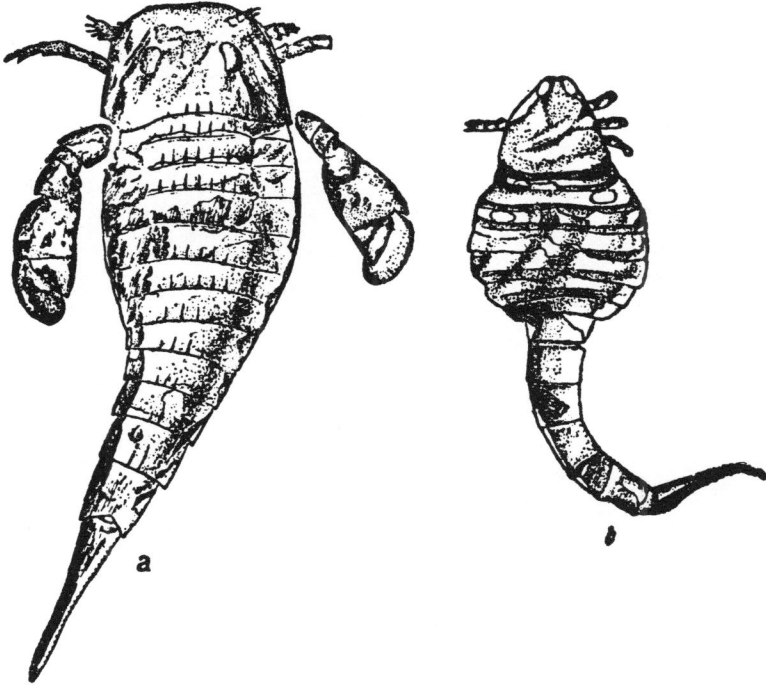

Fig. 61. (a) *Eurypterus lacustris* Harlan. Specimen of immature individual in natural position on the rock. (Reduced.). Cement quarry, Buffalo, N. Y. (b) *Eusarcus scorpionis* Grote and Pitt. Characteristic specimen as preserved on a water-lime slab; showing position of eyes, parts of the appendages and the scorpion-like telson. (Reduced). Williamsville, N. Y.

In Shropshire, England, the interpulsation hiatus is obscured by the relative uniformity of sediments. The end of the Silurian sedimentation is, however, marked by the abrupt disappearance of the graptolites of the late Silurian (*sens. strict.*) (Ludlow or Clunian), and by the Ludlow Bone Bed, which marks the destruction of many of the river fish of the interpulsation period when the salt-sea returned at the opening of Siluronian time.

CHAPTER XVIII

THE SILURONIAN PULSATION PERIOD

1

THE Salinan deposits are primarily restricted to the North American region, where, in the Appalachian geosyncline, they are represented by continental clastics of red loess-like sediments underlain by coarse torrential deposits. These were formed before the climate had become wholly arid. On the marginal platform they are represented by deposits of salt, leached from old, disintegrating, marine rocks, and carried as brine into concentration basins, where re-evaporation separated out the sodium chloride.

In the Caledonian geosyncline, on the other hand, which had become confluent with the Appalachian trough, after the Taconic orogeny, a pluvial but colder climate existed. Erosion was active, and the products of this erosion were carried by the streams into the geosynclines, where they accumulated as continental strata.

That continental sediments, formed during the Salinan interval, but entirely under pluvial or moist climatic conditions, though not recognized so far, must exist somewhere within the area of the old Caledonian geosyncline, or its marginal platform in Europe, and in some of the other geosynclines as well, can not be doubted. We may not recognize them as of Salinan age, unless the enclosing formations stamp them as such. They will have no marine fossils, not even graptolites, since the sea had withdrawn to the edge of the pangæa, but we may expect to find river animals, such as eurypterids (Figs. 61-63), or their terrestrial relatives, the primitive scorpions (Fig. 49, p. 169), and perhaps primitive armored fish. There rocks may be represented by the Caithness flagstones of Scotland which are generally referred to the Upper Silurian, *i. e.* Siluronian, but may be, in part at least Salinan.

We may even expect to find the remains of vascular

land-plants of simple organization in deposits of this age. Such plants have been obtained from beds of approximately this horizon (so-called Upper Silurian), at Alexandria, Victoria (Australia), while others (cf. *Hostimella*) have been obtained from Warrentinna, North Tasmania, from a horizon referred to the Ordovician, though this is questioned by Dr. Cookson.[1]

Early fresh-water pelecypods and gastropods may also be looked for as characterizing such interpulsation beds, but marine animals were entirely excluded.

Finally, we should expect to find, in some parts of the geosynclines of the earth, interpulsation deposits of volcanic rock, for the interpulsation periods were usually periods of volcanism.

For the most part, however, especially in the regions that have been investigated, the contact between the Silurian and Siluronian marine deposits is marked by a simple disconformity, and the sediments on either side are so similar that they have commonly been mistaken for a continuous series.

2

When the sea returned to the geosynclines of the pangæa at the opening of Siluronian time, it found that the old sea-channels had suffered in various ways during its long absence. The most profound modification was of course due to the translation of the pole to its new site in Rhodesia, a distance of over 45 degrees on a great circle. This meant of course an equivalent northward (p. d.) pressure of Africa, and this resulted first in continued sinking of the geosynclines and finally in extensive folding of the older strata of Europe.

Great changes were also experienced by the region which lay in the 'rain-shadow' of the Taconic range of mountains, that is, the Appalachian geosyncline and its marginal platform. In the narrowed Appalachian geosyncline, deposits of continental strata had formed, these consisting of piedmont sands and of gravels near the mountains, while loess and desert salts entered largely into the composition of the

---

1) COOKSON, I. C.: *Fossil Plants from Warrentinna, Tasmania*, 1937.

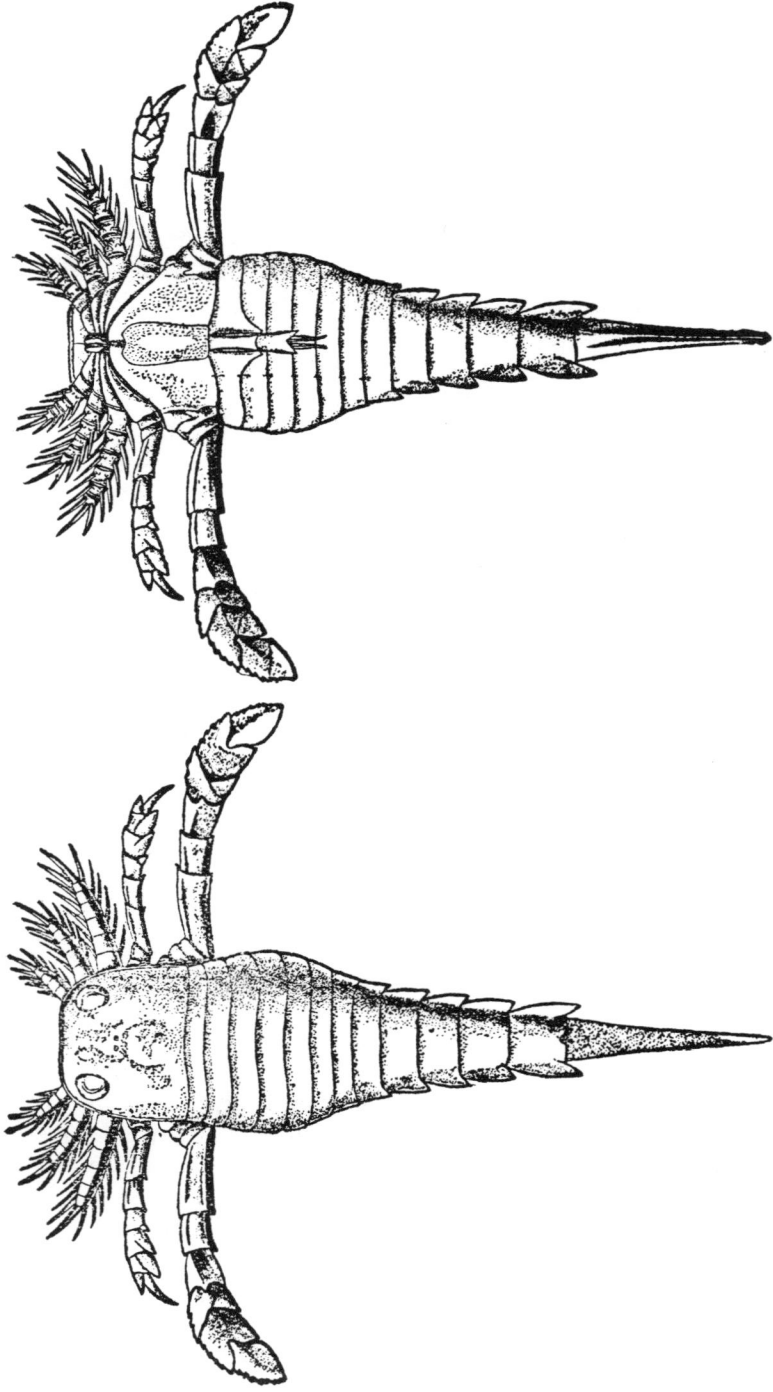

Fig. 62. *Dolichopterus macrochirus* Hall. A typical eurypterid restored with all appendages. (a.) dorsal aspect. (b.) ventral aspect. One half natural size. (After Clarke and Ruedemann).

deposits farther out in the geosyncline or over the marginal platform beyond. These had covered up the older marine strata. Only in the more distant regions, beyond the desert salt basins, were the old rocks still exposed at the surface, but many of these had suffered grievously as the result of erosion. As the country was too dry to maintain streams powerful enough to carry on erosion, the task of rock destruction was mainly undertaken by atmospheric agencies, which worked chiefly by disintegration and deflation. Even the limestones of the older series crumbled away,[2] and their lime-sand and lime-dust were removed by the wind and deposited in the salt basins, where they form dividing beds to the evaporation salts which accumulated there. In fact, when we remember, that the salt in the basin was the old sea-salt imprisoned in the pore spaces of the older marine rocks, that it was set free when these rocks crumbled into sand and dust, and that it was carried in solution by intermittent streams into the salt-basin, where on evaporation it was left as rock-salt, then we can realize that a vast amount of dust and disintegration sand must have been left behind.

Blown about and sorted by the winds, some of the finer material was spread as loess-like clay, some as lime-flour, over distant regions of the 'American Sahara.' The old salt beds of the central area were covered with a protecting mantle of this dust, and so preserved from re-solution when the sea returned. Elsewhere this dust accumulated in heaps, or in sheets, where it fell, and where it could subsequently be reworked by the waves of the returning Siluronian sea and be spread in stratified layers. These would serve as a medium for the burial of the new fauna that came with the encroaching sea.

Other accumulations of this lime dust and clayey loess on more elevated ground were discovered by the rivers which developed with the renewed pluvial conditions that accompanied the sea transgression. These dust deposits were washed into the shallow sea, or spread as huangho deltas on their borders. Such lime-muds and clay-muds

2) An illustration of this in a non-arid region was formerly seen in the crumbling of the crystalline Inwood limestone, in the northern part of the city of New York, into lime-sand. These lime-sands could be dug with a shovel.

abound in the deposits of the Siluronian sea of America, perhaps to a greater extent than elsewhere, because of the special conditions that existed here in the Salinan interpulsation period. The redeposited lime-muds are to a large extent banded limestones, which at intervals have their surfaces covered by dead shells of brachiopods brought in by the flood tide from the purer waters, where these

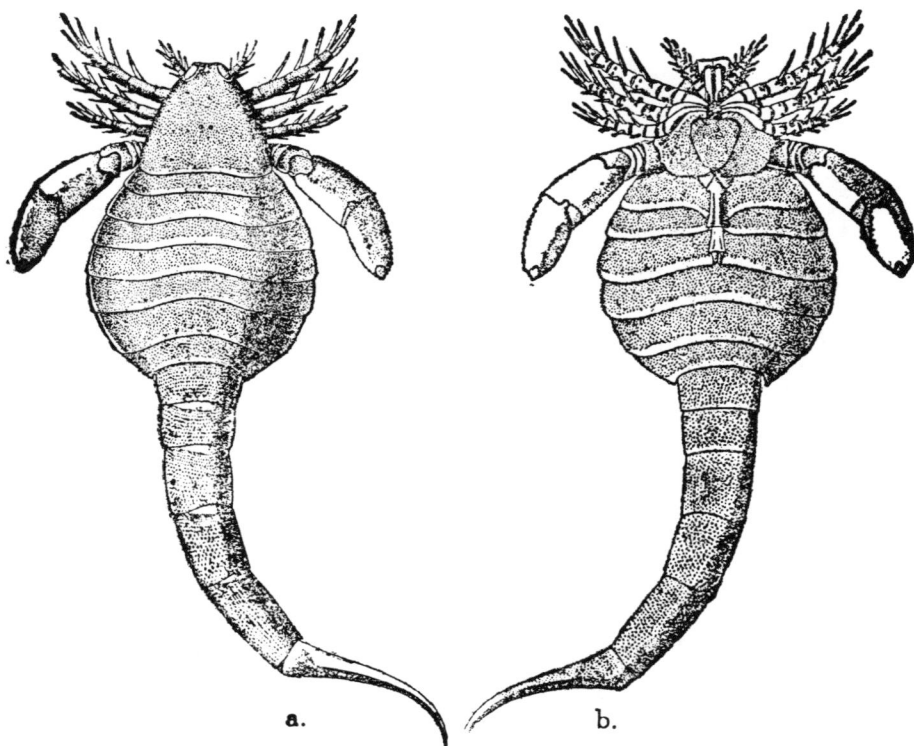

Fig. 63. *Eusarcus scorpionis* Grote and Pitt. A specialized eurypterid restored with appendages; a. dorsal aspect; b. ventral aspect. One half natural size. (After Clarke and Ruedemann).

animals lived. Likewise the planktonic pteropods and other flotsam of the pelagic district were brought by the waters, which were driven on to those mud-flats. Here they are sometimes found swept into "wind rows" of vast numbers of shells, mostly of one kind, and left to lie in close juxtaposition over wide areas. The MANLIUS LIMESTONE

of the State of New York, and other limestones of this
type in New Jersey, Pennsylvania and Maryland, represent
such mud-flat deposits, now consolidated. Pelagic pteropods
sometimes cover the surfaces, all disposed in parallel ranks,
with their large ends pointing down the current, a normal
arrangement, as determined experimentally by Dr. Kindle.[3]

The mud-flat surfaces exposed at ebb-tide were sometimes divided into polygons by desiccation fissures, which
under certain conditions are permanently preserved, and
are a not uncommon structural feature of these rocks.

                                3

Where the rivers brought fine lime-mud into basins
not accessible to the sea, beds of finely and beautifully
banded 'water-limes' accumulated, in texture not unlike
the much younger lithographic limestone. The banding
is suggestive of the varves, or 'annual layers' of Pleistocene glacial clay deposits, though these Siluronian varved
lime-muds were deposited in the subtropical region. On
rare occasions some favored basins received, along with
the fine lime-muds, the inhabitants of the rivers, or their
cast-off skins or exo-skeletons, and these were embedded in
the mud, in an almost perfect condition, and now constitute
nearly the sole fossil content of these rocks. Such are the
famous water-limes of Buffalo, N. Y., from which hundreds
of beautiful specimens of *Eurypterus* and related Merostomes have been obtained; superb specimens, which now
grace the palæontological museum not only of the Buffalo
Society of Natural Science, the collection of which is un-
parallelled, but of most American institutions of learning,
as well. They are also found in foreign museums from
Paris to Peking, and from London to Melbourne. And
wherever they are exhibited, they stand as a monument
to the late Lewis J. Bennett of Buffalo, who, while a
pioneer in the natural cement industry, had the understanding and the foresight to preserve these priceless
treasures of a pre-historic past, and the generosity to
distribute them were they would be of permanent scientific
value. (Figs. 61-63).

_____

3) *Journal of Palæontology*; 1938.

Much has been written [4] and argued about this famous eurypterid burial ground. For that it was a burial ground and not the habitat of living organisms is attested by the nearly, if not complete, absence of any sign of active movement, or of struggle on the part of these organisms. They were dead bodies or cast-off exo-skeletons that were entombed in this mud, or if a spark of life remained, the individual was too feeble to crawl far or struggle much

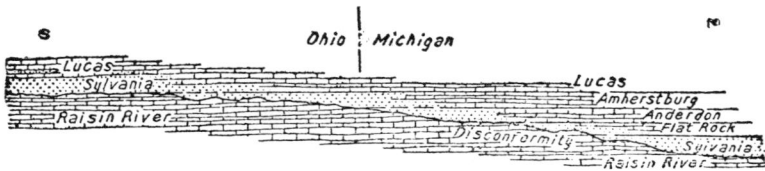

Fig. 64. Section from Michigan (north) to Ohio (south), showing the pronounced disconformity and intrasystemic hiatus and erosion sandstone characteristic of the marginal platform of the geosyncline, due to local withdrawal of water because of rapid sinking of the geosyncline, and readvance due to renewed filling. The Lower Monroe (Raisin River) and Upper Monroe (Flat-rock to Lucas) are both marine Siluronian. The Sylvania is a continental dune sand. (After Grabau and Sherzer.)

and quickly gave up the ghost. Nor could these animals have come into this burial ground from the sea, though that may not have been far away, since scarcely ever a vestige of marine organisms is associated with these animal remains. Moreover, and this is perhaps of the greatest significance, eurypterid remains have never been found in abundance and good preservation in normal marine strata, associated with a typical marine benthonic fauna.

The BERTIE WATERLIME of Buffalo, the last of the Siluronian strata which escaped erosion in pre-Devonian time in western New York, was the burial ground of the inhabitants of the rivers of the period. The region of their burial was probably far from their normal habitat.

---

4) These animals have been described by John M. Clarke and R. Ruedemann in *Memoir No. 14*, of the New York State Museum, with an atlas of 88 superb illustrations, from which our figures are taken. Dr. M. O'Connell, in her work *The Habitat of the Eurypterids* (1916), lists, besides 9 species of eurypterids in 4 genera, one species of *Ceratiocaris*, two cephalopods, one brachiopod, one ostracod, one pelecypod, two pulmonate gastropods, one graptolite (*Buthotrephis* formerly thought to be a plant) and one sea-weed (?) *Chondrites*. The cephalopods and brachiopods probably came from the immediately overlying Akron dolomite, which marks the renewal of marine transgression.

This was also true of the shells of the small brachiopod (*Spirifer vanuxemi*) and of the pteropod (*Tentaculites*), which are found crowding the surface of the layers of Manlius limestone, where this is preserved in Eastern New York and elsewhere. These animals did not live where this rock accumulated as fine lime-mud, but in more normal environment in the not too distant sea, from which the dead shells could be washed at intervals on to the limy mud-flats that preserved them.

The Siluronian deposits of Aemrica are unique in that the material which forms the sediments, especially in the interior between the Appalachians and the Mississippi River, was prepared during the long interpulsation period, when much of eastern North America was a barren desert. This differed from the modern Sahara and from the Takla Makan desert only in details, one of which was that the foundation rocks were nearly all marine sediments with a minimum of sandstone or of crystalline rock. The modern deserts have not only a foundation of crystalline

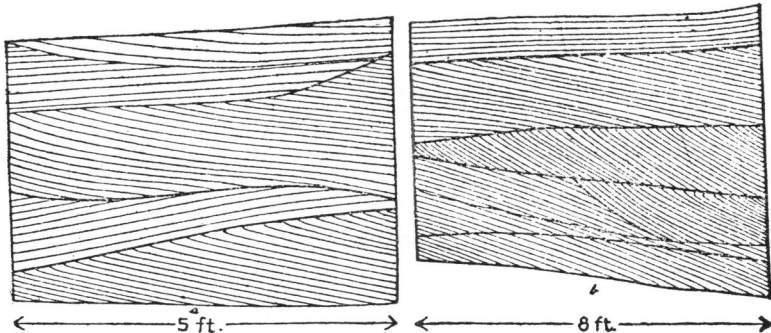

Fig. 65. Sylvania sandstone showing characteristic æolian cross-bedding, as exposed in quarry wall. (Drawn to scale.) (After Grabau and Sherzer.).

rocks, which on disintegration furnishes an abundance of quartz for the production of desert sands, but they have also had their loess-like products (the invariable accompaniment of such disintegration of crystalline rocks) blown out of the desert basin, so that only sand remains. This loess in the case of some of the Asiatic deserts was carried far away and some of it was deposited in North China.

Of course, there were some sands formed in the American desert, especially near the mountains on the east, where some of the older sandstones were uncovered by erosion, and in the region of crystalline rock, in what is now Canada. More probably, however, the source of the Sylvania sandstone, the most noted of these Siluronian sands, was the desert sand of an earlier period, the St. Peter sandstone, which may have been locally uncovered by erosion. This is suggested by the remarkable correspondence in purity, size of grain and roundness of the two. That the Sylvania was exposed to renewed wind activity is shown by the cross-bedding (Fig. 65), but this exposure may not have been of long duration ; since the essential characteristics of the sand were already developed.

The area covered by the Sylvania sandstone, the pure round-grained quartz-sand that occurs in the middle of the Monroe formation, is known to extend from 80° to 90° W. Longitude and from 40° to 46° N. Latitude, or roughly over the area enclosing that within which the Great Lakes are situated, except Lake Superior and Lake Ontario. [5]

Aside from these local sources of siliceous sediment, almost the only rocks subjected to disintegration under the dry atmosphere that prevailed, were the coral reef and shell limestones of the Silurian, or the older calcareous rocks of the Ordovician, and these, in the absence of water to effect solution, yielded only lime-flour, locally mixed with the clay or even sand of the siliceous source rocks.

But the winds, which kept the disintegration products constantly on the move, sorted the material according to size and gravity, and so besides the loess and sand, a vast amount of lime-flour was made available for the returning Siluronian sea from which its marine sediments could be formed, or which the rejuvenated rivers could spread as calcareous muds or huangho deposits in playa-basins and in mud-flats along the sea front.

This, then, accounts for the unique characteristics of

5) It need hardly be said, that these lakes, or even their basins, were not in existence at that time, though for the sake of orientation they are inserted on the map. They began in Tertiary time as river valleys, which were gradually widened and became lakes only in the Pleistocene or even the modern period, when their old outlets were either raised by warping, or blocked by drift. (See GRABAU: *Geology of Niagara Falls*, and literature there cited).

the deposits that were formed during the early part
of the Siluronian pulsation period, and which are hardly
paralleled by any other Palæozoic deposit, since at no
time were the physical conditions quite similar.

4

It must not be concluded, however, that the Siluronian
deposits of North America are all of this type. These are
only the most distinctive and unique, and therefore the
most intriguing. Normal marine deposits were also wide-
spread, especially in the southern part of the Appalachian
geosyncline and the marginal platform adjacent to it.
Moreover, the sea-shore extended close to the more exposed of these mud-flats, which, indeed, were frequently
inundated, though never for long.

The transgressing marine waters from the Caribbean
epi-sea were alive with marine organisms, for the cradle
of their evolution lay near the equator. It was there that
the Silurian survivors lingered, and since the environment,
though limited, was as favorable as any sub-tropical environment could be, they continued in their progressive
evolution. The sedentary brachiopod element was re-
enforced by the pelagic pteropods; there were crinoids
with floating bulbs (*Camarocrinus*); pelagic orthoceratites
were common, and perhaps even some pelagic trilobites
were able to extend their habitat far into the bordering
ocean. The surface waters of the ocean, the panthalassa
in these latitudes, no doubt swarmed with pelagic life.

It is from this region that the normal marine Siluronian
sediments derive their faunas, and these are embedded in
great variety in a series of limestones and shales now
found in Maryland, Virginia, and elsewhere in the Appalachian Mountains.

The Silurian (McKenzie shale) of this section, 300
or more feet thick, and richly fossiliferous, is followed disconformably by another fossiliferous shale, 450 ft thick
(Wills Creek shale),[6] which carries near the middle a fauna
with *Spirifer vanuxemi*. This, as we have noted, was
frequently cast ashore in abundance on the Manlius mud-

---

6) The Wills Creek and Tonoloway were formerly referred to the Salinan,
which is absent in this region.

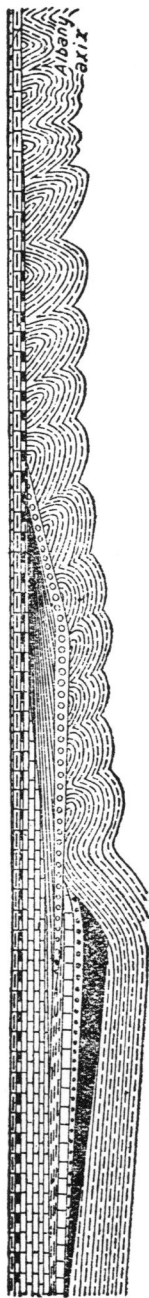
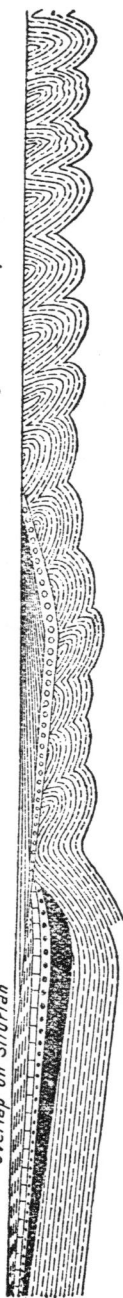

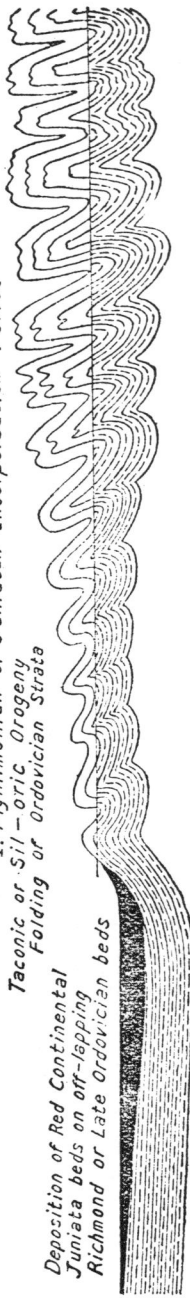

IV. Silurian Pulsation Period
Transgression of Marine Early Silurian (Cayugan) from South° and Helderbergian from North°

Albany axis

III. Salinan Interpulsation Period
Deposition of Continental Salina (Shawangunk and Blumsburg) on Peneplane
Overlap on Silurian

II. Silurian Pulsation Period
Erosion of Folds of Sil-oric Orogeny to Peneplane

Deposition of Marine Silurian
(Tuscarora, Rosehill, Mckenzie, etc.)
on continental Juniata

I. Plynlimonian or Juniatan Interpulsation Period
Taconic or Sil-oric Orogeny
Folding of Ordovician Strata

Deposition of Red Continental
Juniata beds on off-lapping
Richmond or Late Ordovician beds

Fig. 66. A series of sections, showing progressive development in the Appalachian geosyncline.

I. Erosion of Taconian (Ord-oric) folds during Juniatan interpulsation period and deposition of red continental Juniata (Queenston) beds on offlapping Richmond.
II. Early Silurian deposits on Juniata.
III. Salinan interpulsation deposits in geosyncline Shawangunk, etc. on eroded Taconic folds.
IV. Transgression of two faunal phases of marine Siluronian from opposite directions.

flats of the New York region. The two shales, the Niagaran McKenzie and the Siluronian Wills Creek, are separated by a hiatus and erosion disconformity, which alone represents the long time period of the Salinan salt deposits. Above the Wills Creek shale of Siluronian age follows a limestone (Tonoloway), which presents a good example of a lime mud-rock (calcilutyte), like the Manlius above described, but with a thickness of 615 ft. Evidently the transgressing sea encountered a rich deposit of the old lime-dust or mud, formed in the interpulsation period, and reworked this into a marine limestone, just as it would have reworked a previously formed mass of glacial clay into a stratified shale deposit with included marine fossils.

Fig. 67. Disconformity marking the Oriskanian interpulsation hiatus, between the Siluronian (Akron dolomite and Bertie water-lime) and the Devonian (Onondaga limestone) in the cement quarry at Akron, N. Y. The Oriskany sandstone is absent.

These fine mud-deposits have all the ear-marks of a huangho delta of lime-mud, over which at every inundation are spread the flotsam and the plankton of the sea. The latter consisted largely of minute bi-valve crustacea (Ostracoda). At least once, this lime-mud deposit was submerged for a longer period, for, near the middle occurs a sponge bed (*Hindella*), and the zone of the brachiopod *Camarotœchia*, probably living in colonies, was established for a time on that bottom. But by their contrast these normal marine deposits emphasize strongly the abnormal character of the rest of that thick limestone series, and show it to be reworked lime-mud in shallow water, either too shallow to permit permanent occupancy, or too muddy with suspended lime-mud to enable the fauna of the purer outer waters to adventure into this region except at great peril.

But, after more than six hundred feet of this kind of sedimentation, continued transgression finally effected the establishment of the purer water in the parallel of modern Maryland, where 110 ft of shell-limestone (Kayser) follows upon the mud-rock. In the parallel of central Pennsylvania, however, this, too, is mostly a lime-mud-rock (calcilutyte), the total thickness there being 820 ft.

Northward, the thickness of these deposits rapidly decreases, until in New Jersey, at Decker Ferry on the Delaware River, the section shows from below upward: 200 feet of unfossiliferous clay-shale, followed by 100 feet of unfossiliferous banded lime-mud-rock (Bossardville limestone), evidently of purely super-marine origin, and only towards the last is there a deposit, 52 feet thick, of normal marine fossiliferous limestone (Decker Ferry).

5

In the cement region of Ulster County, New York, on the Hudson River and elsewhere (Rondout, Rosendale), may be seen two beds of cement rock, or fine lime-mud-rock of reworked lime-dust or lime loess, a residue of the products of disintegration of the Silurian limestones, during the Salinan desert period. These are separated by a bed, 15 feet or more in thickness, of normal marine fossiliferous limestone, while another thin bed of much purer limestone precedes the water-limes and rests directly on the folded and smoothly eroded surface of Ordovician sandstones (Fig. 58, p. 193)

In Eastern New York (Appalachian geosyncline), the Manlius limestone is succeeded by the Helderbergian series, commonly referred to the Lower Devonian, but here separated and united with the preceding formations, as Siluronian. The general succession is thus in descending order:

DEVONIAN PULSATION SYSTEM

Regressive Series   (3,500 ft)±

    Oneota, and Ashokan Flags [7]   3,500 ft

---

7) Mostly huangho deposits. They are followed by 1,725 ft. of Catskill beds, part of which many be interpulsation continental deposits.

Transgressive Series   (1,035 ft)±
   Hamilton Shales (highly fossiliferous)   600 ft
   Onondaga Limestone   75 ft
   Schoharie Shale and Esopus Grit   300 ft
      (mostly a huangho formation)
   Oriskany Sandstone   5-60 ft
      (highly fossiliferous basal sand)

Masked Hiatus and Disconformity

ORISKANIAN INTERPULSATION PERIOD

(Mostly an erosion period, with locally accumulations of quartz sands, afterwards incorporated as basal beds by the transgressing sea.)

Erosion Hiatus and Disconformity

SILURONIAN PULSATION SYSTEM

Regressive Series   (preserved)   (35-150 ft)
   Port Ewen [8] or Alsen Shales and Sand
   Rock, with fossils at intervals   35-150 ft

Transgressive or Helderbergian Series
                 (extended)   (164-342 ft)
   Becraft Limestone   10-40 ft
   New Scotland Shaly Limestone   60-100 ft (?)
   Coeymans Limestone   30-60 ft
   Manlius Calcilutyte (with fossiliferous
     layers)   20-42 ft
   Rondout Cement Bed   20-31 ft
   Cobleskill Limestone   14-18 ft
   Rosendale Cement Bed   10-21 ft
   Wilbur Limestone   0-8 ft
   Binnewater Sandstone (basal bed)   0-22 ft

Masked Hiatus and Disconformity

SALINAN INTERPULSATION PERIOD
   High Falls Red Shale (continental)   0-25 ft
   Shawangunk grit and conglomerate   0-210 ft

---

8) The Oriskany sandstone was formerly included with the Port Ewen, which is partly Oriskanian and partly overthrust New Scotland. (See GRABAU : *Geology and Palæontology of Schoharie:* pp. 302-314).

(Great hiatus cutting out the entire
salt series and the whole of the
Silurian marine series, as well as
part of the Ordovician.)

Disconformity or Unconformity

ORDOVICIAN (MIDDLE)
  Normanskill graptolite shales (huangho beds)

The Helderbergian series records the invasion of the
Bohemian fauna from the region around Prague, which
was then in 15° to 20° S. Latitude, *i. e.* in tropical waters.
This was also the latitude of the Helderberg mountain
region of eastern New York, where these formations first
attracted attention in America by their wealth of species
of exotic character.   Indeed the pathway of migration,
though covering a distance of at least 25 degrees Longitude
(1,200—1,500 miles) was in a geosyncline, which nowhere
departed widely from the interval between 20° and 30° South
Latitude. The fauna of the Caribbean epi-sea, on the other
hand came form colder waters (40° to 50° S. Lat.) and also
encountered an unfavorable abundance of lime-mud supplied to the southern sea.   This prevented more pronounced
intermingling of these faunas.   As it is, there is a constant
interfingering of the retarded faunas from the south, which
retained many of the ancestral Silurian characteristics held
over from the Niagaran fauna, and the more specialized
fauna from Bohemia, which was a more progressive type, and
set a new and distinctly Bohemian fashion for invertebrates.
    Professor Schuchert, who has made a thorough com-
parison of the Helderbergian with the Konjeprussian of
Bohemia, in 1900 announced his conclusion that the two
faunas are equivalent: "in fact part of one fauna". However, there is no similarity between the fauna of the
Oriskany and that of any southern European time-
equivalent, though there are some faunal elements suggesting relationship to the Coblenzian fauna.
    Coblenz is on the Rhine, where that river is joined
to day by the Moselle.   It lies in approximately 50° N.
Latitude, but before the Alps were formed, it, together
with Bohemia lay within 10 or 20 degrees of the equator
of the time.

CHAPTER XIX

THE SILURONIAN OF EUROPE AND THE
LOWER OLD RED SANDSTONE

1

WE must now take a brief glance at the Siluronian rocks which were deposited in other parts of the world. The geosynclines of pangæa from which Devonian rocks are best known are the Appalachian and the Caledonian, and their strata and fossils have extensively occupied the attention of scientific men. This was a matter of environment, for the west of Europe and, after the voyage of Columbus, the eastern sea-board of America and its neighboring districts were the chief regions that clamored for attention.

In America the rocks are so little disturbed that they have formed the favored training grounds of palæontologists, while the European rocks, though much disturbed and more difficult of analysis, lie in the heart of the older culture, and present problems not only of stratigraphy, but of structure as well, which called for study and solution. The rocks exposed in the picturesque gorge of the Rhine, and the strata of the Ardenne Mountains, vied with the "Old Red Sandstone" of Britain in forming attractive regions and in presenting unique problems for investigation.

As was perhaps inevitable, each political unit divided and subdivided its formations, and these European subdivisions of the Devonian became the standard for comparisons with other regions. Such standards, because founded on incomplete succession, became with time inadequate when sections of greater magnitude were found, and the history of the development of the Devonian in Europe had to be repeatedly revised.

The sediments of what was originally the continuation of the Appalachian-St. Lawrence geosyncline, in what is now North-Western Europe, are well illustrated in Scotland on both sides of the old crytalline axis, the only region in which the relationships of the Siluronian, or older, and the upper "Old Red Devonian" strata are well shown.

Two divisions are recognized: the Caithness flags and the Upper Old Red or Dura Den beds. The latter rest unconformably on the folded and truncated Caithness flags, which, in turn, lie unconformably upon various crystalline schists, granites, etc., and were apparently deposited on the uneven bottom of a sinking basin, or geosyncline, for occasionally even some of the higher portions are found resting against the old rocks.

2

It has long been held that these Old Red Sandstone beds, in common with those of other parts of Great Britain, were deposited in a series of fresh water lakes, which have been given individual names, the one in North Scotland being designated "Lake Orcadie", and that of South Scotland "Lake Caledonia". This interpretation is, however, open to question, for the alternative one, that these deposits are remnants of a series of great river-plains, or huangho deposits, on the border and near the axis of the geosyncline, has scarcely been given adequate consideration. And yet, looked at from this point of view, they fall into line with other normal series of continental sediments, and the red color, the origin of which has given rise to so much speculation, becomes revealed as the normal result of dehydration of a thoroughly oxidized river-plain, formed under prevailingly semi-arid conditions.

Then, too, the Caithness flags, of gray and bluish color, are normal deposits of rivers, characterized by periodic strong floods during an earlier period, when the peculiarities of topographic and climatic conditions which characterized the later Upper Old Red, had not yet appeared in full force. In this respect their physical character would be quite comparable to the Upper Devonian "Catskill" type of sedimentation, on the north-west side (p. d.) of the old-land of Appalachia, but within the Appalachian part of the geosyncline. As we shall see, however, the Caithness flags are of much earlier age than the Catskill.

In the Catskill Mts., the Catskill facies extends through more than 4,000 feet of continental beds, which in point of age range through the entire Upper Devonian. This series shows a constant alternation of gray, greenish and red beds, oscillating between shales, sandstones, sometimes cross-bedded, and conglomerates, and with only plant remains, bone-beds of river fish, and other evidences of non-marine sedimentation. They are preceded by 2,000 ft or more, of similar sandy and shaly beds, with marine fossils in the lower part (Hamilton group), while there are also some black shales (Marcellus). The whole rests on the Onondaga limestone.[1]

5

The Caithness flags of North Scotland, Moray Firth, and the Orkney Islands consist of similar red and gray sandstone-flags, which reach a total thickness upwards of 16,000 ft. Remnants of these old river-plain deposits (huangho type) are found at several places, notably at Oban and elsewhere in Lorne (Argyllshire). Here the series is interbedded with trachytic and andesitic lavas and tuffs, while the shales contain fish, eurypterids, ostracods, myriopods and plant remains. There can be little doubt that these formerly constituted a part of the extensive river flood-plain, to which the beds of Moray Firth and the Orkneys belong, as well as those of South Scotland.

Sir Roderick Murchison referred these beds north of the crystalline axis to the Old Red Sandstone, and since it was thought that the *Cephalaspis* type of fish, characteristic of the Lower Old Red or Arbroath flags of "Lake Caledonia" of Southern Scotland, was absent in the higher part of the Caithness flags north of the Highlands, he referred only the lower portion of that succession, which was characterized by the eurypterid *Pterygotus*, to the lower, and the remainder of the Caithness flags to the Middle Old Red. As remarked by Geikie, however, isolation on opposite sides of the barrier (or as he preferred to call it "in separate lake basins"), would tend to distinctness of development, while the similar tectonic relations between the Caithness flags and the Upper Old Red, *i. e.*, the very marked angular unconformity, which is similar to that

1) See Sherwood's detailed Catskill section, reproduced in GRABAU: *Guide to the Geology and Palæontology of Schoharie*; Bull. 92, New York State Museum,. pp. 275-278.

between the Upper Old Red and Arbroath flags on the south of the old-land barrier, would argue for the identity of age of the two series. Moreover, *Cephalaspis*, *Mesacanthus* and other genera have been found with the eurypterid *Pterygotus* in the rocks on Moray Firth.

Furthermore, the present isolation is not an original one, for the dividing old crystallines, which form the Scottish Highlands, have become exposed as the result of age-long erosion. This erosion removed both Upper and Lower Old Red (Caithness flags), which must have once formed continuous deposits across them.

It must be emphasized that the arrangement of the geosynclines of the later periods was distinct from that at the beginning. The former St. Lawrence geosyncline once extended across the north of Scotland, and in it were deposited the early Palæozoic sediments of that region, as well as those of north-west Newfoundland and the western Vermont region down to the Albany Axis across the geosyncline. But this geosyncline was no longer open in its northern part, where the region between Great Britain and Greenland had become dry land. Likewise the eastern Massachusetts part of the Caledonian geosyncline was blocked by glacial deposits, as well as by folding of the strata. After Ordovician time the Caledonian geosyncline of Great Britain included the whole of Scotland, but it was occupied by marine waters only in the southern area (Devonshire and Cornwall). It had now become continuous with the St. Lawrence geosyncline, the northward extension of the Appalachian geosyncline, which lay to the north (west on modern maps) of the folded Taconic belt of the foreland of the Appalachian old-land, on the geosynclinal side.

Great Britain and Ireland, as well as Scandinavia, were now united with Greenland by the Barents shelf region, and the whole had practically been worn into a peneplane, the height of which above the sea was very moderate. It was on this surface, that the Old Red rivers deposited the 16,000 feet (over 3 miles) of continental clastics of the Caithness flags, this being on the border of the Caledonian geosyncline, of the rate of sinking of which these strata form a measure.

Where the strata of these deltas entered the submerged part of the geosyncline, as in south-east England (De-

vonshire and Cornwall), and the region now included in
Belgium, the Rhineland and France, — they enclosed
the shells and other remains of Siluronian animals which
lived in those waters. In the sections on the Rhine, these
comprise no less than 3,000 meters[2] (10,000 ft) of sandy
and clayey beds, entirely non-calcareous. Even the fossils
occur chiefly as impressions and internal molds, because
of subsequent solution, by percolating waters, of the original
calcareous shells. They occur most frequently in distinct
layers or banks, separated by great thicknesses of practically barren rock, which was formed when the sediment-
bearing waters were so loaded that they either built up to,
or above the surface, or the waters were so muddy, and the
salinity so reduced by influx of fresh water that organisms
could not exist in this region. Often the beds are oxidized,
vari-colored phyllites, with interbedded coarse conglomerates and sandstones. In the unique section cut by
the Meuse in the French Ardennes, these beds have an
estimated thickness of 7,000 meters, or over 4 miles.
This shows the extent to which progressive subsidence
of the geosyncline was proceeding, and the amount of
sediment worn from the drainage basin of the Caithness
River, which was responsible for these deposits.

They virtually blocked the narrow throat, which connected the Caledonian with the St. Lawrence-Appalachian
geosyncline, and for the greater part of this period the
Caledonian fauna had no access to the American region.
During this time the Monroan-Manlius fauna of Silurian
derivatives held sway in the American Siluronian waters,
where also the lime-muds, formed during the preceding
interpulsation period, were deposited. Only in the later
Siluronian, when the maximum transgression was achieved
by the Siluronian sea, was the Caledonian fauna able to
enter the St. Lawrence geosyncline and establish itself as
the Helderberg fauna, which, in the Maryland region,
mingled with the Kayser fauna of Caribbean origin.

4

The piling up of 3 or 4 miles of mud or sand-rock,
layer upon layer, of essentially uniform character, implies,

2) In the Siegerland estimates range up to 10,000 meters.

of course, a corresponding slow sinking of the region, so that the surface would always be at essentially the same elevation above the waters of the Caledonian geosyncline into which these rivers flowed. I hold that the sinking of the geosyncline was the primary factor, and the deposition the consequence, instead of sinking taking place in response to the loading. The sinking of the Caledonian geosyncline, as well as of the Appalachian and of the west African geosynclines, may be regarded as a response to the movement of the African block towards these geosynclines and away from the polar region of Silurian time, so as to bring the new location over the pole. (Plate VII).

Accordingly, the deposition of these strata falls into the Siluronian period,[3] and the final momentum, by which the Siluronian and later the Devonian position was attained, resulted in the folding of these sediments in the British and other European regions, though the Appalachians were not folded until much later time.

It thus appears that the Caithness flags (together with the Arbroath flags or Downtonian, on the south side of the present Highlands) will have to be referred to the Siluronian, and perhaps in part Salinan, of our classification. That would leave the Oriskanian interpulsation period and at least a part of the Devonian time, as here restricted, to represent the interval during which the erosion of these folded Siluronian rocks was accomplished. Considering the great extent of the erosion which these folded strata have suffered, it seems more likely that the greater part, if not the whole, of the Devonian pulsation period, was occupied by this process. This would make the overlying Upper Old Red a late, or even post-Devonian deposit, referring it to the Devono-Tournaisian interpulsation period which might then be called the Old Red interpulsation period. In that case, these deposits would correlate with the red Mt. Pleasant interpulsation beds, between the Upper Devonian Catskill and the first of the Lower Mississippian beds in the Appalachian geosyncline.

This would also be in harmony with their thoroughly oxidized character, since they represent the deposits of a falling sea-level, and therefore more intensification of

3) A part may be cf Salinan age (interpulsation), as suggested in the preceding chapter

aridity, while at the same time it explains their in-
timate relationship to the overlying basal "Lower Carboniferous" (Tournaisian) rocks, into which they seem to
pass conformably, or which overlap against them where
the younger beds are marine.

5

If, however, the Upper Old Red Sandstones are still
to be considered as of Devonian age, they can only represent Upper Devonian, and probably only the later part
of that. On the Island of Hoy, in the Orkney Group,
farther north, there was also an interpulsation outpouring
of lava, marking a period of igneous activity, and the
volcanic necks which penetrate the Caithness flags, have
been definitely located (Fig. 68). These diabase-flows and

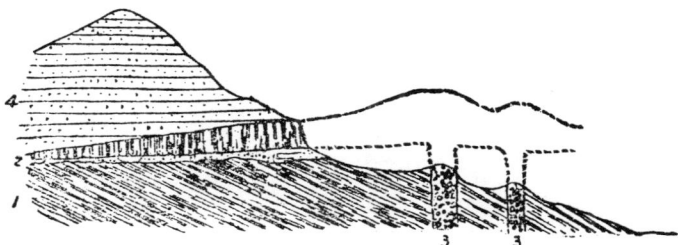

Fig. 68. Unconformity between the folded and eroded Caithness flags
(Siluronian) (1) and the late Upper Devonian or post-Devonian continental
Old Red Sandstone (4), with interpulsation volcanism (2-3). Island of Hoy,
Orkney group, Caledonian geosyncline. (After Geikie).

tuffs rest upon some residual sands, deposited upon the
tilted and eroded Caithness beds, though the flows
occurred and the volcanoes themselves were again destroyed, before the bulk of the Upper Old Red beds were
laid down. Thus, if the erosion period occupied the whole
of Devonian time, these volcanic rocks would mark the
end of the Devonian period.

As might be expected from their position in the
same old drainage basin which was supplied by sediment
brought by the rivers from Greenland, the sandstones of
Northern Scotland, and those of the outliers of Old Red
Sandstone known from the north and west side of Norway,
partake of the same relationship and characteristics, and
attain a thickness of 1,000 to 1,200 ft.

CHAPTER XX

THE ORISKANY SANDSTONE,
A RECONDITIONED INTERPULSATION DEPOSIT

1

ONE of the best marked and most widely recognized interpulsation deposits of the American Palæozoic is the Oriskany sandstone, which marks the dividing line between the Siluronian and the typical Devonian. In the days when belief in continuous series was the orthodox attitude, the Oriskany sandstone was thought to be merely a shallow-water formation, which marked the passage of the upper beds of the Helderbergian series, formerly classed as Upper Silurian, into the Lower Devonian of the older classification, i. e. the Cauda-galli (Esopus) and the Schoharie, and united all of them and the Onondaga limestone into one indissoluble whole, albeit a whole of diverse faunal origins. All the more was this belief held, because the Oriskany, though often a nearly pure quartz sand, was found to be richly fossiliferous in its typical outcrops.

It is now, however, generally recognized that the Oriskany, which represents a reworked æolian, or dune sand, accumulated on the old erosion upland of the Helderbergian, or earlier formations. The highest of the Helderbergian beds are the sandy and shaly limestones originally called the Upper Shaly series, but now familiar to every American student as the Port Ewen. It must be said, however, that this interpretation of the Oriskany as a reworked æolian deposit is not favored by all students of American stratigraphy, though an unbiassed consideration of all the characters of these formations in the light of lithogenesis should leave little room for doubt.

2

The sections exposed on both sides of the international boundary between New York and Canada, here formed by the Niagara River, can leave no doubt as to the great length of the interpulsation period which culminated with the return of the Early (Middle) Devonian sea.

Here the entire Helderbergian, which has a thickness of 240 feet in the Hudson Valley, is absent, having in large part been eroded away during the interpulsation land period, to which the name ORISKANIAN INTER-PULSATION PERIOD is most properly attached.

Fig. 69. Oriskany disconformity between the Siluronian, A (Bertie Water-lime) and the Devonian, C (Onondaga limestone) with the Oriskany continental sand, B (later reworked by transgressing sea), occupying an old erosion channel. Decewville, Ontario (near Buffalo, N. Y.). (After Kindle)

A similar relation is found 110 miles east of the Niagara River at Union Springs, on Cayuga Lake, Seneca Co., N. Y. (Long. 75°42' W., Lat. 42°50' N.). Here the Manlius limestone of Lower Siluronian age and characterized by stromatoporoids is disconformably followed by 4 feet of Oriskany, and this in turn is followed by the heavy-bedded Onondaga limestone. The Helderbergian is absent over this entire region. At the type locality, Oriskany Falls, 75 miles farther east (Long. 75°20' W., Lat 43°10' N.), the Oriskany is 20 feet thick, and rests disconformably on limestone, which represents the lowest of the Helderbergian (probably the basal beds), the higher beds, if they were present, having suffered erosion in pre-Oriskanian time. In the Jamesville section (Long. 76°5' W. Lat. 43° N), Kindle, in his study (1913) of this disconformity, records 15 ft of Oriskany; while in the Manlius section near by (Long. 76° W., Lat. 43° N.), less than half that amount remains, mostly reworked material.

Stauffer[1] writes: "The outcrop in the gorge just west of Manlius shows a stratum between 2 and 3 feet in thickness, in the lower part of which ..... 'the sand predominates over the lime, but toward the top the lime increases at

---

1) STAUFFER, C. R : *Oriskany Sandstone of Ontario*; pp. 371-376, 1912.

the expense of the sand and gradually the top of the stratum becomes entirely lime'. [2] This stratum of sandstone and arenaceous material carries an Onondaga fauna, and forms the basal layers of that formation. It rests directly on a sandstone which contains the characteristic Oriskany fossils. The stratigraphical evidences of unconformity [disconformity] in this case are apparently wanting, but the palæontological evidences are rather marked." From Manlius westward the Helderbergian is absent, and the Oriskany rests directly on an undulating erosion surface of the Manlius or other Lower Siluronian rocks.

This disconformity was recognized by that great pioneer American geologist, James Hall, a hundred years ago, and in his geological classic *Geology of New York*, [3] he accurately described it and clearly recognized that it signified an emergence, for only in that manner could erosion be accomplished. After that came resubmergence, the spread of the Oriskany sands, and finally the formation of the coral reefs of the Onondaga limestone.

In the Appalachian geosyncline the lowest of the Helderbergian beds follows concordantly upon the Manlius limestone, and although there is a most remarkable influx of new organic forms of the Helderbergian fauna, there is no recognizable indication of a hiatus between the Manlius and the succeeding Coeymans. But that, as we have had occasion to note repeatedly, is not a reliable criterion, and even the slight mixture of faunas, which seems to constitute a transition bed, must be regarded with reserve. The fact of the matter is that the faunal change is too abrupt and too pronounced to permit stratigraphic evidence of continuity to dominate our judgment entirely. While not an impossible succession, it is an unusual one, and one highly suggestive of a masked hiatus and a lost interval.

3

There is, however, one significant fact that must not be lost sight of in the study of these Helderberg sections, and that is the difference of source of the two faunas

---

2) HARES, C. J.: *Letter of February 9, 1912* (cited by Stauffer, p. 376).
3) *Summary Report on the Fourth District*; issued in 1843.

which meet and commingle to some extent in the Vlight-
berg and neighboring sections of the Helderberg mountains to form the Siluronian fauna. The new Helderberg
fauna from Europe became established as the water became
freed from fine sediments, and the mud fauna of the
Manlius retired before the change in the physical character of the formations. Thus this is not to be regarded
as a hiatus, but a physical and faunal overlap, a displacement of one type of sediment and fauna by another
adapted to the new conditions.

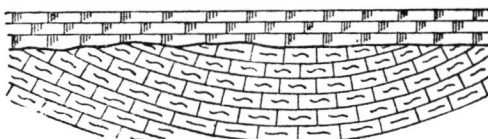

Fig. 70

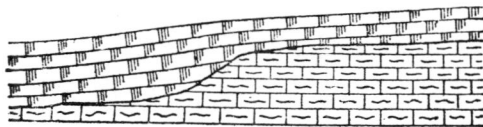

Fig. 71

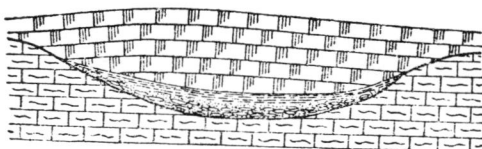

Fig. 72

Figs. 70-72. Sections in the Buffalo Cement quarry showing the erosion
disconformity between the Bertie-Akron below (slightly undulating in Fig. 70)
and the Onondaga limestone above. In Fig. 72 a rubble and sand bed in
the old channel marks the Oriskany interpulsation deposit.

However, the pre-Oriskany disconformable contact
and hiatus is so clearly marked throughout the western
half of the state of New York, as we learn from the
sections already described, and especially in the Niagara
region, that no one can miss its significance. Some years
ago one of the best sections was to be seen in the Bennett
quarry of the Buffalo Cement Co, from which the

illustrations (Figs. 70-72) are drawn. The Akron dolomite, a seven-foot layer of fossiliferous rock, rests upon and is continuous with the Bertie Waterlime, which, though famous for its rich fauna of eurypterids, has scarcely any other organisms and, as outlined in Chapter XVIII,

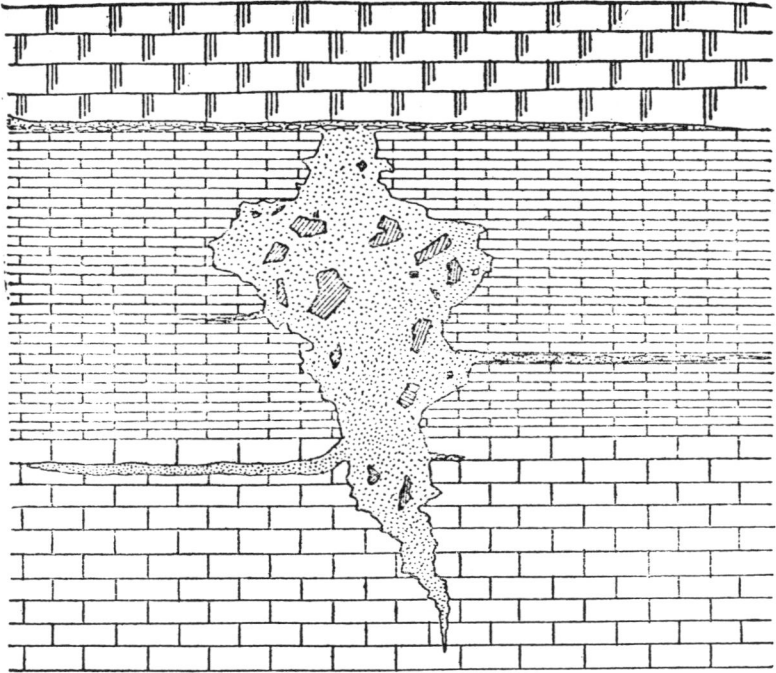

Fig. 73. "Sandstone Dike" formed during an ancient Oriskanian earthquake, which opened the fissure in the Cobleskill (Siluronian) and injected the overlying Oriskany sand together with the broken-off fragments of the wall-rock, which were caught in all positions in the injected sand. The extent (30 ft in some cases) to which the sand was forced between the beds is indicated. The overlying Onondaga limestone is a post-earthquake deposit, being entirely unaffected. (Illustrations after Clarke and Ruedemann).

represents a burial place for dead organisms rather than a place of habitat for the living Merostomes. The Akron dolomite is exposed in many sections in western New York (Fig. 67, p. 209), covering a considerable area. Nevertheless the fauna obtained is a sparse one, the number of species, so far recorded, comprising one plant, one coral, six brachiopods (five specifically determined), two specifically indeterminate gastropods, one cephalopod and one ostracod.[4] It is disconformably succeeded by the richly fossiliferous Onondaga limestone, a formation primarily of coral reef origin. Where the Onondaga lies on the Akron, the latter shows an irregular erosion surface with channeling, and commonly a thin layer of intervening limestone-conglomerate with pebbles of the Akron rock, mingled with quartz grains, which at times may form a nearly pure quartz sand (Fig. 72).

But the most significant feature in this quarry is an ancient earthquake fissure, which cuts vertically through the Akron and enters the Bertie below, and which is filled by the same pure quartz-sand that is found at the contact (Fig. 73). Along the sides of the fissure the country rock (Akron) has been frequently broken off in fragments, and these have been caught in the sand and prevented from dropping to the bottom by the simultaneous injection of the sand, while even stringers and veinlets of the quartz-sand have been forcibly injected into the fragments (Fig. 74). At intervals horizontal films or rootlets of the sand have been forced between the strata, as shown in the illustration (Fig. 73).

Fig. 74. Fragment of limestone caught in the earthquake fissure, with sand injected into the limestone (white). Natural size. Cement quarry, Buffalo.

These features clearly indicate the violence of the force that injected this sand between the strata, which separated with every vibration. The distance to which the interstratal sand injections extend has been found to be 20 to 30 feet from the 'dike.'

Though only 2 feet in average width, this varies greatly from above downward, and the depth to which the fissure was exposed was over 10 feet. It begins

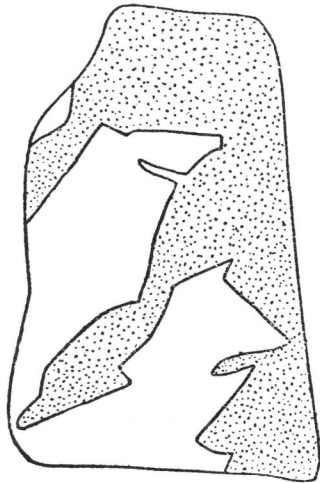

4) GRABAU : *Siluro-Devonic Contact in Erie County*, N. Y., Bull. Geol. Soc. of America, Vol. 11 ; 1900.

abruptly at the eroded top of the Akron dolomite, extends
entirely through it (7 feet), and for at least half that
depth into the waterlime. The 'dike' has been traced
for more than 30 ft, always cutting the country rock.

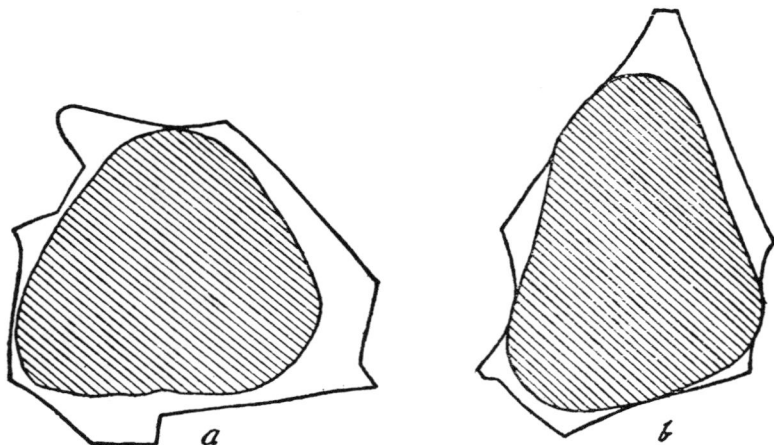

Fig. 75. Sections of sand grains from 'dike', showing the original rounded character of the grain and the secondary enlargement by crystalline silica,
which bound the grains into a solid rock. (Enlarged).

Microscopic examination of the quartz grains shows
them to be well rounded, though now enlarged by the
deposition of secondary silica (Fig. 75 a, b) in crystallographic
continuity with the old quartz.[5] The latter shows by its
inclusions and the outline of the original grains that it
was a wind worn sand, derived from old crystalline rock,
probably with many reworkings.[6] Where the sand grains
are embedded in a lime matrix they show no secondary
enlargements. The sand, of course, covered the surface
of the Akron dolomite as a loose deposit, probably of
aeolian origin, and the injection took place before the Onondaga limestone was deposited, for that is entirely unaffected. This is in fact the record of an earthquake during
the Oriskany interpulsation period, and of course affected
only the older rock, which shows many minute faults.

---

5) Shown by the fact that under crossed nicols they extinguish together.
6) Older pure sand deposits which may have furnished this quartz, are the
   Sylvania of the Silurian and the St. Peter of the Ordovician. For further
   details see GRABAU: 1900, *op. cit.* pp. 347-376.

After the earthquake, the sea was due to return, though there may have been considerable erosion of the older rock; a part of the sand was removed again from the surface, leaving only that which filled erosion depressions, or old river channels, in the surface. In some of these limestone pebbles were embedded in the sand.

4

With the return of the sea, there appeared a luxuriant development of coral polyps, and reefs were built such as had not existed in this region since Niagaran time. As usual the mounds of the Onondaga reefs developed on the marginal platform, while the geosyncline was subjected to more or less clastic deposits. The old reefs of *Favosites*, *Syringopora*, etc., with marginal interfingering of clastic lime-deposits, are known from many localities. Those of western New York and of Michigan are typical, while in the region of the modern Falls of the Ohio the Devonian reefs grew on the surface of a dead Silurian reef, so that it is impossible, except by an appeal to the fossils themselves, to determine the boundary-line between the two formations.

On the opposite side of the Niagara River in Canada (Ontario), the disconformity is equally well marked. Here the erosion surface sometimes cuts down to the waterlime, in which broad channels have been excavated. In Fig. 69 (p. 221), taken from Kindle's article, one of these is shown near Decewville, 20 feet deep and about 300 feet wide. This is filled with Oriskany sandstone, with its characteristic fossils, and covered by Onondaga limestone. As might be expected from the widely different characters of the rock, there is an abrupt change that suggests a mild hiatus,[7] but this again is probably of very minor importance, as in the case of the Manlius-Coeymans contact previously described. The faunal difference[8] is also explained as in

7) Ten miles farther west in Wallpole county, Stauffer reports: "No evidence of unconformity [disconformity] was discovered at the top of this sandstone, and certainly there are no masses of sandstone included in the limestone, although much sand is mingled with the calcareous sediments for 2 or 3 feet above the true (Oriskany) sandstone." (STAUFFER: *op. cit.* p. 375).

8) This was formerly thought to be a transition fauna and called "Decewville type of Oriskany," but as Stauffer has shown in 1901, the collections from the Oriskany and Onondaga were not kept separate by the early collectors.

the earlier case, by the meeting of two antipodal faunas, the Oriskany from the Caribbean (if not the Austral or Ross) epi-sea, and the Onondaga from the Boreal[9] epi-sea.

Within 200 feet of this section the Oriskany is reduced to a few inches, or is absent altogether, and the Onondaga lies directly but disconformably on the Bertie Waterlime.

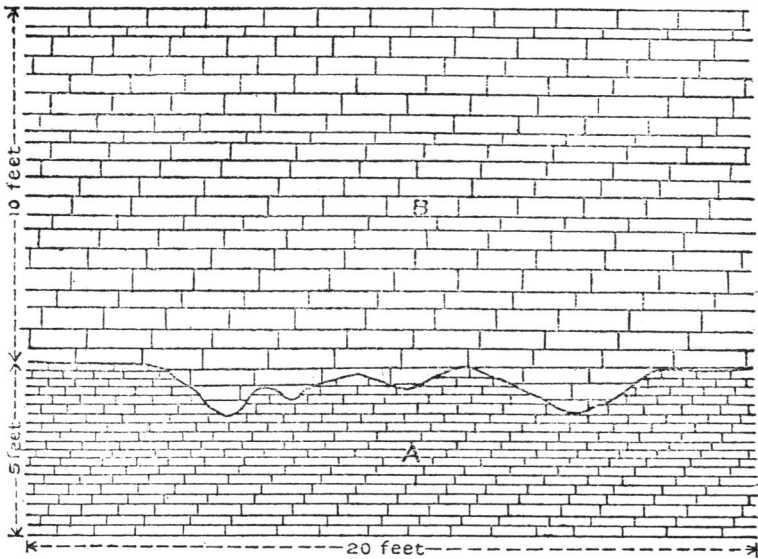

Fig. 76. Disconformable contact between the Silurian (Niagaran limestone A) and the Devonian (Jeffersonville limestone B) on the banks of the Wabash River, Georgetown, Ind. The Salina interpulsation, the Siluronian pulsation and the Oriskanian interpulsation periods are all included in the disconformity. (After Kindle)

In central Ohio the Onondaga is known as Columbus limestone, and in Indiana as the Jefferson limestone, both showing by their fossils that they belong to the same marine transgression as the Onondaga of New York. The former rests upon an erosion surface of the Monroe formation (Lower Siluronian) and begins with a conglomerate, the pebbles of which are worn pieces of the underlying Manlius limestone. In Indiana the older formations, deposited in Siluronian time, were cut down again

---

9) Austral and Boreal here refer to location, not to climate.

to the Niagaran, on which the Onondaga (Jefferson) limestone rests directly. (Fig. 76)

Of course this boundary line is really representative of a hiatus of long duration. The whole Siluronian pulsation period, with its two flanking interpulsation periods, the Salinan below and the Oriskanian above, are all absent in the Louisville reef area, and yet no one except a student of fossil succession would suspect that here is represented a lost interval, covering untold thouands, perhaps millions, of years.

No one can, of course, blame the earlier investigators of these regions for concluding that here was continuous coral growth throughout Silurian and Devonian time, and for using this as a proof that at that region at least, the sea lingered continuously from the early Silurian to the Middle Devonian.

5

In the Appalachian geosyncline the Oriskany sands, after being well sorted by the wind during the absence of the sea, which, in the interpulsation period, had withdrawn to the Caribbean epi-sea, were reworked and rendered fossiliferous by the returning waters of Devonian time. That these waters transgressed from the region of the Caribbean epi-sea is shown by the presence of the Oriskany fauna in the section on the Rio Mæcuru, Para province, Brazil, and in South Africa and the Tassili Mountains of West Africa, all of which localities the sea could reach from the Caribbean epi-sea as a center. At the same time this fauna migrated into the Appalachian geosyncline, where it is typically developed.

The section on the Rio Mæcuru is instructive, as it shows the beds with the Oriskany fauna lying unconformably upon tilted and eroded Silurian strata. The unrepresented interval here corresponds to the Siluronian. The higher beds carry a Hamilton fauna.

The transgressing sea entered the geosyncline underlain by Siluronian limestones, partly of shell rock (Helderbergian) and partly of the old Silurian lime-muds produced by disintegration, as we have seen, during the Salina interval. This had again suffered disintegration during the

Oriskany interpulsation period, when the quartz grains from exposed quartz rocks were mingled with the residual lime-mud to form a mixed rock of lime-mud and quartz-sand.

In Maryland and Pennsylvania, this mixed material reaches a thickness of over 400 feet. The transgressing Devonian sea used this for its first material to shape into stratified beds and to fill with the dead shells of the animals which were the first southern immigrants that came with the invading waters.

Where uncovered again in the Appalachians by erosion, the process of dissociation of the sand and lime constituents by solution of the latter has frequently been observed. In such a process the residual material becomes again a more or less pure sand, from which the fossil shells are readily separable. In some localities these shells have been secondarily silicified while embedded in the rock, and these are found, on weathering out of the matrix, to retain in lasting preservation the form of the shell, though not its finer structure.

In the Maryland region the Oriskany fauna comprises the number of species shown in the following table, those occurring in lower and higher horizons being indicated.

TABLE IV
Oriskany species from Maryland

Phylum	Total number	Also below		Also above		Both below and above	
		number	percent	number	percent	number	percent
Corals	4	1	25.00	—	—	1	25.00
Brachiopods	67	14	20.90	5	7.46	4	5.97
Pelecypods	10	3	33.33				
Gastropods	20	7	2.86				
Pteropods	3	3	100.00				
Trilobites	10	3	33.33	1	10.00		
Ostracods	10	—	—	1	10.00		
Total	124	31	25.00	7	5.65	5	4.04

Out of a total of 124 species 43, or 34.68%, occur either above or below or both, but only 31 or 25% occur only in lower strata, and these are either hold-overs in the epi-sea from which the fauna came, or were deposited in the sands of the retreating seas the last of the Helderbergian survivors.

It was probably the long exposure of this, the last of the Siluronian formations, that favored first, the silicification of the calcareous organic remains which belonged to the retreatal division, by percolating waters rich in silica, and thereafter the gradual removal of the lime from the formation by solution and the consequent setting free of the silicified fossils and sand grains.

If we were to search the Palæozoic formations of the Appalachian geosyncline, it would be hard to find another formation of exactly this type, for none other had precisely this combination of insoluble and soluble material while being fated at the same time to pass through such a cycle of change, because of exposure during a whole interpulsation period to the effects of percolating waters in regions of tropical climate. For the region lay, throughout, near the equator of the time. There were other regions of the pangæa, where the last retreatal deposits of a pulsation system remained exposed for the whole of the succeeding interpulsation period, yet did not suffer the same alteration. This was due in the first place to the more uniform composition of the strata. The last of the series in the Caledonian geosyncline were the Downton sands, already purely continental and largely oxidized sands, so that subaërial exposure could produce little change, except dehydration of the iron oxide. This probably resulted in a change of color from what was presumably the ochery color of loess to red. [10]

Other terminal formations, representing retreatal phases of either earlier or later pulsation periods, were primarily shales or limestone, and though the fossils of the limestones sometimes were silicified where these represented the surface rocks exposed, as in the Niagaran on the marginal platform of the southern states (Kentucky, etc.), there were no sand grains to be left behind on solution.

---

10) This may, however, have been the effect of subsequent dehydration after deep burial.

It is interesting to note that the fauna of this period had a very wide distribution, being found in North Africa, in South America, in South Africa and above all in Tasmania. These regions were part of the "Samfrau" geosyncline (Du Toit) and its branches of that time, the 'feeder' of which was the Ross epi-sea in part (See Plate VIII), but more especially the Caribbean. This raises the question respecting the source of the fauna, a part of which may well have been of Austral origin.

There are some almost unique features about the distribution of these Oriskany sands. Though as a rule of slight thickness in New York, the formation reaches the astonishing thickness of 2,880 feet in the Moose Lake region of Maine. It overlaps the Helderbergian series of formations, which are found so well developed in the Gaspé Region nearer the St. Lawrence River. These beds are absent in northwest Maine, and the Oriskany lies directly on the gneiss of the Appalachian old-land.

6

In Eastern New York, east of the Hudson River, there is another intriguing deposit, which has been handed back and forth among the geological periods between Silurian and uppermost Devonian, seeking a period where it can remain at rest, and finding none. Indeed some recent writers would dispossess it of all rights to a post-Cambrian age.

This is the RENSSELÆR GRIT, a purely continental formation, 1,400 ft thick, which rests on the folded and peneplaned rocks that were involved in the Sil-oric orogeny of the Taconic Mountains. This relation precludes the possibility of a Silurian age for the grit, for that would leave no time for peneplanation. On the other hand, the purely continental character of the Rensselær grit, which contains pebbles of coarse and fine gneiss from the Appalachian old-land, and pebbles of Taconian (Lower Cambrian) from the Taconic Mts., and the alternation of such pebble beds with red and greenish slates, negatives the supposition that it may be a product of the Siluronian transgression, and certifies its membership in the group of deposits characteristic of a post-Siluronian interpulsation

period.   This purely continental deposit may have formed
during the absence of the sea in the Silurono-Devonian
interpulsation period.   Since, however, there is no means
by which we can ascertain its relationship to the Siluronian
on the one hand, and the Devonian on the other, it
remains unavailable as a representative of that interpulsation period.

CHAPTER XXI

THE DEVONIAN PULSATION OF THE APPALACHIAN
AND ST. LAWRENCE GEOSYNCLINES

1

IN seeking for a natural delimitation of the Devonian, and
one based on the phenomena of pulsation, that is of
general marine transgression, followed by equally wide-spread
sea-retreat, we find it exemplified in the formations which
remain after the separation of the beds previously classed
as Lower Devonian. These pre-Devonian formations in the
Appalachian and St. Lawrence geosynclines comprise the
Helderbergian series, which terminates with the Oriskany
retreat. This series is separated from the basal transgressing portion of the restricted Devonian by a long interpulsation period, which is characterized by erosion in some
sections (Fig. 76, p. 228), by continental deposits in others,
and by volcanic activity in still others (Fig. 68, p. 219).
With the recognition of the existence of this interpulsation
period, and the separation by it of these old "Lower"
Devonian rocks, as a part of a distinct earlier pulsation
system, we find that what remains constitutes a single
and complete pulsation system. To this the name DEVONIAN
PULSATION SYSTEM is here restricted.

When the sea returned to the Appalachian geosyncline,
transgressing from the region of the Caribbean epi-sea,
it spread over the marginal platform of the Southern States
and, in Kentucky and Tennessee as well as in southern Ohio,
covered the previously exposed Niagaran limestone, the
corals and other fossils of which were mostly silicified.
The waters were pure, since no mud from Appalachia
reached this region; corals once more began to flourish
and, as had happened at an earlier date in Sweden,
the later corals settled upon the projecting mounds of the
older reefs, from which all residual sediment, formed

during the long interval of exposure, had been swept. Thus Onondaga corals came to grow upon an ancient Silurian reef, and united with that as its basement-rock to such an extent that none but the palæontologist can draw the line between them.

The silicified corals of the "Falls of the Ohio" are known wherever Palæozoic fossils are studied; because of their silicification they weather out of the enclosing matrix in great perfection of form, whatever may be said of their structure. Both Silurian (Niagaran) and Devonian (Onondaga) corals, the latter silicified probably during the later post-Devonian exposure, are here found weathered out, side by side, and fine specimens of both were once available to the zealous collector. Those days are past, but their memory lingers.

Fig. 77. Disconformable contact of the Manlius limestone, with *Stromatopora* (Siluronian) below, and the Onondaga limestone (Devonian) above. The Oriskany interpulsation horizon is absent, but marked by the dark contact zone. Tunnel Creek, Union Springs, N. Y. (After photograph by G. D. Harris).

2

From the compound reef region of Ohio, the waters spread northward, but here they encountered rocks of Siluronian age ; *i.e.* the Monroe formation, including the

Sylvania sands, and conditions were less favorable for reef building. Instead many strata of the early Onondaga, or Columbus, or Dundee, limestone, as the beds are variously known (each State wishing to commemorate one of its localities), consist more commonly of bedded limestone, formed in part from the reworked lime-debris of the disintegrated older limestones. (Fig. 80, p. 242).

These deposits contain only scattered fossils, though some portions of the beds may be made up of dissociated parts, or of ground-up masses of calcareous organisms of the reefs. In the Buffalo region, however, and especially at Williamsville, these limestones again present the aspect of well developed coral reefs, *i. e.* massive mounds of coral-heads, sometimes 30 ft in diameter, growing side by side, the newer above the older. The bedded layers on their flanks interfinger with the coral fringe, and dip away steeply at first, then change to normal horizontal layers, until they reach the next mound, where the same inter-fingering and normal steep dip is repeated. Usually each quarry is opened in a single coral mound (except the larger quarries), and this coral rock furnishes the purest of lime. Thousands of good specimens of *Favosites, Prismatophyllum*, stromatoporoids and other compound forms, and many beautiful single zaphrentoid corals, from these reefs, including among them such giants as *Zaphrentis gigantea* which may reach a length of several feet, and a diameter of several inches,—all these beautiful potential museum specimens (not to mention brachiopods, trilobites and other forms) are fed to the fiery jaws of the limekiln moloch, and when they come through the ordeal they are bleached and crumbling masses of lime, only fit for use as plaster or whitewash.

3

Nearer the geosyncline of the period the limestone loses its purity and becomes interfingered with the muds, that the rivers have brought to the geosyncline from Appalachia. For throughout Devonian time the geosyncline knew little, if any, respite from the constant supply of muddy and sandy sediments brought by the rivers from Appalachia. It was a race between the sinking of the geosyncline and the filling efforts of the rivers, and the latter won in the end. At no time was the water deep enough and the shore far enough removed to permit organic sediments to predominate. Shell-heaps there were, but only such as are seen today on sandy beaches. They made thin limestone streaks in the final record of sedimentation, unless they were ground to pieces in the ceaseless ebb and flow of the tide, or the occasional pounding of the waves. When corals, or stromatoporoids, essayed to build up a small colony of pure organisms, an oasis in the waste of sand and mud-flats probably exposed at every ebb-tide, they soon succumbed to the discouraging environment and gave up the attempt.

There were a few huangho deposits among these sediments, notably, at the beginning, the Marcellus bituminous mudrock. This is characterized by scattered brachiopod shells cast up during periods of marining, and by layers of floating pteropods (*Tentaculites*), which took the place that the now vanished graptolites had held in the grudging hospitality of the older huangho deposits. Only occasionally, and that chiefly in the later part of the series (Genesee), were the more distant submerged parts of the platform subject to freshets, so that the planktonic animals were destroyed in such numbers that their little shells made limestones of appreciable thickness. Such is the *Styliolina* limestone which is exposed in the sections of Eighteen Mile Creek, and there forms a projecting layer in the cliffs on the shore of Lake Erie. It is only four to six inches in thickness, but it is almost wholly composed of the fine, needle-like shells of a pteropod *Styliolina fissurella*, which is so small, that only a sharp eye can detect it on the weathered surface of the rock, while on a fresh fracture only crystalline or compact lime is visible. But in thin sections under the microscope this rock is seen to be almost wholly made of *Styliolina* shells. It has been estimated that a cubic inch of this rock contains 40,000 individual shells, and when it is realized, that this limestone bed has been traced in outcrop for at least 60 miles east and west, and that the present north-south extent is only a fraction of the original, which was parallel to the margin of the platform, it is seen that a staggering holocaust must have taken place here. (Fig. 78 b, c.)

4

Why such millions upon millions of floating organisms should have been abruptly destroyed, so that they formed a pure accumulation of needle-like shells, is a question not easy to answer. Did an earthquake shock destroy them, or were they drifted by the tide and waves into the broad open estuary of one of the streams from the old land, which during an unwonted freshet brought such an abundance of fresh water into the sea as to be destructive to its floating population?

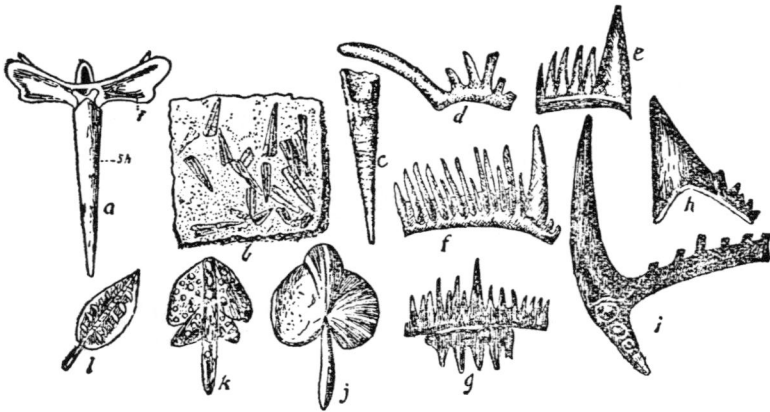

Fig. 78. *Styliola, Styliolina* and *Conodonts*. (After Hinde and others).
a. *Styliola recta* modern species; animal with shell and with wing-like foot expanded (enlarged); b-c. *Styliolina fissurella*; b. A fragment of slate with numerous individuals $\times$ 2.5; c. A shell much enlarged.
d-l: *Conodonts* and associated plates : d. *Prioniodus erraticus*; e. *P. armatus*; f. *P. acicularis*; g. *Polygnathus dubius*; h. *Prioniodus* (?) *alatus* $\times$ 5.6; i. *P. panderi*; j, k. *Polygnathus tuberculatus*; l. *P. pennatus*.
All except h enlarged $\times$ 10.4.

There is some corroborative evidence which supports this theory, for remains of land-plants are not uncommon in some part of this limestone. These are the trunks and branches of coniferous trees (*Dadoxylon* etc.) which grew on she wooded slopes of Appalachia, and which, uprooted by tuch freshets, were swept into the geosyncline and floated to the quieter western part, where they settled with the needle point shells of the plankton, already killed by this freshet. The wood of these trees is often so beautifully preserved that even after these millions of centuries its cell structure is shown in the greatest prefection, though all is replaced by calcite or pyrite. Among the specimens of wood some have been found which, in the words of Sir William Dawson, who named these trees *Syringoxylon mirabile*, "indicated existence in the Devonian period of trees of a higher grade than any that are known in the Carboniferous system". This was written in 1862; perhaps types equally advanced have been found since in these higher strata.

These petrified pieces of wood which can be had for the trouble of picking them out of the rock by any visitor to this region, accessible from Buffalo, are supplemented by the occasional occurrence of the bony armor of Devonian river fish (plates and jaws of placoderms, spines of river sharks and scales of ganoids) — a further illustration of the temporary advent of the river-current in this part of the sea. And with them were preserved the "annelid jaws", or conodonts, which have given the Eighteen Mile Creek rock international notoriety.

The English geologist, G. J. Hinde,[1] discovered them in this region and described some of them. Zittel and Rohon[2] also discussed them in 1886, and pronounced them the horny jaws of annelids, and zealous local students have since made extensive collections of these intriguing little objects. (Fig. 78 d-l).

5

Meanwhile, as the geosyncline was sinking, the river-sediments were laid down in the shallow waters, and wherever the sediments permitted it, the littoral fauna of the Devonian sea was included. Thus, the Hamilton and Early Portage rocks of New York and Pennsylvania have become the favorite collecting ground for American students of fossils, and more American palæontologists were probably trained in the field among these prolific sections than in any other environment, not even, perhaps, excepting the wonderfully prolific but rather uniform sections around Cincinnati.

---

1) HINDE : *On Conodonts.* etc., pp. 351 *et seq.*
2) ZITTEL and ROHON : *Ueber Conodonten.*

But, the full interpretation of the real significance of these strata as great river deltas, washed by streams from Appalachia, into the shaoling Devonian sea, remained for the younger generation to demonstrate, and they have accomplished this in a most satisfactory manner. [3]   It was only after laboriously drawing section after section, often in regions accessible with difficulty, measuring the thickness of individual beds, collecting and studying the fossils preserved in them, and then painstakingly joining section to section, matching beds of one with those of the neighboring section, and carrying this work over the whole of the states of New York, Pennsylvania, Maryland and the regions beyond ; noting the change in type of sediments along lines extending from the Appalachian old-land westward to the purer water ; and discussing and arguing the significance in the scientific journals in hundreds of pages, that make the dullest of reading except to the initiated, — that it became possible to establish the final outline of the history of the American Devonian, and to sketch the picture of the lands and rivers, the seas and their life, and the succession of events which combined to make American Devonian history (See Figs. 79-81).

And this is the story in outline.   From Appalachia, the old-land mass that separated the Appalachian and Caledonian geosynclines in the days when pangæa was still intact and there was no Atlantic Ocean for even Palæozoic voyagers to sail, there stretched away, into what we now call the west, a broad erosion plane.   This was cut near the old shore from the folds of the old Taconic Mountains, which by this time had become wholly bevelled across by a peneplane.   From the products of this early erosion the clastic beds of the Silurian and later periods were formed, as already outlined in preceding chapters.   Twice the sea returned after the folding, and each time it retreated again, and remained away during a long interpulsation period.   During its first return, after the birth of the Taconic Mountains, the Silurian sediments accumulated.   After its retreat the Salina desert developed, where Niagaran coral reefs had grown.   During its next return, the sea found much

---

3) See the papers by CASTER and others cited in the bibliography.

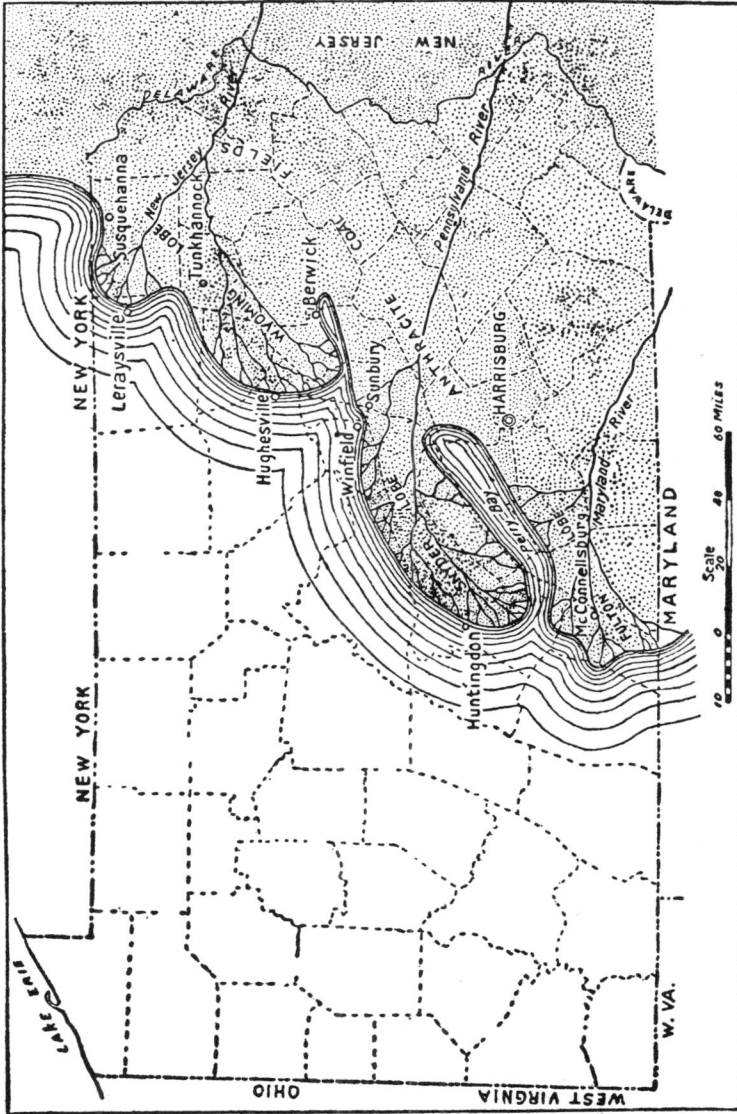

Fig. 79. Map of the Upper Devonian (Chemung) Delta, in New Jersey, Pennsylvania, and its extension into Maryland on the south, and New York on the north. (After B. Willard).

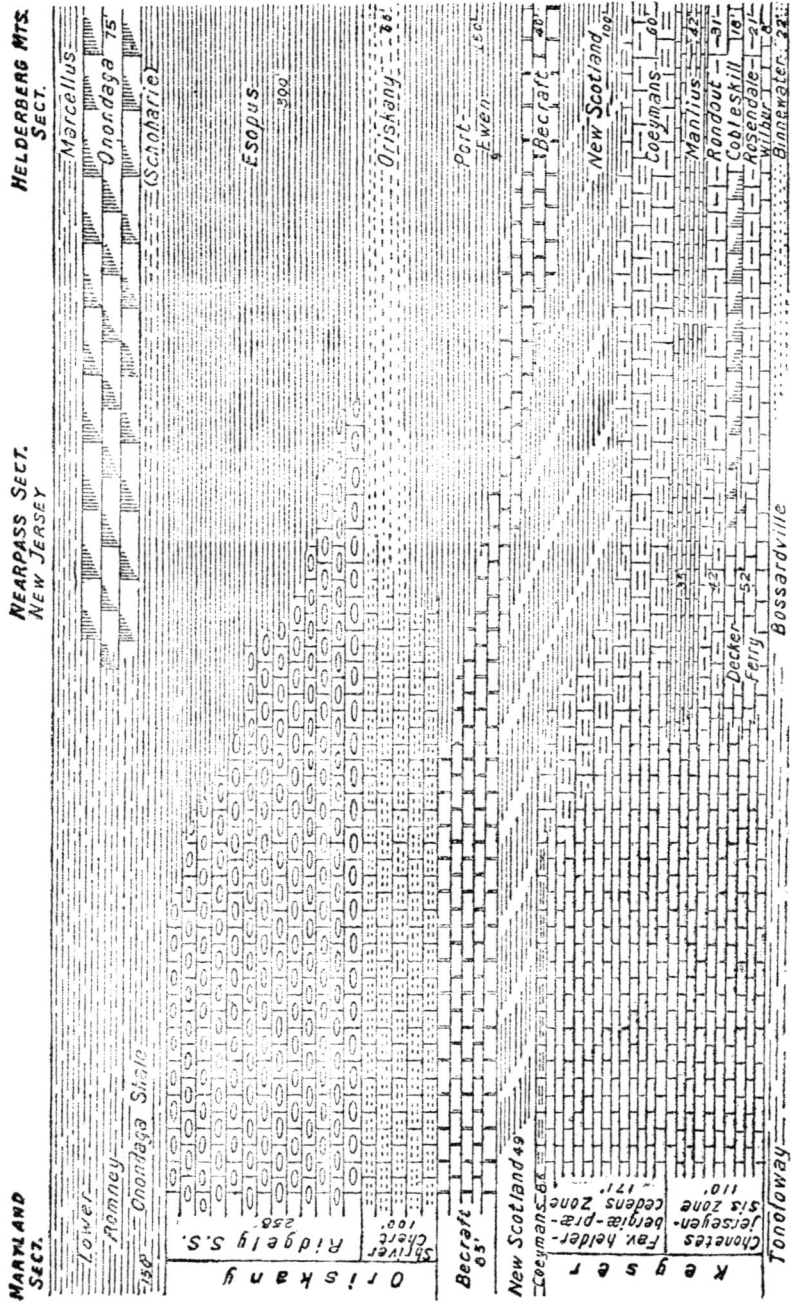

Fig. 80. Diagram to illustrate the interrelations of the strata of the Siluronian pulsation period, the Oriskanian inter-pulsation period, and the Devonian pulsation period, from the Helderberg Mts, through New Jersey to Maryland.

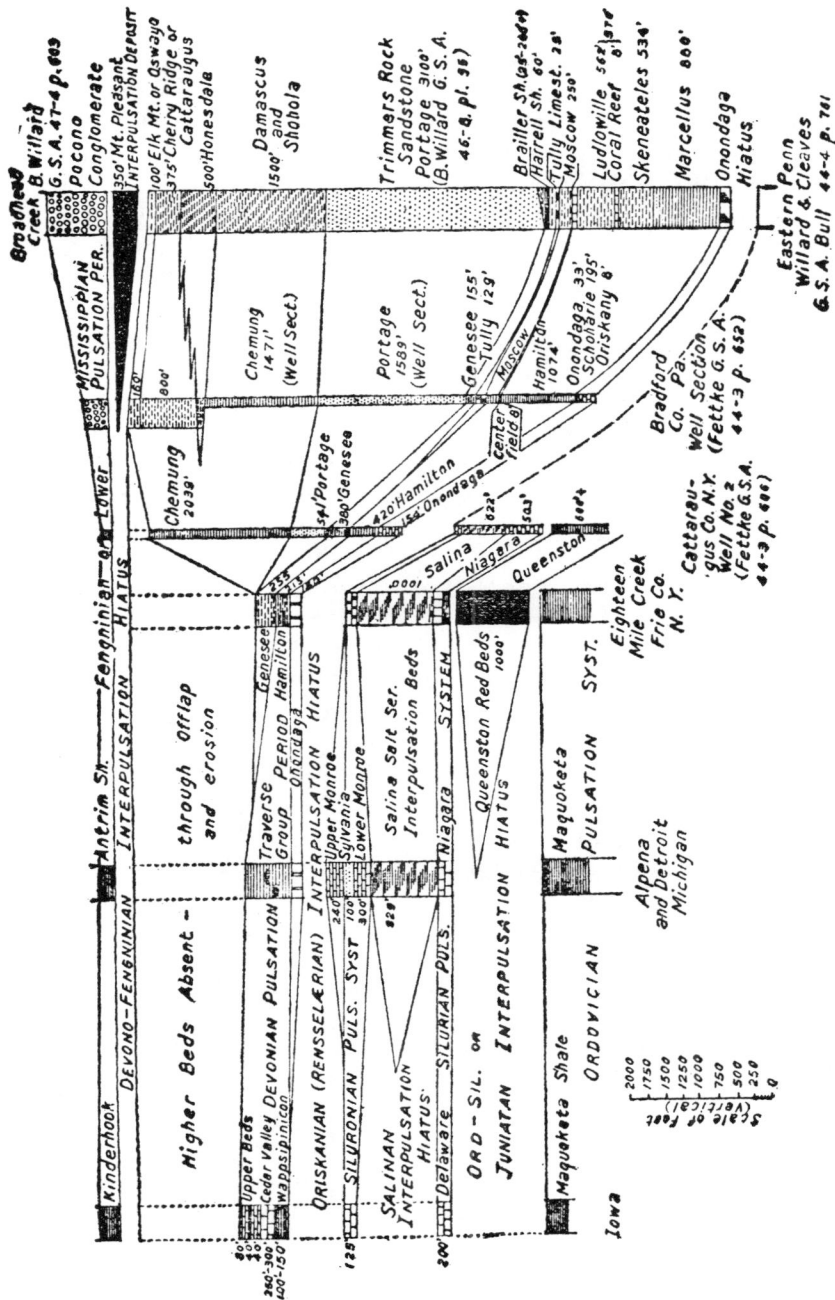

Fig. 81. Columnar sections, showing thicknesses and relationships of the beds from the Ordovician to the Mississippian, their alternations and separating intervals. (Compiled from sources indicated.)

loess and other dust deposits to arrange into stratified beds, and there was no life, such as existed here before the salt desert, until the St. Lawrence sea transgressed over the Albany Axis and brought in the Helderbergian fauna. Again the sea left the geosyncline, and the interpulsation Oriskany sands were spread after its retreat. Then, at the end of this long interpulsation period, the Devonian sea gradually returned to the geosyncline and spread westard over the old platform, which had constantly remained as overflow basin, whenever the sediments from Appalachia prevented the waters in the geosyncline from becoming more than of moderate depth. Near the old-land (on the east p. d.) the deposits were coarser sands and pebbles, but these have long since been swept away again during later cycles of erosion, together with the beds that once covered them in the succeeding periods. But even in eastern New York, in New Jersey and in eastern Pennsylvania, many of these sediments were received by the sea, which then washed the shores of the Appalachian old-land on what is now its western side. But this sea-shore was slowly pushed outward by the growing river plain; huangho muds were formed, which suffered periodic but temporary submergence and are now represented by such deposits as the black Marcellus shales, the Romney shales, and after an interval, by the Genesee and black "Portage" shales. All of these occasionally suffered temporary flooding, or marining, when the floating population of the sea, the pteropods and goniatites, (holoplankton), and the shells of dead brachiopods (pseudoplankton) were spread over the temporary submerged surface where they were buried by the next layer of sediment washed outward by the streams.

Repeatedly the sediments were so voluminous that, in spite of the constant sinking of the geosyncline, the sea-shore was pushed outward, and pure river sands and muds, often with entombed land plants, were spread out between the sea-margin and the old-land. At such times, the waters would be forced to spread over the platform, and though from the first this was submerged, and had become the site of coral reef development, the constant rise of the sea in the geosyncline and the constant urge of the outward pressing shore of the river-plain involved progressive encroachment.

The early lime deposits accompanying the first of the Devonian reefs of western New York came to rest on the old erosion surface which was cut on the Silurian strata. These coral reefs, the first of the Devonian period, began to grow and flourish for a time, while the lime sediments extended beyond the reef area in diminishing thickness for a distance of several hundred miles. Their final edge now lies buried somewhere deep under the later coal beds of Michigan.

6

Then came a period of retreat. The huangho muds had advanced as far as Buffalo (Marcellus shale), while the lime sediments still accumulated beyond; as the muds advanced they killed the coral polyps of the western New York reefs, and sometimes filled their calices with mud, as can frequently be seen. However, a more rapid sinking of the geosyncline drew the waters back over the mud-flats, and the early Hamilton brachiopods and trilobites flourished where formerly the surface of the delta-plain had presented the environment of death and burial, because the waters did not remain there permanently.

But what proved to be rejuvenation for the submerged geosynclines of New York and Pennsylvania, which now supported an abundant shallow water fauna, became disastrous for the marginal platform of the geosyncline in Michigan, Ohio and other States of to-day, for this deepening of the geosyncline drew the waters from the platform, and the reef-builders and other organisms perished by exposure, as their immediate predecessors of the Buffalo reefs had perished by an influx of black Marcellus muds.

For some parts of the platform the emergence proved only temporary, though long enough to put an end to reef-building. Such was the fate of the reefs in eastern Michian. Farther away, in parts of Ohio, the exposure was longer, and here the surface-waters brought silica, which replaced the lime and silicified the corals.

In Michigan the returning sea, however, revivified the organic world by giving it room to expand, and once more pure waters permitted coral-reef formations. But this time most of the corals were new and of different

types.   The Hamilton or Traverse reefs of Alpena and
Petosky are examples, [4] and this time the platform was
submerged to a greater extent, so that Wisconsin, Illinois
and Iowa also received a share of the calcareous sediments.
Those of the former state are only lime-muds (Milwaukee
cement stone), but those of Iowa and Indiana presented
a more normal development, with purer water faunas.

In Iowa, 100-150 feet of carbonaceous shale, with
bands of impure limestone (Wappsipinicon formation), are
followed by 250-300 ft of pure limestone (Cedar Valley
beds), with an abundant fauna.   Although these beds are
the lowest Devonian strata of Iowa, resting with a pronounced disconformity on the Ordovician beds (Maquoketa
shales), they are of late Hamilton age, all the earlier
horizons being overlapped by the slowly transgressing sea.
Moreover such Siluronian and Silurian beds as had been
previously deposited in this region were largely eroded
again during the long period of exposure.

But again the influence of the shoaling sea, because
of withdrawal of the waters into the geosyncline, is seen
in the abrupt succession of black shales (Sweetland Creek
shales) upon the fossiliferous limestones; and the abrupt
change in sediment, coupled with the extinction of the
older forms of life, indicates a period of local emergence,
before the shallow sea, in which these muds were laid,
covered the region again.

7

It is not possible to make detailed correlations between
the succession of events in the geosyncline and those in
the marginal platform, but this much may be taken for
granted: that an emergence on the platform corresponds to a
deepening of the geosyncline, and conversely, a submergence
and purer water conditions on the platform indicate rapid
advance of the continental sediments in the geosyncline.
The waters were constantly rising, but they could not
keep the geosyncline submerged, and so had to spread
out over the platform.

Applying this principle, we may consider that the

4) These lie some miles north of our section line, but that is only a minor
   detail.

change from conditions favoring the formation of the fossiliferous Cedar Valley limestone to those indicating practical
emergence followed by black muds, may correspond to the
deepening of the water in the geosyncline, which permitted
the formation of a fossiliferous limestone (Tully limestone),
in the region, where sediments previously indicated shallower
water, while the succeeding beds of limestone and fossiliferous muds of Iowa mark the advance of the river-plain
in the geosyncline, the *Styliolina* bed, above referred to,
being a result of the destructive effect of the river-water
on the marine plankton.   At the same time, the advancing
river sediments forced the purer-water organisms to spread
again in the region of the platform.[5]

How far the sea extended at the period of the acme
of the pulsation, that is, its maximum rise, has not been
ascertained, but when the reversal came, the falling sea-
level resulted in a rapid drawing off of the water from
the platform, and an exposure of the previously formed
rocks to erosion.   Only in the geosyncline is the gradual
conquest of the river-plain over the marine beds indicated, but even this was an oscillatory one.   At the
last, however, the river-borne sands conquered, and over
a large part of New York and Pennsylvania were spread
the Catskill beds now of red color, because the iron-
oxide, which originally stained these sands with an ochery
color, has lost its water of combination during the long
period that has elapsed since then, and red rocks, the
American Old Red Sandstone, were the result.

These are the rocks of the Catskill Mountains, where
Rip Van Winkle slept for 20 years, as told by Washington
Irving, and portrayed so delightfully by Joseph Jefferson
on the stage of half a century ago.   These rocks had
their counter-part far away, on what is now the same
side of the severed part of old Appalachia, but separated from
the American part by the whole width of the Atlantic
Ocean.   This is in North Scotland, the home of Hugh Miller,
the quarryman.   No ordinary workman was he, but an
observer and a thinker as well.   And to this we owe

---

5) These correlations of beds must not be taken too literally, as there were,
no doubt, oscillations and local warpings of the platform surface, and
other factors, which coöperated with the still rising sea-level. But the
principle holds as a general guide to the sequence of events.

his delightful work, "The Old Red Sandstone, or New Walks in an Old Field", a book that still repays reading, even at this time.   Hugh Miller, under the influence of the elder Agassiz speculated about the fish remains in the Old Red Sandstone, and correctly held that they were fresh water fish.   But under the influence of the orthodox ideas on sedimentation prevalent in his day, he attributed these sites of accumulation of the sediments to a series of fresh water lakes.   That mistake was inevitable, for no river-plains and deltas comparable in size exist to-day in the British Isles.   Had Hugh Miller visited the Western United States or the Great Plain of North India, or had he come to the Huangho River delta of China, he might perhaps have formed a more correct diagnosis of the history of these rocks, and the habits of its extinct fish.

But that error does not matter much, it was an inevitable preliminary step to our present understanding of these rocks.   The old homestead of Hugh Miller at Cromarthy on the Moray Firth, stands as a memorial to the man who, occupied with daily toil in the earning of his bread, had yet time and enthusiasm enough to study and to record in delightfully simple language, his wandering among the *Vestiges of Creation.*

CHAPTER XXII

POST-DEVONIAN PROGRESS

1

THE progressive withdrawal of the sea from the Appalachian geosyncline at the end of Upper Devonian time
induced the type of sedimentation known as Catskill in
America, and as Old Red Sandstone in the north of
Europe. Both are deposits of oxidized sands, formed under
more or less semi-arid conditions, and both are now represented by red beds, because with further ageing the
hydrous oxide of iron, which stained the sands at first,
perhaps, an ochery color, had upon dehydration, changed
to red.[1] Continued withdrawal of the sea, due to the
sinking sea-level (negative pulsation), ended with complete emergence of the geosyncline, and since the country
now stood above sea-level, such rivers as existed continued
to deposit continental strata in the geosyncline, while on
the marginal platform they turned their energies towards
erosion. As a result, it is probable that a considerable
portion of the previously deposited sediments was removed
again.

We do not know how much erosion was accomplished
during the interpulsation period of exposure, because we
do not know the extent to which the platform was covered by Devonian strata, but that the amount was considerable scarcely admits of doubt. The interval between the
top of the Middle Devonian Traverse group and the succeeding Antrim black shale of Michigan, which is shown in
Fig. 81 (p. 243), may not be the true measure of interpulsation erosion; a part of the later Upper Devonian may be
absent as the result of off-lap and non-deposition in the retreating Upper Devonian sea. So, too, the widening of this
interval, between the eastern parts of the states of Michigan

---

1) The process is illustrated in brick-making, red bricks resulting from yellow
   clay, unless substances are added to prevent the chemical change.

and of Wisconsin, was perhaps, in part due to non-deposition of the later strata in the Milwaukee region, because of sea-retreat.   Considerable erosion, however, certainly took place, thus bevelling the strata of the retreating depositional series.

But when we examine the country along a north-south section (Fig. 82), the evidence for post-Devonian erosion becomes pronounced, for we note that the hiatus increases southward, where presumably in the Cincinnati and central Tennessee dome-regions there had been pure water with coral growth in Onondaga time, and this presupposes at least a considerable distance from the original shore.

On the western flanks of the Cincinnati dome the Silurian is directly overlain by a thin bed of white Devonian limestone, from 12 feet in the west, to 3 feet in the most eastern outcrop.   It contains brachiopods and blastoids, characteristic of Onondaga or of early Hamilton time.   This limestone is succeeded by a darker, more sandy limestone, while elsewhere some thin coral limestone is also recorded (Foerste).

At Louisville, Kentucky, and elsewhere, the Onondaga reef-limestone lies directly and disconformably on the Louisville limestone of Niagaran age.   The whole Silurian of Indiana is not over 150 feet thick and often thinner. It rests with a pronounced disconformity on Ordovician beds.   Siluronian is absent, except locally, where it may reach a hundred feet in thickness (Linden limestone), and is followed by representatives of the Oriskany (Camden chert, etc.).   The Devonian is everywhere disconformably overlain by the Chattanooga black shale of Fengninian (Lower Mississippian) age.

As a result of the post-Devonian erosion, the newer formations that had been deposited, were cut away again in Ohio, Kentucky, and Tennessee, until in these regions the erosion plane had reached and bevelled across the Siluronian (Monroan) limestone, and in some parts had even exposed the Silurian (Niagaran) limestones beneath. Over this erosion plane was left a residual soil, produced in part from the weathering of the underlying limestone, and this product of weathering has long been known in local stratigraphy as the Olentangy shale.   The name is

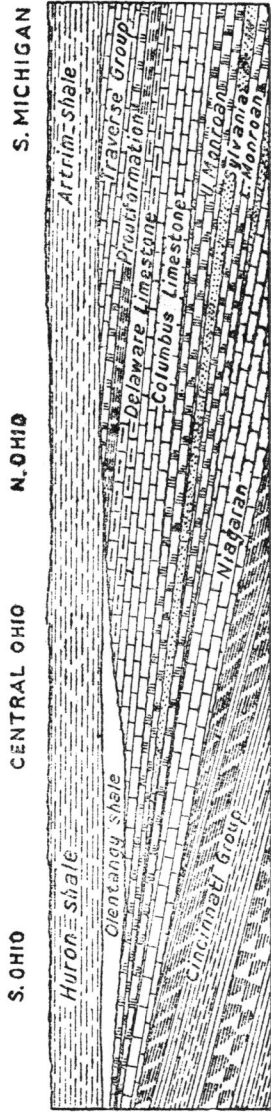

Fig. 82. Generalized north-south section in Michigan and Ohio, to show the original relationship of the formations from the Cincinnati group (Ordovician) to the Black Shales. The interpulsation periods are represented by the post-Cincinnati erosion; the Siluronian (Upper Monroan) Devonian (Columbus) erosion and the post-Traverse pre-Antrim erosion and residual Olentangy shale.

The Olentangy shale an interpulsation residual deposit must not to be confused with the Prout fossiliferous shale as was formerly done. The latter occupies a position between the Delaware limestone and either the Upper Traverse group or the Huron shale, as shown in the section and may seem to be the continuation of the Olentangy shale, but is really of Lower Traverse age. (For details see GRABAU: *Olentangy Shale*, etc., 1917 pp. 557 *et seq.*

from the outcrops on the Olentangy river in central Ohio. This is a light-gray soapy shale, a typical residual clay, the bedding of which is marked chiefly by the occurrence of thin bands of dark shale in the upper part, and by more or less continuous layers of concretions, some of them rich in iron pyrites, others calcareous, with fish scales as the only remains.

This residual clay is evidently the end-product, formed during a period of exposure of considerable length, and before the overlying black Huron shale was deposited. It is essentially an interpulsation residual product of subaërial weathering, covered by the Black (Huron) shale, which elsewhere in northern Ohio lies on the eroded surfaces of beds of Hamilton (Middle Devonian) age, on the Prout limestone of Ohio and on the Traverse limestone, etc., in Michigan. The relationship is shown in the north-south section, on the proceding page. (Fig. 82.)

2

The overlying black shale is the Huron of Ohio, and the Antrim of Michigan, and while neither is typical-ly marine, each represents renewed river-activity, and the spreading of a huangho type of deposit. They have generally been referred to the Upper Devonian, but they are chiefly non-marine, and rest disconformably on a pronounced erosion, or solution surface, cut on typical marine Devonian (Middle), or older formations. This erosion interval is shown to be of interpulsation value by its wide-spread occurrence, i. e., wherever Devonian and Lower Mississippian (Fengninian) strata are in contact. These black shales most likely represent residual deposits of interpulsation age, more or less reworked by the transgressing Fengninian (Tournaisian) sea.

The Ohio shale is the repository of the remarkable "Devonian" fish remains (*Dinychthys*, etc.) which have made this rock famous in the annals of palæontology. These fish may have been inhabitants of rivers and of lakes, but this is not the generally accepted interpretation of their habitat. Their great size is generally taken to imply that they were inhabitants of the sea, but if that was the case I see no escape from the logical conclusion that they belong to the readvancing sea, and that would make them Lower Fengninian (early Lower Mississippian) in age.   To regard these shales as Upper Devonian marine deposits, takes no account of the long period of erosion which preceded their deposition.

It must be emphasized that, although the black Devonian shales of New York and other Appalachian regions carry many large fragments of fish, there is no unquestioned specific identity between these and the Ohio shale fish, though the genera are related.   The only remains reported from the Upper Devonian of New York, of species also found in the Ohio or Cleveland shale, are small jaws of *Callognathus serratus,* and a form referred to *Dinichthys curtus* Newberry, a Cleveland shale species, said to have been found in the Chemung of Warren Co. Penn., but of these Eastman [2] says, that he is "unacquainted with any evidence which will enable one to state positively that the [Pennsylvanian] species is identical with any of the Ohio forms".   Detached plates have, however, been obtained from Pennsylvania of a species fully as large as, or even larger than *D. curtus.*

The commonly assumed ichthyological evidence for the correlation of the Ohio shale with the Upper Devonian, is, to say the least, unconvincing. [3]   The black shale generally rests with a sharp contact upon the limestone surface, whether that surface is Ordovician or Silurian, or even Devonian. It passes upward into, or is abruptly, but apparently without hiatus, succeeded by the limestones of Lower Mississippian (Waverlyan) age.

Whether we regard these black shales as more or less reworked residual shales of interpulsation age, or as typical huangho deposits of the early Mississippian or Fengninian pulsation period, they are undoubtedly to be correlated with the Chattanooga black shale of Tennessee and with others of this type.   This shale has all the characteristics of an old residual soil, formed during a long period of exposure and disintegration of the underlying limestones of various ages, weathered-out fossils of which

---

2) EASTMAN: *Devonian Fishes, etc.,* pp. 139, *et seq.*

3) For a fuller discussion, see the section on "Progressive overlap and the Black Shale Problem" (pp. 593-615), GRABAU: *Types of Sedimentary Overlap.*

it occasionally includes. It is comparable to the modern black soil or "tchernozem" of the Russian steppes, modified by being more or less reworked by the transgressing Fengninian sea. The normal deposits of this sea rest conformably upon the black shale, while progressive overlap is shown by successive members for which it forms a basal bed. The contact may be a sharp one, but no evidence of a hiatus is known between the shale and the overlying marine beds.

In Pennsylvania and other bordering regions of Appalachia, the Devono-Fengninian interpulsation period is marked by a series of red beds in the vicinity of the old-land, and by an erosion hiatus between the marine beds in the north-west.[4] The red beds are known as the Mount Pleasant red shale, which has a thickness of 500 ft in Wayne Co., but this shale thins out north-westward and disappears in the Devonian-Mississippian disconformity. This marks the complete emergence of the Appalachian geosyncline, and the red color of the Mt. Pleasant shale suggests the prevalence of at least semi-arid conditions during its formation over the region of its occurrence.

Such general aridity might be the result of the withdrawal of the sea from the whole of the Pangæa, and the relative rise of the old-land because of sinking base-level. In the region several hundred miles distant in southwestern New York and north-western Pennsylvania, where the adjacent beds are marine, there is a distinct hiatus and disconformity, where these red beds should here been between the Devonian and basal Mississippian. Although this missing interval has been thought to be of only very brief duration (Caster), it is probably much more important than is apparent, for there is every reason to believe that it represents an interpulsation hiatus.

The Upper Devonian beds of the Appalachian region of Pennsylvania likewise represent continental deposits. The highest of these formations, next below the red interpulsation formation, is the Elk Mountain sandstone of the Catskill group. This is a green to grayish-green, flaggy sandstone, with most extraordinary development of cross-bedding, "which", according to Willard:[5] "is more

4) CASTER : *Bull. Amer. Palæontology*, p. 104.
5) WILLARD : *Continental Upper Devonian*; p. 576.

strongly developed and more persistent than that observed in any of the other non-red Catskill formations although all these show a tendency to cross-bedding". A few interbedded shales occur. The formation averages perhaps 200 feet in thickness and passes westward into marine beds (Oswayo), with the Salamanca quartz pebble conglomerate of Rock City fame at its base.

Below this Elk Mt. sandstone lie other (red) beds of the Catskill series. Thus, the sections indicate first, repeated red beds alternating with non-red of continental origin, and second, a progressive replacing westward, or overlap of the continental on the marine beds. Or, to put it otherwise, a progressive restriction of the marine area of sedimentation and an advance of continental sediments is indicated, and this ended in final complete emergence and the development of an interpulsation series of red beds, the Mt. Pleasant series. Elsewhere erosion features testify to the reality of emergence.

How is this to be correlated with the deposits on the marginal platform of the Devonian geosyncline? One of the marked characteristics of Michigan, Wisconsin, Iowa, Illinois, etc., is the extensive erosion which the marine Devonian (Hamilton, Onondaga, etc.) beds have suffered before the deposition of the black shales known variously as Antrim, Huron, Chattanooga, etc., and which because of their direct, though disconformable superposition on the eroded Middle Devonian beds are usually referred to a late Devonian age.

The erosion surface, then, can be paralleled with the retreating marine phase, during which the Catskill continental strata were spread in the geosyncline owing to falling base-level, which increased the grade of the streams, and spelt erosion of the rock, with solution of the limestones over the stationary platform, and continental deposition of sands and muds worn from the old-land in the slowly sinking geosyncline. When finally the sea-level had sunk so low that the old-land became a rain-barrier, only the oxidized muds, swept into the geosyncline by intermittent streams, would build up the interpulsation deposits, which now are represented by the red Mt. Pleasant Shales. At the same time the residual product from the dissolved limestones of the marginal platform would be spread as loess over the old erosion surfaces, augmented, no doubt by the finer material blown by the winds from the surface of the geosyncline. All this fine mud would accumulate as a mantle spread over the scars formed during the preceding moister period, where it would await the transformation which the returning sea in the next pulsation period was to effect.

4

The first of the Mississippian (Dinantian) pulsation systems, the Fengninian, is apparently fully represented in the south-eastern region by a quartz pebble conglomerate and siliceous sandstone, the POCONO sandstone, which marks a rejuvenation of strong river activity. If, as I hold, the red beds mark aridity, owing to general withdrawal of the sea, as well as to the height of the wind barrier, because of falling sea-level, then the Pocono indicates a relative lowering of the barrier partly by erosion, but chiefly perhaps, on account of the renewed rise of the sea-level in the new positive pulsation. The Pocono has a maximum thickness of at least 1,500 ft along the Alleghany front, and rests everywhere with a disconformity on the Upper Devonian continental red and green beds, or on the interpulsation Mt. Pleasant red beds.

This great pebble and sand delta (or better, piedmont river-plain deposit) carries some thin coal beds and a varied flora of land plants; and this clearly marks resumption of moister control of continental sedimentation. This is to some extent analogous to the vegetation of the piedmont belt of gravel deposits at the base of the mountains which enclose the Takla Makan desert of Central Asia. These piedmont gravels encroach on the oxidized sands of the desert, and cover them with a layer of stream-washed sand and gravel, which enclose remains of plants that will be converted into sporadic beds of coal of no very great thickness or extent. Just so the Pocono fan was built as a purely continental deposit over the older beds of a preceding retreatal series, and during this moister interval the incipient coal beds of the Pocono were formed on the flat swampy plain, and trees, washed perhaps from the upland, were buried in the sands of the river plain.

Thus the maximum river activity, indicated by the

coarse deposits of the Pocono, following upon a period of
aridity, coincided with the commencement of the new
pulsation period, when the sea-level again rose, thus relatively lowering the height of the old-land barrier, which
reduced its toll of moisture on the winds.   This permitted
precipitation of some of this moisture where it could
form streams, to descend on the leeward side of the old-
land, there to build the Pocono delta of stream-washed
sands and gravels.   This return of moisture affected the
lœss-covered plains of the interior in a different manner.
The old wind-borne residual soil became covered with
vegetation, the annual decay of which added carbon to the
soil, and the erstwhile light-colored mud became converted
into a mantle of black earth, a veritable Palæozoic "tchernozem", like the black soil of Russia.   This was locally
reworked by the transgressing Lower Mississippian sea
and became the more or less fossiliferous black Huron,
Antrim or Chattanooga Black Shale, one of the most intriguing formations of post-Devonian age in the United States.

5

If the Pocono represents the continental phase of the
deposits of the Lower Waverlyan marine beds of the
Mississippi Valley, the succeeding Mauch Chunk certainly
represents different conditions of sedimentation.   It is
primarily a great series of red beds, the appearance of
which suggests a return of aridification, at least for the
region in which their deposition as continental sediments
took place.   Such a condition is difficult to associate with
rising sea-level of a positive pulsation, for that would
imply a comparative decrease in height above sea-level
of the old barrier, *i.e.*, the Appalachian old-land, which
would not be high enough to deprive the winds of their
moisture, and change them into drying winds, so that
after crossing the barrier they would again expand, and
descend greedy for moisture.

But this is what would happen with a sinking sea-
level, for that implies a relative rise of the old-land, and
this would restore the conditions which were responsible
for the red bed series of the pre-Pocono interpulsation
period, the Mt. Pleasant.

Can we then regard the Mauch Chunk as an interpulsation red series, representing the mid-Mississippian (Fengninian-Visemurian) interpulsation hiatus?

I believe that this is indicated for at least the lower part, while the higher part represents the post-Visemurian interpulsation deposits. Then the Visemurian ought to be represented by the middle portion of the Mauch Chunk, *i.e.* the Greenbrier limestone.

These considerations may suggest to the students of these Appalachian formations problems to solve, problems that involve not only the organisms, but also the physical history of the formations in the light of the geography of the earth as a whole.

CHAPTER XXIII

THE MISSIPPIAN OR DINANTIAN SYSTEMS

1

THE Carboniferous period of the older geologists once
included as Lower Carboniferous the strata now separated
as MISSISSIPPIAN in America, DINANTIAN on the European
Continent, and AVONIAN in England. These are now
generally regarded as a distinct system, but it is only
beginning to be recognized that they do not represent a
unit. In other words, the two divisions, into Tournaisian and Viséen, which are commonly recognized, do
not represent respectively the Lower and Upper Dinantian,
but rather two distinct and separable pulsation systems:
the WAVERLYAN (Fengninian) and VISEMURIAN, divided
by an interpulsation hiatus from each other and by similar
interpulsation intervals from the preceding Devonian on the
one hand, and from the succeeding Moscovian formation
of the Donbassian pulsation period on the other. In many
sections the interpulsation period is marked simply by
erosion, but in other sections thereoccur continental, even
coal-bearing, interpulsation deposits. Typically each pulsation
system has a positive and a negative phase as follows:

B. VISEMURIAN PULSATION
SYSTEM
$\left\{\begin{array}{l}\text{Retreatal, or negative} \\ \quad\text{phase, NAMURIAN} \\ \text{Transgressive, or positive} \\ \quad\text{phase, VISÉEN}\end{array}\right.$

Interpulsation Hiatus or Lanarckian Interpulsation Period

A. WAVERLYAN OR FENGNINIAN [1]
PULSATION SYSTEM
$\left\{\begin{array}{l}\text{Retreatal, or negative} \\ \quad\text{phase, CHIUSSUAN} \\ \text{Transgressive or positive} \\ \quad\text{phase, TOURNAISIAN}\end{array}\right.$

---

1) The Chinese Fengninian is a more complete marine representation than
the Waverlyan of America.

TABLE V.

THE FOLLOWING TABLE PRESENTS
THE PRINCIPAL AMERICAN WAVERLYIAN AND VISEMURIAN FORMATIONS.
(CORRELATIONS APPROXIMATE)

	Mississippi Valley, etc.	Ohio, etc.	Eastern Pennsylvania	Va. & Alabama (Southern Appalachia)	Mid-Continent
					Bend of Texas
Interpulsation	*Hiatus*	—	UPPER MAUCH CHUNK	*Hiatus*	*Hiatus*
	KASKASKIA	*Hiatus*	MAUCH CHUNK (in part)	PARKWOOD Formation (2,000 ft) Shale and sandstone (fossils very rare)	MORROW
	ST. GENEVIEVE Limestone		*Hiatus*	PENNINGTON Fossiliferous Shale sandstone etc. (60—300 ft.)	Floyd Group 1,000
	ST. LOUIS Limestone SPERGEN Limestone	MAXWELL Limestone	GREENBRIER Limestone or *Hiatus*	BANGOR Limestone (700 ft.) (Highly fossiliferous)	
	(Overlap)	(Overlap)			
Interpulsation	*Hiatus*	*Hiatus*	? MAUCH CHUNK	*Erosion Hiatus*	*Hiatus*

Meramecian Chester

Retreatal Namurian — Transgressive Viséen

Visemurian Pulsation System

	Fengninian Pulsation System		Interpulsation System	Devonian Pulsation System
	Retreatal phase	Lower or Transgressive phase		
	BOONE	KINDERHOOK (Overlap)	Hiatus	Hiatus
	Hiatus	FORT PAYNE Chert 200 ft. (Keokuk) / Hiatus	CHATTANOOGA Shale	Hiatus
	MAUCH CHUNK	POCONO Conglomerate	(Mont-Plaisantian) PLEASANT HILL	CATSKILL
	Hiatus	LOGAN Sandstone / CUYAHOGA Shales / SUNBURY Shale / BEREA Sandstone — Waverley Group	CHATTANOOGA Shale	Hiatus
	WARSAW Shales / KEOKUK Limestone	BURLINGTON Limestone / KINDERHOOK Beds / (Overlap)	CHATTANOOGA Shale	Hiatus

If I have correctly interpreted of the origin of the black Chattanooga shale as redeposited 'tchernozem', as the product of long exposure and weathering of the older limestones (including the Middle Devonian limestones) in America under essentially tropical conditions (See Maps on Plates VIII-IX) then the interpulsation interval is indicated to be one of long duration.

2

In Ohio and Michigan the oldest black OHIO SHALE is followed by gray shales and sandstones, often beautifully

NORTHERN ARKANSAS   SOUTHERN MISSOURI

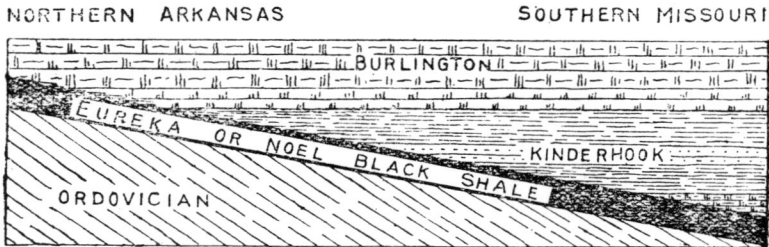

Fig. 83. Section showing the southward overlapping of the Lower Mississippian strata (Kinderhook and Burlington), with the black Noel (Eureka) shale forming a basal bed which rests with a hiatus upon the Ordovician. (Grabau: *Text-Book of Geology*).

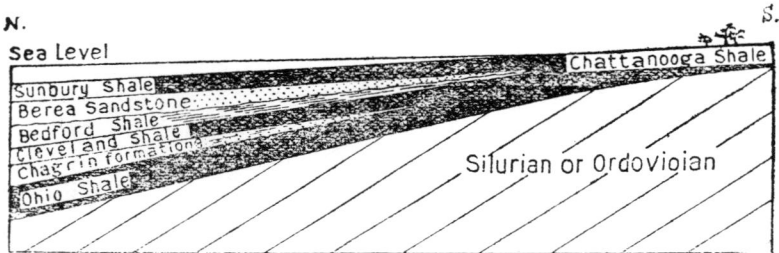

Fig. 84. Diagram showing the relation of the black shales to the other formations. The source of the black mud is the land on the south from which it was repeatedly washed into the sea, interfingering with clastics derived from the east. The Berea, Sunbury, and Chattanooga formations are Lower Mississippian, the others Upper Devonian. (Grabau: *Text-Book of Geology*).

ripple-marked and indicating shallow water conditions (Chagrin formation). These pass upward into black and then red shales (Bedford), which represent the extension of the red beds of the Appalachian region and contain fossils at a few restricted levels. All these are regarded as retreatal Devonian beds, with repeated interfingering of black shale, an early outwash from the 'tchernozem' forming on the surface of the land.[2] The diagram (Fig. 83) illustrates the general transgressive character of the sea, and the overlap of successive strata of the Tournaisian division of the Waverlyan, or Fengninian, pulsation system on the old black Ohio shale (known variously as Eureka, Noel, Antrim or Chattanooga shales). A more complex relation is shown in Fig. 84.

The Berea sandstone which follows, is as a rule devoid of marine fossils; it is mostly a continental, possibly sea-beach, deposit. In some sections there is a suggestion of erosion, between it and the Bedford, which may represent the interpulsation hiatus.

Southward, this passes laterally into Chattanooga black shale, which locally carries thin seams of coaly matter, while a backwash extension of this shale (Sunbury) covers the Berea, and contains a depauperate fauna. This is followed by normal fossiliferous shale (Cuyahoga).

The Waverley series of Ohio terminates with a sandstone (Logan), indicating regression of the sea, after which a hiatus of unknown length intervenes. In the Mississippi Valley the series is more normally marine, beginning with the Kinderhook shales, etc. which are overlapped by richly fossiliferous limestones (Burlington), which in Arkansas rests upon the black Eureka or Noel shale, as upon a basal bed. This shale, in turn, lies on an old erosion surface of Ordovician, or elsewhere, on Silurian (rarely Middle Devonian) beds.

"In some sections the weathering of the older (Ordovi-

2) There may be allowed some difference of opinion in the interpretation of these shales. If, as argued in a previous chapter, the black color of these shales is due to the vegetation which began with and flourished as a result of the moister conditions which returned with the pulsation responsible for Pocono sedimentation, all the black shales must be referred to the same pulsation period as the Pocono, *i. e.* the Waverlyan pulsation system. If, however, we may consider that moister conditions prevailed in the interior during the interpulsation period, when the Mt. Pleasant red shales were forming in the Appalachian geosyncline, or even during the retreatal phases of the Upper Devonian period, when erosion was in progress, the age of the older beds (Ohio and Cleveland black shales and Chagrin and Bedford formations) must still be regarded as Devonian, the Waverlyan beginning with the Berea sandstone.

cian) rock has gone so far that extensive residual accumulations of the shells, formerly scattered through these rocks, were formed. In western Tennessee these shells (mostly the young pelagic stages of gastropods) were highly phosphatic, and this concentration has produced important beds of phosphate of lime at the base of the Chattanooga shale. Such evidence of prolonged exposure to the weather, as well as the presence of the black earth—the ancient tchernozem, which accumulated during a long period of time as a more or less loess-like deposit, [subsequently becoming] highly charged with decaying plant material,—clearly indicates that all of this southern region was exposed land." [5]

3

There is no direct evidence of the connection of the interior Mississippian sea with the Caribbean epi-sea, unless it is under some area covered by Cretaceous, or younger strata. In most regions where accessible, the beds seem to overlap on an old-land surface covered by the residual black soil. In southern New Mexico (Hillsboro, Long. 107° 45′ W., Lat. 32°51′ N.) some 200 feet (65 + meters) of Mississippian limestone rests disconformably on an eroded surface of Devonian calcareous shale, though elsewhere the disconformity is masked. But the exact age of these beds is still unknown. They may be of Waverlyan (Fengninian), or of Visemurian age, or may enclose parts of both period.

Farther south, however, in Guatemala and Chiapas in Central America (90°-95° W. Long., 15°-16° N. Lat.) the whole of the Tournaisian is overlapped by beds of Pennsylvanian (Moscovian) age (Santa Rosa formation) 200-300 meters thick, which rest directly on the Archæan. These consist of basal quartz conglomerate with pebbles of crystalline rock, and are followed by sands, graywackes and slates, all unfossiliferous and passing upward into limestones with *Sinocladia* (*Septopora*) *biserialis* Swallow, and *Lonsdaleia floriformis* Fleming. These are followed by 600 to 800 meters of *Fusulina* beds of Late Carboniferous or Permian age.

Thus so far as the Waverlyan, and probably the

5) GRABAU: *Text-Book*, II, p. 446.

Visemurian as well, are concerned, there was essentially no connection with the Caribbean epi-sea. This suggests that the source of the fauna was from the Boreal region.

It was probably not the East Boreal sea that was the source of this fauna, because: "The known deposits in the northern end of the Appalachian trough, i. e. in the Acadian area, are muds, sandstones and conglomerates, much of them probably of continental origin although gypsum and limestone beds occur. These limestones often consist wholly of marine shells, mainly of species unknown elsewhere either in America or Europe." [4]

This leaves the West Boreal or Alaskan epi-sea as the most likely source of these faunas, though that is by no means established. We only know that the Devonian beds are followed, with no recognized evidence of a disconformity or marked hiatus, by a great thickness of marine limestone (Madison limestone of Montana, Upper Ouray limestone of Colorado, etc.); but the faunas of these beds are still insufficiently studied.

The Burlington is followed by the Keokuk, which in many places indicates the admixture of impurities, and this is followed by the Warsaw Shales, a retreatal series, after which the hiatus, marking the next interpulsation period (the Visemurian), is represented by an erosion plane. To this succeeds the Upper Mississippian (St. Louis) limestones, which locally show striking indications of eolian cross-bedding in their basal beds. [5]

"The top of the Lower Mississippian, or Waverlyan, series is marked everywhere in the interior by an erosion surface, which separates it from the next succeeding series. This indicates that the sea again withdrew and the land rose. The withdrawal was either northward, or northwestward; or both, and the elevation of the country is further indicated by the extensive development of continental sediments. Such deposits appear to have formed in central Arkansas in an east-west trough which lay at the foot of a highland on the south, from which the sediments were derived." [6]

The State of Arizona in south-western United States,

---

4) GRABAU: *Ibid*, p. 447.
5) *Ibid*, p. 450.
6) *Ibid*, p. 449.

where the Colorado River has cut its stupendous gorge, offers some interesting suggestions on the relationship and origin of both Devonian and Waverlyan (Fengninian) faunas. In this State the Devonian formation rests either disconformably or unconformably on Cambrovician, Cambrian or older rock, as described by Stoyanow.[7]

In the southern districts (Pima County) it begins with reworked sandstones (sub-æolian), followed by limestone with Stromatoporoid-masses, small goniatites, etc., and terminates with a thin layer of sandstone with fish-teeth in abundance. These beds are called the PICACHO DE CALERA formation, and may represent a local invasion of Siluronian strata, though the fish fauna (*Ptyctodus calceolus*) is regarded by Stoyanow as being somewhat suggestive of correspondence to the Cedar Valley limestone of Iowa. If that correlation is valid the break indicated by this two foot fish-bed is not a serious one.

The MARTIN LIMESTONE, which follows (261 to 271 feet), has a rich fauna, among which the corals *Acervularia davidsoni* and *Pachyphyllum woodmani,* and the brachiopods *Schizophoria striatula, Stropheodonta demissa, S. perplana, Cyrtia cyrtiniformis* and the *Spirifer hungerfordi, S. orestes, S. whitneyi* and *S. euryteines* deserve mention. In the Ural Mts., beds with *Spirifer whitneyi* overlie the typical *Stringocephalus* beds (Givetian). Closely related species occur in South-West China,[8] an intermediate locality in which this fauna occurs is the Ukrain, *i.e.,* the Volhynian district of Poland (Long. 27° E., Lat. 50° 40′ N).[9]

The Martin is abruptly and disconformably succeeded by the ESCABROSA LIMESTONE 700-950 feet thick, in the Bisbee-Paradise region near the southern border of the State; but nearer the center (Lake Roosevelt) this formation has thinned away to 150 ft, this being in part probably explained by development on the marginal platform. The fauna of this limestone represents Kinderhook and Burlington (Osage) age.

---

7) STOYANOW: *Correlation of Arizona Palæozoic,* p. 484 *et. seq.* See also correlation tables in DEISS C.: *Cambrian Stratigraphy* etc of Montana G. S. A. Special Papers, No 18, 1939.

8) GRABAU: *Devonian Brachiopoda of China.*

9) ALEXANDER von KELUS: *Devonische Brachiopoden von Petzca.*

4

In the region of the Grand Canyon of the Colorado in north central Arizona, the Devonian is known as the JEROME FORMATION and the Lower Mississippian (Waverlyan) as RED-WALL.

At Jerome in Yavapai County (Long. 112°11′ W., Lat. 34°50′ N.) [10] the Jerome has a thickness of 505 ft and rests disconformably on (Middle Cambrian) Tapeats sandstone. It is prevailingly arenaceous, the limestone showing frequent interruptions. About 150 feet above the base is a four-foot bed of sandstone with plates of *Arthrodire* fishes, often forming bone-beds, and filling pockets in the underlying limestone, which itself is made up basally of fragments of the next lower limestone. The thickness of the *Arthrodire* sandstone increases south-eastward, until 55 miles from Jerome a combination of fish-bearing sandstones, with intervening limestone attains a thickness ranging from 50 to 75 ft.

Again, in the northern region at Temple Butte on the Canyon, where the Devonian limestone is called Temple Butte limestone, a thickness of only 77 feet rests unconformably(?) on an erosion surface of Middle Cambrian limestone (Muav limestone), and is followed disconformably by the Mississippian Red Wall.

In some places, deep solution pockets, extending to a depth of 50 feet, were formed in the surface of the Muav (Cambrian) limestone, during the long exposure before the Devonian sands were washed over these beds and filled the pockets. These sands contain the remains of the fish *Bothryolepis*, and are covered by the Red Wall limestone, thus showing that no typical marine Devonian is preserved. "These pockets," Stoyanow says, : ". . . . tempt one to visualize a Devonian river, meandering along the channels cut in the Muav limestone, the course of which has later been dissected by the formation and sculpturing of the Grand Canyon." [11]

During the Devono-Waverlyan interpulsation period, all the higher part of the Jerome formation that had been formed here was eroded again. But it must be kept in

10) STOYANOW : *Op. cit.* 495.
11) *Ibid :* p. 503.

mind, that the original thickness deposited was not necessarily equal to that which formed in the geosyncline, for the retreating sea would uncover the landward areas of the platform first.

From the middle hundred feet of the Red Wall limestone, a considerable fauna of brachiopods, mollusks and a trilobite have been obtained, and from these "the inference may be made that the bulk of its fauna ranges from the later Kinderhook into the Keokuk, with the greater number of forms confined to the late Kinderhook and Burlington." [12]

In the Grand Canyon region there is a great hiatus above the Red-Wall limestone, the next formation being of Permian age (Sepai). This hiatus represents at least two or three pulsation periods, with the intervening and bounding interpulsation periods, in all an enormous length of unrecorded time, though some of these beds representing this time-period were probably deposited in the Grand Canyon area, and later again eroded.

5

The Visemurian pulsation system is represented by the "Upper Mississippian" in south-east Arizona, which follows disconformably on the Waverlyan (Fengninian), or Escabrosa limestone, and is disconformably succeeded by Moscovian (Naco limestone), as will be noted later.

In the type region, at Paradise, in south-east Arizona, the Escabrosa formation is succeeded by 150 ft of limestone, with *Productus marginicinctus* and other species, and *Spirifer bifurcatus* Hall, and other fossils in the upper 30 feet. This has been referred to the base of the Upper Mississippian, *i. e.* Visemurian. Then follows the PARADISE FORMATION, which begins with fragmental limestones and shale and continues as alternating limestone and shales for 195 feet, thus making a total thickness of 345 feet for the two post-Escabrosa series. The fossils of the Paradise formation indicate its Merameccian and Chester age, the beds from 80 to 125 feet above the base representing St. Genevieve, and the higher beds probably St. Louis age. Evidently, then, these Upper Mississippian beds re-

---

12) STOYANOW : *Ibid.* p. 514.

present a distinct transgression, and inferentially, a distinct pulsation system.

Farther north, in east central Idaho, the Devonian is represented by the Jefferson dolomite, 1,150 feet thick, followed by Upper Devonian Grand View dolomites, 1,170 feet thick, with the *Spirifer whitneyi* fauna, these beds evidently representing the accumulations in the geosyncline. They are followed by the Milligen formation, mostly of black Carbonaceous argillite, with some impure coal and a few intercalated limestone beds. It has a thickness estimated at 3,000 ft, or less, and contains mainly plant remains in fragmentary condition.

These continental beds are followed upward, and are in part replaced, by the marine Brazer limestone, 2,000 feet, or more, thick. This carries a marine fauna, called Upper Mississippian, and probably represents the Visemurian pulsation system.

6

Turning now to the Caledonian geosyncline, we meet with an instructive display of these strata in the southern Uplands of Scotland, where, as we have already seen, late Upper Devonian, or more probably Devono-Fengninian interpulsation beds in the form of Old Red Sandstone, lie unconformably on tilted and eroded Siluronian (Downtonian) strata, which contain fish remains, etc. The orogeny here is Dev-oric, and the missing beds are the true Devonian. In Great Britain the divisions of the so-called Carboniferous limestone series, are as follows in descending order.

SUPERFORMATION: MILLSTONE GRIT

Hiatus and Disconformity

AVONIAN SERIES ("Carboniferous Limestone")
3. Yoredale Group;
2. Scaur or Main Limestone;
1. Lower Limestone or Calciferous Sandstone Group.

Hiatus and Disconformity

SUBFORMATION: UPPER OLD RED SANDSTONE (Interpulsation).

The Calciferous sandstone consists, according to Geikie, of-"thick courses of yellow and white sandstone, dark shale, and seams of coal and -ironstone, among which only a few thin sheets of limestone are to be met with."[13] It grades down into the Upper "Old Red", which itself apparently represents the continental deposits of post-Devonian interpulsation age.

The lower division is the cement-stone group, and consists of thin-bedded white, yellow and green sandstone, gray, green, blue and red clays and shales, with thin beds of argillaceous limestone or cement-stone; seams of gypsum occasionally appear. Organic remains are few and limited. Some limestones, composed chiefly of ostracod shells and containing fragmentary plant remains, are also found. On the whole these beds suggest mud deposits, on a broad low river plain covered occasionally by overflow from the sea.

Coming southward to the Welsh borderlands, we find fossiliferous limestones replacing the sands and muds. A. Vaughan[14] subdivided the "Avonian" rocks on the Avon River (near Bristol) as follows:

DIVISION ε

VISÉEN

    DIBUNOPHYLLUM ZONE

        D_2 *Lonsdalia floriformis* Subzone
        D_1 *Dibunophyllum* Subzone

    SEMINULA ZONE

        S_2 *Productus cora* (Mut. $S_2\alpha$).
        S_1 *Productus semireticulatus* (Mut.)

Usual Location of Interpulsation Hiatus

DIVISION δ

TOURNAISIAN

    Horizon C_2 absent
    SYRINGOTHYRIS ZONE
        C_1 *Syringothyris* aff. *laminosa* Zone } *Caninia* Zone

---

13) GEIKIE: *Text-Book*, p. 980.
14) VAUGHAN: *Palæontological Sequence*; p. 181 *et seq.* 1904.

DIVISION γ

ZAPHRENTIS ZONE

Z₂ *Schizophoria resupinata* Subzone
Z₁ *Spirifer* aff. *clathratus* Subzone

DIVISION β

CLISIOPHYLLUM ZONE

K₂ *Spirifer octoplicata* Subzone
K₁ *Productus bassus* Subzone

DIVISION α

MODIOLA PHASE (ZONE)

M   *Modiola lata* Subzone

Interpulsation Hiatus and Disconformity

SUBFORMATION

Etroeungtian, with mixed Devonian and Tournaisian faunas or *Old Red Sandstone* (Interpulsation continental beds)

Belgian geologists place the Etrœungtian division of
light colored crinoidal limestone, with primitive species of
Tournaisian corals and brachiopods, including *Productus
niger*, at the base of the Lower Dinantian. The fauna
is a mixed Devono-Tournaisian fauna, such as one would
expect to find after an interpulsation period, where an
area of weathered Upper Devonian, with loosened fossils,
was transgressed by the Tournaisian sea. It is not a
transition fauna, as it is so often called, but a true mixture
at a disconformable contact between successive systems.
The same mixed assemblage of Devonian and Tournaisian
species is found in North Devonshire, where marine
Devonian is capped by Early Tournaisian (Pilton Series).
Where, however, as in the South Western Provinces, these
beds, which Vaughan correlates with the Etrœungtian,
directly overlie unfossiliferous Old Red Sandstone, they
have a pure Lower Avonian fauna.

These Etrœungtian reworked beds are followed by
higher pure Tournaisian, with the fauna of the *Zaphrentis*
Zone, as in England, with *Z. delepini* and *Caninia cylindrica*, and other forms, indicating a transgressing sea and
overlap of zones. On the other hand, in Russia, the

beds which have been correlated with the Etrœungtian
of Western Europe are also of marine calcareous type, and
are closely related faunally to the underlying marine
Upper Devonian.   They contain almost exclusively Upper
Devonian forms, with *Spirifer verneuili* Murch. predominat-
ing, while Rhynchonellids and Productids also represent
Devonian groups.   The fauna differs sharply from the
Tournaisian.   The last of the *Clymenias* also occur in
these Russian beds.

The explanation apparently lies in the fact, that the
Russian "Etrœungtian" beds are not the true Etrœungtian
beds of the Belgian and other Western European regions,
but are still part of the Upper Devonian.   That they lie
between beds with a "Chemung" fauna and beds with
a "Kinderhook" fauna has no significance in this connection, once it is recognized that an interpulsation hiatus
of world-wide extent separates Devonian and Tournaisian.
Then it will be seen that marine Devonian beds in Russia,
younger than any known in Europe or America, simply
indicate a later retreat of the Devonian sea, or a less
extensive erosion in Russia than in Western Europe or
America, or in both.   Again, the absence of true Etrœungtian beds in Russia means a later post-hiatus transgression, than the transgression in Europe, so that the
typical European Etrœungtian was not deposited there.
The fact that certain species regarded as index fossils of
the Etrœungtian are present in both regions is likewise
not significant, if it is remembered that an unknown
amount of Upper Devonian, probably of the Russian
'Etrœoungtian' type, was available in weathered form with
exhumed fossils, ready to be re-interred by the Tournaisian
sea with its own species, to form the West European
Etrœungtian, which, according to Bisat,[15] nevertheless is
of basal Tournaisian age.

7

One of the characteristic rock types of the Tournaisian
is the "Waulsortian" or "Knoll" facies, which consists of
an irregular accumulation of powdery limestone (calcilutyte)

---

15) BISAT: *Faunal Stratigraphy*, p. 531.

often brecciated and seamed with calcite-veins. Its most striking aspect is a massive limestone mottled with bluish blotches and abounding in Fenestellids. Locally thin masses swell out into mushroom-like "reefs" with the marginal beds often containing nests of brachiopods arching over them; foraminiferal limestones also occur.

Tournaisian		Tournaisian
Lower Avonian: True Etrœungtian[16]		Hiatus
Interpulsation Period Hiatus		
Marine Upper Devonian	Weathered Upper Chemung	Unweathered Upper Chemung (False Etrœungtian)
	Unweathered Lower Chemung	Chemung (Lower)

A disconformity between Tournaisian and Viséen is suggested by the brecciated condition of the black marble of division δ or the beds lying between C_1 and S_1, i. e., essentially the horizon C_2, that is the horizon of faunal change, whereas in south-western England the hiatus occurs earlier, i. e. between C_1 and C_2. Vaughan holds that "Horizon δ bridges over the passage from Tournaisian to Viséen, and that no single line within it satisfies all the requirements of a dividing line between [these divisions]..."

Interpreted in terms of pulsation periods, horizon δ marks the last of the Tournaisian beds remaining after an unknown amount of erosion, and these beds were brecciated and reworked so that the returning Viséen sea mingled its fauna with the dead fossils of the preceding Tournaisian. For, of course, the presence in the same bed of both Tournaisian and Viséen species does not imply their co-existence, nor can we assume that Viséen species suddenly appeared, and the surviving Tournaisian forms gracefully yielded the ocean floor to them. Such a "change of administration," familiar enough in politics, does not occur in nature; the older "conquered" fauna is not exterminated, except locally. And, of course, the older fauna

16) EXPLANATION: *Etrœungtian*, a formation containing a mixture of marine Upper Devonian fossils weathered out during exposure in the interpulsation period, with earliest Tournaisian. *Chemung*, a general term (as here used) for marine fossiliferous Upper Devonian. (See also PÆCKELMANN, W., and SCHINDEWOLF, O. H.).

does not change directly and rapidly into the newer — of that there is no evidence; transitional species are absent. Therefore, the only rational explanation is that after a long interpulsation interval, when, because of restriction to the epi-seas, the Tournaisian fauna had largely become extinct, while the few survivors in the epi-seas had changed to new types, the sea returned to the old geosynclines, and the "new" fauna established its home over the graves of the old, in a measure its distant ancestors, though not recognizable as such. Moreover, the upper layers of the Tournaisian graveyard were disturbed and the sediments reworked, so that Viséen S_1 types were buried with Tournaisian C_1 types, the two forming the mixed fauna of the C_2 or the "δ" horizon.

It is the old story over again: a mixing of distinct types of adjoining horizons by reworking of the upper weathered layers of the older by the incoming sea, which brings the new fauna. But mingling of old and new is not change of one into the other; each is distinct, and represents its own closed chapter in the evolution of life. Mixed faunas abound, but transition faunas have yet to be discovered, and, when it is recognized that the mixing process is the all-prevailing one, the idea of continuous change of one fauna into the other *in the place of burial* will receive no more credence than would an assertion of contemporaneity of *Homo sapiens* with *Sinanthropus pekinensis*, whose bones may be mingled together in the soil of the Peking plain.

CHAPTER XXIV

OROGENIC PERIODS AND POLAR SHIFT[1]

1

THAT mountain making, or orogeny, is directly associated with polar shift, *i. e.* the displacements of the poles (not the axis) of the earth, through drifting of the continental sial mass, is abundantly illustrated by the history of mountains formed in Palæozoic time. This applies primarily to folded mountain chains, though block mountains and volcanic cones also bear a genetic relationship to the events inaugurated by shiftings of the polar site.

Sial shifting may be thought of as a rotary movement, either clockwise or counter-clockwise,[2] and the apparent shifting of the pole as taking place in the opposite direction to the movement of the sial crust. The pivotal points of the movement from one station to the next can usually be placed at the apex of an isosceles triangle, of which the path of migration of the polar site forms the base. What the apical angle of the triangle is, whether it is the same in all cases, and what determines its location (for it must be obvious that the location of the pivotal point cannot be the same for all shifts), these are among the unsolved problems to which attention might profitably be directed.

It might be said that an apical angle of 30° serves for an experimental basis, but a smaller angle, which means of course a longer radius of rotation, is probably indicated. As for the position of the pivotal points, these must for the present be located experimentally, until something more is known of the cause of the shifting of polar sites. It may be that it is the displacement of the center of gravity that, as suggested in Chapter I, was the cause of the sial shift under the influence of a rotating earth. But if so,

---

1) See also Chapter V.
2) Clockwise rotation when viewed from the South Pole corresponds in direction to counter clockwise movement as seen from the North Pole.

the location of the gravity disturbing intrusion, as well as the determination of the pivotal point, that is to say, the point at which the sial is anchored to the sima by a deep-seated intrusion—require to be ascertained. These are problems for future investigation.

When we visualize polar shift in terms of land-block or sial movement instead of actual displacement of the pole as a whole, we gain an insight into the process by which such movement may influence orogeny. For example, if the pole seems to move from the region of the Nile delta in Egypt to the vicinity of Taodini in the western Sahara, it is really a movement of the crust in the opposite direction. Thus, instead of the pole moving in a south-westerly direction (speaking in terms of present day geography)[3] across North Africa, from its original position on the pangæa to its location in Ordovician time, there was in reality a movement of the African land block in a north-easterly direction (p. d.), to the same amount, the actual position of the pole being to all intents and purposes stationary. Such a north-easterly movement of a huge mass of crystalline sial would have comparatively little effect, if, as today, the sial were essentially a distinct continental block with little contact with the other continental masses. It might suffer marginal disturbances, owing to the fact that it must plow its way through the more resistant sima in which it is submerged for the greater part of its thickness, but that would probably have no recognizable effect on other land masses.

But in Palæozoic time Africa was not a separate mass, but a part of a single continuous sial cap, one of its most stable and perhaps thickest portions, and it was separated from other continents, if at all, only by shallow depressions, the geosynclines. These, though constantly sinking, were also constantly being filled by sediments, and, so far as we have any reliable knowledge, were never of an abyssal nature, the repeated suggestions to the contrary notwithstanding.

It is a very different matter, when such a land mass is set in motion even for a rotation of limited arc, for it must inevitably pull some of the other continents with

3) This is indicated by the letters (p. d.) in the text.

it and push others before it. Thus, if the movement of the African part of the Palæozoic pangæa is eastward (p. d.) it would pull the Americas after it, while pushing Eurasia before it. But moving away from a resistant sima-mass on its outer margin, would have no compressing effect such as would characterize the mass that is pushed against the resistant sima mass on the other side of the pangæa. The dragging pull might, if it acted too rapidly or spasmodically, open rifts in the weaker portions (the geosynclines) and so inaugurate localized volcanism. Only in regions beyond the pivotal point, where the sial cap rotates in the opposite direction would there be pressure exerted, which might culminate in an orogeny. Thus, if the pivotal point of a rotation is situated within or near the mid-margin of an extended geosyncline, one part, that on the polar side moving away from it would suffer tension and perhaps volcanism, as a result of rifting or normal faulting; or at the least, a cessation of subsidence would result, which, as seen in section later, would suggest a local elevation, because of the absence or slight development of certain sedimentary formations found elsewhere. But the part of the geosyncline or of other geosynclines beyond the pivotal point on the opposite side, would suffer compression, which at first would become manifest as a more rapid subsidence or down-bowing and finally in a buckling up into mountain folds and the formation of the thrust faults and overthrusts, perhaps of great magnitude.

The most profound compression, however, would be suffered by the land massess pushed before it by the moving African colossus. Such compression would be strongest near the moving mass, but would be translated to the opposite side of the pushed continent, and, though at first weaker, would increase in strength as the weakness of the intervening part was removed by compression into folds of greater resistance. Thus the immediate result of the moving mass would be first to cause more effective subsidence of the geosyncline across the path of its movement, and secondly, to crush the strata into a folded mountain mass or relieving overthrust. With the continuance of the pressure by the moving mass, the momentum would be transmitted in perhaps wave-like effects across the continent. This eventually would begin to move through the retarding sial-mass, a movement which would be manifested by the development, or at least rejuvenation of the marginal geosyncline. This belated transmission of the compressive force would increase with time, as the intermediate weaker spots were eliminated by crushing, and with the accompanying strengthening of the strata through increased compactness. Eventually, if the pressure were continued long enough, the marginal geosyncline of the pangæa in the direction at right angles to the movement, woulld also suffer an orogeny. The observed migratory nature of orogenies is thus accounted for.

2

During the closing stages of the Ordovician period and the Plynlimonian interpulsation period the pole experienced relatively a very extended migration. From the site at Taodini in west Africa, in Skiddavian or early Ordovician time, it moved relatively to its Silurian location in the Levant, east of the Mediterranean, that is to within a relatively short distance of its pre-Cambrian location. This was a distance measured today by 2,700 miles. The movement of the African land mass was westward, the pressure being against the American land blocks (chiefly the North American), while Asia was dragged-westward (p. d.), suffering tension, which should be manifested by further decrease of geosynclinal subsidence or actual rifting and volcanicity. The east American region and its European extension suffered compression of the Taconic geosyncline, which at last culminated, during the Plynlimonian interpulsation period, in the Sil-oric or Taconian orogeny, a folding which extended from Virginia or further south to Great Britain or further north, with its greatest intensity near the middle in the Hudson River-Quebec region where the pressure was most pronounced. But the effects of this compression were transmitted only slowly across the continent of North America, and only geosynclinal depression was experienced on the West Coast and in the geosyncline from which very much later the Rocky Mountains were born.

In the next few pages we will review some of the better determined and more accurately dated of the Palæo-

zoic orogenies and their relation to the shifting of the
polar site.   In this connection it must be emphasized that
the dating of unconformities, the visible indications of older
orogenies, by the use of the time-honored subdivisions
of geological history into its broader periods, is not
sufficient.   To state that an orogenic act occurred between
the Ordovician and Silurian is not sufficient, because it
leaves only time for the orogeny itself during the interpulsation period, and does not allow for the much longer
time required for the erosion of the folded rocks and the
ultimate production of a peneplane, across which the sea
might eventually transgress, or the rivers spread their
deposits of sand and mud.   We must first of all recognize
the existence of interpulsation periods, when the sea has
entirely withdrawn from the geosynclines owing to the
lowering of the ocean level, concomitantly with subsidence
of the ocean bottom, through contraction of the sima, or
swelling upwards of the land through expansion of the
sima under it, or to both combined.

But recognition of the existence and importance of
interpulsation periods is not enough.   Our old divisions
of Palæozoic time must be rendered more precise.   Even
today in continental European literature we find the old
practice persisting of classing all pre-Devonian post-Cambrian formations as Silurian.

In America, Great Britain and China and a few other
progressive countries, the Lower Silurian has long since
become recognized as an independent Palæozoic unit, under
the name Ordovician.   More slowly has the separtion of
the old Carboniferous formation into its three most obvious
components taken root, and even now we meet with
terms like Lower Carboniferous (Unterkarbon) and Upper
Carboniferous (Oberkarbon), while Permian is more often
grudgingly given recognition to independence, though not
by everyone even now.   But we must go farther than that.
The Ordovician is not a unit, neither is the Silurian, nor
the Devonian, nor the Mississippian nor, the Permian, to
mention only a few.

3

We must recognize the independence of the Lower
Cambrian or Georgian, of the Acadian (Middle Cambrian)

and of the Cambrovician (Upper Cambrian and Tremadoc). The Ordovician must be divided into the Skiddavian (or Canadian) and Ordovician proper. The Silurian must be restricted, and the Siluronian (Upper Ludlow and Downton and Lower Devonian of Europe, Cayugan plus Helderbergian of America) given independent rank, while the name Devonian should be restricted to the old Middle and Upper Devonian. Mississippian (or Dinantian) must be divided into the Fengninian (Tournaisian plus Chiussuan) and Visemurian (Viséen and Namurian), each an independent pulsation system. The Carboniferous, or Carbonic deprived of the Namurian below and shorn of the Uralian above, now appears to constitute a single system, to which the name Donbassian has been applied; [4] while the Uralian and Artinskian, usually tossed back and forth among the Carboniferous and Permian, must be recognized as an independent pulsation system, the Uralinskian. Finally, the surviving members of the upper part of the Palæozoic may retain the name Permian, this being now recognized as an independent pulsation system. Whether future discoveries will demand further subdivisions remains to be seen, those enumerated must be recognized if we wish to date Palæozoic history accurately, and especially if we wish to gain a true concept of the time relations of the Palæozoic orogenies.

Palæozoic orogenies are best named after the period of erosion, though the orogeny itself occurred during the immediately preceding interpulsation period. Eventually, when these interpulsation periods are better known, and their final names established officially, the orogenies may be named after them, by adding the ending "oric". At present, however, they can be designated by the name of the missing period, for that is the most reliable evidence of their age. If two or more periods lack representatives the first one absent may usually be regarded as the one during which erosion at least commenced. The orogenic periods thus recognized are: 1. Sin-oric; 2. George-oric; 3. Acad-oric; 4. Cambrov-oric; 5. Skidd-oric; 6. Ord-oric; 7. Sil-oric; 8. Sild-oric; 9. Dev-oric; 10. Feng-oric; 11. Visem-oric; 12. Donb-oric; 13. Ural-oric; 14. Perm-oric; 15. Appala-

---

4) See: V. K. TING, and A. W. GRABAU: *The Carboniferous of China,* p. 16 etc.

chian orogeny.   Only this last orogeny affected all the strata
of the geosyncline.

Table V (p. 282-283) summarizes these orogenies.

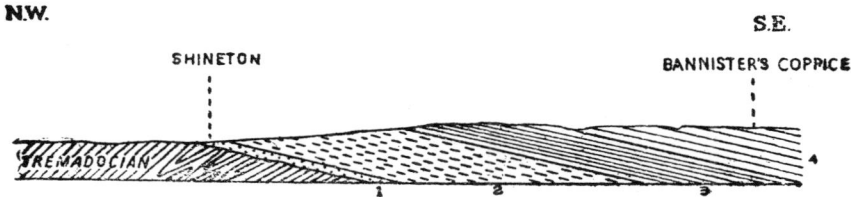

Fig. 85.   Skid-oric Orogeny in Great Britain.   North-west south-east section through Shineton, showing the unconformable overlap of the Silurian
strata of the typical region of Shropshire on the strongly folded Tremadocian (Shinetan or Cambrovician).   The section lies about 2 miles north of
Much Wenlock and an equal distance south of the Wrekin.   This illustrates
the Skidd-oric orogeny.

The dips range from 85° to 30° frequently in opposite directions.   As
here shown the section is incomplete for elsewhere in the vicinity the Ordovician beds (Bala) rest unconformably upon the Shineton shales, while the
Arenig (Skiddavian) is absent.   This is therefore an example of Skidd-oric
orogeny, the Skiddavian being the period of erosion and peneplanation of
the Shineton (Cambrovician, or Ozarkian) folds.   The period of folding must
be placed in the Rhobellian interpulsation period.   Finally, during the Plynlimonian interpulsation period, the region was slightly deformed chiefly by
doming, this being a mild echo of the Taconic (Sil-oric) Orogeny of America.   The domes were eroded during the later part of this period so that the
Silurian transgression brought various early Silurian beds to rest disconformably on the Ordovician *Trinucleus* shales (Onny River Sect. Map. fig. 46, p. 164)
and in the Shineton Section (16 miles north-east) directly and unconformably on
the Shineton shales, the whole of the Ordovician here having been eroded
in the late Plynlimonian interpulsation period.   This, however, is a gentle
bevelling not a peneplanation as on the Shineton shales.

1-3.   Valentian: 1. Basal Sandstone (40 ft); 2. *Pentamerus* Beds (320
ft); 3. Purple Shales (220 ft).

4.   Salopian: Wenlock Shales. (Redrawn with additions from Stubblefield and Bulman after Whittard, Q. J. G. S. Vol. 83, pl. 56. fig. 2).

TABLE VI.—PALÆOZOIC OROGENIC PERIODS

Orogenic Periods of the Palaeozoic	Latest Formation involved in Folding	Period of Erosion	Age of Oldest Covering Formation	Typical Locality	Geosyncline Affected	
					Caledonian	Appalachian St. Lawrence
1. Late Sin-oric	Torridonian	Post-Torridonian (Late Sinian)	Taconian or Lower Cambrian	North Scotland	—	St. Lawrence
2. George-oric	Sinian	Georgian	Acadian	—	—	—
3. Acad-oric	Taconian or Lower Cambrian	Acadian or Middle Cambrian	Cambrovician	Shropshire	Caledonian	—
4. Cambrov-oric	Acadian	Cambrovician	Skiddavian	—	—	—
5. Skidd-oric	Cambrovician (Shineton)	Skiddavian	Ordovician and Silurian	Shropshire	Caledonian	—
6. Ord-oric	Canadian (Skiddavian)	Ordovician	Silurian (Llandovery)	N. W. Ireland Hope Valley Shropshire	Caledonian	—
7. Sil-oric (Taconian orogeny)	Ordovician	Silurian	Siluronian	Eastern New York	—	St. Lawrence Appalachian
8. Sild-oric (Ardennian)	Silurian	Siluronian	Devonian (Late Siluronian?)	Ardennes S. E. Scotland	Caledonian	—

		Devonian (Early)	Late Old Red or Fengninian	N. Scotland Greenland		St. Lawrence
9. Dev-oric (Erian)	Silurionian				—	—
10. Feng-oric (Bretonian)	Devonian	Fengninian	Visemurian	Brittany	Caledonian	—
11. Visem-oric (Sudetian, Selkian?)	Fengninian	Visemurian	Donbassian	Saxony	Caledonian	—
12. Donb-oric (Asturian)	Visemurian (+ Moscovian?)	Donbassian (Upper or Donetzian)	Uralinskian (Ottweilian part)	Asturia	Caledonian	—
13. Ural-oric (Saalian)	Donbassian (+ Lower Uralian?)	Uralinskian (Upper?)	Permian, (Zechstein)	Halle on Saale	Caledonian	—
14. Perm-oric	Uralian	Permian etc.	Triassic	—	—	—
15. Appalachian Orogeny and Migration of Geosynclines	Permian and Older End of Geosyncline affected	Migration of Geosyncline into Old-Land of Appalachia Deposition of Continental Triassic on former old-land surface, followed by mixed continental and marine Jurassic, Comanche Cretaceous and Tertiary.				Appalachian and St. Lawrence

CHAPTER XXV

OROGENESIS IN PALÆOZOIC TIME

1

BECAUSE of the necessary limitations of space, I will select here a few of the better known and most fully studied Palæozoic orogenies, in which exact dating is possible of: a) the latest formation involved in the orogeny, and b) those formations that overlie the folded series unconformably, so that it can be readily fitted into the frame-work of the newer classification. In a few cases orogenies, that have been dated in terms of the older more inclusive general terminology have been traced down to the detail of the classification here adopted, and these may serve as an illustration of the manner of treatment necessary to a correct understanding of all orogenies.

The following illustration (Fig. 85) shows a relationship which might be misleading unless compared with other sections in the neighborhood.

2

One of the most definitely determined is the Sil-oric or Taconic orogeny of eastern North America, for here the true Silurian beds are absent, the Siluronian lying unconformably upon the strongly folded and peneplaned 'Hudson River' slates, which include not only Deepkill shales (Beekmantown or Canadian), but Normanskill shales as well (typical Ordovician). This orogeny is recognized from Canada south to Maryland or Virginia.

Schuchert summarized the successive deformations in the Acadian region as follows. During the Sil-oric orogeny, the Ordovician and older strata of the Hudson Valley region (known formerly as Hudson River beds, in the Albany region) were "intensely folded and then profoundly overthrust; there are at least two major nappes and many minor overthrustings, besides imbricate structure. The shale formations are as a rule closely folded and overturned to the west, with the result that in the shale belts all the strata appear to dip about 70 degrees east, and strike north 20 degrees east. The more competent quartzites and grits, however, show broad anticlinal and synclinal symmetrical folds. The folding dies out gradually toward the west."[1] These folded strata are followed unconformably by Siluronian beds ranging in this region from the Binnewater, or Wilbur, to the Cobleskill, or Manlius, or more rarely Helderbergian beds.

In Vermont and southern Quebec these older beds are known under the collective name 'Quebec Series'. These are also close-folded and crumpled, with the folds isoclinal, and in many places overfolded toward the north-

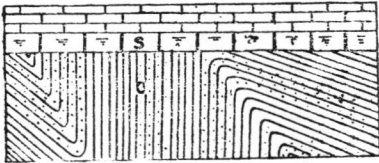

Fig. 86. Sil-oric or Taconian orogeny. Section at Kingston, New York, showing the strongly folded and peneplaned Ordovician graptolite shales (O) unconformably overlain by the Siluronian (Cobleskill and Rondout waterlime) the equivalent of the Bertie. The peneplanation represents the Silurian pulsation period which is unrepresented by sediments. The series was again folded during the Appalachian disturbance (Fig. 58, p. 193).

west. To what extent these strata in Quebec are actually overthrust in sheets or nappes, is mostly undetermined, though such faulting has been mentioned by authors and proved for the vicinity of Quebec City, Levis, and in north-western Vermont. The whole Quebec series is of course widely thrust to the north-west, as demonstrated first by Logan,[2] and by others since, making the well-known Saint Lawrence-Champlain fault.

"Structurally," says Schuchert, "one of the more complicated areas of the Saint Lawrence geosyncline is that of the Shickshock-Notre Dame-Green Mountains, bounded to the north-westward by the great overthrust Saint Lawrence-Champlain fault. This fault may be traced from Kingston, New York, and Lake Champlain to Quebec city, and thence it follows the estuary of the Saint Lawrence, undoubtedly continuing to the south of Anticosti. To the west of this line are the flat-lying Ozarkian (Potsdam)

1) SCHUCHERT: C. *Orogenic Times,* etc., p. 711
2) *Geology of Canada,* 1863.

and Lower to Upper Ordovician formations of the Saint Lawrence plain, chiefly limestones, which dip about five degrees south-east and doubtless rest on a very stable shelf of the Laurentian nucleus at no great depth. Against this stable edge, the eastern strata have been folded and widely overthrust (Logan's fault line). To the overthrust formations Logan applied the name Quebec group, which embraces in general Lower and Middle Ordovician, with the addition of Lower and Upper Cambrian, known *in situ* only in western Vermont and northern Newfoundland, but surely to be found in the Province of Quebec."[3]

On Gaspé Peninsula (Mt. Albert Area) the very thick Quebec series consists of slates, quartzites, limestone conglomerates, etc., overlain by volcanic rocks of complex basic character, and long since altered by metamorphism in places to chlorite schists with the intruded ultra basic rocks (peridotites and amphibolites) largely altered to serpentine. All these rocks are thought to belong to the Ordovician pulsation system.

Sixty miles south-east (Port Daniel—Gascon region) several thousand feet of shales with much volcanic material and serpentine (Mictaw series) carry graptolites in the upper part, which, according to Ruedemann, represent late Trenton age. These deposits have been traced to the Lake Memphremagoc area on the one hand and into New Brunswick (Bathhurst region) on the other. "It is clear" says Schuchert, "that southern Quebec in late Middle Ordovician time had extensive outpourings of lava (now serpentinized) and that clastics accumulated in intervals between the eruptions. Over these Middle Ordovician formations lies disconformably another thick fossiliferous series, dominantly of shales with limestones, and of Richmondian age. All of these formations were greatly deformed by the Taconian movement that took place toward the close of the Ordovician".[4]

It is possible that this outpouring of lavas in mid-Ordovician time marks the culmination of the tension before the Taodini polar position was attained, while the movement of Africa was still north-eastward (p. d.). Then, during the reversal of the movement the depression of

3) SCHUCHERT, C.: *Orogenic Times, etc.* pp. 711-712.
4) *Ibid*: p. 712

the geosyncline for the deposition of Richmond beds went
forward. These beds were all folded during the post-
Ordovician interpulsation period, though the intrusion and
flows are older, being also affected by the disturbance.
They may represent torsional strain in the Quebec region,
while compressive force was applied in the 'Southern
Appalachians' because of the position of the pivotal
point. If so, the compressive force later migrated, so as
to affect the region of earlier stress and volcanism. The
erosion was then accomplished during the Silurian pulsation period, followed by transgression in Siluronian time
as in the Albany region. Schuchert calls these 'Middle
Silurian'[5] and 'Lower Devonian' (Gaspé marine series).

3

The Gaspé peninsula forms the south shore of the
modern St. Lawrence estuary in Quebec and is separated
from New Brunswick on the south by an inlet of the
Gulf of St. Lawrence known as Chaleur Bay.

Around the head of this bay and elsewhere on the
peninsula, the Ordovician beds, which here include the thick
Matapedia series of Richmondian age (with a fauna of
European affinities) are strongly folded, bevelled by erosion and
covered by Siluronian strata ('Middle Silurian' of Schuchert—
Upper Silurian and Helderbergian, which is our Siluronian).
These marine 'Silurian' (Siluronian) beds have according to
Schuchert, a thickness of more than 3,000 ft, and are
succeeded "by a similar thickness of Lower and Upper De-
vonian clastics and conglomerates".[6]

That the 'Silurian' beds cannot be the true Silurian
will be evident, when it is realized that the older folded
and peneplaned Ordovician beds include several thousand
feet of Richmondian or uppermost Ordovician. If the
folding occurred during the post-Ordovician (Plynlimonian)
interpulsation period, the whole of Silurian time was probably taken up with the erosion of these folds, as was
probably also the Salina interpulsation period. Then the
first marine beds deposited on the erosion surface could

---

5) Probably Upper Silurian as formerly used is equivalent to the lower
   part of my *Siluronian*.
6) SCHUCHERT: *Ibid:* p. 715.

only be Siluronian, *i. e.*, beds of the general age of the Rosendale of New York or perhaps the Mackenzie of Maryland, but not much older.

The entire series was again folded and much faulted during the Feng-oric orogeny at the end of the Devonian, probably during the Mont-plaisantian interpulsation period. The biotite-granite dikes, which cut these strata locally, may belong to the same period of disturbance. Much overthrusting during this and the later Appalachian orogeny further occurred to complicate this region.

Typical Silurian certainly occurs in some part of the St. Lawrence area, as on Anticosti Island, where it rests disconformably on the Ordovician. It also occurs at Notre Dame and White Bay, etc., in Newfoundland, from which region Silurian fossils are reported,[7] but the strata which contain these fossils rest unconformably on the crystalline schists, the Ordovician being absent. Moreover, there are many faults in this region. These areas were not affected by the Taconic (Sil-oric) orogeny, and Silurian could be deposited while the folded ranges were undergoing erosion. If typical Silurian strata are found to rest on late Ordovician beds with more than the slightest difference of dip, the contact cannot be a sedimentary one, but must be a fault contact. From this general conclusion there can be no dissent.[8]

4

A very similar problem confronts us in the mountains of New Jersey, Pennsylvania and Maryland (Fig. 53, p.

---

7) SCHUCHERT lists typical early Silurian fossils from Herring Neck, New World Island, Newfoundland.

8) SCHUCHERT and DUNBAR write: "The very fact that Murray and others say that the Silurian lies unconformably upon Ordovician, and that the older Silurian has much conglomerate and sandstone, *implies an orogenic episode* here — "in Notre Dame Bay, more certainly than on the west coast of the island (1934; p. 102, The italics are mine). This conclusion is here challenged. It must be remembered that in the days of Murray and "others" no distinction was made between unconformity and disconformity — the latter being included under, and designated by, the same name as the former. Disconformities are normal between all systems, unless their place is taken by interpulsation continental deposits; unconformities imply an orogeny — *and the lowest Silurian beds cannot follow upon folded and eroded upper Ordovician.* The time of the interpulsation period is not long enough for both folding and peneplanation.

179). I have already referred to the fact that the Shawangunk conglomerate throughout this region rests unconformably on the strongly folded Ordovician shales. No one will dissent from the statement that this implies a long unrepresented time interval, first for the folding of the strata, which I doubt whether any one would consider catastrophic, though it may not have occupied the whole of the succeeding interpulsation period.

But someone will say : "Folding began earlier in some sections than in others, and while upper Ordovician beds were still forming in some localities, folding followed by erosion was going on in others." The possibility of this is not denied, though I maintain we have at present no reliable evidence for it. Certainly such deformation cannot be expected along the axial line of the geosyncline while sedimentation continued unabated in other axial portions, unless the pivotal point of rotation is situated within the geosyncline (*ante*, p. 277). Folding *might* go on in the geosyncline, while the marginal platform remained undisturbed and continued to be submerged, and the erosion of the folds proceded. I say it might, but I do not believe that it ever did happen. Certainly such an orogeny as the Taconian (Sil-oric) was a powerfully disturbing interruption, affecting the whole geosyncline, though the folds themselves may die out in various directions. Most certainly marine sedimentation cannot continue for long in the shallow geosyncline or on the marginal platform, whatever may happen in the deeper sea, should folds arise there, — but such folds are not eroded into a submarine peneplane as was once assumed.

I have already discussed the age of the Shawangunk conglomerate, and demonstrated that, where it rested on the folded and peneplaned Ordovician beds, its age could not be pre-Niagaran (Tuscarora) as has been claimed.

But it may be asked, "What about the Maryland section, where the conglomerate with Eurypterid beds rests unconformably on the folded and eroded Ordovician beds and is followed with apparent concordance by beds with a typical Niagaran fauna (Rose Hill)?"

This reported Silurian settlement or oasis in the wilderness of the surrounding areas of folding and erosion demands accounting for. Either we have here an area outside of the regions of Taconic orogeny, brought into association with it by the great Taconic overthrust, an area which could be reached early by the returning Silurian sea, needing no peneplanation to make it available, and where there is in consequence no unconformity between the Ordovician and later beds; or else the region is one of thrusting and faulting, and the puzzling relationship is of purely tectonic origin. [9]

The force which developed the Taconic geosyncline as well as the Acadian, and which afterwards destroyed them again by the folding of their strata was generated by the persistent westward movement (p. d.) of the African colossus, with the resulting eastward shifting of the south polar site across the entire width of that continent, from the western Sahara to the Nyassa location and on to Mozambique by Devonian time. The geosyncline, an unstable region, lay at practically right angles between these two powerful stable masses, the agressively mobile one (Africa) crowding the resistant static one (North America), which moreover was backed by the resistant sima mass on the west of North America.

Thus the thrust was entirely from the south-east to the north-west (p. d.), as the folding and overthrusts indicate, and as is shown by the north-east south-west strike of the strata and their cleavage planes.

5

Not until the end of the Devonian, when the polar site moved relatively westward again, (p. d.) was there a change in the direction of pressure. At first thought one would suppose that the movement of the land was reversed, so as to bring the pole to the western side of Africa again, and that this would cause tension in the west. But the location by now of the polar stations near the lower (outer) end of the rigid African mass suggests that the movement was rather one of rotation or of torsion. In the present case

9) The trouble with the description of these sections is that the authors forget to mention the nature of the contacts,—especially to discriminate between unconformities, which are the evidence for orogenies, and disconformities, which are merely records of withdrawal of the sea during interpulsation periods, and demand no more than a normal time interval between systems.

a counter-clockwise rotation of about 30° is indicated, with a pivotal point somewhere in North Africa.[10] The movement would cause the European mass (Spain, France, England, etc.) to impinge against the Acadian mass, with the result that the Acadian area, perhaps as far as Massachusetts, suffered folding and thrusting with the force apparently coming from the Atlantic of today (Acadian orogeny).

Of course, a similar effect would be experienced by the adjoining margin of the actively pressing block — and this we find in eastern France, where the Laval Syncline in the Maine Province (Long. 0° 46′ W., Lat. 48° 4′ N.) shows well this unconformity.   This is known in Europe as the Bretonian orogeny.

The Devonian, which forms an apparently continuous series with the Upper Gotlandian in the Finistere basin [11] is, represented by the Gedinnian and the Coblenzian, that is the old Lower Devonian, equivalent to the American Helderbergian.   This, with the Upper Silurian forms the SILURONIAN PULSATION SYSTEM.   The terminal beds of sands and shales contain calcareous lentils, which enclose *Phacops potieri* associated with *Uncinulus orbignyanus, Nucleospira lens,* etc.; they were thought by Haug to represent Eifelian, but they are still referable to the Siluronian as uppermost Gedinnian. [12] The typical Eifelian forms do not occur in these shales.

These beds are folded into a series of synclines and overlaid by Dinantian, with a basal conglomerate or pudding stone containing pebbles of Devonian rocks, and with a strong angular unconformity.   The pudding-stone is followed by shales with *Cardiopteris polymorpha, Sphenopteris elegans* and *Rhodea gigantea,* plants indicating late continental Dinantian.   Then follow black limestones, with *Productus giganteus* and other typical Viséen fossils.

Finally it should be noted that the Dinantian is transgressive and passes from the Lower Devonian to Silurian or even Algonkian. [13]

---

10) This would place it about at the 30° apex of an isosceles triangle, with the two polar points at the opposite angles. (It is best to use the apparent direction on the map, rather than translate it into its opposite for south polar-position. That will avoid confusion).

11) This is 175 miles farther west in the Department of Finistere 48° 13′ N., 4° 4′ W. (p. d.),

12) *Phacops potieri* characterizes the higher greywackes of the Upper **Gedinnian** of the Rhine region below typical Eifelian.

13) HAUG :   *Traité*, p. 771.

Thus the statement which satisfies the tectonic geologist that "the Lower Carboniferous rests with a strong angular unconformity on the folded Devonian" is misleading and must be modified and amplified as follows:

In the Laval syncline, the Siluronian rocks (old Lower Devonian and Old Upper Silurian, so far as present) are strongly folded, this folding having taken place during the Oriskanian interpulsation period, and perhaps in a part of the early Devonian [14] time. Then the remaining time of the main part of the Devonian pulsation period, that of the succeeding interpulsation period (Etrœungtien, etc.), and the time of the duration of the entire Fengninian pulsation period (Tournaisian, Chiussuan and inter-Dinantian interpulsation period) were all used up in the erosion of these old folds, and it was only later that Viséen plant-bearing continental beds were laid down with a basal conglomerate. Then came the transgressing Viséen sea with the *Productus giganteus* fauna. The orogeny must be regarded as Dev-oric or early Bretonian (Erian?), but it may have continued until post-Devonian time, and so include Feng-oric.

In Stille's statement of the tectonic relation as quoted above, he suggests that Dinantian follows with a violent unconformity on folded and truncated Devonian, but no time is apparent for the folding and enormous erosion of the 'Devonian' strata. The correct statement of the relationship is that folded and eroded Siluronian strata are followed unconformably by Visemurian. This at once implies that two entire pulsation periods — the Devonian and Fengninian, and three interpulsation periods (Oriskanian, Mont-plaisantian and Mauchchunkian) are absent, an interval of more than usual length thus being allowed for all the folding and erosion observed.

Probably more exact analysis of other orogenic discordances would show an equal discrepancy in the time element. As the Acadian orogeny was the American counterpart of the Bretonian, it follows that both should show essentially the same character and age of strata involved; and this is what we actually find.

According to Stille the Bretonian (Feng-oric) folding is very pronounced in the Tien Shan, as might be expected

---

14) Early mid-Devonian in the older classification.

from the fact that the same rotary (counter-clockwise on map) movement of the African mass that caused the Bretonian folding of France and the Acadian of North-East America also exerted a northward pressure via India on the Asiatic land-mass, which was then expressed in the folding of Devonian and older strata north of the Tarim Basin (See Map, Plate VIII). If these Devonian strata are involved in these folds, this orogeny must be referred to the Feng-oric.

The first Andean folding of western South America was the direct result of the westward pressure of the African mass, when it moved from the south polar location in southern Brazil (approx. Lat. 30°, p. d.) in Visemurian time to its location in Natal in Moscovian time,—a trek of about 1,500 kilometers (about 950 miles), — pushing the old mass of South America before it westward. This was submerged, as were all the land-masses for seven-eights of their sial-mass in the resistant sima, and it was this static resistance that crumpled up the pre-Andean geosyncline, as it did the Andean at a later period, though under different influence.

In a discussion of the history of the pre-Andean geosyncline Dr. Kirtly Mather [15] expressed the opinion that the crumpling of the Devonian and older strata of this geosyncline occurred at the end of Devonian time, these strata being unconformably overlain by Permian and younger terrestrial strata. When it is recognized, however, that the motivating force of the crumpling, that is the African colossus, was moving east from Devonian to at least Tournaisian time and dragging South America with it, until the south-eastern part of Brazil was brought under the pole, it will be seen that the Devono-Fengninian (Monteplaisantian) interval was one of tension in the South American region rather than of compression, though Bretonian (Feng-oric) orogeny occurred in the north (p. d.). During the South Brazilian Uruguay location of the south pole, the Andean part of the Samfrau geosyncline was blocked by ice (Map Plate IX) as was also probably that part of the Amazonian geosyncline near the confluence with the Samfrau geosyncline, both regions

---

15) MATHER : *Bull. Geol. Soc. America, 1921.*

lying between 60° and 75° S. latitude. Thus, fossiliferous Mississippian (Fengninian and Visemurian) sediments could not accumulate in these areas, but were restricted to the more distant ice-free regions of lower latitudes.

At the end of Visemurian time, if not before, the westward movement of the blocks was inaugurated, and it was this irresistible force which was pushing the western border of South America against the resisting sima, that crumpled-up the pre-Andean geosyncline. There was thus ample time left for the erosion of these folded strata during Donbassian time and the long interpulsation periods, before and after the Donbassian, and it was only after this, in the early Permian or the Uralian period that deposition of continental sediments began again, this time on the eroded surfaces of the folds.

In the Argentine, however, where Devonian and Fengninian occur in sequence, they are perfectly concordant (though, of course, separated by the usual interpulsation

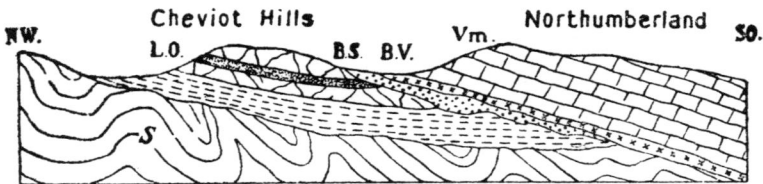

Fig. 87. Sild-oric, or Ardennian, and Dev-oric, or Erian orogenies. Caledonian geosyncline, Great Britain.
Cheviot Hills in south-east Scotland. Folded and peneplaned Silurian Beds (S), overlain unconformably by continental 'Late Lower Old Red' of Cheviot Hills, cutting out marine Siluronic (formerly Upper Silurian and Lower Devonian). The continental beds of the Cheviot Hills most likely represent post-Siluronian (Oriskanian) Interpulsation deposits. Another unconformity (Dev-oric) separates the Oriskanian from the overlying Visemurian (Vm), Devonian (and Tournaisian?) are absent, Basal Sand (B. S.), Basal Volcanic (B. V.). (After Goodchild).

disconformity). Moscovian, however, is absent in the Andean geosyncline, that being the period of folding and erosion, before the Uralian transgression. [16]

The main folds of the pre-Cordilleran Andes are found in the provinces of San Juan and Mendoza (Long. 67°30′ to 70° E., Lat. 28° to 38° S.) (p. d.) (Long. 165-180° W., Lat. 60°-75° S. on the map of the period), where they

16) STILLE, H: Grundfragen, p. 114.

XXV OROGENESIS IN PALÆOZOIC TIME 295

are strongly developed and steepened on the east (towards the under-thrust Brazilian mass). It is accompanied by sheet-thrusting (*Deckenbildung*) and involves horizons up to and including the Ecca formation of early Permian (Uralinskian?) age. Stille would correlate it with the Alleghany (Appalachian) folding. But this is a later pre-Andean folding than the Moscovian (early Donb-oric or Mosc-oric orogeny), while the folding of the main Cordilleran, *i. e.*, the Andean orogeny, did not occur until the Tertiary. The later pre-Andean (Donb-oric) orogeny is in harmony with the continued westward (clockwise) movement of the continent, which allowed an eastward migration of the polar site for many thousands of miles, first to east Antarctica near Ross Sea (Donetzian location), then to central Australia (Uralian, Pl. XI), and finally to India (Permian, Pl. XII).

The disturbances due to the shifting of the polar site from south-eastern Brazil to the Transvaal is known in Europe as the Sudetian (Visem-oric) orogeny, from the fact that the Sudetic mountains on the north-east border of Bohemia were formed at that time (Long. 17°-18° E., Lat. 50° N., p. d.).[17] This was due to the pressure of North Africa against Europe, which was translated into a north-eastward pressure of the Bohemian mass against the unstable land of Germany, and this resulted in the raising of the barrier by the folding of the strata into the mountains, the remnants of which are today, after prolonged destructive erosion, seen in the Sudetian Mts. of Bohemia.

The Carnic Alps of North Italy,[18] 46° 34′ N., 12° 50′ E. (p. d.) received what was probably the initial folding in post-Dinantian time, all the beds from the Ordovician or older, to the Viséen being strongly folded by north-eastward pressure from the direction of S.S.W., some of the synclines becoming closed and overturned folds. These are unconformably succeeded by conglomerates, sands and shales of Uralian age, above which the *Fusulina* limestones (Uralian) lie concordantly, though disconformably. All these were cut by later thrust faults in the same direction.

---

17) There was, however, an older, probably Sild-oric, folding, which is considered the main orogeny of these mountains. The newer is a rather weak one, because the northward pressure was not pronounced.
18) HAUG : *Traité*, p. 777.

The earlier folding may have been brought about by the shifting of the polar station from southern Brazil to the Transvaal, this resulting in a moderate pressure of Africa northward against the east-west Alpine geosyncline. This is the same pressure that affected the Sudetian Mountains. A much more powerful and longer continued pressure in the same direction was exerted by Africa during the next period, while the east Antarctic region (Station No. 9.) was moved on to the polar site.

Such pressure would involve the disturbance of the Moscovian beds, which are wholly absent from the Alps. This suggests that the period of folding began in the post-Visemurian interpulsation period, and probably was continued with intensification through the Moscovian, while erosion peneplaned the folds during the Donetzian and the succeeding interpulsation period, after which, in Uralian time, Stephanian beds, followed by marine Uralian (*Fusulina* limestone) covered the truncated edges.

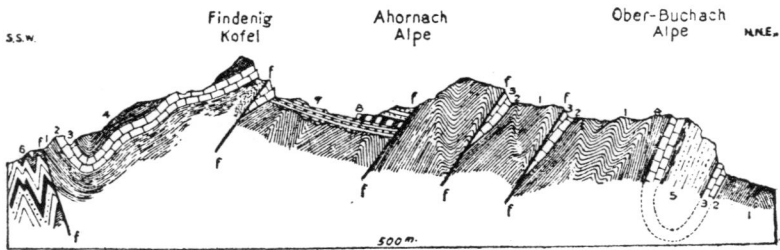

Fig. 88. Donb-oric Section across the Carnic Alps, showing the folded strata and the unconformities and faults (F).
1. Ordovician lutytes etc.
2. Black graptolite shales.
3. Silurian limestones.
4. Argillaceous Silurian (or Siluronian) shales.
5. Devonian limestones.
6. Dinantian graywackes and shales.
7. Conglomerates, sandstones and shales of the Uralian, unconformable on Ordovician.
8. *Fusulina* limestone. (After Geyer, from Haug, p. 777).

Stille hesitates between the Sudetian and Asturian orogeny of his classification (Visem-oric or Donb-oric in our terminology). Since, as we have seen, it is probably the longer movement which was more effective, we may place it in the latter, though the beginning was in the former, so that no Moscovian could be deposited.

The same pressure which carried Africa north against south Europe, drove Asia south against India. Here lay the great east-west geosyncline of North India, from which the Himalayas were born long after; more than 12,000 ft of Palæozoic rocks, the Dravidian Series (which includes a concordant series from the Cambrian to the Donbassian and possibly some Uralian), accumulated here before the polar center reached this region, and throughout Uralinskian (especially Artinskian) time kept the sea out of this geosyncline, because of the great development of the polar ice-cap.

The Aryan era, which succeeded, began with the chaos of the polar ice-sheet invasion, shown by the Talchir boulder bed, whose center of accumulation, for a time at least, lay in the region of the Aravalli Hills of North-West India today (Long. 73°40′ E., Lat. 25°30′ N.,) This condition of glacial chaos was, of course, only temporary, like all such disturbances in the orderly development of the earth. When it was over, the Himalayan geosyncline was much as it had been before, and the interrupted late Palæozoic development was resumed and continued under conditions similar to those in other parts of the world.

The Himalayan geosyncline is probably the longest continued geosynclinal depression on the surface of the earth. It began in Cambrian time, if not before, and continued to the end of the Eocene, when it collapsed, to rise in the Himalayan Mountains. Whether pressed from the south by India, during counter-clockwise rotation or from the north against India during clockwise rotation of the sial-mass, it continued to sink steadily, and throughout its history great thicknesses of sediments accumulated, either in shallow water or just above the level of the sea, which was kept out by the abundance of the sediments. But if the Himalayan geosyncline did not respond to the continued pressure of the various rotary movements, except to continue its subsidence, the weaker areas near its northern border did respond.

In South China, according to the present state of our knowledge, the Asturian or Donb-oric (Pennsylvanian) orogeny is a most important one in Yunnan and is widespread, occurring in Szechuan and in the Nan-Shan —

Tsingling-Shan and is also reported from Indo-China. Part of these may be referable to the Uraloric or Saalian.

The Perm-oric orogeny, which was also a clockwise movement of the sial, involved the folding of the Uralian and older formations, but not the Permian. It originated in the shifting of the polar station from the central Australian location to the Burma site.

This resulted in a powerful eastward pressure (p. d.) of Europe against Asia, and the most notable effect was the folding of the north-south (p. d.) Ural geosyncline into the Ural Mountains (Long. 60°, p. d.). It forced Spain, with Africa backing it, northward against France, and the Pyrenees, an east-west range, received its first

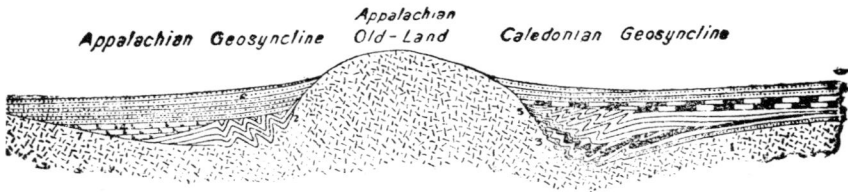

Fig. 89. Cross-section of the Appalachian old-land and its bordering geosynclines, the Appalachian (St. Lawrence) on the west and the Caledonian on the east.

The diagram represents the end of Permian time. The numbers indicate the formations and the periods of folding in the two geosynclines.

compressive folding. In the great ring formed by the Donetz and the lower course of the Don River (joining in Long. 41°, Lat. 47° 30′ N., p. d.) lies the renowned Donetz coal basin of South Russia (N. E. of Sea of Azov). The beds are strongly folded into sharp anticlines and synclines, which extend east and west, and thus imply pressure from the south. This was also the direction of the pressure required for the development of the geosyncline in the first place, with its more than 4,500 meters (nearly 15,000 ft) of clastic deposits, part shallow water marine, and part continental and coal-bearing (Devonian, Visemurian, Moscovian, Donbassian). This pressure was developed in part during the passage of the polar-site from the east Brazilian (Visemurian) to the Transvaal (Moscovian), but especially during the change from the latter to the east Antarctica location.

The clockwise rotation which brought about this change also developed a powerful northward pressure by the African-Arabian masses (there being as yet no Mediterranean sea). The folding was probably brought about by the pronounced pressure in the same direction during the rotation from the Donbassian to the Uralian site in Central Australia), and possibly during the next stage in the shifting of the Indo-Burma site over the pole. If it seems strange that northward, north-westward and eastward, and even southward pressure can be applied by the same rotation, it must be remembered that the rotation about a pivotal point forces different parts of the pangæa to move in different directions. This can easily be verified by experiment.

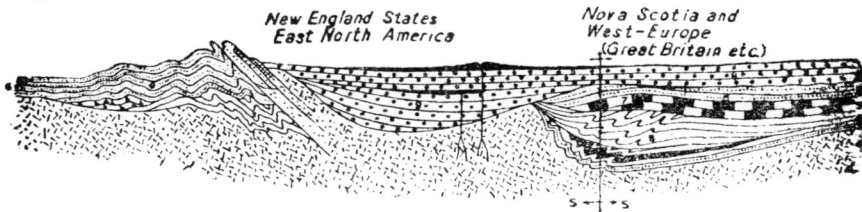

Fig. 90. The same Cross-section as fig. 89. after folding of the strata of the geosynclines to form the Appalachian Mountains, i. e., the new old-land. The sinking of the old land of Appalachia formed the Triassic geosyncline, while the Caledonian geosyncline suffered no folding, but became the Marginal Plain of the new geosyncline, which at first, however, was entirely filled by continental strata. The later (Muschelkalk) invasion from the southeast is not represented. Note that the Triassic beds rest concordantly, that is without unconformity, on the beds of the old Caledonian geosyncline in Western Europe and Nova Scotia, while they rest directly on the crystalline old-land in the East American region, that is the part which now forms the coastal plain of the United States, and the western part of the New England States. The vertical line indicates the approximate point at which the continents subsequently separated, this being essentially along the line of the old Antillean geosyncline, shown on the Map, on Plate I. The following is indicated by the numbers.

Appalachian Geosyncline of America	Caledonian Geosyncline of East Canada (Nova Scotia, etc.) and Western Europe
1. Archæan Crystallines	3. Folded Silurian to Viseen strata
2. Ordovician (Taconian folding)	5. Uralian-Stephanian (Rotliegendes)
4. Silurian-Devonian	Saalian Folding
6. Mississippian to Permian (all folded into the Appalachians)	7. Zechstein
9. Triassic on old-land	8. Tartarian followed by Disconformity
	9. Triassic, or New Red Sandstone concordant with the Permian

The last of the great orogenies that owed its inception to the clockwise rotation (p. d. reading) of the sial
was the Permian, or the Appalachian orogeny of America.
The latter must be placed here because none of the subsequent movements of the pangæa could have a compressing effect on that part of the earth, such as most of the
earlier movements had.   This first affected the continued
sinking of the geosyncline, allowing the deposition of
thousands of feet of strata, after which the folding came as
the final act.

This was the most significant orogeny of the eastern
and southern United States, and in fact of the whole
Palæozoic.   From the time of the first great folding, the
Taconian or Sil-oric, the geosyncline had been undergoing
continued sinking under the uninterrupted pressure from
the south-east (p. d.), mainly effected by the African mass,
especially from Mississippian time onward.   Every stage
of the clockwise movements that brought about a polar
shift effected a powerful pressure of this ancient land-mass
directly against the North American coast, keeping the
geosyncline constantly active to the last.

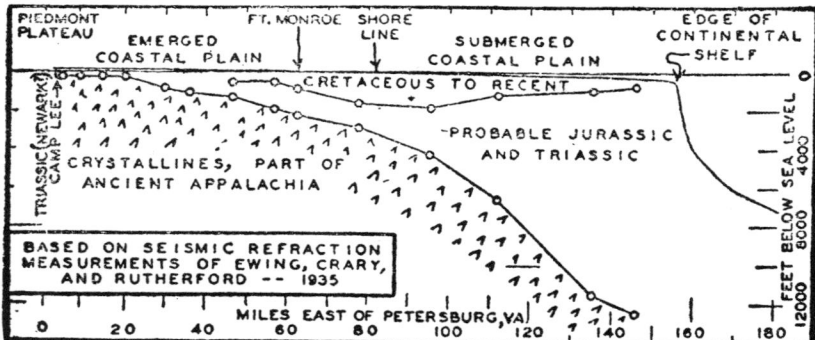

Fig. 91. Diagrammatic cross-section of submerged Appalachia and its
Mesozoic deposits; from the coast at Camp Lee to a point 180 miles out to
sea, showing character and deposit from sea-level to a depth of 12,000 feet
below it, as determined by the records of seismic refraction. (From Gutenberg)

But by Permian time the subsidence had become so
great, as measured by the amount of detritus deposited in
it (40,000 ft), that the floor of the geosyncline was brought
within the region of temperature high enough to sufficient
ly reduce its strength, so that the continued pressure

finally effected the buckling of the strata and the formation of the Appalachian folds. This may have begun in mid-Permian time, which is the age of the last division the Dunkard group (1,300 feet thick) with the Washington coal at the base. But it may be later, for the later Permian beds may have been removed again by erosion. No later movement, however, could effect this buckling, for the post-Permian movement was a counter-clockwise one, which resulted in the release of pressure against North America by Africa, permitting the collapse of the old-land of crystalline rock, and thus producing the Triassic geosyncline. It resulted, in effect, in the migration of the geosyncline into the old-land. (Fig. 90).

The totality of this pre-Appalachi-oric movement involved a clockwise (p.d. reading) rotation of the pangæa, through an arc of nearly 180° and over a distance of roughly 9,500 miles, fully one half of which was accomplished by Donbassian time, the remainder during the Uralian and Artinskian. During all of these movements the brunt of the pressure was borne by the Appalachian-Caledonian and the pre-Andean, pre-Cordilleran and Franciscan geosynclines, all of which suffered folding at one time or another during this interval, an interval which lasted from late Visemurian until after the close of Thuringian (Zechstein) time.

After that, began the tremendous movement in the opposite direction, that northward pressure movement that caused the fan-wise collapse of Asia. So great was this last movement that it carried Antarctica over the South Pole, removing that pivotal point some 4,000 miles (55°) at right angles to the axes of the folds, perhaps in two separate stages. At the same time it caused the continued sinking of the Himalayan geosyncline in Permian time and later periods. This culminated, in the Oligocene period, in the final northward move of India, which resulted in the elevation of the Himalayas, while at about the same time the northward pressure of Africa was instrumental in the final folding of the strata of the Alps, etc. But by that time the Atlantic rift was well under way.

CHAPTER XXVI

THE LATER PALÆOZOIC PERIOD

1

WE must now go back and briefly glance at the conditions in the various regions of sedimentation on
the Pangæa during post-Visemurian Palæozoic time. The
remaining pulsation and interpulsation periods can be
grouped as follows in descending order:

14′. Post-Permian Interpulsation Period
14. Permian Pulsation System
13′. Post-Kongurian or Lebachian Interpulsation
Period
13. Uralinskian Pulsation System
12′. Lochinvarian Interpulsation Period
12. Donbassian Pulsation System
11′. Lanarkian Interpulsation Period
11. Visemurian Pulsation System, etc. —

The primary control in these later as in the earlier
periods is the polar location and the influence which the
polar ice-cap had in modifying sedimentation in the geosynclines. As heretofore, Africa makes the chief claim to
furnish the polar sites, though towards the last, east
Antarctica, Australia, and India qualified as temporary
polar hosts, each during a limited portion of a pulsation
period. Except for minor modifications due chiefly to the
successive orogenies, the essential location of the geosynclines
and of the epi-seas remained unchanged, but different portions
varied in their ratio of subsidence, primarily because of
varying pressure due to the rotary motion of the continental
masses, which resulted in the shifting of the polar site.
Such increased depression also generally affected the
marginal platforms, which then became more widely
flooded and covered by marine sediments or by river-
spread continental clastics. The effects due to isostatic

readjustment are not to be overlooked, but these probably lag far behind those due to lateral pressure which was induced by rotary motion of the sial cap.

Our survey will begin with the polar region as the center of control. It will be remembered that in Dinantian (Fengninian and Visemurian) time, the pole lay somewhere in the region of the Rio Grande do Sul in southern Brazil, [1] on the mid eastern border of South America (approx. 32° S., 52° W. p.d.). It is possible that in Visemurian time the eastward shifting of the site had begun and a location in West Namaqualand, on the adjoining African coast, had been achieved. This is indicated by the direction of the striæ on the glaciated rock floor beneath the Dwyka tillite of South Africa, as recognized by Du Toit. It must, of course, be understood that the glacial erosion and scouring of this floor antedated, probably by long periods, the deposit of this tillite, for after it was spread over any part of the rock surface that surface was preserved from further scour.

The scouring of the rock floor beneath the Dwyka tillite may have commenced in Fengninian time, when the ice-center was in the Urugayan location, though during its translation from the Devonian to the Fengninian some parts of the old floor, now covered by the Table Mountain series, may have been scratched. The glacials in the Table Mountain beds are not associated with a striated basement, and may represent, in part at least, ice-rafted material, as suggested by Du Toit.

From the time of development at the Dinantian center, however, until its passage from South Africa to Antarctica in post-Moscovian time, the scouring of the old rock-floor, and the subsequent deposition of the tillite must have been in progress in South Africa. Du Toit gives some intermediate locations for the ice-center during Dwyka time, these being in succession, after Urugay: 1. Southern Brazil, in South America, 2. Namaqualand, South-West Africa (at that time adjoining the South American locality), 3. Griqualand West; and 4. Transvaal, in Moscovian time. Thence it apparently passed to ice-center D(5) on Du Toit's map, in or near Natal on the south-east coast, or perhaps on

---

1) It may have been farther south, in Uruguay.

the adjoining west coast of Antarctica.   Throughout this
time, and until the polar point had moved eastward
beyond the center of Antarctica, the western margin of a
20 degree ice radiation would impinge on the South African
coast and contribute to the morainal material there deposited.

During all this time the Samfrau geosyncline was
ice-blocked, or at least infested by icebergs, and thus
this section became a barrier between the South American
and the Australian portions of the geosyncline.

2

In the East Australian part of the Samfrau geosyncline the Dinantian is represented by the Burindi Series,[2]
which includes both the Fengninian and the Visemurian
pulsation systems of our classification.   During this time
the polar ice-cap centered in the Uruguay-Rio Grande do
Sul area, but was probably moving eastward to its Transvaal
location in Moscovian time.   This was a period of heavy
glacial scouring in South Africa, though deposition of
morainic material (except ground moraine) came much later.

In South Africa the Cape System underlies the glacial
beds of the Karoo (Dinantian).   The lowest division of
the Cape System is the unfossiliferous Table Mountain
series, 2,000-4,000 ft thick, which rests with a strong
discordance on the folded older series of pre-Devonian
age.   It also contains a glacial band formed by an extension of the Nyassa lobe in Siluronian time, while the
general source of the sediments lay on the north.   The
Bokkeveld series, next younger, follows conformably and
consists of 2,500 feet (maximum) of shales and sandstones.   The lower three-fifths of the series yields marine
fossils of early Devonian age, with relationships to similar
faunas in the Falkland Islands off the coast of South
America.   These at the time were within the same
geosyncline, as were also the Argentine, southern Brazil
and Bolivia, which have the same fauna (See Pl. IX).
More distant relationship is seen with the North American

2) Schuchert correlates the Burindi with the Burlington and Keokuk, while
the Australian geologists mostly correlate it with Viséen.   The series is
thick enough to cover both, as well as a dividing interpulsation period
between, though this has not been located.

(Oriskanian) and the European (Coblenzian).

The higher beds carry only poorly preserved plant remains and indicate purely continental deposits, perhaps induced by the oncoming glacial condition.

The succeeding Witteberg series about 2,500 ft. thick, is probably of Upper Devonian age and certainly represents deposition before the oncoming of the ice [3]. Some of the fine-grained beds may represent deposits of fine sands and muds brought by streams from the still distant ice and deposited on river flood-plains or in lakes. All the beds are strongly folded into isoclinal folds, but this folding is post-Karoo.

Surprise has often been expressed at the remarkable similarity, often amounting to specific identity, between the marine faunas of the Bokkeveld and those found in the Falkland Islands, Argentina and Southern Brazil. Indeed, as Du Toit says, the description of the formations of the Cape System : "applies almost word for word to the beds in the Falklands, the three divisions in those islands being all but identical in their lithology, thickness (so far as can be judged), and fossils with those of the Cape, and are overlain by the Lafonian Tillite, the undoubted equivalent of the Dwyka Conglomerate".[4]

In Argentina the relationship is equally close, even to the isoclinal folds of the strata. In Brazil and Bolivia the succession and faunas are the same, while the lithology varies only to a moderate and predictable degree. Similar conditions continued through the Permian and Triassic.

When it is remembered that these regions are separated by 80 degrees, or about 3,500 miles, of open Atlantic Ocean, one is forced to admit that only conjunction in a continuous geosyncline will explain this close relationship, unless one has recource to the imaginative extreme of a land bridge across these 3,500 miles of deep water, or to the equally imaginative connecting continent which has since been swallowed by the sima of the ocean bottom.

---

3) The plants are all poorly preserved, but the genera identified suggest early "Carboniferous" affinities. A Eurypterid (*Hastimima*) is represented by fragments of body segments.
4) DU TOIT : *South Africa* p. 199.

3

While the south polar ice-center occupied the Southern Brazil site the continuity of the Samfrau geosyncline was interrupted as a pathway for faunal migration, because it was partly filled by the polar ice masses. The influence of this ice probably did not extend above 20° or 25°, *i. e.* to 65° or 70° South Latitude of the time. This left both the Ross and the Caribbean epi-seas open (each then in about 20° to 30° South Latitude), and these could continue as faunal feeders, the one to the Australian-Burindi fauna, the other to the American-Mississippian fauna, unless these latter were wholly supplied from the "Boreal" part of North America. This emphasizes and explains the difference of those faunas in the temporarily dis-united part of a peripheral geosyncline.

The Burindi series is extended northward across Tonking to Yunnan and the Yangtze Mouth near Nanking, as elsewhere discussed in detail. [5]

The general extent of the transgression is marked by the convex side of a southward curving broadly V-shaped lobe, extending from Nanking south-west to Changsha and Kweiyang, then north again through Weining (approx. 105° E., 27° N., p.d.), thence north and north-west to Lat. 35°, Long. 100°, and onward to the Nan-Shan Mountains.

In the Nanking region the Fengninian is represented by nearly 60 meters, the Viséen (Shangssuan) by about 10. Then follows the Weiningian, 90 meters, above which lie Uralian beds. Southward all these increase, until they reach their maximum [6] in Kweichow (Tushan Section). The Viséen of South China indicates a renewed transgression, and rests disconformably on the Chiussuan. [7] It contains such well known European brachiopod elements of the Viséen an *Daviesiella comoides* (Sowerby), *D. llangolliensis* Davidson, *Productus giganteus,* etc., *Chonetes papilionaceous*

---

5) TING and GRABAU: *The Carboniferous of China* pp. 555-571.
6) Kolaohoan, or transgressive Lower Fengninian, 450 m.; Chiussuan or retreatal Fengninian, 237 m.; transgressive Viséen or Shangssuan of the Visemurian pulsation. 440 m., followed by Weiningian, 145 m., and this, after, a disconformity, by Uralian, 115 m.
7) This is mostly a thick continental series without fossils, except in Kweichow province, where it is characterized by the unique *Cryptospirifer* fauna. Elsewhere extensive erosion marks the pre-Visemurian emergence and interpulsation hiatus.

Phillips, and many others, besides characteristic compound corals.

The succeeding Weiningian follows closely the northern outcrop of the Viséen from Nanking to Weining in Kweichow, where, however, it turns sharply southward and westward, thus clearly showing the termination of the geosyncline on the north-west and its severance from Central Asiatic waters by the relative rise of the Tibetan mass. In other words regression of the sea had commenced, and the Weiningian represents the falling sea level of the Visemurian pulsation system.[8] This nowhere reached the region in which the typical Moscovian sediments (Penchi Series) of North China were deposited, while, moreover, the typical Russian Moscovian did not reach the Samfrau geosyncline, or if so, only temporarily, any more than it entered the American geosyncline as a glance at the map (Plate X) will show. From Kwangsi in southern China the Weiningian (Namurian) series has been traced into Indo-China (Laos, approx. 18°-20° N., 102°-104° E. p. d.), where at Cammon, Fromaget found Namurian fossils. In a general way the line of the geosynclines passes west of Sumatra.

4

The Moscovian is represented in the East Australian part of the Samfrau geosyncline by the Wallarobba con-

---

8) This was tentatively suggested by us (Ting and Grabau) in our 16th International Geological Congress article, on the basis of the Weining fauna (including Laokanchai and Huanglung—local names in other provinces), and of the striking distinctness of the Weiningian from the Moscovian (Penchi) fauna of North China. Although the preserved margins of the two formations in the Nanking region approach to within 50 miles or less of each other, the beds, judging from their thickness and character must at one time have been much closer. Taking our faunal lists of 1933 as a basis, we find that of the 27 species of corals described from the Weiningian, not one occurs in the Penchi, while of the 11 species listed from the latter, not one occurs in the Weiningian; of the 23 species of brachiopods described from the Weiningian (including Laokanchai) of South China, and 28 species from the Penchi of North China, not a single species was common to both. Only among the Foraminifera is there any relationship. Of the 15 species described from the Penchi and the 27 of the Weiningian only 8 are common to both. Moreover, the Weiningian has such characteristic Viséen genera as *Daviesiella, Cryptospirifer, Striatifera striata* (a Namurian form in Europe) and *Athyris trigonalis*, while the Penchi has a typical Moscovian fauna including such index fossils as *Enteletes lamarcki, Spirifer mosquensis, S. strangwaysi*, etc.

glomerate, which rests disconformably on the Burindi and attains a thickness of 2,300 ft. It merges upward into tuffs with conglomerates, and in the upper part contains a ten-foot plant-bearing bed, with *Lepidodendron osbornii* as the dominant form. By this time the glacial center had probably advanced some degrees eastward (on Antarctica) and a lobe of the ice-sheet extended to south-eastern Australia, where a glacial bed 100 feet thick overlies the great conglomerate and testifies to polar influence. But now volcanic phenomena awoke anew as the result of the rifting of the crust by tension, and 2,600 feet of lava were poured out over the older sediments, interspersed with tuffs and conglomerates. Glacial conditions vied with volcanism for supremacy, as in Iceland to-day.

The glacial stage of this period has been called a rare and remarkable example of glaciation for Carboniferous time, but on our reconstruction of Pangæa it was to be expected. For by this time the south polar center of glaciation had reached its station in east Antarctica, from which point it not only influenced much of Australia, but by slight westward extension may have been responsible for giving the final touches to the great Dwyka tillite of South Africa.

The contemporaneity of the Dwyka glacial tillite, that most remarkable and best known example of fossil glaciation, with other late Palæozoic tillites has frequently been suggested. But the Dwyka is itself a compound deposit. Glaciation probably began to touch South Africa in late Devonian time when the ice-sheet shifted to the east coast of South America, as already outlined[9]. With continued eastward translation of the ice-center an opportunity was furnished for deposition of the old morainic material on the glaciated rock surface (in so far as this had not been covered by ground moraine).

The polar ice was inexorably advancing under the clockwise rotary movement of the crustal mass. This, translated into pressure on the west and north, became a drag in the Australian region. That is why volcanic activity and glaciation were both concentrated in this

9) For an exceptionally fine photograph of the Dwyka glaciation see: DU TOIT: *Geolgy of South Africa*, plate XXIII (opposite p. 112).

region, the former to an unprecedented degree. Both were consequential phenomena, and both were due to the persistent clockwise rotation of the crustal mass.

The major glacial stage of the Donbassian period is known as the Kuttung series. According to the Australian geologists : "The beds of this stage, which have a maximum thickness of about 4,700 feet, consist mainly of tillites, fluvio-glacial conglomerates, varved shales and tuffs. The tillites and fluvio-glacial beds at the base of the series contain an abundance of large boulders as much as 12 feet in diameter, many of which have been derived from far distant sources and some of which are ice-scratched and facetted. At one locality these beds rest upon a glaciated pavement. The varved shales. . . . contain fine examples of contemporaneous contortion."[10]

That a region, which was subjected to such violent and conflicting phenomena, as the alternation of glaciation and of volcanism must have produced, and to the devastating floods that must have resulted whenever the glaciers succumbed to the volcanic heat, should still be able to support life, is a telling illustration of the phenomenal power of adjustment and tenacity of resistance to extinction, which had been developed, especially by the plant life of the period. Embedded in both glacial and volcanic deposits we find the record of the flora of this volcanic 'Iceland' of the past, a flora which has no relation to that of the tropical world into which this ice and lava-scarred land block has since glided.

Even in the deposits from the ice-sheets we find the remains of nine species of ferns, among which the genus *Rhacopteris* predominates. This gives the flora its name and distinctive character.

Nevertheless this flora was doomed, for, as the ice center finally became established in Central Australia in Uralian time, practically the whole of that continental mass— still a part of the Pangæa, was buried under ice. The glacial beds deposited by the ice-sheet at this time are known as the Lochinvar tillites, and are only 300 feet thick, most probably because they were wholly formed under the ice-sheet.

---

10) DAVID and SÜSSMILCH: XVI *Int. Geol. Congr.* p. 633.

5

The typical Russian Moscovian fauna originated in what is today the frozen north, but was, in the time we speak of, within the zone of mild temperate climate. Its pathway of migration is now marked by the Ural Mountains, which today form a north and south dividing line between Europe and Asia, but in late Palæozoic time were a sinking geosynclinal region flanked by broad marginal platforms which, when submerged, supported a wealten of growing organisms. What is today Central Europe was then in the tropical belt, and the seas which covered it supported a rich lime-secreting fauna, among which the bryozoans were the chief reef builders.

The Moscovian sea extended over South Asia and into China as far as Peking. Its deposits are known as the Penchi series, which consists of less than 100 meters of shales, sandstones and thin limestones, with coal seams in some sections, and rest disconformably on Ordovician limestone. It is covered disconformably by beds of Uralian age (Taiyuan series), which mark the second invasion of China from the Ural geosyncline. Between these two invasions there was a pronounced retreat of the sea, while the deposits that were formed in the retreating sea were entirely eroded again from North China and from the principal region of the Ural geosyncline, during the succeeding long pre-Uralian interpulsation period. Only in South Europe (Donetz Basin) and in an arm that extended to England and the north of France, were these deposits preserved.

In the Donetz basin the Moscovian is followed by the Donetzian, comprising 500 meters of shales and limestones with marine fossils and coal beds (Flora VI), and this by 885 meters of arkosic sandstones, shales and, rarely, limestones, partly marine, but also coal-bearing beds. At some horizons these beds carry species of brachiopods and pelecypods and *Fusulina* (*F. gracilis*), also characteristic of the North American later Pennsylvanian, showing that some connection between the neighboring epi-seas was maintained. In the arm of the Russian sea which extended south-westward across Germany to England and northern France and the Pyrenees, the Visemurian is followed by the Millstone grit, a quartz pebble conglomerate largely, if not wholly, occupying the interpulsation period. This is followed by continental plant- and coal-bearing beds, the Westphalian or Moscovian and the Staffordian and Radstockian, which together represent the Donbassian.

Since Uralian beds everywhere in North Europe and Asia follow directly but disconformably upon the Moscovian, or Middle Carboniferous beds, they have been generally but mistakenly classed as Upper Carboniferous, because frequently the great erosion hiatus between them and the underlying Moscovian beds is not very apparent physically, and is commonly overlooked.

When, however, it is realized that in the Donetz basin the place of this disconformity is taken by over a thousand meters of partly marine and partly continental coal-bearing beds, and in Great Britain and adjoining districts by a similar, though more completely continental coal-bearing series, it must be apparent that the reference of the Uralian to the Upper Carboniferous is erroneous.

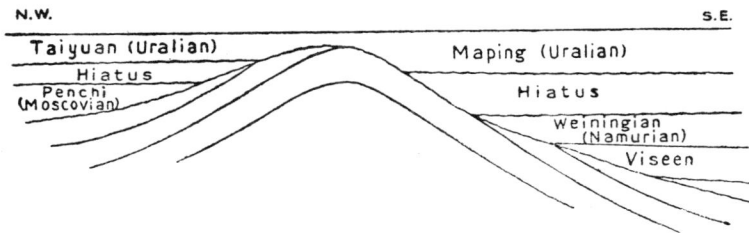

Fig. 92. Section across East China, showing on the North-west the Ordovician followed disconformably by the Penchi (Moscovian) and this, separated by a hiatus and interpulsation disconformity, by the Uralian. On the S. E. the Maping (Uralian) is preceded by the same interpulsation hiatus, below which is the Weiningian (Namurian) and Viséen (=Visemurian). Here the Moscovian is also absent, i. e. the hiatus covers the whole of the Donbassian. (Owing to exaggerated vertical scale the Ordovician beds dip too steeply).

The Uralian, then, represents a renewed transgression of the sea from the Ural geosyncline, which spread as far as the Peking region of North China on the one hand, and over South China on the other, extending to the Yangtze near Nanking. The South China deposits are known as the Maping limestone, a richly fossiliferous white limestone, which covered essentially the area occupied by the Weiningian sea, upon the deposits of which it rests disconformably. Thus, while in the North China region the Uralian (Tayuan) rests with a hiatus and disconformity on the Moscovian (Penchi), in South China the same Uralian, represented by the purer highly fossiliferous Maping limestone, rests with an even greater hiatus upon the Weiningian, the equivalent of the European Namurian. [11]

6

The late Palæozoic rocks of Japan are still very incompletely studied, but it would appear that in Viséen time the North Asiatic geosyncline had established a connection with the Pacific coast, where a marginal epi-sea may have been forming, from which, later on, faunas were supplied to some of the geosynclines. This Okhotskian epi-sea, as it may be called, probably covered Sakhalin and Korea, where, in the western part of that peninsula, still little studied formations of the late transgressions are known. Practically the whole of Japan was under this epi-sea, this being indicated by the great thicknesses of limestones with only intermittent clastic beds and some coals.

The Omi limestone, 2,000 meters thick and typically developed on the west coast of the main island of Japan, includes characteristic Dinantian fossils (*Lonsdaleia floriformis*, *Syringothyris*, *Productus edelburgensis*, *Retieularia lineata*, etc.), which suggest that this limestone includes both Tournaisian and Viséen. The interpulsation hiatus still awaits location.

On the east coast the lower part is chiefly a clastic graywacke-sandstone, with limestone lenses, suggesting the presence of a land-mass on the Pacific side, from which those sediments were derived, and which has since entirely disappeared. These beds are overlain by limestones with subordinate clastics comprising Upper Viséen (Visemurian) (*Lonsdaleia* zone) Donbassian (Moscovian, *Fusulinella* zone), Uralinskian (*Shwagerina* zone), and Permian (*Neoschwagerina* zone), the entire fossiliferous series being more than 1,000 meters thick. They also include the interpulsation gaps.

---

11) GRABAU : *Maping Limestene*

In the southern part of the Kitakami Mountains on the northeast coast of Honshu Island (approx. Long. 40°), the following succession has been recognized, according to Endo's approximate correlation: [12]

Middle Permian (Permian pulsation system).

7.	Slates and sandstone	200 meters
6.	*Fusulina* limestone	60 meters

(Hiatus) and Disconformity

Lower Permian — Upper Carboniferous (Uralinskian and Donbassian(?) pulsation systems)

5.	Slates and sandstone	400 meters
4.	Crinoid limestone	400 meters
3.	Slates and sandstones	600 meters

(Hiatus) and Disconformity

Lower Carboniferous (Visemurian pulsation system)

2.	Coral limestone	300 meters
1.	Coarse sandstone	80 meters

The coral limestone contains *Lithostrotion, Lonsdaleia* and *Syringopora,* etc. of British types. Between the Carboniferous and Permian a strong disturbance has recently been discovered, suggesting a profound disconformity.

Thus this section, by its abundance of clastic material testifies to a land-mass on the east, which enclosed this region as a part of the East Cathaysian geosyncline. Westward, limestone seems to become more prominent in the sections.

Post-Palæozoic volcanism and intense disturbance by folding and faulting characterize all the outcrops, and frequently have metamorphosed the deposits.

The metamorphic biotite and mica-gneisses and schists of the Loochoo Islands (Long. 124°-128°, Lat. 24°-28°) as well as the quartz schists, crystalline limestones and metamorphosed sandstone and slates of these islands are thought to be metamorphic derivatives of these 'Anthracolithic' beds, but may be in part fragments of the old

---

12) ENDO in HAYASAKA: XVIth Intern. Geol. Congr. Report. p. 587

land-mass, that bounded the Cathaysian geosyncline on the east. To this Formosa may likewise belong.

What were the conditions in the western part of the Pangæa at this time? The Samfrau (pre-Andean) and Franciscan geosynclines of South America were buried under the ice of the Uruguay ice-sheet, and this probably interrupted the Amazonian arm as well, though north-west Africa was probably submerged.

7

There is no direct evidence available to show that the Caribbean epi-sea at this time functioned as a feeder of the North American geosynclines, these being largely supplied from the West Boreal (Canadian) faunal epi-center, in Lat. 30°-40° of the time. The interior North American seas were grouped in the vicinity of the equator, and the coral and other warm water elements of their faunas were indigenous. But if distinct during the greater part of Mississippian time, that distinctness came to an end in late Viséen time, when connection with the Caribbean epi-sea was established, and the late Visemurian fauna invaded western America through the western Samfrau geosyncline, as it invaded the Cathaysian during the same time through the eastern part of the Samfrau geosyncline. But the eastern (Weiningian) and the western (Morrow) remained distinct, because the influence of the ice-sheet was still dominant in the central geosyncline, even though it had begun to move towards its Transvaal station.

We have already seen that the Mississippian of America includes both the Fengninian and Visemurian pulsation systems, and that the evidence of the interpulsation retreat is just as marked as it is in Europe and Asia, though locally it is masked. But we are now in a position to appreciate the fact that the limits usually set for the system are not the true limits. This view was also formerly held by Moore, though he subsequently departed from it.

In the northern mid-continent (Missouri, Iowa, Nebraska and Kansas), i. e. the northern and western flanks of the Ozark Mts, the mid-Mississippian disconformity is very pronounced: "the entire upper half of the Middle Mississippian being generally absent".[13]   An even greater disconformity, or unconformity, occurs at the base of the Pennsylvanian, and this becomes even more marked in the central mid-continent district (Arkansas, Oklahoma and North Texas).   Here the Morrow formation shows evidence of offlap and retreat, followed by transgressive overlap of the Pennsylvanian, the hiatus gradually increasing from east to west in Oklahoma, from west to east in central Texas, and from south to north in the intermediate region.

The retreatal or off-lapping members in the first example are known as CANEY SHALE and SPRINGER and WAPANUCKA FORMATIONS, and in the latter as BEND. Collectively they form the MORROW DIVISION, which, with the underlying transgressive Chester forms the Visemurian pulsation system.   Though a disconformity is postulated between the Chester and the Morrow, this is much subordinate to that between the latter and the Desmoines, where, moreover, the Arbuckle orogeny intervenes.   The DESMOINES, which, with the MISSOURI, forms the Pennsylvanian (Donbassian) system, is separated from the Permian by the Arbuckle orogeny in the mid-continent, and by the Marathon orogeny in West Texas.

The fauna of the Morrow group is as markedly different from that of the underlying formation as is that of the Weiningian from the Viséen.   But this is to be expected, if both mark an invasion from new centers of origin and mainly represent the retreatal phase with offlap, and therefore progressive restriction in distribution. Also the Morrow carries the flora of the Pottsville conglomerate, which underlies the Allegheny.   This too is understandable, if we regard the Pottsville as a part of the interpulsation deposit, with the upper Mauch Chunk, as I think we may.

There is no great violence done to our conception of floral progress, if we consider the flora—which had come into existence during the retreatal phase of the Visemurian if not before, and had temporarily been barred by aridity of climate from the Appalachian geosyncline—as returning thence during the interpulsation period or even at

13) R. C. MOORE: *Carboniferous Rocks of North America* p. 600

the beginnging of the new transgression (Moscovian), when moister conditions had returned to the region, as is shown by the deposition of pebble deltas by the rivers.

CHAPTER XXVII

THE PERMIAN — LAST OF THE PALÆOZOIC SYSTEMS

1

TO understand the conditions in Permian time and the immediately preceding periods and the varied deposits formed during this interval, we must keep in mind that throughout the typical Permian region in North Russia, the Uralian sea had transgressed across a country in which the preceding Upper Carboniferous (Upper Donbassian, or Donetzian) had never been deposited, because of retreat of the sea, or, if it had been formed in part, was removed again by erosion during the interpulsation period which followed. In some localities only the Upper Carboniferous or Donetzian deposits are preserved from erosion, and of these only the typical region of the Donetz basin in South Russia and the sections in Great Britain are well known. There may be many others, but up to the present time these remnants of the deposits formed by the retreating Carboniferous or Donbassian sea in Upper Carboniferous or Donetzian time have not generally been separated from those which mark the return of the sea in Uralian time.

It must be remembered that disconformities between formations of similar lithic character are not easily recognized, except by very careful examination of the sections and detailed study of the faunas, such as has so far been made in only a few localities.

It signifies little that the interval we are looking for represents an interpulsation period, for if the country was low-lying during such a period of exposure, and especially if it was distant from the sea, so that the grade of the rivers was not a steep one, not much erosion could be accomplished. In such cases the Upper Carboniferous (Donbassian) beds, which had previously been deposited by the retreating sea, would only be partially removed, if at all, but may have become subject to disintegration, so that when the sea returned in Uralian time after a long absence, it found much disintegrated material to be reworked. This material it would incorporate, with the new sediments, into what would apparently be a continuous series, but one which nevertheless enclosed within it a masked hiatus.

An example of this is found in the northern Timan. The typical Uralian limestone, characterized by *Schwagerina*, rests on oölitic limestone and flagstone with *Spirifer supramosquensis*, a typical Donetzian fossil. This rests on coral limestone with *Spirifer marcoui*, also a Donetzian form, and this in turn rests upon the Moscovian. Elsewhere in the Timan the *Spirifer marcoui* (Gschellian) beds are separated from the overlying basal Uralian, by from 1 to 2 meters of arkose sandstone with cross-bedding, which is a subaërial bed marking a distinct hiatus. Farther west in the region of the Onega and Dwina Rivers, Uralian (*Schwagerina* and *Cora*) beds rest disconformably on Donetzian beds with *Spirifer supramosquensis*.

It is chiefly the failure to note the hiatus and disconformity between the transgressing Uralian and the

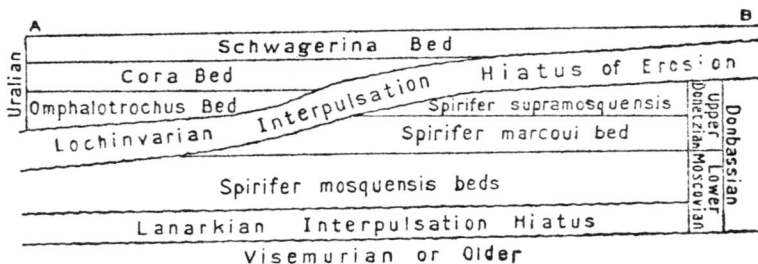

Fig. 93. Diagram, showing the relationship of the Donbassian and Uralian beds in North Russia.

eroded surface of the Donetzian or even the Moscovian, and to appreciate its significance, that has led to the mistaken assumption that the beds above the Moscovian represented a unit deposit of Upper Carboniferous age, all referable to the Uralian series. The truth is that a long erosion hiatus separates the true Uralian from any remnants of the Donetzian such as the *Spirifer supramosquensis* beds and the *Spirifer marcoui* beds, both typical Donetzian, and from the Moscovian beds beneath, this erosion hiatus representing the interpulsation period. This relation is shown in the diagram on the opposite page.

Thus, at A all these divisions of the Uralian rest on Moscovian 'Middle Carboniferous' or (Lower Donbassian). At B the *Schwagerina* limestone (of Upper Uralian age, and only in part present by overlap), rests on Donetzian (Upper Donbassian). Here all the divisions of the Donbassian are present from the Moscovian (which alone is represented at A) to the *Sp. supramosquensis* bed of Upper Donbassian (Donetzian) age. Failure to recognize the disconformity and overlap leads to identifying the whole of the Uralian at A with the Upper Donbassian followed by Upper Uralian at B, and to identify the whole as Uralian and refer it to 'Upper Carboniferous'.

2

The most complete succession of these formations is seen in the 'Ufa Plateau'[1] region of the Southern Urals, to the south of the city of Perm, from which the name Permian was derived. The Ufa Plateau is a table land formed of the great limestone beds to which the name Uralian is applied.

On the western border of the plateau, which is here bounded by the Ufa River, the limestones are succeeded by the Artinskian shales and marls, which are named from the small village of Artinsk on the Ufa River. These Artinskian beds form the retreatal series of the Uralinskian pulsation system, of which the Uralian limestone forms the transgressive division.

The Artinskian beds are succeeded by the Kungurian dolomites and limestones, with which they form a continuous series. The fossils of the latter are closely related to, and often identical with, species in the Uralian. The Kungurian is in fact a calcareous phase of the Artinskian, which is sandy and shaly and frequently plant-bearing. The two together represent the retreatal phase. A characteristic feature of the later Kungurian is the presence of gypsum and potash salts, and these represent deposits in a lagoon which was cut-off from the sea by

---

1) Present day location of Ufa : Long. 56°8' E., Lat. 54°46' N.; of Perm Long. 56°32' E., Lat. 58°1' N.; of Artinsk Lat. 56°25' N., Long. 58°25' E.

a bar, with the evaporation of the water and the deposition of the salts, as outlined in a previous chapter. For the deposition of the potash salts, however, special conditions are necessary. Not only must the lagoon be finally and completely isolated, but the climate must be dry enough to permit complete evaporation.

These conditions are induced by the complete withdrawal of the sea during an interpulsation period, leaving a cut-off portion behind to evaporate. Thus these salt deposits must belong to the interpulsation period which followed the Artinskian-Kungurian retreat. The fact may be noted that the Burma ice-center, responsible for the Talchir glacial deposits of India, occupied the Himalayan geosyncline before the marine Permian beds were formed there; at that period the Ufa region lay on the equator.

However, even during the Permian pulsation period, when the south pole had moved to the Aravalli Hills in Rajputana (26°30' N., 75°0' E., p.d.), and the centre of Antarctica was beginning to move pole-wards, the Ufa Plateau region was still near the equator, as was also that other lagoon on the opposite side of the Ural geosyncline, in which the Zechstein salts of west Europe were forming at a later period.

If, then, the Kungurian salts were deposited from the waters of a cut-off, left by the Kungurian retreat and separated out during the succeeding interpulsation period (post-Kungurian (Lebachian) interpulsation period), the succeeding beds must represent the true Permian pulsation system. These comprise the Kama series[2] as the transgressive part, and the Tartarian series as the regressive and continental final division.

The series begins with the Ufa beds well exposed around the town and on the river of that name. This comprises, about 200 meters of limestones, marls, gypsum, clays of various colors and red sandstones, but all apparently without fossils, though these may occur. Farther north these are replaced by fossiliferous beds like those of the overlying Samara series. Upon the Ufa beds follow the Samara beds, 75-100 meters, of marine marls and limestones, with brachiopods, pelecypods and gastropods.

2) GRABAU : *Permian of Mongolia,* p. 385.

Among the brachiopods two species are most characteristic: *Productus* (?) (*Strophalosia*) *horrescens* and *Productus cancrini*. Typical pelecypods are *Macrodon, Allorisma* and *Pseudomonotis*, while the gastropods are represented by *Murchisonia* and other genera. Intercalated with these are beds with plant remains and fish fragments.

Farther north, as already remarked, both the Ufa and the Samara beds are replaced by a richly fossiliferous series (Kostroma dolomite), and these in the Arctic region of modern Russia lie directly but disconformably on Donetzian, or even Moscovian beds, the entire Uralinskian being absent here.

All these fossiliferous Permian beds, which represent the transgressive phase, are succeeded by the regressive phase. This is represented by the Tartarian series, which forms the last of the deposits. This reaches a thickness of 175 meters or more, and begins with rose-colored marls, with limestone layers with fresh water pelecypod shells, and ends with red colored clays, and marls, and with sandstones.

3

During both Uralinskian and Permian time, a deep embayment,—the Zechstein Bay,—extended from the western side of the Uralian geosyncline as far as Great Britain.

In Great Britain and adjoining regions the Donbassian or 'Carboniferous' is represented by the (chiefly) non-marine coal measures of the Westphalian. The lower and middle Westphalian represent the Moscovian with a *Sphenopteris-Alethopteris-Lonchopteris* flora, which includes early species of *Neuropteris* as well.

In the north of England and in Westphalia these beds are further characterized by goniatites, the lower by species of the genus *Gastrioceras* and the middle by species of *Anthracoceras*. The Upper Westphalian represents the post-Moscovian Donbassian, *i. e.*, the Donetzian, but has neither goniatites, nor other marine faunas in the Franco-British Basin, where it is represented by coal measures (Staffordian, Radstockian), with a *Pecopteris-Mixoneura* flora, also containing later species of *Neuropteris*.

The Uralinskian pulsation system is represented in

France and Belgium by the Stephanian division, with 21
coal seams in the lower (coking-coal division), and 47 in
the upper (gas-coal division).   This contains a *Pecopteris-
Mixoneura-Linopteris-*and *Odontopteris* flora, but no goniatites.

In Great Britain the Stephanian is not developed in
its plant-bearing facies, and is either absent or possibly
replaced by the lower sandstones of the succeeding 'Permian' series.   These are 3,000 feet thick, of red and
variegated color, and with some marls and calcareous
conglomerates and breccias.

In Shropshire, on the other hand, the Uralian
(Stephanian) time was the period of an orogeny.   The
Westphalian and older beds were strongly folded, probably
during the interpulsation period, and the folds eroded
during the Uralian (Stephanian) positive phase of the
succeeding Uralinskian pulsation period.   Then the early
Permian continental sandstones of the 'Rothliegendes' type
were deposited upon these.   This Ural-oric orogeny, then,
accounts for the absence of Uralian strata in Great Britain,
even in the continental coal-bearing phase (Stephanian).

The succeeding Permian pulsation is only in very
small part represented by marine beds, the greater part
of the sedimentation being of the continental type.   The
sections are as follows : [3]

	W. of England	*E. of England*
Red sandstone, clay and gypsum	600 ft	50-100 ft
Magnesian limestone } Marl slates }	10-30 ft	600 ft
Lower red and variegated sand- stones, reddish brown and purple sandstones, and marls with calcareous conglomerates and breccias	3,000 ft	100-250 ft

The greater thickness of the calcareous rocks in the
east, and their progressive replacement by clastic rocks
westward, is evidence of the transgression from the east.

The lower series attains its greatest development in

3) GEIKIE : *Text-book of Geology*, p. 1070.

the Vale of Eden, where it consists of brick-red sandstones, with some beds of calcareous breccia, locally known as 'brockram', derived principally from the waste of the Carboniferous Limestone.

In South Scotland these beds lie unconformably on Ordovician formations, which generally have furnished the material of the breccia. In this region much volcanism occurred at the time, numerous volcanic vents and sheets of diabase, picrite, olivine basalt, andesite and tuff being associated with the red sandstone. Geikie says that though these volcanic phenomena were on a feeble scale, yet: "They are interesting as marking the close of the long continuance of volcanic activity during Palæozoic time. Neither in Britain nor, save at one or two places on the continent, has evidence been found of renewal of eruptions during the lapse of the Mesozoic ages." [4] No fewer than eighty such vents have been observed in a space 12 miles long by 6 or 8 miles broad, between St. Andrews and Largo, in Fifeshire, piercing the coal measures. Others have been found in Devonshire.

Some of the breccias have been thought to be of glacial origin (Ramsay), but if this were substantiated they must have been the moraines of mountain glaciers, for the salt deposits in the next higher beds clearly indicate the position of this embayment in the tropical zone.

The only fossils found in these red beds are plant remains (*Ullmannia, Lepidodendron, Calamites, Sternbergia, Dadoxylon*) and fragments of coniferous wood, foot-prints and the crania of a labyrinthodont (*Dasyceps*).

"The Magnesian Limestone group of the north of England has yielded about 150 species belonging to some 70 genera of fossils, a singularly poor fauna when contrasted with that of the Carboniferous system below. The brachiopods include *Productus horridus, Spirifer alatus, Camarophoria humbletonensis, C. schlotheimi, Strophalosia goldfussi, Lingula credneri,* and *Terebratula (Dielasma) elongata*. Of the lamellibranchs *Schizodus schlotheimi, Bakewellia tumida, B. antiqua, B. ceratophaga, Mytilus squamosus* and *Parallelodon striatus* are characteristic. The univalves are represented by 10 or more genera, including

4) GEIKIE: *Ibid*, p. 1070.

Pleurotomaria and *Turbo* as common forms. Nine genera of fishes have been obtained chiefly in the Marl Slate, of which *Palæoniscus* and *Platysomus* are the chief. These small ganoids are closely related to some which haunted the lagoons of the Carboniferous period. Some reptilian remains have been obtained from the Marl Slate, particularly *Proterosaurus speneri* and *P. huxleyi*, while the amphibian *Lepidotosaurus duffii* has been found in the Magnesian Limestone". [5]

On the continent the basal red beds reach a thickness of up to 6,000 feet and include breccias, which like those of Britain were formerly thought to indicate glacial origin, but for reasons already given cannot be regarded as such. They are probably referable to accumlations of talus fragments. There is also much associated volcanic activity, and it is possible that a part of this series (known as Rothliegendes, etc.) may represent the interpulsation period.

In the Halle region of Germany the Stephanian (Ottweilian) is followed by the coarsely crystalline Landsbergian quartz-porphyry lava, which is known to be more than 800 meters thick, and which undoubtedly represents interpulsation volcanicity. This is followed by continental beds, with remains of conifers *(Walchia piniformis* and *W. filiciformis)* also including layers of porphyry lavas. The series is terminated by another porphyry flow (Petersberg porphyry), this time finely crystalline and reaching a thickness of more than 60 meters, over which lie 8 meters of clays, with non-marine fossils and 71-78 meters of decomposed tuffs.

This series, which is probably all referable to the interpulsation period, was folded in the closing stages of that period and eroded during the early part of the Permian pulsation period.

Eventually the transgressing Permian sea occupied the greater part of the embayment, and the first of the marine limestones was laid down upon the old continental sands. Usually it begins with a limestone conglomerate, which is in many regions followed by half a meter or more of the so-called Kupferschiefer (copper-shale), a

---

5) GEIKIE: *Ibid*, p. 1071.

bituminous marl-shale, with disseminated grains of copper sulphides, etc., which is the chief copper ore of Germany. But the most significant feature of this deposit is the vast number of fossil fishes which it contains, these being often preserved in great perfection.

Two forms predominate above all others, and are in large part responsible for the bituminous character of these shales, which is sometimes so pronounced that pieces will continue to burn after ignition. Agassiz long ago described and named these fishes, the ganoid : *Palæoniscus freieslebeni* and the broad-fish — *Platysomus gibbosus*. *Callipteris*, regarded as the fern typical of the Permian, is also found.

4

The first invasion of the sea produced the basal beds, often conglomeratic, which are characterized by brachiopods (*Productus cancrini, Strophalosia,* etc.) and pelecypods (*Bakewellia ceratophaga, Pseudomonotis speluncaria,* etc.), species which we have met with in the basal marine Permian (Samara beds) of Russia, and in the equivalent Kostroma dolomites of the far north. With these occur remains of *Ullmannia*. This invasion was premature, because stability had not been established, and the basin was cut off again. The normal marine organisms died out, and later the fishes and other vertebrates succumbed, their bodies sinking into the bottom, which their decomposition changed into a foul slime, or sapropelite.

When they were all dead the sea broke in again, this time successfully, and the Permian organisms came with it, but not all of the older forms returned. The obese, finely striated and wrinkled, *Productus cancrini* had apparently died out and a larger form, bristling with long spines (*Productus horridus*), had taken its place. With this are associated other characteristic brachiopods and pelecypods, while locally bryozoa (*Fenestella retiformis*) grew in reef-like masses and built up local centers of settlement for other organisms.

After more than 50 meters of limestones, dolomites and dolomitic marls had accumulated — in the slowly sinking basin, encroached upon by the rising sea in the positive phase of the Permian pulsation, — further transgression ceased, and the lagoon was again cut off by the building of a bar, through the inlet in which sea water could be supplied. Then evaporation came into play, the salinity of the water increased to the point of saturation, when deposition of salt commenced.

The first salts to separate out were the lime-sulphates, probably in the form of anhydrite, but now commonly changed to gypsum, especially in the upper layers.

The main mass of the older rock-salt, which ranges in thickness up to 670 meters, is interrupted by thin layers of anhydrite, indicating repeated temporary overflow, with sufficient dilution of the water to substitute anhydrites for the salts, the separation of which, however, recommenced after a while. In such manner were formed the so-called annual rings in the salt layers the older salt bed containing some 3,000 of these, each a seasonal interruption and change. Then follow the mother-liquor salts. [7]

A generalized section of the succeeding beds which compose this upper Zechstein series, is as follows in descending order :

---

7) The chief natural mother liquor salts are :
Sylvite (KCl) *Kieserite* ($MgSO_4$, H_2O), *Carnallite* ($KCl.MgCl_2.6H_2O$) *Kainite* ($KCl.MgSO_4.3H_2O$) and *Polyhalite* ($K_2SO_4.2CaSO_4.MgSO_4.2H_2O$). Some of these are secondary derivatives from the original salts by the formation of new combination. The detailed succession in the salt deposits of Stassfurt is as follows in descending order (See GRABAU: *Principles of Salt Deposition*, p. 60) : —

7. Salt clay.
6. Schönite layer ($MgSO_4.K_2SO_4.6H_2O$).
5. Sylvinite or Hart salt layer, with mixed Sylvite (KCl) and rock salt (NaCl), some Kieserite ($MgSO_4.H_2O$).
4. Kainite layer ($MgSO_4.KCl.3H_2O$).
3. Carnallite zone, containing Carnallite   55 per cent.
   ($KCl.MgCl_2.6H_2O$). Halite (NaCl)   26 per cent.
   Kieserite ($MgSO_4.H_2O$)   17 per cent.
   Anhydrite ($CaSO_4$)   2 per cent.
   Also minor quantities of Kainite ($MgSO_4.KCl.$
   $3H_2O$) Sylvite (KCl) and other minerals.
2. Kieserite zone, containing Halite (NaCl)   65 per cent.
   Kieserite ($MgSO_4.HO$)   17 per cent.
   Carnallite ($KCl.MgCl_2.6H_2O$)   13 per cent.
   Anhydrite ($CaSO_4$)   5 per cent.
1. Polyhalite zone, containing Bischofite ($MgCl_2.6H_2O$) small quantity.
   Polyhalite ($2(CaSO_4)$ $MgSO_4.K_2SO_4.2H_2O$.)
Subformation: Rock salt with anhydrite layers

ZECHSTEIN SALT SERIES:

Zechstein clay-rock (lutytes)	20	meters
Younger rock salt (with about 400 "annual rings" of polyhalite)	120	,,
Main anhydrite	40	,,
Gray salt clay	6-12	,,
Mother-liquor salts (main potash salts)	30	,,
Older rock salt, with 3,000 annual anhydrite rings, and a basal anhydrite	670	,,

In some parts of the old bay region occur shore facies of conglomerates and sands, which suggest the underlying Rothliegendes, but are believed to be a facies of the Zechstein. The lingering of the sea in the neighbourhood is shown by the presence in the salt-clay of shells of characteristic pelecypods and gastropods. This raises the question, whether the complete evaporation followed the final withdrawal of the sea during the Tartarian interpulsation period, or whether these salts were separated during the latter part of the Zechstein period, because of exceptional aridity. Whatever their age in detail, these salt and potash beds and their American representives, give the last account of the influence of the retiring Palæozoic sea, which left to the eras that were to follow a record of such intricacy and detail, that we who struggle to read it, marvel when we obtain a glimpse of the wonderful harmony that pervades it all.

CHAPTER XXVIII

A WITNESS FOR THE DEFENCE

THE folding of the strata of the Franciscan geosyncline
of Schuchert also belongs to the later Permian period,
and received its "crushing blow" or series of blows by
the impact of the African land mass. This is also a
meridional geosyncline, or nearly so in terms of the modern
map, though on the pangæa its position varied with its
relationship to the pole.

As described by Schuchert this geosyncline is supposed
to deal a "crushing blow to the displacement hypothesis",
but since, apparently, the hypothesis cannot be crushed, the
blow became a boomerang and by crushing the geosyncline
testified not only to the proximity of the great African
mass, the "crusher", but also to the force engendered
by the shifting of the pangæal cap in obedience to gravitational readjustment. Schuchert says:[1] "This Franciscan
geosyncline of eastern Brazil is a long and narrow marine
trough present at least since the early Silurian." Its center
or axial line extends from east of Maranhao (Long.
44°5' W., Lat. 5°0' S.) to somewhat west of Sao-Paulo
(Long. 46°33' W., Lat. 23° 30' S.) and finally along the
coast to Rio Grande do Sul (Long 52°18' W., Lat. 32°6' S.).
It has thus a length of almost 30° of latitute, or something over 2,200 miles. Schuchert continues: "The
Palæozoic sediments are essentially sandstones and shales,
having a united thickness of less than 6,000 feet, though
there may also be present strata older than the Silurian.
The Silurian and Devonian are marine deposits, about
2,000 feet thick, while those of the Permian, 2,400 to 3,400
feet thick, are in the main continental, although marine
and brackish water zones occur almost throughout the
whole series. All the Palæozoic detritals appear to have

---

1) SCHUCHERT: in Symposium: *Theory of Continental Drift*, p. 126.

come from the east: to the west they are buried under the plateau lavas."

This is as it should be, for Africa was competent to furnish an abundance of detritus. Moreover, in both Siluronian and Devonian time sediments brought by streams from the ice sheet of the period were carried into the geosyncline, especially at the end of the Siluronian, when the "western" border of the ice was melting, while the center was shifting farther "east" in conformity with the relative shifting of the pole. Thus in early Siluronian time the southern end of the geosyncline received much drainage from the melting ice, just as the north end did from the Taodini ice sheet when that began to melt during an earlier period. But during the Devonian, while the southern end still lay close to the south polar ice cap the northern end, where i t joined he Amazon geosyncline, lay in 60°.

Schuchert does not tell us whether or not Mississippian strata (deposits of Fengninian and Visemurian pulsation systems) and Donbassian (Pennsylvanian) are present in the geosyncline, and his failure to mention them may be regarded as implying their absence. [2]

This is entirely in harmony with my palæogeographic maps, which were drawn originally without reference to this geosyncline. In Mississippian time the polar ice cap was squarely over the Franciscan geosyncline and completely smothered it, while it supplied glacial drainage debris to the Amazonian geosyncline. When the ice-cap finally melted with the removal of the polar center to south-east Africa during Moscovian-Donbassian time, it was still potent enough to endow the geosyncline, newly emerged from its icy baptism, with the certificate of its submergence, the tillites of the "areas of proved glaciation", for which Professor Schuchert is sponsor.

The Permian beds in this geosyncline "are in the main continental", with marine and brackish water zones. This is quite in harmony with the further progress of

---

2) This is also the conclusion to be drawn from the thicknesses which Schuchert gives: "The Palæozoic sediments have a united thickness of less than 6,000 ft. The Silurian and Devonian marine beds are about 2,000 feet thick, while those of the Permian [are] 2,400 to 3,400 feet thick". This leaves no space for intermediate beds.

Palæozoic history, for the continued westward pressure of the moving lands (which was the cause of the polar shift) for a long time still effected only the deepening of the geosynclines and the corresponding elevation of the old-lands which supplied the sediments, the latter lying on or within the African border.   There was now a mass of coarse and fine morainic material available, left after the recent occupancy by glaciers, like the debris left by the retreating army after its short but destructive invasion. This the streams sorted and swept into the geosyncline, which could hardly sink fast enough to assimilate it all, and only occasionally could admit the sea to place its seal of legitimacy and rehabilitation on the geosynclinal sediments.

But finally the geosyncline had suffered such extensive subsidence, and was filled by such thick masses of clastic sediments, that collapse resulted from further pressure, so that in middle Permian (Uralinskian?) time the strata were folded, apparently on the west and north.

In late Triassic time further sedimentation of river-laid sands took place here, after which came the stupendous outpouring of the Plateau basalts, which covered "an area at least 300,000 square miles in extent between the Amazon, Paranà and La Plata Rivers, [the lava] averaging about 1,000 ft in thickness" (Schuchert).   Such an outpouring might well accompany the beginning of rifting at the end of Triassic or during Jurassic time, an outpouring which was duplicated in the Appalachian geosyncline, along the potential rift, in the Hudson and Connecticut Valley.[3]

The same thing happened much later in Polycene time, when the Atlantic rift had reached the further end between Spitzbergen and Greenland and when the Iceland volcanoes burst out on the rift-line. (Chapter XXXIX).

The last of the deposits was the Franciscan sandstone, believed to be river-spread, and of Cretaceous age.   This is probably of purely South American origin, belonging to a post-rift period, while the continents were still drifting apart.

---

3) Such eruptions need not be on, or even near to, the line of separation, for the tension which was finally relieved by the separation, would develop many parallel rifts, by which the igneous matter could reach the surface.

But what shall we say to the charge of disjunctive disharmony between the Franciscan geosyncline and the African mass? Let us review the facts.

Following Schuchert's lead we turn to Lemoine, since Krenkel has not yet reached this part of Africa in his second volume. Lemoine maps the Gold Coast region and the area abutting against the granite terrane of Dahomey as Palæozoic, and continues this over wide areas of the Niger Country, and northward as far as the Tassili Region. In the central Touareg region are Silurian graptolite shales much restricted and partly metamorphosed it is true, yet unmistakable. In the Niger Valley, in the Timbuktu basin, and farther south, disturbed quartzites, with limestone lentils, are referred to Silurian.

In the Atacora range north of Dahomey, a region eight to ten degrees west of the coast of Cameroun, at the head of the Gulf of Guinea,[4] strata of Silurian age are recorded. There are quartzites, sometimes extremely metamorphosed by intruded granites, and maintaining either a horizontal or a vertical position, but always strongly folded.

Both Siluronian and Devonian are wide-spread in Western Africa. They can be traced northward nearly or quite to the Moroccan border, and south-east to the Tassili plateau (about 25° N., 4°-5° E.). This is a region famous for its Devonian fossils, which unite elements of the North American Devonian with those of west Europe (as might be expected from the geosynclinal connection shown on our map, Plate VIII). Twenty degrees farther south, within the elbow of the Niger, south (p. d.) of Timbuktu, flat-lying sandstones and grits with intercalated limestone beds are held to be of Devonian age. They closely resemble in character and topography the Devonian of the Ahnet mountain region and the other north-Saharan beds and they are also very like the sandstones of Table Mt. in South Africa, which are likewise of Devonian age. Until fossils are obtained from the Sudan rocks, however, all age determination must be provisional; but so much appears

---

4) Maranhao, east of which Schuchert places the center of the Franciscan geosyncline in South America, is 9˚17′ west of Cape San Roque, which on our reconstruction joins the Cameroon coast. The northern end of Schuchert's Franciscan geosyncline would therefore abut against the south coast of Dahomey.

certain that Schuchert's assertion, that west Africa shows no relation to eastern South America is wholly unwarranted. Furthermore, there is good reason for believing that the Amazon geosyncline continued across what is now the Sahara, and that the Franciscan geosyncline joined it near the passage from South America to Africa.

It would thus appear that this missile hurled by the opponents of continental drift, in the expectation that it would smash at a blow the whole structure built by Taylor and Wegener and by Du Toit and his followers, has missed its mark and may now be picked up by the protagonists of the new view and incorporated as one of the corner-stones of the rebuilt scientific edifice devoted to earth evolution.

CHAPTER XXIX

ON THE THRESHOLD OF A NEW ERA

1

CASTING our mental vision back over the periods of the Palæozoic we find a constant thread of unity running through the history of the development of the earth and of life. We begin with an earth of concentric spheres, each essentially of uniform, though graded, composition, specific gravity and other characters, but each differing from all the others in these respects. Such an earth revolves on an axis which throughout all the periods of earth history has maintained essentially a uniform inclination to the plane of its ecliptic, revolving about a sun, which during the Palæozoic probably did not vary materially in heat radiation, and on a pathway which need not have varied appreciably from that which it follows today.

From such an earth the sial was withdrawn by the attraction of some extra-telluric body, at present undetermined,—almost the only unidentified force in earth history. This outside agency acted in a polar direction, *i. e.* one coinciding with the earth's axial direction. This can readily be deduced, since no other direction could have any other but a tidal effect. That our award of the role is made to the South Pole, is the result of the unfolding of the subsequent history of the earth. If it had been the North Pole, as we formerly assumed, it would have been possible to credit Polaris with the enforcement of the command to "let the dry land appear", and "the waters be gathered together *into one place*", as the poetic version in Genesis has it. But since Polaris failed to qualify for the duty by lack of position, some other celestial body, perhaps by now deposed, perhaps no longer existing, must bear the charge of setting the earth on her evolutionary course.

A constant axis, but a shifting sial cap, under polar control; the dominant force that of rotation; a temporary anchoring of the sial-crust to the sima by an intrusion into the former of a deep-seated igneous mass; this was the setting. Then followed a drifting of the sial cap under the dominant force of earth rotation, by rotary shifting on a pivot, which was formed by the igneous anchor rooted in the sima, as a ship at anchor drifts with the tide. Rotating clockwise under one group of positions of the anchor, counter-clockwise with another, the sial-cap shifts so as to bring different regions over the pole, which maintains its constant position with reference to the mass of the earth.

But polar residence, even if only transient, is inevitably accompanied by polar glaciation, milder in perihelion, more severe in aphelion, but *persistent polar* glaciation. That is the kind that different regions of the Palæozoic Pangæa and different fragments of the dismembered Neozoic sial-crust experienced. Excepting only mountain glaciation, we may confidently assert that there is no other than polar glaciation, and the doctrine of a universal, or even semi-universal ice-age, at one or several particular periods in the history of the earth, must be exorcised, like Banquo's ghost, to prevent its periodic return. But polar glaciation we have always had with us, and probably always will have, as long as the earth maintains its present axial relation to the sun.

2

But the drifting of the crustal cap of sial brought different regions of the pangæa under polar control, and an ice-sheet varying in radius from 15 to 20 degrees, developed at each polar end. The North Pole was in the Pacific throughout Palæozoic time, though approaching at times sufficiently close to the margin to make its influence felt when the South Pole was at the antipodal margin.

Polar shift also influenced deformation of the sial by pressure. This is exerted against adjoining lands, which were slow to respond to the rotary movement of the time, because of the resistance of the sima in which the sial-crust was partly submerged. Continued subsidence of the geosyn-clines was the first effect of this induced pressure, followed by orogenic disturbances. On the other hand, areas of tension were areas of rift, of volcanism and of normal faulting.

All these diversified effects were influenced, if not induced, by the rotation of the earth. But the rise and fall of the sea-level (except where due to tidal influences) were probably caused by the radioactive processes, producing heating and swelling followed by cooling and contraction under the ocean bottom and in the sima of the land blocks. Thus the process became a rhythmic advance and retreat of the sea in the geosynclines, a regular pulsation: the positive phase of rising sea-level causing the advance of the waters and the spreading of the faunas from the several marginal centers, while the negative effect was marked by progressive withdrawal and final complete drainage of the geosynclines. This was, of course, accompanied by extermination of marine organisms, until finally only a few survivors were segregated in the concentration camps of the marginal epi-seas, there to gather new strength and develop into a new host, which on improvement of the environment with the return of the waters, was destined to occupy the re-flooded areas.

3

It is of course obvious that these relationships, dependent on, and induced by, the existence of the pangæa as a whole, can endure in their totality only as long as the pangæa exists, and that modifications and new conditions would arise when the pangæa broke into its several component continental blocks, which have endured down to the present day, though they have undergone marked changes in relative positions.

The period of commencement of fragmentation is by universal testimony referred to post-Palæozoic time. Wegener placed the time of the rupture between South America and Africa in Cretaceous time, and the final breaking away of Greenland from North Europe he visualized as beginning in Tertiary, and as continuing through Quaternary time, the drifting apart being in progress even today. In this he has been generally supported by others.

In Triassic time the Newark geosyncline, along the Atlantic border of the United States, could not have been separated from the African mass by any water bod of any

Atlantic type, since the sediments are wholly continental, and indicate an arid climate by their red color and the arkosic character of many of the sandstones. (Fig. 91, p. 300)

4

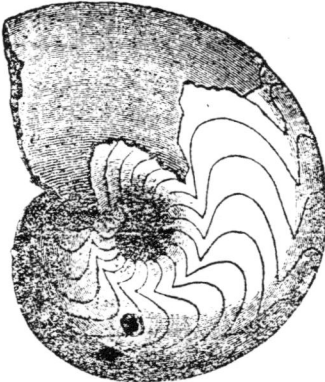

Fig. 94. *Goniatites* (*Manticoceras*) *intumescens* Beyr. Upper Devonian; with most of the shell removed to show simple sutures characterized only by lobes and saddles. (After Zittel).

Schuchert's St. Franciscan geosyncline of eastern South America, which had suffered virtual demise under the influence of the Dinantian and Moscovian glaciers and, in early Permian time, had suffered folding by the pressure of Africa against South America, became sufficiently revived in Triassic time to permit the deposition of continental strata. Then, during the later Triassic and earlier Jurassic time the plateau lavas, averaging about 1,000 feet in thickness,[1] spread over 300,000 square miles of country.

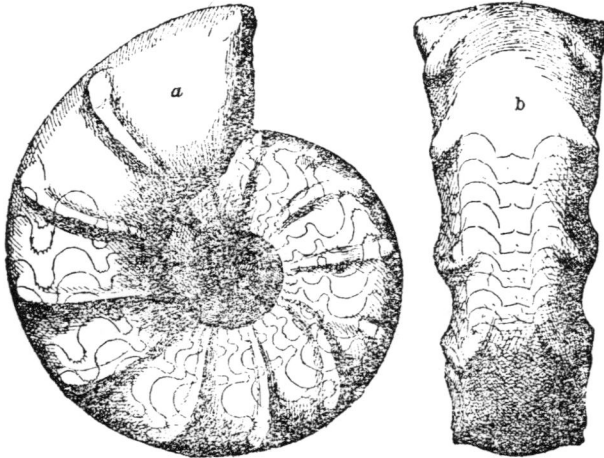

Fig. 95. *Ceratites nodosus* de Haan; Triassic internal mold of hardened mud, with shell removed to show the sutures, i. e. the edges of the septa. The saddles (forward bending) are smooth, the lobes backward bending) are crinkled. (After Zittel).

1) SCHUCHERT: in *Theory of Continental Drift.* p. 126.

Here then, we have the first indication of the tension and rifting that culminated in the separation of the continents. Schuchert says, that these plateau lavas: "are overlain by fresh water sandstones thought to be of Cretaceous age [and that though] none of these post-Permian formations are folded they are more or less normally faulted." What is this other than a premonitory warning of the impending rift, for normal faulting, like volcanism, implies tension, and tension, if prolonged, will lead to rifting and breaking apart.

The evidence for rifting indicated by the faults which bound the Newark systems of terrigenous strata in the Atlantic Coast region, and the volcanism indicated by the Palisade trap and the other lava-masses of late Triassic age, likewise suggest the tensional strains which preceded the period of ultimate rifting and complete separation.

5

If we now consider the question of the opening stages of the Mesozoic era, we are confronted with a remarkable phenomenon, which, though generally passed over with little or no comment, deserves the most careful consideration. Stated in the briefest terms we have :

1. Complete and absolute extinction of Palæozoic marine animals.
2. Apparent sudden appearance of a wholly new fauna in the ocean.
3. Such pronounced modification in the land flora and land fauna that the age of any formation characterized by either can almost as readily be determined as Palæozoic or Mesozoic, as it can by the marine fauna.

Obviously the complete extinction of marine faunas cannot be a matter of a short time period. Only an absence of the sea from the geosynclines and from the epi-seas as well, for a period of exceptionally long duration, can account for such extinction. Likewise, the appearance of a wholly new fauna, already of high specialization in its own members, demands a long period of development, and of this we have no record whatever. If the sea-level should fall low enough to lay all the marginal epi-seas bare, as well as all the shelf-seas of the pangæa that had developed, draining all the geosynclines of the earth, extinction of all bottom-living, shallow water forms would naturally result, except where a precarious foothold could be maintained by some of them on the abyssal slopes of the panthalassa. For pelagic animals, such as the goniatites of the late Palæozoic are believed to have been, and for fish and other swimming (nektonic) types, such a fall of sea-level would, of course, have no such deleterious results;

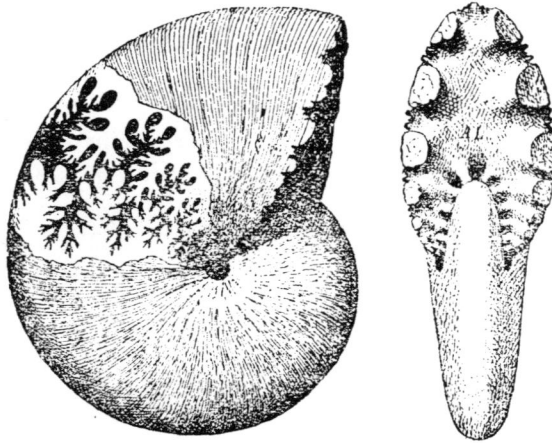

Fig. 96. *Phylloceras heterophyllum* (Sow.). Upper Lias of Yorkshire. (After Zittel).

they would merely float outward with the falling sea-level, as with a falling tide, and continue their existence in the surface waters of the open sea. But — and this is the point of importance — unless the interval was a long one, the return of the sea to the geosynclines (new or

Fig. 97. Suture of *Pinacoceras metternichi* (Hauer), showing the characteristic ammonite type of suture (Noric of Alps). (After Zittel).

old) would bring only specifically, or at best generically, new types, not types which represent a wholly new order, such as the ceratites are. But the goniatites, which are typically Palæozoic, had become almost extinct.

In our standard texts,[2] we find the arrangement shown in the following table :

TABLE V.

NUMBER OF GENERA IN TRIASSIC AMMONOID FAMILIES

Family	Number of Genera in:		
	Palæozoic	Upper Palæozoic and Lower Triassic	Triassic
Clymeniidæ	5	0	0
Goniatitidæ (Fig. 94)	45	0	3 (Lower)
Noritidæ	4	0	2 (Lower)
Medlicottidæ	2	2	5
Ceratitidæ (Fig. 95)	2	3	61
Cyclolobidæ	10	1	7
Total	68	6	78

In addition to this there are 5 exclusively Triassic families, with 81 genera, and 13 Jurassic and Cretaceous families, of which only one has a Triassic representative genus.

8

How can all this distinctness be explained ? It is so pronounced, so much greater than that which we have found between any two other successive pulsation systems, that we can only understand it if the surface of the sea sank so far below the normal level to which it had fallen during the Palæozoic pulsation periods, that no shallow bottoms remained for animals to attach themselves to, or to live upon, so that practically all died out. But, as we have seen, that should not affect the pelagic forms, those that float or swim in the surface waters ; and yet these

2) ZITTEL : *Text-Book of Palæontology.*

too were changed so greatly during this interval that they represent a new world of life. Can it be that the waters sank so low that during the next pulsation period they could not rise sufficiently to enter the geosynclines, and so left no record of that period? If we may postulate an unrecorded pulsation period, with its two interpulsation periods, one preceding and one following, this might give sufficient time for the pelagic forms to change to the degree shown by their Triassic descendants.

If that is the true explanation, the main part of the Bunter-sandstein of Central Europe, as well as the New Red Sandstone of Britain, may be the continental representatives of a part or the whole of this long interval, representing both the pulsation periods, which could not reach land, and the interpulsation periods, when land was normally out of reach.

If this reasoning is sound, we must add a new system to our table, one between the Permian and the Triassic, which bridges the interval between Palæozoic and Mesozoic. This, provisionally, we may name the METAZOIC.

It is probable that during this Metazoic period the lands stood relatively so high that erosion was the chief activity of the rivers. Only in the interior of Cental Europe would aridity be so marked that no rivers could exist, and disintegration of the old rocks would furnish a vast mass of debris, which during later periods of moister climate could be reworked into the continental bedded series of the Mesozoic. Such a long period of aridity would explain the absence of plant life of intermediate stages, such as probably characterized the marginal portions of the pangæa, where there was, however, no opportunity for preservation.

Thus, on the threshold of a new era, we find a lost interval, one of which the record may never be found, because it is still at the bottom of the sea, which does not give up its secrets. These are the lost volumes that the sibyl destroyed, and they leave for ever a gap in the historic record of the evolution of our earth.

CHAPTER XXX

THE LOST INTERVAL

1

WE have seen that the marine record is hopelessly mutilated and we have no means of determining the extent of the missing interval, because we are not able to ascertain what is the length of time required for the great change in the marine fauna. This change is most typically displayed among the ammonoids, but is equally clearly seen in the all but complete extinction of the other Palæozoic marine organisms, and the appearance of a totally new fauna in the Triassic. For the development of this, from the few survivors, a long unrecorded time interval must be allowed.

But changes that resulted in destruction of the old forms and enforced development along new lines of adaptation for the marine fauna, proved providential for the terrestrial fauna and flora, because of the greater land expansion, which provided exceptional opportunities for their development and diversification, and, moreover, provided them with a wide expanse of land over which to roam, and a variety of habitats to which to adapt themselves. All of this made, of course, for enormous expansional evolution.

2

As shown on the palæogeographic map, much of the land of the period lay in tropical, or at least temperate, climates. The south pole had at last become more or less stationary with reference to the land, or to put it more accurately, the successive slidings of the pangæa had finally brought Antarctica upon the south pole, and there it remained essentially fixed during the Mesozoic and, with slight excursive strayings, to the present day. Meanwhile, however, the other continents were destined to drift away from it in due time.

During all this time the north pole had not yet come into the picture, although it approached within sufficient distance of the margin of the pangæa to make its influence felt. Indeed, we may deal from this period onward with a definite *Boreal Realm*, not only geographically but climatically as well. If the pangæa was symmetrically disposed, the north pole lay approximately in Long. 160° of today or 30° Long. of Triassic time, and between Latitudes 75° and 80° N. *on the opposite, or panthalassic, side.* This would bring the nearest point on the coast of the pangæa to within about 25° or 30° of the north pole on the landward side of the globe.

The borders of Siberia were all within the Arctic or sub-Arctic regions (45°-60°), reaching approximately to Latitude 65° N. on the border of the East Boreal epi-sea, while Kamtchatka ended in approximately Lat. 30° N., and Anadyr Peninsula of Asia, and Alaska, both lay in Lat. 40°, but on opposite sides of the globe. Behring Straits of today was then a broad belt of open ocean, covering more than 150° of Longitude on the parallel of 40° north Latitude.

The equator of the period passed through southern Asia and central Europe, where lay the great median geosynclines. These were once radial on the pangæa when the pole was in North Africa, but now had become truly equatorial. They were the dominant sites of sedimentation during Mesozoic time, the other geosynclines having become mere feeders. Westward, this line of geosynclines was continued as the Newark geosyncline of present-day eastern North America, which was produced by the migration of the new post-Appalachian geosyncline into the Appalachian old-land on the folding of the strata of the Palæozoic geosyncline towards the close of Permian time.

This Newark geosyncline of eastern North America was throughout Triassic and Jurassic time a region of non-marine sedimentation. Not until the process of rifting had begun, which tore America from Eur-Africa, did the Atlantic have access to this geosyncline. It then became a coastal plain region over which were deposited the early Cretaceous and the later Coastal Plain strata.

3

When the sea had withdrawn at the end of the Permian, and the long unrecorded period of exceptional fall of sea-level began, the opportunity for the development of extensive vertebrate life upon the land was given. [1]

At that time vast areas of level land were accessible to the moisture-bearing winds from the sea. Africa, Australia and India were without pronounced mountain chains to intercept the winds, and only in the Asiatic region was there an extensive system of folded mountains. The Himalayas, Alps, Carpathians and other mountains had not yet arisen, and in their places were low-lying lands or shallow geosynclines flooded by the sea. Our reconstruction, it is true, shows parts of South Africa still on the border of the Antarctic ice sheet, and extensive areas lay south of 60°. But Central Africa lay within the moist belt (30°-60° S. Lat.), and probably produced sufficient plant and animal food for large amphibians (labyrinthodonts, stegocephalians) and primitive reptiles (theromorphs) to thrive on and grow to great bulk.

A part at least of the interval unrecorded in the sea is found to be included in the strata of the continental Karroo formation of South Africa.

---

1) Every now and then a case is reported where Permian and Triassic species are said to come from the same deposit, and are therefore supposed to represent transition faunas. I confess to an entirely skeptical attitude of mind with reference to such cases. Anyone looking at plates 36 and 37 of Kayser's Lehrbuch, 5th ed., needs to have strong faith in the infallibility of the collectors to accept such a mixture. On the one hand, typical Permian, not necessarily uppermost Permian; on the other, pure Triassic or younger types are mingled together. *And these are supposed to have lived together at the same time.* I cannot accept this without more convincing evidence.

This does not mean that I refuse to believe that they may be found in the same stratum or even in the same rock fragment. But that does not mean that they lived at the same time. As well might the bones of *Megatherium*, found in a modern grave-yard, make it the contemporary of modern man. There was ample opportunity for the commingling of fossils of an older with the remains of a newer fauna, even without tectonic disturbance, for these horizons are separated by at least two long interpulsation periods, and a missing pulsation period besides. Thus there was a super-abundance of time for the old rock to weather to a deep regolith full of loose fossils, and thus to furnish material for re-incorporation in the newer formation.

This begins with the Dwyka tillite or glacial series of late Palæozoic age. It probably began when the ice stood at the southern Brazil center (Pl. II, Sta. 7) in Fengninian time, the lower shales of which are conformable with the Witteberg series. These shales, which Du Toit,[2] with much reason, would regard as belonging to that series, were deposited at that time. During its location at the Transvaal center (Sta. 8) the main mass of the interglacial tillite was probably formed, though as this continued, frontal morainal material must have been added until the ice-center had reached East Antarctica (Sta. 9) when the only material that could reach the South African region was that transported by glacial streams. This was probably therefore the period of formation of the upper shales, 650 feet in thickness.

4

The Dwyka is followed by the Ecca series, 6,000 ft thick, which represents the remainder of the deposit during Uralinskian and Permian time, while the ice-centers stood respectively in Central Australia (Sta. 10), Burma (Sta. 11) and the Aravalli Hills region (Sta. 12). At the end of this period *Gangamopteris* had died out and the flora had taken on a Mesozoic character.

To what extent the Ecca continental series of South Africa represents later Palæozoic must remain for the present undetermined. Perhaps a part should be referred to post-Permian. The succeeding Lower Beaufort series, however, is without much doubt referable to the long transition period of which we have no marine record and during which the land-life developed to a previously unknown extent and diversity.

Perhaps the preceding Ecca also belongs to the lost interval, which, it must be remembered, includes at least one pulsation period and two interpulsation periods, if not double that amount.

If this interval includes the whole of the Beaufort, as well as the preceding Ecca, there will of course be no deposits to represent the succeeding Triassic. But, except to make the record everywhere complete, it is not necessary

2) DU TOIT: *Geology of South Africa.*

to consider that Triassic representatives are found in areas of continental sedimentation. The Triassic was a period of rising sea-level and, in consequence, of diminishing river grade, whereas the interpulsation periods and the lost pulsation periods were times of low sea-level and therefore of greater river activity, and hence of active erosion. On smaller, more or less mountainous, lands this product would be swept out to sea and so lost, but on large continents, with inland drainage, continental deposition is most active during the period of low sea-level.

It is thus not incongruous to consider that, on a broad expanse such as Africa and perhaps North America, especially if it is more or less rimmed by highlands, extensive continental deposits should form during periods of low sea-level, while during periods of normal sea transgression very little if any continental sedimentation goes on, though weathering of exposed rocks may provide abundant material to be borne into the continental basin when the sea-level sinks again.

Thus, unless cogent reasons appear that would demand the reference of any part of the Karroo series to the Triassic period proper, it is more rational to refer the whole of the series, Beaufort as well as Ecca, to the pre-Triassic interval. The succeeding Molteno beds of the Stormberg series are generally referred to the Rhætic, that is to the beginning of the Liassic pulsation period, but they may represent the Keuper interpulsation period, formed when the sea-level had fallen again, this time only for a normal interpulsation interval.

5

I follow Du Toit in summarizing the characteristics of the several continental deposits, which I here refer to the pre-Triassic post-Permian time interval.

The post-Dwyka shales overlie the tillite concordantly, though they may be separated from it by a long unrecorded time interval. They reach 700 feet in thickness in the south, but decrease northward to 300 feet and are absent in Natal, where the Ecca directly overlies the tillite.

The beds are carbonaceous shales and sandy beds.

They are terminated by a 'White Band' from 50 to
100 ft thick, thinning to half that northward. These
sediments were apparently an original deposit of black
carbonaceous muds, strongly charged with organic matter
and with sulphuretted hydrogen, which resulted · from
its decomposition. Du Toit compared them to the black
sulphuretted muds formed at the bottom of the Black
Sea, or in the Gulf of Bothnia at the present time.

These Upper Dwyka shales evidently mark a temporary incursion of the sea, probably in Uralinskian or
Permian time and thus we may correlate these beds with
the similar *Eurydesma* beds of Australia. Other marine
fossils recorded here, besides *Eurydesma globosum* of the
equivalent beds of Australia, include *Orthoceras* sp. and
Conularia sp. A number of fishes (*Acrolepis lotzi, Na-
maichthys schrœderi*, etc.) are also characteristic, but may
belong to fresh water types. Of plants, *Lepidodendron
australe* is characteristic of the equivalent beds of Australia;
Gangamopteris cyclopteroides Feist. occurs in the boulder
bed, as previously noted.

The 'White Band' at the top of the series is especially characterized by reptilian remains of the genus
Mesosaurus, of which three species (*M. tenuidens* Gervais,
M. pleurogaster Seeley and *M. capensis* (Gürich)) have
been identified. Besides these, the fish *Palæoniscus capensis* Broom and the crustaceans *Pygocephalus* and *Anthrapalæmon* are characteristic.

6

The moist tradition is carried into the lower Ecca,
which because of its coal beds is known as the Australian
coal measures, but must not be confused with the 'Coal
Measures' (Carbonic) of the northern lands. These deposits
reach a thickness of 1,700 ft at Impofana, but thin away
in most directions. In the center of the basin, where they
pass under the higher Karroo beds, they may be much
thicker.

These coals are very thick, 10-20 feet seams are
common, and are known to have been formed by the
growth of vegetation *in situ*. Therefore the free circulation of the moisture-bearing winds at the time is evident.

The flora is unlike the earlier 'Coal Measure' flora of Europe, and consists largely of the remarkable genera *Gangamopteris* and *Glossopteris* (Fig. 98, a, b), the ordinary genera of the Europe coal beds (*Calamites*, *Callipteris*, *Alethopteris*, *Odontopteris*, *Pecopteris* and *Annularia*) being absent.

The reason for the limitation of this flora to the 'Gondwana continent' (South America, Africa, Australia, India) will be understood when it is recognized, that these countries were always more or less under the temperate or even cold temperate influence of a climate, dominated, even if distantly, by polar influence, and separated from the other countries (North America, Europe, Asia) by an equatorial belt, which this flora apparently could not readily cross. That the *Gangamopteris- Glossopteris* flora (Fig 98) did occasionally cross into the north temperate zone is seen by its occurrence in the Kusnetzk beds of the Novo‑Sibirsk region.

The coal measures are covered by 800 ft of the Upper Ecca, which resembles the Lower Ecca and probably must still be referred to the closing stages of the Palæozoic.

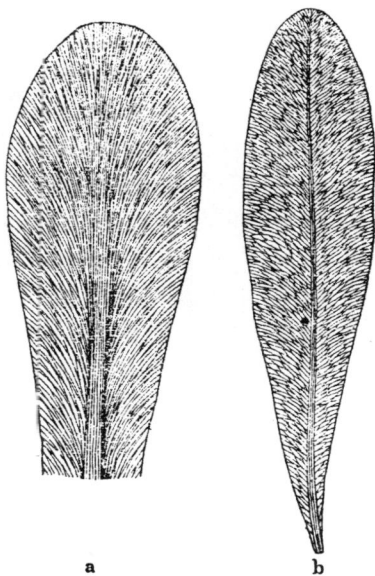

a   b

Fig. 98. Ecca plants: (reduced) a. *Gangamopteris cyclopteroides*; b. *Glossopteris browniana*. (After Du Toit ; Geol. S. Africa)

7

It is the succeeding Beaufort series, the thickest and most interesting of the subdivisions of the Karroo, that was formed during the long interval which separated the Permian from the Triassic and probably includes the whole of the Triassic period as well, in so far as that is represented at all.

The Beaufort substantiates its claim to represent the missing interval by presenting us with the wonderful series of primitive reptiles (Theromorpha[3]) and amphibians (Stegocephalia[4]) which have made the Karroo region famous in the annals of vertebrate palæontology. The following zones have been established:

Fig. 99. *Pareiasaurus baini* Seeley, a theromorph reptile of the Karroo formation, Cape Colony, S. Africa; about 1/25 nat. size: (After Seeley).

III.	Upper Beaufort:	150-2,000 ft
	6. Zone of *Cynognathus* (Fig. 100),	
	5. Zone of *Procolophon*.	
II.	Middle Beaufort:	500-1,000 ft
	4. Zone of *Lystrosaurus*.	
I.	Lower Beaufort:	6,000 ft
	3. Zone of *Cistecephalus*,	
	2. Zone of *Endothiodon*,	
	1. Zone of *Tapinocephalus*.	

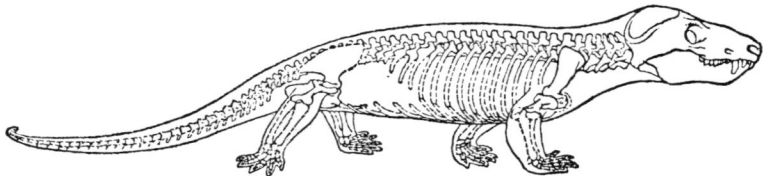

Fig. 100. *Cynognathus*, a theraspid, or mammal-like, reptile from the Karroo formation of South Africa; 1/22 nat. size. (From Romer).

3) Wild-beast-like reptiles.
4) Covered-head amphibians.

By far the most interesting is the Lower Beaufort with its rich reptilian and amphibian fauna: *Pareiasaurus* (Fig. 99), etc.   It consists of bands of fine- to medium-grained sandstone, yellowish and feldspathic, alternating with thick bodies of blue, green, red or purple, poorly lithified lutytes and shales, sometimes with calcareous nodules.

The series abounds in disconformities and 'wash-outs' and other evidences of irregular stream 'cut-and-fill' activity, — such as characterizes modern river-plain deposits.

The Middle Beaufort is only about 1,000 ft in maximum thickness and consists of bright-colored lutytes, alternating with those of more somber color and some feldspathic sandstones, while the Upper 2,000 ft, or less, is very similar, but red and maroon lutytes are more common.   The fine sandstones are all feldspathic and represent disintegration products under conditions of prevailing aridity.

8

The fauna of the Beaufort as a whole is especially noteworthy, because of the great number of reptiles, only a few of which are known outside of this region.   Those that have been recorded from Europe occur in the Buntsandstein, which, as we have seen, represents deposits of this same interval of low sea-level, or in the Keuper, which is the Triassico-Liassic interpulsation deposit.   They must be regarded as individuals that wandered north during the period of continuous land connection, but coming into a region of tropical climate and intense aridity could not long survive.   Among these migrants to the desert regions of Europe are the amphibians *Trematosaurus*, *Cyclotosaurus* and *Capitosaurus*.   The richness of the basins of the Beaufort beds is shown by the number of genera in each division, as given in the table on the next page.

Among the reptiles the earliest known *Mesosaurus* from the Lower Ecca White Band, above the Dwyka, is known elsewhere only in the *Iraty* shales of southern Brazil, associated with an allied form, *Stereosternum*.   The fishes of these post-Dwyka shales are with one exception well known Carboniferous and Permian genera.[3]

---

[3] DU TOIT: *Geology of South Africa.*

TABLE VI

DISTRIBUTION OF BEAUFORT AND STORMBERG VERTEBRATES

	Number of Genera in:				
	Beaufort			Stormberg	
No. of Genera of	A Lower	B Middle	C Upper	D Red Beds	E Cave Sandstone
1. Reptilia	77	10	24	14	8
2. Amphibia (Stegocephalians)	4	3	8	—	—
3. Fishes	3	2	9	1	2

The reptiles and amphibians of the Beaufort beds are for the most part unique and have no close relationship with those of other regions. They were excluded from the North American interior basin by the recently formed mountain chains, though the eastern piedmont belt, the new Triassic geosyncline, was probably open to them. But their remains are still unknown.

9

A marked change in flora occurs after the *Ecca* period of deposition, which probably marks the end of Permian time. Most significant perhaps was the disappearance of *Gangamopteris* (Fig. 98-a), while *Glossopteris* (Fig. 98-b) continued in the Lower Beaufort with many species. *Cordaites* is doubtfully present, but *Sphenopteris* and *Sphenophyllum* have been obtained. The equisetales are represented by *Phyllotheca whaitsi* and *Schizoneura africana*.

A similar assemblage is known from the Newcastle stage of New South Wales in Australia and the upper half of the Damuda Series of India. In the Middle Beaufort, the flora is even more impoverished, being restricted to *Glossopteris* (2 species) and fragments of *Schizoneura*. This is paralleled by the Panchet stage of India and the Narabeen stage of New South Wales.

During Upper Beaufort time a few new forms made their entry (*Odontopteris, Danæopsis,* etc.), but on the whole the flora was destined to become extinct. By the end of Triassic time, however, a new flora had developed, and this appeared in full force in the lower Stormberg-Molteno beds. But between these two periods, if we have read the history aright, there extends a long intervening pulsation period, the first of the new succession of marine invasions recorded on the pangæa.

This was separated from the last of the older pulsations that rose sufficiently to flood the geosynclines, by a long period of depression, when, despite pulsatory rise, the sea could not reach the geosynclines. So long, indeed, was this period that the marine fauna had undergone a complete change, while the land fauna had developed into a new and rich assemblage of organisms, chief among which were the reptiles. These were now dominant, and during the succeeding periods of the Mesozoic became the rulers of the land, until they had reached the limit of possible growth and were replaced by the smaller but more intelligent mammals.

10

Since it is probably in the moderately elevated parts of the lands that opportunity for forest growth is given, while the high-lands are actively eroded and the low-land basins are the sites of inland deposition, we cannot expect to get a complete and perfect record of plant development during the interval of land expansion. For only such remains as are accidentally swept into the basins of deposition, will be preserved. The plants of the uplands flourish, die and decay and leave little record behind, since these uplands are not places of deposition, but at best only places of slow rock decay. This product of decay will later be swept away and reworked by the transgressing sea and only accident makes for preservation of plant-remains in exceptional cases.

The same holds true for terrestrial vertebrates, which, when death overtakes them, serve as food for the living, and if their bones remain behind, there is no chance for burial except by accident, and so they will crumble to dust and never leave any record.

Basin deposits like those of the Karroo are exceptional, and because of their history, preserve for us a modicum of the land-life, which would otherwise remain wholly unknown.

But they do more than that. They give us a hint of the imperfection of the record which we are accustomed to rely upon. And this hint will bear fruit if we recognize the significance of these deposits as of *interpulsation origin* and recognize them as a piling up of inter-pulsation and of missing intermediate pulsation systems, instead of attempting to correlate them with the normal record of the marine series, — in other words, if we grant the terrestrial world of life time enough to develop and an expansion of lands sufficient to develop on.

In the future, then, we must recognize the following types of periods:

I. PULSATION PERIODS, or periods of expanding marine waters; — periods of widening marine habitats, and the diversification of marine faunas, — but also periods of narrowing land conditions and of contracting habitats for terrestrial organisms, both plants and animals. This implies periods of fierce struggle for existence, with survival of the fittests and destruction of the great mass of the land fauna and flora, — but of their preservation as fossil records only under exceptional circumstances. Thus most of the land record is lost.

II. INTERPULSATION PERIODS, or periods of with-drawing seas, until only the marginal epi-seas form a habitat for the survivors,—extinction of all others with adaptation of a few to a new environment, some to fresh water habitat, others even to terrestrial life. But the remains of the dead will mostly be preserved in the marginal deposits. Finally, when the sea has completely withdrawn to the epi-seas, the struggle for existence among marine animals will be fiercest there and modi-fication of form under duress will take place. But these developing types, the transition forms, though their remains may be preserved, are practically lost to us, since epi-sea strata may never enter into the preserved geological record of the lands. Meanwhile, the land biota flourishes and expands, but it, too, mostly leaves no permanent record of its expansion except in a few favorable regions like those of Karroo sedimentation.

III. Finally, we must recognize the exceptional periods, when the sea-level falls so low that, for a succession of pulsation periods, it cannot flood the land. Then nearly all marine forms, except pelagic types, will become extinct and when, after a long unrecorded interval, the sea is able to return, a wholly new creation, with only the faintest flavor of the past, will appear with it.

That the land faunas make tremendous strides in development during such an interval is, of course, to be expected, but only an occasional volume of the history of this expansive evolution will be preserved on the land. As long, however, as we persist in trying to fit the continental volumes into the same cover with that of an adjacent marine volume, we hopelessly confuse the record.

CHAPTER XXXI

THE SEA RETURNS

1

WITH the reference of the main portion of the Bunt-sandstein to the lost interval, or to the interpulsation period that followed it, the remainder of the Triassic becomes an orderly record of a pulsation period initiated by a transgression of the sea into the geosynclines. It begins with the Rhöt, or Upper Buntsandstein (the probable equivalent of the Scythian of the Alpine region),[1] and culminates with the Muschelkalk, which is followed by the regressive stage of the Keuper.

The movements of the sial caps, *i.e.* the pangæa, had at last brought Central Antarctica over the South Pole, and as a consequence, the Himalayan geosyncline and its westward continuation, the Mediterranean geosyncline, lay in the vicinity of the equator.

The Mediterranean, or Tethyan, geosyncline ended near Spain and the Balearic Islands, transgressing across these and for some distance beyond, in mid-Triassic time. It also extended over parts of south Europe of today.[2]

All this region had been a desert basin in the time of the Lower Buntsandstein, which represented the inter-

---

1) The Scythian stage in the marine Lower Triassic, marks the first transgression in north-west Europe. It is represented by the higher Buntsandstein, or Rhöt, 20 to 150 meters, with the *Myophoria costata* fauna, and other bivalve molluscs, some of them characteristic of the succeeding transgressing beds. The first representation of this is at the base of the Middle Buntsandstein.

2) This was a normal geosyncline of not very deep water, and sinking as the sediments were built up in it, to a totality of nearly 5,000 meters, or over 16.000 ft (approximately 3 miles). Probably the depth of water was never more than 50 feet, and most of the time much shallower. The present deep basin of the Mediterranean is a rift-valley formed at the time of the Alpine folding, or shortly after. The strata of the old geosyncline now lie in a folded and crumpled mass many miles to the north, where they form the modern Alps.

pulsation period that preceded the Scythian transgression, and locally, perhaps, also the lost Permo-Triassic interval.

The Alpine and Himalayan geosynclines lay mainly to the south of the equator of the time (0°-20°), while the north German and British regions lay in approximately the same relation north of the equator.

The normal transgressive and regressive marine phases of the Triassic pulsation are divisible into the following stages or series, which have their more open water representatives in the Alpine Trias (in ascending order) :—

TRANSGRESSIVE SERIES :

1. SCYTHIC STAGE (or first transgressive series) zone of a few to .... — 1,000 m (3,280 ft)
   a. *Tirolites cassianus* Quenst. (Werfen beds)
2. ANISIC STAGE : (Virglorian) zones of : — 1,000 m (3,200 ft)
   b. *Ceratites binodosus*
   c. *Ceratites trinodosus* Mojs.
3. LADINIC STAGE :
   comprising the zones of : .... — 1,000 m (3,280 ft)
   d. *Protrachiceras reitzi* (Buchensteiner beds)
   e. Volcanic flows of augite prophyry and tuffs :
   f. Zone of *Daonella lommeli* Wissman (Wengen beds)
   g. Zone of *Trachyceras aon* Münster (St. Cassian beds)

REGRESSIVE SERIES :

4. CARNIC STAGE : beginning of undoubted regression .... — 700 m (2,300 ft)
   Zone of *Trachyceras aonoides* (Raibler beds); to this is generally added as a terminal member :
5. NORIC : (Renewed transgression? = Rhætic-Liassic pulsation system)... — 800-1,000 m (2,625-3,800 ft)
   (Haupt dolomite below)
   Dachsteinkalk includes Hallstätter kalk above.
   i. Zone of *Megalodon (M. gümbeli* and *M. complanatus)*

The Alpine Trias has a maximum thickness approaching 4,700 meters, or a total of over 15,400 ft of marine sediments; or, excluding the Noric, an aggregate of 3,700 meters or more than 12,000 feet.

The lowest beds are transgressive over the old erosion surface formed during the Metazoic interval, and its various members overlap one another progressively in the direction of transgression. Frequently there is a basal conglomerate. Needless to say, the lithic character of these beds varies very greatly, as is to be expected in a transgressive series, and this variation in character of sediments is accompanied by a variation in the littoral types of organisms which it contains, for these are often a sensitive index of the physical conditions prevailing in their environment. While pelagic types like the ceratites may become stranded in any kind of sediment, bottom-living animals are more choice in the selection of their habitat, or perhaps it would be more correct to say that unless they happen to settle on a favorable bottom they are unsuccessful in the struggle for existence and will languish or perish.

The earlier transgressive stages represent the battle of the sea to get possession again of the lands which had been for so long beyond its reach. This is indicated by the development of great gypsum and salt beds in the lower series, which implies that the transgression was an intermittent one, punctuated by local return of the desert climate, under the influence of which cut-off portions of the Scythic sea became subjected to evaporation.

2

The Scythic, or Werfen beds range from slight representation to thicknesses of several thousand feet, or about 1,000 meters. They begin with thin-bedded sandy strata (Seiser beds), followed by the red clayey Campiler beds, which include the salt layers from which the region in the Austrian Alps has received its name of "Salzkammergut". Here again we have a good illustration of the deposition of salts in cut-off lagoons, while gypsum beds and fossiliferous layers accompany the series and testify to the derivation of the salt from marine waters.

As this was an interrupted transgression rather than the deposits of retreatal series, the areas of salt deposition were mostly overwhelmed again before the process was completed, and this is reflected in the comparative scarcity of the mother liquor salts, though these are not wholly absent.

3

It is generally held that in the Alpine orogeny there was a northward displacement of more than 15°. In Triassic time that region was a geosyncline of first magnitude, characterized by continued sinking and filling by marine strata. It was, as we have seen, in the equatorial belt and therefore one of luxuriant life in the sea, but never a region of deep or abyssal waters, such as is the Mediterranean today. Furthermore it was a region of transgressing seas.

On the marginal platform of the geosyncline, where now is North Germany, the transgressive character of the Muschelkalk is clearly indicated. The transgressing sea converted the upper layers of the desert sand into a fossiliferous series, the *Rhöt* or Upper Buntsandstein, with its many gypsum layers deposited in drying pools and its shells of *Myophoria (M. costata, M. vulgaris, M. ovata,* etc.), and other pelecypods and its rare ammonoids. The clearer water conditions, which were gradually developed, permitted the entrance of a richer fauna from the geosyncline, including some of the species of *Ceratites*, as well as representatives of *Encrinus*.

4

After the first transgression of the sea, which culminated in the formation of the Lower Muschelkalk (90-100 meters), retreat from the platform set in, or at least stationary conditions, which induced the development of lagoons. These, being cut off from the main water body, became concentrated until sea-salts were deposited. This is the so-called Middle Muschelkalk, or anhydrite group, consisting of 30 to 100 meters of porous dolomites, marls and limestones with lens-shaped or tower-like masses of gypsum and anhydrite, as well as rock-salt masses, from 8 to 22 meters in thickness, followed by dolomite, either solid or leached to porosity.

Fossils are comparatively rare in this division, since it represents drying lagoons. Those found are mainly pelecypods (*Myophoria, Mytilus, Monotis*) and a *Lingula*, etc.

The upper or main Muschelkalk comprises 40-120 meters of limestones, abounding in crinoid stem-joints (Trochiten), and occasionally complete calices of *Encrinus liliiformis*. It carries in the upper part the well known *Ceratites nodosus* (Fig. 95, p. 336), and some other species (*C. compressus, C. evolutus* and *C. spinosus* in the lower, *C. semipartitus, C. dorsoplanus, C. intermedius* in the higher part), besides many other fossils.

It would appear that these fossiliferous portions of the Muschelkalk are all referable to the Virglorian or Anysic division of the Alpine (geosynclinal) succession, and that for some time after that period, the sea did not flood the North German basin again. Of course, a long time interval may be included in the formation of the middle gypsum and salt beds of the Muschelkalk, which is then bounded by a great hiatus below as well as above. But there is no palæontological evidence for this. The nearest relationship, as pointed out by Tornquist, is in the Upper Buchenstein beds or Lower Ladinic of the Vicentin Alps, which contain ammonites of the group of *Ceratites nodosus*, a species that in the Alpine succession belongs to the horizon of *Protrachyceras reitzi*. That the withdrawal of the sea followed immediately is shown by the fact that the lower, or Kohlenkeuper, which is the only part that carries marine fossils (the Rhætic is here referred to the Lias) contains a depauperate Muschelkalk fauna, but all normal marine types such as brachiopods, oysters and cephalopods are virtually wanting. A *Ceratites* however (*C. schmidti* Zimmerman) and a *Nautilus* are of rare occurrence.

The Keuper begins with clays, sandstones and some intercalations of dolomitic layers, indicating the dwindling marine conditions. Impure clayey coals (Lettenkohle) and a variable bed of dolomitic limestone terminates the lower part and carries mainly a depauperate relict fauna, which consists largely of the pelecypods *Myophoria* and the brachiopod *Lingula*, while some surfaces are covered with thousands of the little bivalve crustacean *Estheria*, which lived in the desert lakes of the Buntsandstein at the beginning of the Triassic period. With these occur teeth of a Dipnoan fish *(Ceratodus)* and fragments of labyrinthodonts and saurians *(Mastodontosaurus* and *Nothosaurus)*. The whole of the lower series ranges from 10 to 50 meters or more, and indicates the disappearance of the sea and the return of desert conditions.

Then follow sands, clays and gypsum beds, with salt pseudomorphs, aggregating in some sections 450 meters, and representing enclosed basin deposits, though *Myophoria*, *Corbula* and *Estheria* still characterize some of the beds. Many of the higher beds also contain the characteristic fresh-water fish *Semionotus*.

Over the emerging lands roamed the early dinosaurs *(Plateosaurus)*, and the giant stegocephalians that found their food in the fishes which abounded in the ponds and streams.

The Buchensteiner beds of the Alps (the larger representative of the Upper Muschelkalk) are followed by the Wengen beds, which are especially characterized by abundant lava flows of augite-porphyry; these lava flows cover the older rocks, and are associated with tuffs that pass up into sandy beds, and then are followed by black, platey, marl limestones. *Daonella lommeli* is the chief fossil of the lower and *Posidonia wengensis* of the upper part. Various ceratitic ammonoids *(Protrachyceras, Proarcestes, Joanites, Asperdites,* etc.) also occur as well as a rich flora (flora of Corvara).

The Wengen beds are followed by the St. Cassian series, tuffaceous marls, with a rich and beautifully preserved, though diminutive fauna, for which these beds are internationally famous.

A remarkable type of deposit of this same time interval is seen in the great 'Dolomites' of the Tyrolian Alps, which sometimes reach a thickess of 1,000 meters. They may extend from the Anysic through the Ladinic, and they have been interpreted as ancient reef masses, though generally the reef building organisms were not corals so much as coralline algæ *(Diplopora)*. They, of course, include very many other organisms. Some students of these isolated masses of dolomitic rock have, however, regarded them as infaulted fragments of a once continuous and uniform bed.

5

The next higher series, the Carnic beds, are referred to the Upper Triassic for purely systematic reasons; structurally the lower beds still belong to the transgressing series.   There is evidence of partial withdrawal of the sea at the end of Carnic time. This is furnished by shallow water deposits, and evidence of local concentration is shown by the diminutive Raibler fauna, and the presence of gypsum, but this is followed by renewed transgression and even overlap, in Noric time.[4] This means that the transgressive character and pure water type of sediment is maintained throughout the remainder of the Triassic, and that such retreatal beds as were formed at the end of the Triassic were removed again during the long interpulsation period of erosion which followed and which is only indicated by an erosion hiatus.   On the platform, however, it is probably represented by the continental Keuper, which there rests on lower beds.   Still, since it is followed by the transgressive Rhætic, the basal bed of the Liassic pulsation, we may accept the Haupt or Main Keuper as representative of the Triassico-Liassic interpulsation period, and designate that period by the name Keuperian.

The complete withdrawal of the sea allowed the development locally of desert conditions, with the formation of red beds formed of ancient loess.

How long the emergence continued and how much of the Keuper beds are to be referred to the interpulsation period of emergence, deserves to be made the subject of future investigations. Certainly the Lower Keuper alone has the true earmarks of a retreatal series, carrying the last of the relict faunas of the sea.   The Middle Keuper, with its continental beds, characterized by conifers (*Voltzia*), *Equisitites*, cycads (*Pterophyllum*) and ferns, its giant salamander-like stegocephalian, and dinosaurs, or

---

4) The question of referring the Noric along with the Rhætic to the Liassic pulsation system will be considered in the next chapter

marked by desert sands and gypsum deposits, may well be regarded as product of the interpulsation land period. The next succeeding Rhætic, which follows disconformably and with transgressive overlap, marks the return of the sea in the next, the Liassic pulsation period. Both French and British stratigraphers generally make this the base of the Lias, while the Germans, for historical reasons, retain it in the top of the Trias, where it evidently forms an incongruous appendage.

6

Westward the Himalayan geosyncline, of which the Alpine is only a part, becomes shallower and more subject to invasions of land-derived sediments with sparse faunas. In the western Alps of Switzerland and France conglomerates, sandstone and quartzites rest disconformably on older rocks. They are succeeded by massive and by porous dolomites, with gypsum and with few fossils, including pelecypods of the German Muschelkalk. These are succeeded by vari-colored shales and marls, the upper part of which carries limestones and marls with *Avicula contorta*, representing the Rhætic.

When compared with the eastern Alps the thickness is markedly reduced, and it is evident that the marine waters which flooded the Alpine geosyncline did not extend much beyond Central Spain, where in Andalusia gypsiferous and saliferous marls predominate with a few fossils of the Muschelkalk type. Similar conditions in other parts of the western Mediterranean region show that here lay the western shore of the Triassic waters that had filled the Alpine geosyncline. Thus no connection with an oceanic body, like the Atlantic, can be postulated from the characters of the formations, the Atlantic, in fact, not yet having come into existence.

This leaves only two directions from which invasion might proceed, the North Russian, and the Eastern Pacific. Northward the earlier beds are unknown until we come to Spitzbergen, Bear Island and the Siberian region, where, however, we meet with a distinctive West American type of Triassic.

This, then, leaves the eastern region through the

Himalayan geosyncline the only approach to the Alps, and here indeed we meet with an extensive Triassic succession, in which all the divisions are represented by faunas, especially of the pelagic ammonoids.

The most typical section of the Himalayan geosyncline is that in the Spiti region, Shall-Shall cliff, etc. In this region the Triassic beds rest disconformably on the Permian *Productus* shales, and it is sometimes assumed that there was a continuous sedimentation from the one to the other, and that the most characteristic fossil of the lower Triassic, the remarkable ammonoid *Otoceras woodwardi*, continued across the border-line from the one to the other.

The absolute untenableness of such a view must of course be evident from the foregoing considerations (see pp. 337; 343), since, as we have seen, such a view leaves no opportunity whatever for the extinction of the older and the development of the newer fauna. Thus, the apparent occurrence of a typical lower Triassic ceratite, associated with some fossils of the *Productus* shales, can have no weight whatever, since any number of weathered-out Permian fossils could be included in the basal beds of the transgressing Triassic sea, along with the remains of the early Triassic fauna. Unquestionably, there must have been a vast amount of older fossils available, considering the length of time that these rocks were exposed to weathering and erosion. Indeed, the wonder should be that so few 'mixed' faunas are known, and it is safe to predict that in the future many more will be found. But mixed faunas are not transition faunas, as we have said before, and can suggest continuous development only to the uncritical mind.

7

In the Himalayan geosyncline of India we have another great development of the Triassic as calcareous beds, but it is not comparable in magnitude with that of the Alpine geosyncline. At maximum it reaches only 3,250 feet or something less than 1,000 meters, while that of the Alpine geosyncline is more than 3.5 times that. The most complete Himalayan development is at Spiti.

The Himalayan succession at Spiti is very complete and very fossiliferous including the following zones in descending order :[5]

SUPERFORMATION : LIASSIC

Limestone with *Megalodon* (Rhætic type)

Hiatus and Disconformity

TRIASSIC

E. NORIC : 5. Quartzites, with *Spirigera maniensis* etc.; 4. Shales, with *Monotis salinaria*, etc.; 3. Coralline limestone, with *Spiriferina gries-bachi*, etc.; 2. Limestone, with *Halorites*, etc., very fossiliferous; 1. Limestone, with *Proclydonautilus griesbachi*, etc.

D. CARNIC: 5. Dolomitic limestones, with *Lima austriaca*, etc.; 4. Shales, with *Tropites discobullatus*, etc.; 3. Shales and sandy limestones, with lamellibranchs and brachiopods; 2. Shales and calcareous grits at Spiti, with ammonoids, etc.; 1. Limestones and shales, with *Halobia comata*, represented throughout a thickness of nearly 300 m, and with ammonites (very abundant).

C. LADINIC : represented by a hiatus—or by a limestone, with *Traumatocrinus,*—poor in cephalopods and with *Daonella indica*, etc.; 1. Shales, with *Daonella lommelli*, and ammonoids.

B. VIRGLORIVN : 2. Limestone, nodular, black and extremely fossiliferous (127 species of ammonoids), corresponding to the zone of *Ceratites trinodosus* of which it encloses several species; also other genera; 1. Limestone, with *Ceratites subrobustus*, etc.

A. LOWER OR WERFENIAN: 2. Limestone, with *Hedenstræmia mojsisovicsi*, etc.; 1. *Otoceras* beds; limestones and bluish shales very thick, and with numerous ammonites, the species all special to the Himalayas, distributed over many genera, including *Otoceras* (6 species),

5) HAUG: *Traite*, p. 902-903.

Ophiceras (10 species), *Meekoceras,* etc. Other molluscs and brachiopods are rare.

Long Hiatus and Masked Disconformity

PERMIAN PULSATION SYSTEM of the Palæozoic.

8

We must now enquire, Where does this geosyncline have its feeding epi-sea? For that none of the older epi-seas of the Palæozoic geosyncline could qualify must be evident from a glance at the map. We have already seen that the Boreal region was cut off, and we must now record that the old Samfrau geosyncline was no longer functioning in Triassic time. In the first place Triassic strata are unknown in Australia, and in the next, the old Ross epi-sea of Antarctica was now blocked by ice. But the continuation through Burma, and across the East Indian or Malayan Archipelago is indicated by the development of the Triassic fauna on those islands, which, however, then formed a part of the south-eastern rim of the pangaea.

From Sumatra have been obtained characteristic *Halobias,* and *Daonellas,* and from Borneo the typical *Myalina salinaria.* From Timor, Rotti, etc., and from New Caledonia and New Zealand sufficient faunal evidence is known to make it reasonably certain that those island masses lay on or near the pathway of the Pacific connection. An arm of the Himalayan geosyncline embraced eastern Tibet (p. d.), and extended to Tongking and South China (Yunnan and Kweichow). The Triassic faunas of these regions are all of Himalayan type, but of more or less incomplete development. All lay near the equatorial zone of the time.

On the other hand, the Triassic of Japan is an extension of the Boreal (Siberian) type, as is shown on our map. (Plate XIII).

This Boreal type was more fully at home in the West Boreal or Alaskan epi-sea, as is shown by its development on Spitzbergen and Bear Island, the Arctic archipelago of North America, north of the 75th parallel of today, and down the west coast of North America to Mexico. At the same time another geosyncline indented western North America from the Arctic region, where the Rocky Mountains now rise. In all of these the ceratitic ammonoid *Meekoceras* is typical of the lower division.

The Beaufort series of South Africa is essentially the continental deposit representing the long period of emergence, during which the sea had no access to the land because of the extreme depression, and if we regard the Buntsandstein of Europe as representing the last of the interpulsation periods, before the sea again took possession of the land in the Triassic pulsation, we probably gain a much truer picture of the interrelations of the several formations.

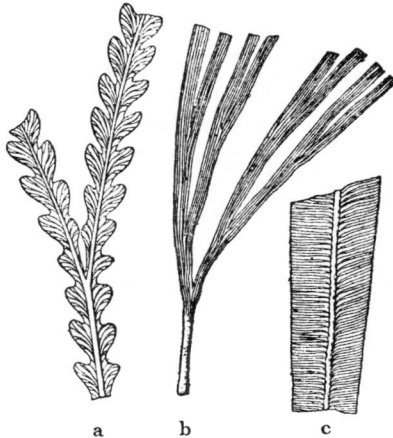

Fig. 101. Stormberg (Molteno) plants (reduced).
a: *Thinnfeldia odontopteroides*;
b: *Baiera schencki*;
c: *Taeniopteris carruthersi*.
(After Du Toit; Geology of South Africa.)

True, we have not solved the problem of the origin of the *Thinnfeldia* flora (Fig. 101), which burst suddenly upon the world in the post-Triassic interpulsation period. The beginning of the development of that flora falls into the late Triassic pulsation period, when the older flora met with a restriction of its environment, by wide-spread sea transgression, and suffered all but annihilation, while only a few individuals survived to become the progenitors of the new flora. That flora developed with the extension of the land in the post-Triassic interpulsation period, but its center of origin has not been found. It would seem that Africa is the logical home-land, where this flora, as well as the Karroo fauna, could develop, and if further exploration brings to light a formation older than the Molteno but younger than the Beaufort, *i.e.* a truly continental representative of the Triassic, the ancestors of the *Thinnfeldia* flora should be found in it.

9

But someone may ask, Why not the Molteno formation, which contains the flora fully developed? That is just the reason; a flora cannot spring into existence fully developed in a formation which follows one from which it is wholly absent. That fact alone signalizes the existence of a hiatus between the Upper Beaufort and the Molteno, and if our reasoning is sound the missing period in South Africa corresponds to the time of the marine Triassic in the Alpine and Himalayan geosynclines. That much is good logic. It does not of course follow that the continental Triassic (in the sense here implied as the time equivalent of the marine Trias) was of necessity preserved, but it is a reasonable supposition that it was so preserved somewhere.

The Molteno beds are probably the representatives of the Keuper interpulsation period, which in Europe is formed by unfossiliferous red sandstones. That would make the *Thinnfeldia* flora appear before the Rhætic, which is the early Liassic transgressive series, to which it has spread after the inhospitable conditions of the Keuper had come to an end. I conceive of the interrelations and age relations of these beds essentially as follows: [6]

---

6) This is, needless to say, an entirely different concept from that current in our stratigraphic literature, where in the first place no interpulsation periods are recognized, and in the second place, the great missing interval between Permian and Triassic is not acknowledged as such, and the question of the cause of the extinction of the old, and the appearance *de novo* of a distinctive fauna, is almost overlooked.

Non-Marine Series (Africa)	LIASSIC PULSATION	Marine Series (Europe)
Red Beds 1,600 ft		Lias-Rhætic (transgressive)
Molteno 2,000 ft		*Interpulsation* Keuper beds
Hiatus	TRIASSIC PULSATION	Marine Triassic
Bunter Continental Beaufort Series	Metazoic Interval	*Hiatus*

PERMIAN PULSATION

The border geosyncline of western North America, now mostly represented by much disturbed and isolated fragments of folded strata, was probably a continuous marine water-way, most likely separated from the panthalassa by a protective border-land. This, however, was not continuous with the similar border-land of the west coast of South America, the persistent Caribbean epi-sea dividing them. As this lay approximately on the equator of the time, the geosyncline in either direction could be supplied with tropical forms, especially pelagic types, many of which no doubt entered from the Pacific. Others, however, were supplied by the Boreal sea, for not all of this region was in the grip of the north polar cold as would appear from the projection in which the map is drawn.

CHAPTER XXXII

THE RHÆTO-LIASSIC PULSATION SYSTEM

1

THE upper divisions generally placed in the Triassic are the Noric and the Rhætic, although as previously remarked, the Rhætic is now more generally referred to the Liassic transgression. And the Noric and the Dachstein may have to go with it.

The Schlern district of the Tyrolean Alps presents an instructive section. It begins with the Werfen beds or the Scythian, followed by Anysic or Virglorian and then by the Buchenstein dolomites of the Lower Ladinic. Then, in the main mass, follows the heavy reef rock. This, in the southwest, rests directly on the Buchenstein dolomites, that is Lower Ladinic, and is followed by flows of augite porphyry. In the north-east end of the section, the igneous beds, much thicker here, abut against the reef-rock, and are covered by beds of Wengen age, which include rafted blocks of the reef-rock.

In the south-western part of the section the porphyry is overlain by bedded Wengen and St. Cassian beds, and these by Raibler beds (Carnic), with some Plattenkalk of retreatal Lower Noric (Triassic) age. Then disconformably above this lies the transgressive phase of the Dachstein (Upper Noric-Rhætic age).

2

A very illuminating description of the 'Dachstein' is given by Suess,[1] from which I quote condensed paragraphs.

The well stratified member of the Alpine limestone series which immediately underlies the Rhætic is known as the 'Plattenkalk'.

---

1) SUESS: *The Face of the Earth*, II, p. 260, *et seq.*

"It has been traced from the Vorarlberg to near Vienna, and through the whole limestone region of the southern Alps, maintaining the same characters in the faulted-in band of the limestone which traverses Carinthia.[2] The pale limestones, however, which overlie the richly fossiliferous beds of the Rhætic stage, the upper Dachsteinkalk of the northern Alps, are only a recurrence of the beds of the Plattenkalk type at a higher stage."

Foraminifera apparently play a large part in the building up of these limestones but these are more likely a shallow water accumulation. Moreover bright red shards are strewn through some of the beds. Sometimes these fragments are angular, as if broken from a hard red bed; sometimes they are thinly laminated in red and yellow layers. The large shells of *Megalodus* are sometimes found filled with red material up to a level line, while above this occurs white limestone or calc-spar. "This red material is the *terra rossa* of the Karst and of the emerged coral reefs of Oceania, the residue left after the solution of limestone, and thus scarcely likely to have been formed beneath the sea."[3]

The corals of these rocks have been greatly altered by crystallization and are generally referred to as *Lithodendron*. The radiating branches present a rosette-like or nodular form; sometimes, encrusting the shell of a *Megalodus*. Sometimes they are scattered through the limestone, but there are some beds which are entirely made up of them; their branches are more or less vertical, the intervals between them being filled up with limestone. These are true coral limestones. They are separated both from the underlying and the overlying limestone by a bedding-plane or continuous plane of division, which may frequently be followed for a great distance along the face of the cliff, and across which the knoll-like reef masses and branches do not extend.

The beds themselves vary in character; gray, very splintery limestone without a trace of organic remains

---

2) Suess suggests here that Plattenkalk and typical Dachsteinkalk are synonymous terms, but this can only apply to reworked Plattenkalk material which is incorporated as basal transgressive Upper Noric.

3) *Ibid*, pp. 260-261.

alternating with pale yellowish or grayish white, even-fractured rock, occasionally rich in fossils.

Sometimes the beds are laminated, this being clearly shown on weathered surfaces; polished surfaces often show larger or smaller fragments of limestone of a different origin enclosed in the matrix. In the Dachstein mountains vast numbers of *Rhynchonella ancilla* are found embedded in fragments of a light gray limestone, which forms foreign inclusions in this rock. Polished surfaces also reveal fragments of a yellowish-white limestone containing other organic remains, and of an unfossiliferous gray limestone, as well as parts which are coloured by red earth in bands of lighter and darker tints.

Not infrequently each of these derivative components has been covered with a thick crust of carbonate of lime, occasionally enclosing several adjacent fragments. This crust, which in section shows a radiate structure, was deposited before the formation of the matrix. This indicates repeated exposure, for, as Suess says: "It is not easy to understand how this process could take place beneath the sea". (p. 263)

The bedding-planes are usually marked by argillaceous layers, while thin streaks of lustrous coal, formed of drifted stems of plants, make a rare appearance here and there in the uppermost beds of the Plattenkalk, not far below the lowest beds of the Rhætic series. The partings are formed of black bituminous shale containing remains of ganoids and numerous scale-like leaves and twigs of *Araucarites alpinus*. These black shales with their fishes and terrestrial plants present a most striking contrast to the light colored limestones.

Higher up, calcareous shales between the beds of this limestone contain bivalves such as in Swabia characterize the development of the Rhætic series. At first these beds occur more or less as shale partings; they increase in thickness, but the pale limestone continues to alternate with them in isolated bands; the whole series grows darker in colour, the clastic elements increase in importance; and finally the typical Rhætic sediments and their fauna predominate. Thus there seems to be a perfect gradation from the Dachstein to the Rhætic, indicating that they form one series.

Elsewhere (in the Piestingthal), "where the thickness of the Plattenkalk is estimated as at least 1,000 meters, reddish marl appears as partings in the beds near the summit; these fill up the trifling depressions in the upper surface of the beds, and sometimes unite together to form thin continuous layers. They contain numerous scales and teeth of *Gyrolepis, Sargodon, Saurichthys, Acrodus,* and other fishes, and represent the bone bed which accompanies the Rhætic series in its littoral development. They are repeated four or five times at least, between the hard limestone beds. Next follows hard limestone, again with *Megalodus,* then a parting nearly two feet thick, with Rhætic mollusca of a littoral character; again limestone, and another reddish layer with a bone-bed, and then, after some closing alternations, the representatives of the rising Rhætic sea, increasing continually in depth . . . . The presence of the marls and red shards, the heaping together of the various kinds of limestone fragments, and perhaps also the encrusting sinter, all combine to show that the surface of some of the beds of the Plattenkalk was uncovered by the sea, exposed for a while to the air, and then again submerged. With the appearance of the Rhætic shells we observe an increasing proportion of detrital sediments".[4]

That this brecciation and intercalation of beds of terrigene origin indicates exposure was already noted by Suess, who considered that the exposure took place after the deposition of each limestone bed. It seems likely that we have here the products of a prolonged exposure, during an interpulsation period, with the production of limestone breccias and much lime-dust, all of which was reworked by the sea in the Rhætic transgression. The first product of this reconstruction consists of a confused mass of rock with Triassic affinities, rearranged by the transgressing Rhætic sea, which also placed its own distinctive impress upon it. Hence the terms Plattenkalk, Dachstein and Noric have been used for different beds, some retreatal Upper Triassic, and some residual beds reworked by the transgressive Rhætic sea, and therefore referable to the latter.

+) SUESS: *Ibid*; p. 265.

3

The transgressive character of the Noric has long been recognized. In the Vicenza district of north-east Italy, where a part of the Ladinic and all of the Carnic is absent, the Haupt-dolomite (Noric) begins with a conglomerate and overlaps the older beds, resting locally even on crystalline schists.

Ordinarily fossils are rare in the Haupt-dolomite or its 'Dachsteinkalk' facies; they comprise primarily the huge bivalve shells of *Megalodon* (*M. guembeli*, *M. complanatus*, etc.), the mytiloid shell *Gervilleia exilis*, the gastropod *Worthenia solitaria*, and a *Gyroporella*, which is often encrusting.

Only in special phases of the rock, such as the limestones of Hallstadt, are .ammonites found, among which may be listed the following:
Pinacoceras metternichi, Cladiscites tornatus, Paracladiscites diuturnus, Arcestes gigantogaleatus, A. intuslabiatus, Megaphyllites, Joanites, Halorites, Juvavites, Nautilus, Orthoceras dubium, etc.

Pinacoceras metternichi v. Hauer (Fig. 97, p. 358), the leading fossil, has a typical ammonite suture, far advanced over that of the *Ceratites*, so typical of the Muschelkalk and the Triassic as a whole. If any evidence, in addition to that of the stratigraphic relation, is needed to prove that these Hallstadt limestones of the Norian belong to a distinct pulsation system, this clinches the argument beyond question. A similar lobation to produce an ammonite suture is seen in the other genera cited above, except the last two *(Halorites* and *Juvavites)*, which are relatively simple-sutured. Among brachiopods *Halorella amphitoma* (Bronn.), and among pelecypods *Pseudomonotis salinaria* (Schloth.) may be mentioned as most characteristic.

4

The Rhætic proper is a further transgressive series. Suess distinguishes several phases according to the lithic and faunal characters.

" . . . . The first of these is the SWABIAN FACIES: a bone-bed, layers with *Mytilus* and *Tœniodon*, with *Avicula contorta*, but without brachiopods. It is succeeded by the CARPATHIAN FACIES with *Avicula contorta, Tere-bratula gregaria, Ostrea haidingeri.* Then comes the KÖSSEN FACIES with numerous brachiopods, such as *Spirifera oxycolpos;* and last of all the SALZBURG FACIES with *Choristoceras marshi* and *Avicula speciosa*". (Suess: p. 265)

The first or SWABIAN FACIES maintains a distinctly littoral character, and extends far beyond the region of the Alps. From Sutherland, North Scotland, where it is only known in scattered blocks, it extends into the north-east of Ireland and England (Nottingham, Warwick, Worcester, Gloucester, and Somerset, to Dorset). Some obscure littoral patches occur on the coast of Scania and the facies is found over a large part of France, and the whole of central Germany.

The CARPATHIAN FACIES is far more restricted in its distribution. It occurs in the Jura Mts., the Alps and the Carpathians, the Apennines, and in Corsica. The KÖSSEN FACIES is still more restricted. It is found in the north-east Alps, at several places in the Carpathians, and extends to the Bukowina in northern Roumania. Sometimes the brachiopods of this zone lie together in dense masses in the light-red limestone (Starhemberg beds of the north-east Alps). This is also the case with the SALZBURG FACIES.

Thus the littoral beds of the Alpine Rhætic occupy the lowest position, beneath all the other subdivisions; while at the same time they attain their greatest distribution over Europe.

During the deposition of the upper beds of the Plattenkalk, repeated withdrawal of the sea resulted in exposure of the limestone, with accompanying brecciation and erosion. With the readvance of the sea, clays with a littoral molluscan fauna were deposited, followed by the purer water facies, the last of which, the Kössen facies, is rich in brachiopods. This transgression, as previously noted, extended far and wide over Europe and completely buried the Plattenkalk under later sediments.

Suess has interpreted this transgression as a progressive deepening of the Alpine region, but a progressive rise of sea-level with normal subsidence of the geosyncline and overlap on the Russo-German platform would readily account for the phenomena.

The *Megalodus* limestone and the Rhætic series are known, outside the European area, only in the Himalayas and the outer chains of the Hindu Kush.

5

An isolated Rhætic fossil was found among the specimens brought home by Payer from east Greenland. With this exception the marine beds of the Rhætic did not extend beyond the Central Mediterranean, or its extension towards the south-west of Europe, which had been covered during the previous (Triassic) transgression.

Outside of the region of marine transgression plant-bearing beds of Rhætic age are known from Siberia, Turkestan, Tongking, Australia, in the Gondwana and Karroo series, and in the Argentine Republic as well as in the Newark beds of Eastern United States.

The Lias represents the final series of this pulsation system, and its lowest beds, of purely terrigenous character, are as a rule most widely distributed, showing continued transgression. The higher beds, representing the result of the regression of the sea, are found to be more and more restricted.

6

The detailed subdivision which has been made of the Lias (14 palæontological zones, chiefly ammonites) is given in the following table. Besides the ammonites and other invertebrates, the Lias is characterized by the first of the great marine Saurians, some of which have been preserved in marvellous perfection in the Liassic shales of Holzmaden in Bavaria.[5]

They illustrate the rapid rise and diversification of that great class of reptilians which ruled the sea, the land and the air during the mediæval period of the history of our earth. Among them, the colossal dinosaurus were the bulkiest and mightiest because of their strength. They sought to dominate their environment by brute force, and so brought about their own downfall. Today we know them only from their bones preserved in Mesozoic rocks.

---

5) For illustrations of these see my Text-book of Geology.

TABLE OF THE SUBDIVISIONS OF THE RHÆTIC-LIASSIC
PULSATION SYSTEM

SUPERFORMATION :   Jurassic Pulsation System
Interpulsation Hiatus and Disconformity

RHÆTIC-LIASSIC PULSATION SYSTEM

LIASSIC OR RETREATAL PHASE

Upper Lias or Toarcian   10 m
  ζ. —JURENSIS MARLS (zone of *Lytoceras jurensis*):
      d) Marls, with *Harpoceras aalensis,*
      c) Limestone, with *Lytoceras jurensis,*
      b) Marls, with *Harpoceras radians,*
      a) Sub-zone, with *Hammatoceras variabilis.*
  ε. —POSIDONIA SHALE [6] (zone of *Posidonia bronni*):
      d) Subzone of *Cœloceras crassum,*
      c) Subzone of *Hildoceras bifrons,*
      b) Subzone of *Harpoceras serpentinum,*
      a) Subzone of *Cœloceras annulatum.*
Middle Lias or Charmoutien and Pliens-
bachien   15-20 m
  δ. —AMALTHEUS CLAYS:
      c) Limestones, with *Amaltheus spinatus* Brug.,
      b) Dark shales, with *Amaltheus margaritatus*
          Montf. Also includes *Pentacrinus basalti-*
          formis, Belemnites paxillosus, etc.,
      a) Light coloured shales with *Lytoceras ilne-*
          atum.
  γ. —NUMISMALIS MARLS
      d) Gray marls, with *Deroceras davoei, Aego-*
          ceras capricornis and *Lipoceras henleyi.*
      c) Marls, with *Phylloceras ibex,*
      b) Zone of *Aegoceras jamesoni, A. pettos,* etc.
          Waldheimia numismalis, Rhynchonella
          rimosa,
      a) Limestone, with *Gryphœa cymbiium.*
Lower Lias or Sinemurien and Hettangien   40-100 m

---

6) In the Swabian type-region these are bituminous paper-shales filled with
*Inoceramus dubius, Posidonia bronni, Belemnites arcuarius, Amm. communis, A.
bollensis,* etc., *Pentacrinus briareus* ; reptiles (*Ichthyosaurus, Plesiosaurus,
Telesaurus*); fishes (*Dapedius*); sepeoid cephalopods (*Geoteuthis, Beloteuthis*)
chiefly as the pen and dried ink. The famous localities of Holzmaden
and Banz are in this horizon.

β. — TURNERI ZONE

    c) Clay shales with *Ophioceras varicostatum,*
    b) Clay shales with *Oxynoticeras oxynotum,*
    a) Limestones with *Arietites obtusus* (Sow.)
    (=*Ammonites turneri* Zieten).

α. — ARIETETES BEDS

    C. c_3) Oil shales, with *Cidaris olifex,* etc., c_2)
    Pentacrinus band with *P. tuberculatus,* c_1)
    Arietetes or *Gryphites* limestone with *Gryphæa
    arcuata* aud *Arietetes bucklandi,* etc.

    B. *Angulatus* beds, b_2) Sandstone, with *Schlothei-
    mia angulata*; b_1) Zone, with *Cardinia listeri.*

    A. *Psilonotus* limestones, with *Psiloceras planorbe*
    series (=*Ammonites psilonotus* Queenst.)

RHÆTIC STAGE.   With the following facies:

    C.   CONTORTA OR KÖSSEN BEDS with *Avicula contorta.*
    B.   DACHSTEIN LIMESTONE (may extend into Lias).[7]
    A.   The succession regarded as typical in the region
    between Lago di Garda[8] in Lombardie (Italian
    Alps) and Lugano[9] across the Swiss border is as
    follows:

        6)—*Conchodon* dolomite, with *C. infraliassicus,*
        5). *Lithodendron* limestone, 4). Azzarola
        beds, with *Terebratula gregaria,* 3). *Contorta*
        marl, with *Avicula contorta,* 2). Bactryllien
        shales, 1). Platy limestone (Plattenkalk),
        which rests directly on

NORIC STAGE—(Transgressive phase)   800-1,000 m

HAUPT-DOLOMIT or DACHSTEINKALK. Local facies are:

ZLAMBACH BEDS — shales, marls and limestones, with
    Choristites haueri and corals.
    (*Thecosmylia, Phyllocœnis, Montlivaultia,* etc.)

HALLSTADT LIMESTONES, with *Pinnacoceras metter-
    nichi* with a highly complicated ammonitic suture
    (Fig. 97, p. 338) and other ammonoids.

---

7) KRONECKER, W: *Centralblatt für Mineralogie,* etc. 1910.
8) Long. 10°40′ E., Lat. 45°40′ N.
9) Long. 9° E., Lat. 46° N.

SEEFELD ASPHALT SHALES or bituminous marls of Seefeld, etc. in the northern Tyrol with numerous fish impressions (*Semionotus*, *Lepidotus*, etc.)

Hiatus and Disconformity or Unconformity

SUBFORMATIONS: Trias retreatal type of Noric (Plattenkalk) (Carnic) or older, even to crystalline schists.

Thus separated from older and younger beds, with which they have long been united for historical reasons, these Rhætic-Liassic beds form a sequential series of a unit pulsation system, which falls into place between the restricted Triassic below and the equally restricted Jurassic system above.

CHAPTER XXXIII

THE WORLD IN THE JURASSIC AGE

1

To the conservative believer in the sanctity of the older institutions, nothing is more distasteful than the dismemberment of the systems founded by the fathers of stratigraphy, especially such widely accepted groups as the Triassic and the Jurassic. European geologists of the present generation will probably resist the breaking up of the so well established systems and the recombination of the members into new units. And they are right in their attitude, so long as the evidence is not sufficient to substantiate the claims of the advocates of the new classification. But conservatism must not be allowed to become a bar to progress, and the longest and most widely accepted classification must give way, if the new view-point has the facts and the logic to support its claims.

We have already given reasons why at least a part of the Noric and the Rhætic should be cut from the Triassic and joined with the Lias in a distinct pulsation system, and now we must adduce additional reasons why the Dogger should be separated from the Lias and united with the Malm in the independent Jurassic pulsation system (restricted).

A number of localities are known which present evidence of a marked disconformity[1] between Lias and

1) In the Dachstein Massif of the Austrian Alps solution funnels or pockets in the Dachstein are filled with Hierlatz limestone of crinoidal, brachiopod, and other fragments sometimes showing bedding. This suggests a disconformity between late Rhætic and Liassic, and would seem to mitigate against uniting the two into a single system. From the position on the eroded slope of the Dachstein this can of course not be considered an ancient structure. It is a comparatively modern solution-hollow, into which the debris of the higher bed has been carried and reconsolidated. If the illustration is correct, there is a vast time interval between the beginning of erosion here and the filling of the hollow by reconstructed Liassic beds. (HAUG: *Traité*, p. 981, Fig. 307).

Dogger, but not all are of equal value as recorded, unless it is borne in mind that absence of certain basal beds may be due to overlap. The zoning generally used for the Dogger and Malm, the restricted Jurassic, is as follows:

SUBDIVISION OF THE JURASSIC SYSTEM (*sens. strict.*)

UPPER JURASSIC (MALM)—Regressive Series in part

    F. TITHONIAN
          3. PURBECKIAN or AQUILONIAN
              Zone of *Olcostephanus subditus* and
                    Perisph. transitorius
          2. PORTLANDIAN (II-ζ)
              Zone of *Oppelia lithographica*
              Zone of *Pachyceras portlandica*

    E. UPPER OÖLITE
          1. KIMMERIDGIEN (restricted)
              b. Virgulien (II-ε)
                  Exogyra virgula Zone
                  Aulacostephanus eudoxus Zone
              a. Pterocerien (II-δ)
                  Pteroceras oceani Zone
                  Aulacost. pseudomutabilis Zone

    D. INTERMEDIATE OÖLITE
          2. SEQUANIEN (II-γ″)
              Astarte supracorallina Zone (of Oppel)
              Streblites tenuilobatus Zone and
              Perisph. polyplocus Zone (of newer zoning)
          1. PRE-SEQUANIAN (II-γ′)
              Diceras arietinum Zone and
              Sutneria reineckiana Zone

    C. MIDDLE OÖLITE
          3. CORALLIEN OR RAURACIEN (II-β)
              Coral oölite, with Zone of *Nerinea visurgis*
              and *Ostrea rastellaris* in N. Germany
              (Oppel's Zone of *Cidaris florigemma*)
              Zone of { *Haploceras wenzeli* / *Peltoceras bimammatum* }
          2. OXFORDIAN (II-α″)
              Zone of { *Peltoceras transversarius* / *Aspidoceras perarmatum* }

1. CALLOVIAN (II-α') Oxford Clay (Kelloway)
    (Quenstedts Zone I-ζ)
    Zone of *Quenstedticeras lamberti*
    Zone of *Peltoceras athleta* (and *P. anceps*)
    Zone of *Cosmoceras jason*
    Zone of *Macroceph. macrocephalus*

LOWER OR DOGGER — Transgressive Series
  B. *Great Oölite*—(Quenstedt's Zone I-ε)
      1. BATHONIAN (I-ε) (extended Bath Oölite)
          Zone of *Oppelia aspidoides*
          Zone of *Parkinsonia parkinsoni*

  A. *Lower or Inferior Oölite*
      4. UPPER BAJOCIEN (I-δ) (Cornbrash a local
              'estuarine' series, terminates this).
          Zone of *Parkinsonia bifurcatus*
          Zone of *Stephanoceras humphriesianum*
      3. MIDDLE BAJOCIEN (I-γ)
          Zone of *Stephanoc. sauzei*
          Zone of *Sonninia sowerbyi*
      2. LOWER BAJOCIEN (I-β)
          Zone of *Lioceras concavum*
          Zone of *Ludwigia murchisonæ*
      1. BASAL BAJOCIEN (I-α) (*Opalinus* beds)
          Zone of *Lioceras opalinum*, etc.
          Zone of *Ammonites affinis*.

Hiatus and Disconformity

LIASSIC PULSATION SYSTEM (including Rhætic)
  Upper Lias with *Lytoceras jurense* or older beds.

2

The evidence for the transgression of the sea in Jurassic time is widespread. In the London region deep borings have shown the Lias and Inferior Oölite to be absent ". . . the beds immediately superposed on the Palæozoic formations begin with the Bath Oölite (Bathonian), which is followed by the Kelloway and the other marine stages of the Upper Jurassic. It is precisely the same with the series superimposed on the Devonian reef at Marquise, near Boulogne, which also begins with the Bath Oölite."

In borings near Memel in Lithuania ". . . the Kelloway beds were met with 95 m beneath the surface, resting on red sandstone, which belongs probably to the Trias."

Over wide regions of western Europe the transgression begins sometimes with the Oölite, sometimes with the Kelloway, and sometimes again with the Oxford stage, but in every case the Lias and Inferior Oölite are missing. In Czechoslovakia (north of Brno) Upper Bathonian and Kelloway [2] beds rest directly upon the Devonian.

These of course are all examples of overlap of Jurassic on older rocks, and have no reference to the Lias and Rhætic of the preceding pulsation system, which either did not reach this region or were removed again during the post-Lias—pre-Jurassic interpulsation period. Finally evidence of transgression is seen in Poland and over wide areas in Russia, as a long and comparatively narrow zone of Lower Callovian, probably a shallow geosyncline, at the border of which Middle and Upper Callovian overlap. Suess summarizes these several transgressions as follows:

". . . with the advent of the Bathonian, which no doubt is not over sharply separated by its fauna from the Kelloway stage, the strand-line advanced far and wide. The Bathonian lies on the down-thrown Armorican arc beneath the soil of London, and on the Devonian at Boulogne; with it commences the great transgression over Abyssinia and the north-west of India. The boundary of the sea enlarged still further during the deposition of the Kelloway; in the extreme north of Scotland this lies on the fluviatile coal-bearing beds of Sutherland, and it extends over the lower Jurassic stages of Pomerania, proceeding towards Memel and as far as Lithuania; and while the Lias of Franconia already comes to an end at Regensburg, and the other stages of the middle Jurassic disappear in their turn, the sea of the lower Kelloway stage extended from Poland past Kiev, forming a long belt on the western side of the Ural, and so reached the Arctic Ocean; at the same time, passing Orenburg, it encroached on the eastern side of that chain. At this epoch the Jurassic sea extended from the Petchora on the one hand to Sutherland on the other, to Abyssinia and Cutch, and much further still to

---

2) SUESS: *The Face of the Earth*, p. 272.

the south and south-east. Withal the transgressive beds have preserved a horizontality as undisturbed on the banks of the Petchora as on the Blue Nile; and *Stephanoceras (Macrocephalites) macrocephalum* maintains its horizon in the Kelloway from Brora in Sutherland to Cutch in India.

"This extraordinary extension did not, however, mark the culmination of the positive movement; . . . . The succeeding stages occupy, as far as can be seen, not only the whole region covered by the Bathonian and Kelloway, but in Europe they even extend beyond it."

Bohemia was submerged in Upper Oxfordian (Lusitanian or Corallian) and Lower Kimmeridgian time, as was also part of Saxony, where remains of these transgressive deposits are preserved in faulted inliers of the mountains. The later Kimmeridge sea covered the southern part of the Russian Platform and extended over the Devonian red sandstone at the same time. In the Dobrudja, horizontal limestone beds of the Kimmeridge rest on the upturned green schists, exposed in some of the mountain fragments, which extend to the Black Sea or beyond.

The Lusitanian beds are known in England as Corallian (Coral rag), because of their abundant coral faunas which testify to the influence of the warm waters from the equatorial regions. North of the Alps, in Bavaria, these Jurassic rocks are also represented by reef masses which rest upon the bedded Jurassic limestones, while hydrocorallines *(Ellipsactinia)* are among the chief reef-builders. In the upper surface of these reefs, huge basins or lagoon-like hollows came into existence, in which the lithographic limestone beds of this region were formed. (Figs. 102-103)

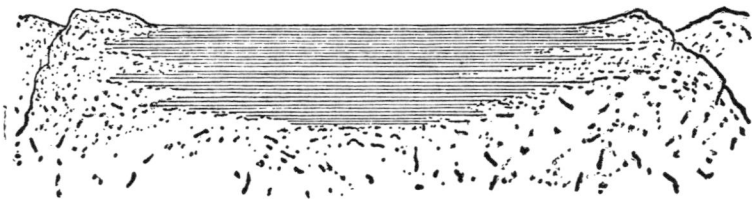

Fig. 102. Diagrammatic section of a lagoon in reef-rock, with lithographic limestone layers of extreme thinness (Plattenkalk), occupying the depressions. (After Walther).

"The basins containing the fine and thin-bedded lithographic rock form the strongest possible contrast to the enclosing reef mass. In these sediments of impalpable lime mud, the most delicate organisms were preserved with a marvelous perfection. The feathers of the ancient bird, *Archæopteryx*; the wing membrane of the flying saurians, the dragon-flies and other insects, with the veining of their wings perfectly retained, are wonderfully well preserved, and even the delicate jellyfish left its impressions in marvelous perfection. Sometimes secondary reefs occur in the midst of the basin, indicating a temporary encroachment of the reef-builders, which later on were again overwhelmed by the fine mud which produced the lithographic calcilutytes. The strata are often very thin, and very uniformly bedded. The heavier bedded ones are used for lithographic purposes, the thin ones for roofing and flagging purposes. At intervals argillaceous layers occur, and some of the beds retain a mud-crack structure. The evidence adduced by Walther goes to show that the clayey beds are due to deposition of terrigenous dust, and in them the terrestrial insect fauna was buried.

Apparently, then, ". . . . these lagoon-like depressions or basins in the coral reefs of the Jurassic sea of that region

Fig. 103. Section of the reef rock (Franken dolomite, Jurassic) of Kelheim, Bavaria. The thin bedded "Plattenkalk" is shown by horizontal lining. Those in the upper left-hand portion show distortion through gliding ("Krumme Lage"). (After Walther).

were slowly filled by the fine lime mud which was derived from the destruction of the reefs, by terrestrial dust brought by the strong winds from the distant land; and by chemical precipitation of lime. Thus were formed the fine-grained lime deposits, which reach in places a thickness approaching a hundred feet. In the more clayey beds and between the layers were preserved the insects and plants blown from the mainland, or the marine types brought there during the flooding of the lagoons. Walther has shown, from the position of these remains, that with few exceptions they were brought there dead, and left as stranded carcasses on the ooze of the lagoon bottom, which was mostly covered by little water, if not altogether exposed. The repeated evidence of shallow water, and even complete exposure of the lagoons to the air, suggests that these deposits were accumulating during a slow subsidence of the region, and that the filling of the lagoons, in general, kept pace with the sinking of the reefs. The absolute uniformity of the entire series of thin-bedded limestones further shows that the physical conditions remained uniform during their deposition. . . . the terrigenous dust incorporated in the thin sediments represents wind-blown material. It is in these dust-bearing layers that most of the fossils are found, especially the insects, of which there are 72 genera and 103 species, 35 per cent of these being dragon-flies. The peculiar character of the fauna, and the indications which it furnishes of having been stranded on surfaces of calcareous mud, suggest that the lagoon was nearly dry and was flooded only during exceptionally high tides or during storms, after which the water soon ran off again.[3] Some such conditions are represented to-day by the lagoons of Lil and Mejt, in the Marquesas Archipelago, where the water is brackish and poor in organisms, while at Jabor and Jaluit the lagoons have become fresh."[4]

The Kimmeridge stage marks the last recognized submergence known to have occurred in Russia before a great change of conditions set in. Pavlow describes its occurrence at several localities in Simbirsk, as well as in the neighbourhood of Orenburg: it also occurs according to Gourow, on the shores of the Donetz, and, according to Levisson-Lessing, at Nishni-Novgorod. The fauna presents precisely the same characters as in the contemporaneous deposits of western Europe; *Exogyra virgula*, the little oyster so abundant in the Jura as to have led to the creation of an independent sub-group, the 'Virgulian', and well known also in the Kimmeridge of Spain, England, the whole of the north of France and Hanover, as well as in the platy limestones of Ulm, now likewise appears in Poland and in the south-east of Russia in company with many other characteristic species of the Kimmeridge fauna of western Europe.

---

3) WALTHER: *Solnhofener Fauna.*
4) AGASSIZ: *Pacific Coral Reefs,* p. 31, 273, 284. GRABAU: *Principles of Stratigraphy,* pp. 439-441.

"With the close of the Kimmeridge a complete change of conditions set in and affected the whole of Europe. Everywhere the sea receded".[5]

3

The presence of a Lower Oxfordian or Callovian stage in the east of Greenland, within ten degrees of the pole, has been proved by the discovery of *Macrocephalites macrocephalus* and other characteristic fossils cited by Suess. Below this stage lies another band containing *Macrocephalites ishmæ* and three species of belemnites, which may perhaps represent the Cornbrash. In the same group of strata a characteristically Jurassic flora is met with, including species of *Phyllotheca, Anomozamites, Zamiopteris, Asplenium*, etc. Farther south on the Greenland coast, Jurassic rocks have been found at Cape Stewart on Scoresby Sound (Lat. 70°25′) (p. d.). From those thirty-seven species have been described, probably indicating a Callovian horizon.[6]

We must take into consideration that we are dealing with simultaneous transgression from two distinct epi-seas; —the Boreal on the north (Volga series); and the equatorial (Kimmeridge and Portlandian, etc.), probably the Caribbean. When they meet and overlap it appears that the northern series transgresses over the southern, which it replaces, giving the appearance of a new invasion. In reality it is a part of the same transgression, which began locally in Kimmeridgean time, and in some sections much earlier. Apparently the *Virgatites* beds with the ammonite *Virgatites virgatus* and the *Perisphinctes giganteus* beds correspond to the Lower Portlandian and the Upper Kimmeridgean. These Lower Volgian beds extend farther south and may be followed by typical continental Purbeckian.

These Boreal types begin earlier in the north-east, while typical Kimmeridgean occupied the west European region. Then, as the retreat began, the Portlandian and Purbeckian (mostly non-marine) beds merged westward into higher and higher Volgian: first *Virgatites* beds, then *Peris-*

5) SUESS: *op. cit.* p. 277.
6) GEIKIE: *Text-book.* pp. 1158-1159.

phinctes giganteus beds and finally the *Craspedites* beds (the essential equivalents of the Purbeckian). Then, after an interpulsation period of erosion, during which the typical Wealden held sway in Poland and farther west, the Neocomian sea transgressed across north-eastern Russia, where the beds with *Olcostephanus hoplitoides* and *O. spaskensis* (often erroneously referred to the Wealden) were the first to be deposited above the hiatus which in Russia represents the Wealden. The fauna is wholly new and marks the re-invasion of the Volgian sea at the opening of the Neocomian, after the Wealden interpulsation period. These basal Neocomian beds are followed by Valanginian, with *Aucella volgensis*, *A. crassicollis*, etc. and ammonites (*Craspedites* etc.).

4

The Boreal influence was extended southward and even made itself felt in the Himalayan geosyncline, where the dark Spiti shales represent the time from Callovian to Tithonian and contain the characteristic Boreal *Aucellas*, *Hoplites*, etc.

This same Boreal sea also influenced the Jurassic deposits of northern and north-western North America (Lat. 40°-50° at that time), where the *Aucella* fauna is found in East Greenland, Alaska, the Aleutian Islands, and Vancouver, and extends south to Dakota and Wyoming.

Since the marginal epi-seas were not re-established to any large degree and did not function strongly in Triassic to Jurassic times, the benthonic fauna was never a strong element of these deposits except locally. Probably it had not become re-established to any marked degree after the complete extinction at the end of the Palæozoic. The pelagic fauna, on the other hand, had become the dominant type, and ammonites, belemnites and nautiloids are the leading elements of the Jurassic fauna.

To summarize, we may stress the fact of the wide transgression of the Jurassic sea after the Lias contraction and final continental coal-forming condition of the interpulsation period, or the re-erosion of some of the earlier deposits of the Lias during the land period. Though

developed in west Europe, the Lias is already wanting at
Regensburg and Passau, in parts of Saxony, in the whole
of trans-Carpathian Moravia, in Upper Silesia, near Krakau,
and over most of the Baltic areas, over the whole of
European and Asiatic Russia, north-west and Arctic North
America, East Greenland, Spitzbergen and Franz-Joseph's
land, and over most of Africa (except in the north and
east) and in India and South America, except the west
coast. At all these points the Jurassic overlaps directly
on to the older rocks, beginning with various stages, the
older being progressively overlapped, until in some sections
the whole of the Dogger is likewise missing.

This was probably the most extensive marine transgression known in the whole history of the earth. It
was made possible by the great extent of the country
which had been peneplaned during the long continental
interval, for this provided extensive flat areas, over which the
sea could transgress, as it rose during the Jurassic period.
Beginning at the several epi-seas the transgression filled
the geosynclines, extended over the marginal platforms
and swept the country in one gigantic flood that spelt
destruction to all animal- and plant-life in its path. It
was the last great inundation; after its retreat, the pangæa
became subjected to the process of dismemberment, beginning with the opening of the South Atlantic between
Africa and South America, widening as the latter land
block drifted westward. This increased, step by step with
increase in size of the Atlantic, until finally the arrangement of lands and seas that we are familiar with came
into existence. This was, and is a process of such extreme
slowness that though we can record the fact for the long
past time intervals, progress of the drifting in a single
human life time is so slight that we have come to believe
in the permanence of continents and ocean basins, and
most of us have hardly yet become conscious of the
drifting movement of the land. For that the land blocks
of our earth do move, that 'continental drift' is no
longer a theory but is entitled to the dignity of an un-
assailable fact, is being recognized by an increasing number
of thinking geologists.

CHAPTER XXXIV

THE WEALDEN INTERPULSATION PERIOD AND THE LOWER
CRETACEOUS OR NEOCOMIAN PULSATION SYSTEM

1

IT is a generally recognized fact that the "Lower Cretaceous" or the Neocomian[1] is sharply separated from
the underlying Jurassic, the close of which is everywhere
marked by a profound withdrawal of the sea. The Neocomian is in turn always characterized by transgression;
the transgressing series begins commonly with a basal
conglomerate or a sandstone, and the successive series
overlaps on the older beds. But whether the time that
intervened between the base of the Lower Cretaceous and
the top of the Jurassic was characterized wholly by an
erosion interval or, whether in some sections interpulsation deposits of a continental character were forming, is
still a matter under debate.

That the Wealden of the type region in South-
eastern England has all the characters of an interpulsation
deposit will not be denied, but because the older marine
Neocomian strata are absent in this region, and the first
marine bed which overlies the Wealden, is usually of late
Neocomian age (Aptian or even Albian), the impression
is produced that the Wealden continental beds are the
equivalent of the older Neocomian, and this is the general
position upheld in practically all our text-books.

In the Schaumburg-Lippe syncline of North Germany,[2]
however, several hundred meters of continental Wealden
lies on Purbeck limestone, and is followed by the Valanginian and Hauterivian stages, which belong next above
the lowest of the marine Neocomian formations. This is
certainly indicative of a pre-Neocomian post-Jurassic age

1) From the Roman name of Neuchâtel, Switzerland.
2) KAYSER : *Lehrbuch*, p. 496.

for the Wealden, and that can only mean that it is an interpulsation deposit.

Fig. 104. *Brontosaurus excelsus*; the Thunder lizard. Comanchean beds southern Rocky Mountain region Texas and New Mexico (From *Natural History*, May 1939) "In bulk, this dinosaur was equal to four or five 6-ton elephants : Man would have reached only to his knee."

Wealden Interpulsation and Neocomian Pulsation

These later Mesozoic formations may be classified as follows :

Classification in general use :	Suggested classification :
	Laramie Interpulsation System
Upper Cretaceous System	Cretaceous Pulsation System (Restricted)
	Dakota Interpulsation System
Lower Cretaceous System	Neocomian or (Comanchean) Pulsation System
	Wealden (or Morrison) Interpulsation System
Subformation Upper Jurassic	Jurassic Pulsation System

2

We shall consider the Lower Cretaceous as a unit with the following subdivisions : (See Fig. 105 p. 405).

SUPERFORMATION : CRETACEOUS SYSTEM

Disconformity

Hiatus equivalent to or represented by DAKOTA INTER-PULSATION SYSTEM (continental)

Disconformity

Europe   (Approximate Equivalence)   America

NEOCOMIAN PULSATION SYSTEM

COMANCHEAN PULSATION SYSTEM

Albian (Gault)..................... }..............Washita
Aptian (Apt in France).. }

Barremian .... }..............Fredericksburg
Hauterivian .... }

Valanginian .... }.............Trinity
Berriasian............................ }

Disconformity

Wealden Interpulsation System, Morrison Interpulsation System
(continental)   (continental)

Disconformity

SUBFORMATION: UPPER JURASSIC   MARINE JURASSIC
OR OLDER SYSTEMS   OR OLDER

Suess calls attention to the fact that the Upper Jurassic shows a gradual change towards brackish water or water of reduced salinity, as we reach the upper limits of the Portlandian, and he adds:

"At this point the great change set in: The sea, which extended beyond the Volga at the time of the Kimmeridge, and at least as far as the Dniester during the Portland, was now so closely restricted that it did not pass beyond the region of the Alps and Carpathians, and indeed did not retain its hold on the Jura. In the territory it had abandoned, lagoons remained in which clay and gypsum, with here and there rock-salt, were deposited.

"In England these gypsiferous beds, which are always very poor in organic remains, are never exposed at the surface; but when in 1874 a boring was made near the middle of the Weald, they were encountered with a thickness of over 100 feet. They lie at the bottom of a basin, probably as a lenticular mass, completely buried out of sight beneath more recent sediments.

"In north Germany they are known as the MÜNDER MARLS, and they are frequently exposed at the surface as a consequence of the folding, which has affected this region. The salt-beds underneath the city of Hanover, which were passed through by a boring, belong to this series. The marls attain a thickness of over 300 meters. The resemblance they bear to the Keuper has been pointed out by [others].

" . . . In the region now occupied by the Jura mountains . . . . the course of events was the same . . . as in England. Gypsiferous marls, but only 3 or 4 meters thick, were deposited above the Portland, . . . chiefly between the Doubs and the lake of Neuchâtel, and somewhat further to the north-east and south-west; but the succeeding deposits have a much wider extension, particularly towards the south-west, and if the region had not been folded, the gypsum would form a lens buried out of sight as in England.

"Far to the west also, on the Charente, the gypsum has been laid down over the Portland.   In this case, we have before us only part of a basin; the rest is concealed, on the one hand by the Atlantic Ocean, on the other by the transgression of the middle Cretaceous. . . . Above the Portland lies first a bed of cavernous limestone, about 1.6 meters thick, then 35 to 40 meters of gypsum accompanied by clay, with fish scales and fragments of wood.   This bed of gypsum crops out in a zone striking to the north-north-west, from Châteauneuf west of Angou-lême for a distance of 40 kilometers, and it may be traced in isolated exposures through Rochefort to the point de Chassiron, the northernmost promontory of the island of Oléron; so that its total length is more than 100 kilo-meters.   The Gironde and the Atlantic conceal the greater part of this basin, and its outline is unknown."

Then, "The basins in which the gypsum was de-posited began to be refilled.

"In the Jura the sediments of this new stage, which consist of limestone and marl, extend . . . through the whole mountain range and further still into the depart-ment of the Isère.   They are sometimes separated from the underlying beds by a cavernous dolomite, and attain a thickness of only 4 or 5 meters.   At their base is a layer containing a mixed fauna of fluviatile and brackish-water shells, then a fresh-water bed, and above that again brackish-water layers.   The upper brackish-water bed is, however, of but trifling extent, and appears, at least according to existing observations, scarcely to pass beyond the region of the gypsiferous clay, which lies deep below it.   The fresh-water limestone-beds reappear outside the Jura near Baume on the Doubs, above Besançon, and near Gray on the Saône, . . . But in the department of the Yonne the beds above the Portland are already absent, as well indeed as every trace of fresh-water de-posits; and the marine lower Cretaceous rests directly [but disconformably] on the marine Jurassic.

"In north Germany, an imposing series of limestone beds, with marine, brackish, and fresh-water shells, was deposited during this period above the marls of Münde: they are known, owing to the abundant occurrence of *Serpula coacervata*, as the 'Serpulite'. The Serpulite also occurs near Boulogne and corresponds, in England, with that part of the Purbeck, which lies above the Gypsum. Oscillations were so frequent that Bristow distinguishes in this zone, as developed in the isle of Purbeck, the remains of four ancient forests, eleven fresh-water beds, four brackish, and three marine beds, succeeding one another in repeated alternations.

"On the Charente, a limestone bed with fresh-water mollusca also occurs, the 'couche de deux pieds'.

"The marine or brackish-water mollusca of these beds fully maintain the character of the lower beds; many species are common to both; we have here *an impoverished Jurassic fauna.*

"The later transgressions show clearly that at this time all the western part of the Paris basin beyond the Pays de Bray was dry land, . . .".[3]

The wide withdrawal of the sea is thus established, though locally it was oscillatory. The non-marine character of the Weald is of course not questioned by any one, though there are differences of opinion in regard to the nature of the conditions of the area of deposition; Mantell in 1822 demonstrated the fresh water origin of the Wealden. Goodwin Austin, as long ago as 1872, thought, that a great lake covered the south-east of England, extending over Kent, Sussex and Hants counties, and that it was continued on the other side of the channel in the Boulonnais in France, and that it also left its traces in Oxfordshire and the Isle of Wight. Other "lakes" were located in the Charantes and in North Germany, in Spain, Portugal, the Balearic Islands, etc. Since that time the lake hypothesis of origin has been gradually replaced by the river plain hypothesis, based on the Indo-Gangetic plain, or the Huangho type, as represented by the Yellow River of China.

It must, however, be pointed out that most students of these formations, though accepting a non-marine origin for the Wealden, have regarded it as contemporaneous

3) SUESS: *The Face of the Earth,* pp. 279-281.

with marine Lower Cretaceous (Neocomian) strata elsewhere, and so have made the Wealden the continental
equivalent of some part of the Lower Neocomian series.

We must now enquire into the validity of this correlation. The Volga stage of North Russia and other
Boreal regions, characterized by the pelecypod *Aucella* in
the upper and *Perisphinctes* in the lower part, is regarded
by many Russian geologists as the marine equivalent of
the non-marine Wealden and contemporaneous with it in
deposition. But the only argument advanced in favor of
this supposition is, that it follows beds with a Portlandian
fauna. If these Volgian beds represent a transgression of
the Boreal sea in Upper Jurassic time, their relation to
the normal Portlandian and Purbeckian is that of replacing overlap. But if, as others of the Russian geologists
hold, these Volgian beds mark the beginning of the Neocomian transgression (Infra-Cretacic of the Boreal sea),
they may with every good reason be referred to the post-
Wealden transgression.

This is equally true of the first marine Neocomian
strata, which overlie the Wealden beds in Great Britain,
in Germany and elsewhere. Different beds rest upon the
Wealden in different sections, but this can be explained
just as readily as a case of normal marine overlap on a
previously deposited continental series. Again, the contact
of the Wealden and the next succeeding bed appears to
be a gradational one, but so is the contact with the
underlying Jurassic. Thus Judd [4] gives the following typical
succession for the British formation to which is added
the revised classification.

NEOCOMIAN PULSA-TION SYSTEM	("Lower Green-sand")	marine
	Punfield Forma-tion	fluvio-marine 160-230 ft
	Wealden Interpul-sation System	fresh water 1800 ft or less
JURASSIC PULSATION SYSTEM	Purbeck Forma-tion	fluvio-marine 406 ft
	Upper Oölite	marine

The details of the minor layers as given by Judd,

4) JUDD: *Neocomian of Yorkshire*, p. 326 et. seq.

show, that the contact is not so much gradational as a normal contact of beds, formed in a sea transgressing across a surface of previous continental sedimentation.

3

The lowest marine Necomian (Barriasian etc.) is the only representative, found in the Alps. Here we meet with some significant sections. In the Central Tyrol (Colle di Muntijella), the Valanginian rests directly but disconformably, on the Triassic Dachstein limestone. The whole of the Jurassic appears to be absent. However, on the Dachstein, rests a breccia formed of Dachstein fragments with dolomitic cement, and this passes gradually into a green dolomite without fossils, which may represent the Tithonian (or Uppermost Jurassic), which is generally transgressive in the Tyrol. These lower beds are followed by a thin bed of siliceous limestones with siliceous concretions, and this by normal fossiliferous Valanginian and higher beds up to and including the Barremian, after which follow unfossiliferous Aptian shales.

In Portugal, the Neocomian begins with Valanginian limestones, with *Trigonias*, corals, etc., resting disconformably on a plant-bearing terrestrial delta deposit, which contains the oldest *Dicotyledonous* plants.

In England the thickness of the Wealden deposits increases rapidly towards the west, and reaches 2,000 ft in the beds above the Purbeck. Possibly the clastic sediment came from that direction.

Where the Lower marine Neocomian rest directly on Upper Jurassic, as in the Balearic Islands, it is separated only by an erosion hiatus, but in the adjoining mainland of Spain and in Portugal, this gap is filled by a terrestrial plant plant-bearing sandstone which corresponds to the Wealden. The Valanginian has dropped out by overlap.

Suess held that the great change in physical conditions did not exactly coincide with the change in fauna, but slightly anticipated it. But here he is in error, for though the Jurassic fauna had not yet wholly disappeared when the gypsum and salt beds of the Portland and Purbeck beds were formed, recurring in depauperate form in the higher Jurassic, he neglects to take account of

the Wealden Interpulsation period, during which the continents were entirely drained and the littoral sea-life,
restricted to narrow epi-seas. Yet he says correctly "The
marine Jurassic fauna, so far as we can trace its history,
did not give rise to the marine fauna which succeeds it.
Some of the pelagic forms of the Alpine area may have
contributed direct descendants, but the fauna of moderate
depths was destroyed, first by the extensive retreat of
the sea, and next by fresh-water conditions; some brackish-
water species may have been derived from marine forms
and have persisted for awhile; but the new marine faunas
came, for the greater part, from elsewhere.

"The Valengian [Valanginian] of the lower Neocomian,
which extends but a slight distance beyond the pelagic
region of the Alps, does not contain a rich fauna, and
shows foreign affinities."[5]

If the Neocomian fauna was not derived from the
few survivors of the Jurassic fauna in the epi-seas, its
origin is entirely unknown. But its development required
a long period of time, during which no marine sediments
were formed in the geosynclines. Therefore a long Interpulsation period separates higher marine Jurassic and
lowest marine Neocomian, and this time interval was
filled locally by the deposits of the continental Wealden.
Thus palæontology and stratigraphy bring us to the same
conclusion, namely that the Wealden, wherever found,
represents the continental Interpulsation deposits, during
the period, when in the distant panthalassa the new
Neocomian marine fauna was developing from the few
survivors of the old Jurassic fauna.

We cannot agree with Suess, when he says: "The
Alpine region remained constantly submerged"[6] for unless this was still in connection with the outer sea, it
would have developed into a stagnant water body, in
which no new marine fauna could develop. Moreover,
the evidence from the Tyrol which we detailed above,
shows, that there was no such continuity, for the Lower
Neocomian rests disconformably on the older Triassic rocks,
or on disintegration products of these limestones, which
were formed during the exposure and remained unfossiliferous recemented lime-sand.

5) SUESS: *Op. cit.* p. 288. 6) *Ibid:* p. 289.

4.

The Upper Volga phase of the Neocomian of Russia, with its distinctive *Aucella* and ammonite fauna represented, as was stated above, an independent transgression from the Boreal epi-sea. There this fauna had developed during the Interpulsation period from the survivors of the Jurassic retreat, which were sequestered in that region. When the new transgressions had effected a contact between the Boreal and the equatorially derived waters, the respective faunas intermingled. Suess adds: "afterwards in the epoch of the Gault the universality of the fauna was gradually re-established even in the most distant regions. Neumayr has arrived at similar results. Phenomena of this kind cannot be explained by oscillations of the continents." [7]

The Atlantic breach had not yet opened, but Lower Cretaceous beds are widespred on the Pacific coast. But on the Atlantic coast, except in the case of Europe, we seek in vain for representatives of any of these deposits. Even in Europe the Wealden advances with its freshwater deposits right up to the brink of the ocean. Nevertheless a communication of the Lower Cretaceous seas of Europe with those of the west of South America is indicated by their fauna. This communication, however, can only have existed in the vicinity of those parts of the coast, which are constructed on the Pacific type, *i.e.* between the existing Mediterranean and the West Indies." [8]

This proves our contention, that the tropical element of the fauna was supplied by the Cordilleran epi-sea, and this, of course, implies that the continents were still in contact.

Thus we arrive at last at the realization that the commencement of the fragmentation of the pangæa into separate continents did not begin until the Cenomanian, that is the Upper Cretaceous, and so was a phenomenon purely of the latter part of the Mesozoic and of Tertiary time.

---

7) SUESS: *loc. cit.* p. 289.
8) *Ibid*, p. 290.

CHAPTER XXXV

THE COMANCHE PULSATION SYSTEM AND THE DAKOTA INTERPULSATION DEPOSITS

1

THE sea, which occupied the equatorial region of Europe in Neocomian time, extended to the west coast of the American lands. This is shown by the faunas of the Comanche series, which is the American equivalent of the Neocomian, and is typically developed in the southern United States and the Mexican and Central American region.

The Comanche there has a threefold division, which is most complete in the state of Texas. It comprises the following members below the Interpulsation formation:

SUPERFORMATION: CRETACEOUS PULSATION SYSTEM
INTERPULSATION SYSTEM: Dakota Sandstone enclosing hiatus
COMANCHE (NEOCOMIAN) PULSATION SYSTEM
    Washita Series
    Fredericksburg Series
    Trinity Series
INTERPULSATION SYSTEM: Morrison Sandstone with variable
    hiatus and disconformity (equivalent to the Wealden)

or Hiatus and Unconformity

SUBFORMATIONS: Folded and peneplaned Older Rocks

In the Gulf coast region, especially in the state of Texas, the lower Comanche rests unconformably on the folded and peneplaned Palæozoic rocks, and begins with a basal sandstone.

In the Sierra de Mazapil in the States of Zacatecas and Coahuila, and in other ranges of north Mexico, the succession is as follows:[1]

---

1) BAILEY WILLIS *Index:* p. 552; also: R. W. IMLAY: *Studies of the Mexican Geosyncline*, Bull. G. S. A. Vol. 49, pp 1651-1694, 1938; and *Upper Jurassic Ammonites from Mexico;* Ibid Vol. 50, pp. 1-78, 1939.

		Thickness in ft
CRETACEOUS PULSATION SYSTEM		5,910-5,963 + ft

SENONIAN

Maestrichtian	absent ?	
Campanian	Difunta formation	
Santonian	Parras shale	2,000 + ft
Coniacian	Caracol formation	3,350 ft
TURONIAN } CENOMANIAN }	Indidura formation	560-635 ft

Dakotan Interpulsation Hiatus and Disconformity

(not generally recognized because of masked character)

COMANCHE (OR NEOCOMIAN PULSATION SYSTEM)		2,284-3,999 + ft
ALBIAN Cuesta del Cura limestone		830-1,025 ft
APTIAN La Pena formation		260-388 ft
BARREMIAN } HAUTERIVIAN }	Cupido limestone	862-1,123 ft
VALANGINIAN } BERRIASIAN }	Taraises formation	332-463 ft

Wealden Interpulsation Hiatus and Disconformity

JURASSIC PULSATION SYSTEM

TITHONIAN	La Casita Formation	300-3000 ft
PORTLANDIAN	*(retreatal and emergent)*; or	
KIMMERIDGIAN	La Caja Formation	140-270 ft
OXFORDIAN—	Zuloaga limestone	1,200-1,800 ft

Hiatus and Disconformity

SUBFORMATION: , Red Beds (Triassic or Older)

The Zuloaga limestone is thick-bedded and forms the main resistant mass of the eastward trending mountain in the vicinity of the 25th parallel. "On the east slope of the mountain, 6 miles southwest of La Ventura, the complete formation above the red beds is about 1,200 feet thick. . . . The underlying red beds, exposed along the axis of an anticline, are about 20 feet thick and consist of red shale, red sandstone and conglomerate. The conglomerate consists of angular masses of red shale, red sandstone and yellow quartzite. The pieces of shale and sandstone range from small particles to more than a foot in diameter. Their size and poor rounding suggests slightly reworked mantle rock." (*Imlay:* 49, p 1659)

There is thus shown here a profound transgressive overlap of Oxfordian and Kimmeridgean beds on the pre-upper Jurassic land surface, which corresponds to the great transgression recognized in Europe.

The next succeeding CAJA formation includes thin-bedded limestones, marls, shales and marly limestones, ranging up to 269 feet in thickness in the type section. Imlay correlates the La Caja formation with the Kimme-ridgean, but of this it can only represent an early part. In the western end of the Sierra de Atajo it is replaced by the gypsiferous shales and sandstones of the LA CASITA formation, this representing in part at least a retreatal and even emergent series. A long hiatus, representing the Wealden Interpulsation period, separates these beds from the Comanchean system. This hiatus, according to Imlay [2], in the Sierra de la Paila on the north, cuts out in addition the Upper Jurassic (Portlandian—Tithonian) and the whole of the Neocomian beds below the Aptian. He, however, does not recognize the hiatus in the more southerly region.

When it is realized that the La Casita formation varies in thickness from less than 300 ft to more than 3,000 ft, it will be apparent that it was subjected to long Interpulsation erosion. This establishes the Wealden Inter-pulsation hiatus for this region.

The intense folding at a later date, amounting even to inversion of the whole series of strata, and the remark-ably abrupt change in the trend of the mountains will be referred to again in chapter XXXIX.

These later Jurassic rocks extended over the Central American mass, of which the island of Cuba was formerly an integral part before the development of the Antillean hair-pin loop.

The completion of this equatorial water-way was, however, only effected in later Jurassic time, though beds of Oxfordian age are known from Cuba, while in Puebla and Vera Cruz, Mexico, black and yellow clay slates with Liassic fossils, and referable to the Rhæto-Liassic pulsation system are disconformably overlain by Upper Jurassic beds.

In all these regions the Interpulsation deposits are absent, these periods being represented by erosion intervals.

---

2) IMLAY: *Studies Mexican Geosyncline*, p. 1689. fig. 6.

This is true also of the Jurassic-Comanchic contact and of the Comanchic-Cretacic, both the Morrison interpulsation (Wealden) deposits and the Dakota interpulsation deposits being absent.

But the hiatus between the pulsation systems is usually larger than this. Not only is the Dogger or lower Jurassic absent, but the highest bed, the Purbeckian, as well, while the earliest Comanche, the Barriasian, is also omitted by overlap, as are also some of the upper beds of the Comanche by off-lap and erosion. The Cretaceous transgression is pronounced, as everywhere else in the Cenomanian.

2

In the State of Texas the formations of the three divisions of the Lower Cretaceous show the following succession in descending order.

COMANCHE SUBDIVISION

C. UPPER OR WASHITA SERIES   175-400 ft
Retreatal, with the following subdivision on the Red River in the north and 250 miles further south at Austin on the Colorado River

Division in Red River Valley: 400 ft	Austin on Colorado 175 ft
8. Grayson lime marls 7. Main street limestone 6. Pawpaw clays and sandy-limestones 5. Weno clays and sandstones 4. Denton marls and shell conglomerate 3. Fort Worth limestone 2. Duck Creek, with arenaceous limestones alternating with marls 1. Kiamitia bituminous shale	Buda limestone Del Rio clays Georgetown Limestone

The Washita division varies considerably in composition and thickness; shallow-water deposits characterize the northern border region. Here extensive formations of ferruginous sands and bituminous clays occur, which cease to the south, where the strata become more calcareous, but even this rule has its exceptions.

"As a whole, the group decreases from about 400 feet
in the Denison section, on Red River to less than 175 feet
in the Austin section, on the Colorado. This loss in
thickness of the Washita division to the south is compensated for by a corresponding gain in that direction in
the thickness of the Edwards limestone of the Fredericksburg division, so that the thickness of the Comanche
series in its entirety is not impaired thereby.

". . . What is a conspicuous clay formation in the
Red River section may become a limestone when traced
300 miles south to the Colorado, or a formation which
is a limestone upon the Red River may be a clay on
the Colorado."[3]

B. FREDERICKSBURG DIVISION   30-350-700 ft
Chalky limestones with the following members:
  c) *Edwards Limestone:*
     Semicrystalline limestones with
     flint nodules, *Requienia* and
     Rudistes; These grade down into: GOODLAND lime-
  b) *Comanche Peak Limestone:*   stone (a few feet)
     Compact, white, easily shattered   on the Brazos.
     limestones with numerous casts
     of mollusks and echinoids.
  a) *Walnut Formation:*
     Calcareous clays intercalated
     with brecciated limestones pass-
     ing up into chalky layers.

"As a whole, these sediments thicken and become
more calcareous seaward, away from the position of the
old peripheral shore-line at the close of the Fredericksburg
epoch. . . . This thickening took place by the addition
of calcareous layers off-shore and horizontally away from
the diagonal basement beds. . . . These changes in thickness were very gradual and can only be noted by comparing widely separated local sections. Hence the rocks
of this division, as a whole, notwithstanding the variations
to be noted, may be considered as presenting a remarkable
example of uniformity of thickness and composition. In
its entirety this division is composed of more calcareous
rocks than the other divisions of the Comanche series.

3) HILL: in B. WILLIS, *Index,* p. 604.

Although its interior margins ultimately pass into the Basement sands, this change mostly takes place beyond the borders of the east-central province, within which it is practically a great limestone formation (EDWARDS LIMESTONE), initiated by beds of clay (WALNUT FORMATION) at its base.

". . . The limestones of the Fredericksburg division thin toward the Rocky Mountains and the Ouachita uplifts to the north, and thicken southward or toward the Rio Grande, being in the neighborhood of 700 feet in the latter region and less than 30 along the southern foot of the Ouachita Mountains of Indian Territory [Oklahoma]. They ultimately pass into clays and sands around the western ends of the Ouachitas in southern Kansas and west of Texas in eastern New Mexico.

"The combined Comanche Peak and Edwards limestone thins out north of the Brazos, where it is no longer separable into individual beds (and is called the GOODLAND LIMESTONE). On the other hand, it thickens south of the Colorado, where several distinct subdivisions could be made." [4]

A. TRINITY DIVISION

3. PALUXY FORMATION $\left\{\begin{array}{l}\text{Upper Sands}\\ \text{Thin Limestones}\\ \text{Lower Sands}\end{array}\right\}$ 190-251 ft

2. GLEN ROSE FORMATION   5 ft to trace

1. TRAVIS PEAK FORMATION $\left\{\begin{array}{l}\text{Hensel Sands}\\ \text{Cow Creek Beds}\\ \text{Sycamore Sands}\end{array}\right\}$ 115-200 ft

"The division includes the lower or initiatory beds of the Cretaceous formations of the Texas region, embracing all the rocks lying below the Walnut beds of the Fredericksburg division. . . . In general these strata consist of sands, clays, marls, and massive limestones (including in the latter shell breccias, agglomerates, and chalks), all of which grade imperceptibly into one another, both vertically and horizontally, according to their proximity to the shore line against which they were deposited. . . ." [5]

---

4) HILL: in B. WILLIS.
5) *Ibid*, pp. 602-603.

5

On the tropic of Cancer the basal sands and conglomerates form the base of the Comanche series, the
overlying beds changing progressively through arenaceous
and calcareous clays to limestones (TEHUACAN LIMESTONES).
From this point northward to Texas and Oklahoma the
basal bed rises progressively, but several oscillations are
indicated. The general advance, however, is shown by
the overlap and change in character and thickness of the
formations. Thus, at Austin, Travis Peak beds are over
800 feet thick, and begin with basal sands and conglomerates. More than two-thirds of the formation is limestone, and it is succeeded by 600 feet of Glen Rose limestone, the Paluxy being undeveloped as a sandstone. At
Twin Mountain in Erath county, Texas, the Glen Rose
is a slightly siliceous limestone 5 feet thick, and is enclosed between 115 feet of basal sands and conglomerates
and 190 feet of Paluxy sands. At Decatur, Wise county,
nearly 100 miles northeast along the strike from the preceding locality, the merest trace of the Glen Rose limestone appears between 200 feet of basal sand and 125
feet of Paluxy. This indicates the uniform thinning
north-westward of the formation, partly by lithic change,
but largely by disappearance, through overlap, of the basal
members. The age of the basal bed at the localities of
the last two sections is clearly Glen Rose, though it is
nearer the middle than the upper part of the formation.

Along the Texas-Oklahoma line the Trinity beds
have disappeared by overlap of the Fredericksburg, which
begins with the basal Antlers sands followed by the Goodland limestone which is only 25 feet thick. This represents the Comanche Peak and Edwards limestones, which
are 350 feet thick in the Austin region and approximately
700 feet on the Rio Grande. In western Texas, in New
Mexico, and in southern Kansas, the upper Fredericksburg
beds are represented only by shore-derived clastics. In
southern Kansas these are the plant-bearing CHEYENNE
sandstones, which rest directly upon the Red beds (Permian)
and have a thickness of 65 feet. They are followed by
the Kiowa shales, with *Gryphœa corrugata*, which have
been found to extend northward into southern Colorado,
where they overlie the Morrison formation. It is not

impossible, that this horizon or a somewhat higher one will be traced north as far as the Black Hills, where a thin limestone band holds the proper position. Only the lowest Washita beds were deposited over this more northern area; the Dakota regression begins in early Washita, if not actually at the beginning of Washita time, and continues throughout that epoch.

The inter-relationships of the beds and the Dakota and higher beds are shown in the following diagrams:

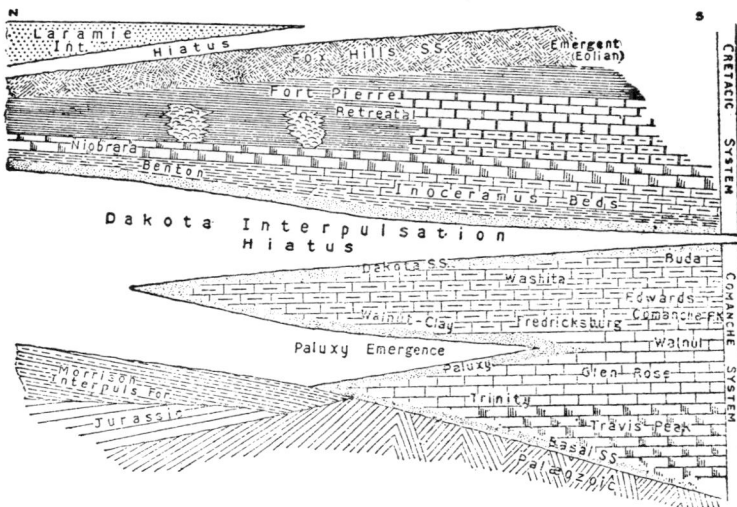

Fig. 105. Ideal section, showing the relationships of the Trinity and Fredericksburg formations of the Texas Comanchean and their transgressive overlaps. The basal sandstone rises in the scale with the advance of the sea; the Paluxy sandstone indicates a local retreat and readvance, and therefore represents a combined sandstone of emergence and submergence. The Washita is mainly a retreatal series; the Dakota is an interpulsation deposit, retreatal in Upper Comanche, readvancing in Cretaceous time, with overlap of marine Cretaceous beds against it as a basal bed. The enclosed hiatus increases northward, where the Upper Dakota is replaced by 'Benton'. The Niobrara marks locally the maximum transgression of the Cretaceous, followed by the Fort Pierre retreatal and the Fox Hills emergent beds. The continental Laramie interpulsation beds follow. (Original).

4

The basal sandstone continues beyond the region of Trinity deposition and underlies the whole Comanchean series in the southern United States. From its character and extent it appears, that we are dealing here with an old residual quartz-sand, deposited in pre-Comanchean time, and reworked by the advancing Comanchean sea, so that it became a basal sandstone for the entire overlapping series.

With the beginning of Washita time the sea again withdrew, and continental sands were spread over the emerging area by rivers and the wind. These formed the DAKOTA SANDSTONES, and they rest on progressively higher members of the Washita series as they are followed southward, thus showing the characteristic off-lap features of an emerging series. While they follow upon the lowest Washita (Kiowa or Kiamitia shales) in Kansas, they are preceded by 840 feet of Washita shales and limestone in southern Oklahoma. The sandstone does not extend as far south as Austin, where the upper Washita limestones show evidence of erosion, a character, of course, quite consistent with the other evidence of emergence.

The sea, however, continued to submerge the greater part of Mexico during Washita time, while it retreated from the southern United States. This is shown by the fact, that the Washita formation is represented by limestones 600 meters in thickness. Finally, however, emergence also took place over all of Mexico, as is shown by the fact, that these rocks were eroded before the next higher formations (the Cretaceous) were deposited upon them.

In Central and South America at this time, the Andean geosyncline received the sediments of the Neocomian series, these representing practically a complete succession. In Colombia, Barremian beds carry species of ammonites, characteristic of the same horizon in Algeria, Spain, Provence and Thuringia (Wernsdorf beds). In the Aptian also, the relationships with the Mediterranean region are very pronounced.

In Peru, beds of Wealden type are succeeded by beds with *Trigonia lorenti* (Valanginian). In the Andes of Chili Valanginian and Hauterivian are known, Barremian and Aptian more doubtfully. Finally, in Patagonia the metamorphic old rocks are unconformably followed by porphyritic tuffs, overlain by Valanginian fossiliferous beds, which are in turn unconformably succeeded by Tertiary beds.

The fauna of the older beds was derived mainly from the Ross epi-sea and is distinctive, showing the influence of the colder waters of the Antarctic region, in contrast with the more tropical types of the Caribbean epi-sea which was nearer to the equator.

Fig. 106. Section showing the relationship of the Cretaceous and Tertiary formations to the older rocks in Patagonia (Cerro Belgrano) (after Hauthal, from Haug)

1. Metamorphic rocks ; 2. porphyritic tuffs ; 3. ammonite beds (Valanginian) with *Holcostephanus, Neocomites, Crioceras*, etc. ; 4. Belgrade beds, with pelecypods, gastropods and ammonites (*Hatchericeras*). These are unconformably succeeded by 5. vari-colored sands, and 6. Tertiary shales etc., and the whole is capped by 7, a basalt flow

In Graham Land, Antarctica, the peninsula west of Weddell sea, by which at this time Antarctica was still attached to South America, beds indicating at least Barremian age have been obtained. This also indicates that this region was at that time far removed from the influence of the South Pole, which then was situated some 25° or 30° south of the coast of Graham Land.

5

It is possible that already, at the beginning of Neocomian (Comanche) time, the South Pole had been displaced in a relatively south-west direction (speaking in modern terms) so that it came to lie perhaps 35° or 40° (if not more), from the southern end of Africa. As the shifting of the poles in the pre-fragmentation stage could only be brought about by the movement of the pangæa as a whole, we must visualize a shift which would leave the stationary South Pole in the panthalassa south of Africa, while the North Pole would lie on the northern margin of Fenno-Scandia. This, while not an absolute barrier to the entrance of the Boreal fauna into Europe, became one between the East and West Boreal seas, the latter being the chief source of the Boreal element in the Shasta (Comanchic) fauna of the west coast of North America.

The new location of the South Pole gave an impetus for the breaking away of South America from Africa, and as Antarctica was then a part of South America and remained attached to its southern end into late Pleistocene time, it also swung out with the southern end of South America and may have caught up with the South Pole, which again became located over it.[6]

The breaking away of South America with Antarctica from Africa opened the South Atlantic rift (Pl XIV) which is a much older ocean than the North Atlantic; the latter did not assume its full width until Holocene time.

6

The chief indication of the Neocomian marine transgression in Africa is the *Uintahage* formation of South Africa. This series is divisible as follows in descending order (location on south coast approximate : Long. 26° E.).

"UINTAHAGE SERIES"[7]

"3. SUNDAY'S RIVER BEDS. — Blue and gray clays, shales, and sandy limestones predominate, with occasional thin greenish sandstones, to a maximum thickness of about 1,000 feet. Marine fossils are abundant, of which the commonest are : — lamellibranchs: — *Astarte, Exogyra, Gervillia, Meretrix, Modiola, Pecten, Pinna, Pleuromya* and various species of *Trigonia ;* gastropods: — *Natica, Turbo,* and *Turritella ;* and cephalopods: — *Belemnites, Hamites, Hoplites,* and several species of *Holcostephanus.*

To this should be added the skeleton of a *Plesiosaurus.*

2. VARIEGATED MARLS AND WOOD BEDS of yellow sandstone and greenish clays passing into red, pink, blue and gray, poorly bedded, sandy clays, with bands of sand

---

6) SUESS ; *The Face of the Earth,* II. pp. 290-292.
    I must here again remind my readers of a fundamental postulate, namely that the axis of the earth, and therefore the poles, always occupied the same position with reference to the earth as a whole, but that the sliding of the thin sial crust brought different parts over the poles, — both north and south, at different times. The variability in position of the lands over the South Pole was comparable in Mesozoic time in kind, though not in amount to that of Palæozoic time, and even later the South Pole frequently lay in the southern ocean, leaving Antarctica mostly free from ice (See Chapter XXXIX).

7) DU TOIT : *Geol. South Africa,* p. 305.

1,000 to 2,500 ft. The plant remains in this division include: ferns (*Onychiopsis* and *Cladophlebis*), cycads (*Cycadolepis, Zamites,* and *Nilssonia*), and conifers(*Brachyphyllum*).

1. ENON BEDS (locally basal conglomerate), yellow to brilliantly red marls and irregular conglomerates, probably originally a product of the shattering of the local rocks without much reworking later, 500 to 1,000 ft or over.

The beds are found at intervals in the folded east-west ranges of South Africa from Port Alfred on the east to within 1 or 2 degrees of Cape Town on the west.

These marine animals "find their closest allies among forms in the Upper Valanginian and Lower Hauterivian of Europe, that is to say the middle of the Neocomian. Some of the lamellibranchs, such as the *Trigoniæ*, are closely allied to European Neocomian species, but are more closely related to, or identical with, forms in the Umia Beds of Cutch in India, for example *Trigonia ventricosa* and *T. holubi*. The fauna has been recognized in the territory formerly known as German East Africa, at Mozambique, and on the south-west side of Madagascar, while a few species are very like forms obtained in Bolivia, Chile, and Argentina. Recent work in the latter country has led to the discovery, in the Nequen district, of such characteristic UINTAHAGE forms as *Trigonia ventricosa, Grammatodon jonesi, Holcostephanus featherstonei, H. rogersi* and *H. uintahagensis,* according to Grœber." [8]

The occurrence at Mozambique and on south-west Madagascar is significant as it indicates the extent of the rifting on the east coast of Africa.

7

Passing now to the west coast of North America, we meet with the Comanche system in California, Oregon and Canada. It is here known as the *Shasta Series*. It rests unconformably upon the Jurassic (Knoxville-Tithonian) and is disconformably [9] succeeded by the Cretaceous *Chico*.

---

8) DU TOIT: *Ibid:* p. 305.
9) The statement in Anderson's text, that these formations are separated by marked unconformities, is not borne out by his sections (Fig. 1) where the beds are all inclined in the same direction to the east, and, so far as can be determined, at uniform angles, varying from 45° to 60° or more within the same formation. The contact is therefore the normal disconformable one, representing only the interpulsation period but not an orogeny.

The Subformation consists of more or less metamorphosed strata, ranging in age from Palæozoic to Jurassic, and complicated by igneous intrusions. A basal sand or conglomerate is generally present and sometimes seems to grade downward into the rocks of the old land, owing to the apparent slight rearrangement of the disintegration soil, formed by the decay of the crystallines. The Lower Shasta or Paskenta beds extend north to the Shasta county line in California, where they are overlapped by the Upper Shasta or Horsetown beds, which extend 125 miles beyond the Paskenta. In this distance the higher beds of the Horsetown progressively overlap the lower ones. Where the Horsetown beds come to an end, the Chico overlap them, resting unconformably on the metamorphics.

The classification of these series for the Palæocordilleran geosyncline is as follows:

CRETACEOUS PULSATION SYSTEM

> III CHICO SERIES (Senonian, Turonian, Cenomanian)

> *Dakota Interpulsation Hiatus and Disconformity*

COMANCHE OR NEOCOMIAN PULSATION SYSTEM

> II SHASTA SERIES ....13,000± to 27,000 ft

>> B.   Horsetown Group   7,660± to 12,500 ft +

>>> 2.   Hulen beds (Albian) (4,600 ft)
>>> Concretionary sandy shales, with
>>> a rich cephalopod fauna;
>>> Sandstones.

>> 1.   Cottonwood beds (3,480-8,400 ft)
>> (Aptian, Barremian and Hauterivian)

>>> Shale with strongly ornamented (ribbed) or with degenerating ammonites (phylogerontic as shown by their loss of power to coil-*Ancyloceras, Acrioceras, Hoplocrioceras*) and sandstones and conglomerates with many cephalopods in some sections; in others with few cephalopods, many *Aucellas* and plant remains.

A.   Paskenta Group . .   5,340$\pm$ — 11,000 ft$\pm$
Valanginian and Infra-Valanginian (Barriasian)

2.   Hamlin-Broad zone etc.
Sandy shales with ammonoids, or elsewhere with *Aucellas* and *Belemnites*

1.   Duncan Creek Zone, etc.
Sandy shales with degenerating am-
monoids (phylogerontic): *Crioceras,*
Neocomites, etc.

Interpulsation Erosion Hiatus and Disconformity
Elsewhere: Unconformity

JURASSIC PULSATION SYSTEM

I KNOXVILLE OR TITHONIAN SYSTEM.

8

The orogeny at the close of Knoxville time affected
the entire Pacific coast from Alaska to Mexico, but it affected
the geosynclines only in so far as it narrowed them.   In
the area of sedimentation the beds are concordant, though
of course separated by the interpulsation hiatus.

The deposits throughout the geosyncline are almost
wholly detrital. Anderson says,[10] "their stream-borne materials were laid down under marine conditions and within
limited embayments or channels.   There is a general absence
of organic deposits, except fossil Mollusca, and they are not
abundant."

Also there are lacunæ: "Faunal evidence of deposition during intervening time (Barremian to early Albian)
is meager or altogether lacking.   Either such sediments
were not laid down, or they have been removed, or are
hidden beneath later deposits by overlap. . . . In many
such areas the absence of Horsetown deposits from the
sequence is one of the most striking facts in the Cretaceous history of the Pacific Coast.   In the Joaquin embayment of the Great Valley trough itself, the upper Cretaceous sequence rests for the most part upon the earliest
beds of the series.   Only at rare intervals have Horsetown

---

10) ANDERSON, F. M.: *Lower Cretaceous of Cal.* etc., p, 10.

fossils been found, whereas in most places no Horsetown sediments occur at all between upper Albian and Valanginian strata on the west border of the Joaquin embayment." (See Chapter XXXV).

Such relationship is especially recorded for Oregon sections and emphasizes the Shasta-Chico interpulsation hiatus and disconformity. The Horsetown represents in large part the regressive series of the Shasta with offlapping strata, and therefore incomplete at the top. It was followed by complete emergence (withdrawal of the sea because of sinking sea-level), and extensive interpulsation erosion, which in many sections removed the Horsetown sediments that had been deposited there. With renewed transgression in Cenomanian (Lower Chico) time, these Cretaceous beds came to rest disconformably on the erosion surface formed during interpulsation time, and which was cut in some sections down to the Paskenta (Valanginian) group.

The source of the clastic sediments, which accumulated in this geosyncline, was the western border-land (old-land), only a part of which is still intact, the remainder having been removed by erosion or by faulting. Within the geosynclines are found deposits of the entire Comanchean system from Barriasian to Albian, and of the Cretaceous from Cenomanian to Maestrichtian, though not occur all within the same general area.

The deposits are most complete in the western part of the area, which therefore represents the axis of the geosyncline nearer the old-land on the west, while the eastern area represents the marginal plain or platform, where the wedge-like portions of the successive deposits thinned away, or were overlapped. Conglomerates, though frequently local, where streams entered the geosyncline, are nevertheless of great thickness; limestones on the other hand are rare.

9

The thickest succession of these deposits is found on the west borders of the Great Valley, which marks the axial portion of the geosyncline. The thicknesses may be summarized as follows:

	Maximum	Minimum	Average
Chico Series	26,600 ft	5,700 ft	16,150 ft
Shasta Series	26,800 ft	11,200 ft	19,000 ft
	53,400 ft	16,900 ft	35,150 ft

Thus the maximum thickness of the two series combined is over 10 miles, and the volume, estimated from the average thickness within a longitudinal distance of 36 miles and a width of not less than 24 miles, is about 13,400 cubic miles of sediment. All this mass of sediment accumulated approximately at sea-level in the geosyncline, that is, just a little below and at other times just a little above sea-level, but never in deep water, and never in greatly elevated alluvial plains.

This, of course, implies that the geosynclinal floor was subsiding as the sediments were added, or to put it more correctly, the sinking was constantly going on, and with it, as constantly, the filling by sediments. There can be no question of primary isostatic adjustment here, for the sediments could not depress the trough; they accumulated as the result of the sinking. Therefore only lateral pressure due to the stress of the westward striving sial block against the resistant sima, in which it was immersed, could have effected the downward bowing of the geosyncline. This westward striving was manifest in Comanchic (Shasta) time, even before the Atlantic rift began, though the sial crust may have suffered preliminary stretching [11].

10

"The Cretaceous stratigraphic sequence in the Great Valley embayments, on the whole, divide themselves naturally into two great sedimentary series, belonging to earlier and later Cretaceous epochs. The depositional relationship of the two is one of clear unconformity [or disconformity]. Wherever their contact has been found, and in fact, wherever evidence has been sufficiently sought, the basal beds of the upper series are found to rest discordantly upon those of

---

11) For much more detail the student should refer to Anderson's monographic paper, published as special paper 16 by the Geological Society of America 1938. Where our interpretations differ, I alone am responsible.

the lower series.   In many places this discordance is marked
by beds of conglomerate which carry convincing evidence
of disturbance, of erosion, and of reworked materials, in part
derived from the older series, and in part brought from
more distant points, all lodged in the basal beds of the
later series.''

The basal beds of the Shasta series "rest in some places
discordantly upon Knoxville beds, with thick basal conglomerates which contain distinct evidences of diastrophism,
erosion, and transported material, some of which is fossiliferous, derived from underlying Knoxville and older formations.   In other places the basal beds rest directly upon
pre-Knoxville (Franciscan, or older) rocks, as in the Yolla
Bolly basin, Morgan Valley, Russian River Valley, Mount
Diablo, and the Diablo Range.   In a few exceptional places,
as in the McCarthy Creek—Elder Creek district, this unconformity is less evident [is in fact a masked disconformity],
and the succession from upper Knoxville to lower Cretaceous has been [erroneously] believed to be continuous.
Similar conditions are rarely found farther south, but for
the most part unconformity [or disconformity] is readily
proved.   In many places the unconformity [disconformity]
of the two series is marked by bodies of conglomerate at
the base of the Shasta series, resting directly upon dark
clay shales of the Knoxville series.   In some places they
contain Upper Jurassic (Tithonian) fossils.   These conditions
of unconformity [disconformity] have been found by N. L.
Taliaferro in southwest Fresno County and in southern San
Luis Obispo County, and are known to extend as far south
as the Santa Ynez River, Santa Barbara County.''

In most, if not all cases the "unconformity" is merely a
disconformity.   "The Shasta series is overlaid at the top by
sediments of the Chico series, having in many places
distinctly basal, or near-basal, conglomerates which contain
evidences of [disconformity] . . . it seems to have been
recognized in Alaska (Yukon and Kuskokwim Valleys,
Chitina Valley), although the definite horizons of the overlaps are not always indicated.

"In the district between Thomes and Elder creeks,
the fauna of the older group, through a vertical range
of 1,800 feet, is dominated by *Aucella*, of which there
are at least 11 distinct species, as determined by Pavlow.

With them are found various species of cephalopods, including Barriasellids, Olcostephanids, and Neocomitids, specifically different from any found in the northern district. Only a few species of ammonites are known to be common to the two areas in the lower group. Many of the cephalopods found at the south are confined to this area. Among Belemnoids the case is somewhat different, and a number of species is common to both districts. The contrasts in faunas are illustrated by the total absence of *Aucella* in the lower group at the north. None have yet been found, although a few, and mostly different, species of *Aucella* have been found in the upper group at the north. Reference to the lists of species [given by Anderson] from the Paskenta group in the two areas will illustrate these facts.

"Similar contrasts of faunas continue almost throughout the Horsetown group in the two districts. The abundant cephalopods in the Horsetown group, in the Cottonwood district, have few representatives in this group in Tehama County, although a few species are common to both districts. There is a general scarcity of ammonites in the Horsetown group south of the delta. The plant remains in the Paskenta group, as far as known, show similar contrasts, in that they are abundant south of the delta and comparatively rare to the north. On the contrary, the Horsetown group north of the delta contains much fossil wood, some leaves, cones, and nuts, especially in the Hulen beds, but at the south these are not plentiful.

"The causes of the faunal contrasts here described, probably lay in the hydrological condition of the embayment already described, namely, the inflow of large volumes of fresh water laden with sediment from the land areas to the west, seasonally cooled, and having a southward current along the western side of the embayment; tidal currents bringing in marine water, that followed the eastern shore northward, and into the areas north of the delta; and lastly the influence of these currents upon the molluscan life, entering the embayment from the sea, but in part coming from opposite directions along the littoral corridors of the time."[12]

12) ANDERSON, F. M. *Ibid.* pp. 37, 40, 72. For much more detail the reader should refer to the original paper. Also for lists of species.

11

Summary: In the west coast region the Dakota interpulsation deposits are not developed as such, unless a part
of the adjacent series of purely non-marine character is
referable to it.   If there is no continental deposit separating the Comanche and Cretaceous (Shasta and Chico), we
must regard the two series as separated by an erosion
interval which, as we have seen, often cuts out the whole
of the Horsetown Series and represents the interpulsation
period, so well marked in all other parts of the earth.
Of course it may be masked by reworking but that should
be ascertainable by careful study of the strata in the field.

12

In the Appalachian geosyncline of the modern Atlantic
border region of North America, the Comanche is purely
non-marine, resting either on the Triassic or by overlap
on crystalline schists.   The succession includes:

DAKOTA INTERPULSATION SYSTEM

  RARITAN SANDS   400 ft

COMANCHEAN PULSATION SYSTEM

  3. PATAPSCO FORMATION   up to 200 ft
      Highly colored and variegated sandy
      clays and sands, with an *Angiosperm*
      flora.

            Interruption of Sedimentation

  2. ARUNDEL FORMATION   up to 125 ft
      Chiefly clays, locally iron-bearing, and
      lignitic flora similar to that of the
      Patuxent, while remains of *Dinosaurs*
      also occur.

            Interruption of Sedimentation

  1. PATUXENT FORMATION   up to 350 ft
      Chiefly arkose sandstones with sands
      often cross-bedded, with small masses
      of clay and an early Comanche flora.

Hiatus and Unconformity or Disconformity

SUBFORMATION: Crystalline schists or Triassic sandstones.

The several breaks in the series, which mark periodic discontinuance of deposition of the continental strata, because of pauses in subsidence of the geosyncline, make up by their time-duration the shortness of the period indicated by the slight thickness of the sediments themselves.

CHAPTER XXXVI

THE BEGINNINGS OF CONTINENTAL FRAGMENTATION
AND THE BIRTH OF THE ATLANTIC

1

WE have now reached a point in our outline of the history of the earth, where we come to the end of the pangæa, for before the post-Dakota readvance of the sea the Atlantic rift had made its appearance, and from this time onward it steadily widened as the Americas continued to drift westward under the influence of the force generated by the eastward rotation of the earth on its axis. Just why the break should occur at this time, since the position of the pangæal mass between the North and South Pole, which favoured such a break, had extended back as far as the beginning of the Mesozoic at least, is not wholly clear, and further investigation would seem to be called for. That there had been westward striving during the Comanchic, and even the later Jurassic, is seen by the thickness of these formations on the west coast of North America, which implies a corresponding sinking of the geosyncline under steady lateral pressure, followed by the folding of these strata during the Dakota Inter-pulsation Period.

There is, however, every reason for believing that the two Americas, with their broad connecting central land-mass, continued to act as a unit, though it was the southern end, i. e. South America, that seems to have broken away first from the pangæa and drifted more rapidly than did North America. However, the southern end of South America formed a continuous land-mass with Antarctica, the two not separating until the end of Pleistocene time. India, too, and perhaps Australia, probably did not break away from the main mass of the pangæa, until Tertiary time.

A direct consequence of this westward drift of a

unit continental mass extending from pole to pole was the deformation of the whole of the western border of the Americas, because of the resistance of the sima, in which the drifting sial-mass was submerged for 7/8 of its depth.

At first the conflict between the westward urge and the resistance of the sima caused only the formation of the marginal geosynclines and bordering old-lands on the west. But finally the geosynclinal floor subsided under this compressive force to a depth of thousands of feet[1], while the depressed region was filled at the same rate with sediments only partly, if at all, consolidated. Then the decreased resistance of the subcrust, now carried down into regions of higher temperature, and perhaps additional lateral pressure due to greater resistance, or to stronger lateral drift, brought about the folding of these strata.

2

If we contrast the nearly 5 miles of Comanche strata, which were forming in the Palæocordilleran geosyncline during this period, with the 700 feet or less in the eastern geosyncline, or the 1,000 feet of strata in the latter, accumulated during the Cretaceous, with nearly 4 times that amount preserved in the Cordilleran region, we have a measure of the difference of subsidence of the two regions. If drifting had begun in Comanche time the deepening of the eastern geosyncline could only be the effect of stretching before final separation, if that began in Cretaceous time.

The interior or Colorado geosyncline of North America, now transformed into the Rocky Mountains by the continued pressure of the drifting land-block against the resisting sima of the Pacific, also contains a succession of strata which accumulated in a steadily sinking but never very deep trough, and the simultaneously slow rising of

---

1) This subsidence is usually explained by isostacy—the sinking under the pressure of the sedimentary load. But since, as argued before, a thin layer of sediment can obviously not force the underlying area to sink to an equal amount, and since a great thickness of uniform shallow water or even continental deposits could obviously not accumulate except under constant and progressive subsidence, we must look for another force to bring about such subsidence, and this force is compression under drifting.

the sea-level.   This resulted in a gradual transgression of
the sea with progressive overlap of the formations and thinning of the whole series in the region of overlap, *i. e,*
the marginal platform.   Thus even in the region of the
Rio Grande of today there was a total accumulation of
some 7,500 ft of strata of which 1,500 ft was white limestone (Austin Chalk).   Further north the beds are thinner
by overlap, so that at Austin Texas, even the limestone has
decreased to 600 ft, and by the time Colorado was submerged, the limestone was less than 100 feet.

While the limestones were forming in the open waters
of the south, river-borne muds were accumulating in the
north.   These formed the Benton shales, which are chiefly of
estuarine character, but with marine limestones in the
center.   In many places the shales abound in fish remains,
but other fossils are rare or absent.   This suggests that
the fish were either river types, killed in large numbers
by the advancing sea, or marine types, killed by the
inpouring of fresh water.   Their decaying bodies were
probably in large part the source from which the petroleum
which is now stored in porous members of this series was
derived.   Similar fish-bearing shales overlie the Niobrara
limestone series.   During the earlier part (Coloradoan) of
the Cretaceous, when the muds of the north were mainly
estuarine, the marine fauna from the south invaded this
trough. The first immigrant was the pelecypod *Inoceramus
labiatus* (middle Benton), and this was followed by ammonites (*Prionotropus, Prionocyclus*) and later by the
coarse thick-shelled *Inoceramus deformis*, easily recognized
by its coarsely prismatic shell-structure.   In Montana time,
however, the Arctic sea transgressed into this trough from
the north, bringing with it *Baculites, Scaphites, Helicoceras*
and other peculiar cephalopods, which are also found on
the west coast of Greenland. [2]

3

We return now to Europe which had been separated
from America by the opening of the Atlantic rift.

Here we meet with a vast transgression of the sea
at the beginning of Cretaceous (Cenomanian) time.   It

---

[2] For illustration of these see GRABAU: *Text-Book, II,* pp. 699-701-733.

continued during the Turonian and Senonian; its greatest extension apparently coinciding with the Senonian.

The Cenomanian transgression is recorded on the summit of the Spanish Meseta from where it can be traced to France and the north of Scotland, while isolated patches of Senonian have been revealed by dredging off the Norwegian coast, up to the highest latitudes. The transgression extended across the Bohemian mass as far as the Bavarian Jura. Further north it is seen in numerous patches in Denmark, in Scania, and to some extent in the Baltic. It crossed Poland, while according to Karpinsky, its northern limit passed somewhat north of Vilna towards Moscow, and north of Simbirsk and Samara towards Orenburg (Long. 55°12' E., Lat. 51°47' N.). The northern half of Russia, however, was not covered by the sea, but the southern outrunners of the modern Urals were submerged. However, Cretaceous beds with *Baculites* do exist beyond the Urals on the northern Sosva (Lat. 62°30' N.), where they rest upon fossiliferous strata, probably of upper Volgian age. The whole succession is horizontal.

The Cenomanian sea extended over the whole of the Caspian region and covered the Kizil-Kum and all the plains of Turania, up to the great mountain chain, which had apparently not been submerged since Palæozoic time. The Mesozoic, in so far as it is represented in these mountain ranges, consists of coal-bearing beds. The Cenomanian sea, however, continued further into the Tarym basin.

The middle and upper Cretaceous transgression covered the whole region of the central Mediterranean and even extended beyond it, its deposits being represented chiefly by hard limestones. These deposits extend from south Europe across Syria, cover the whole eastern half of the Sahara and Arabia, and penetrate into the valley of the Narbada, thus reaching India.

We have already noted the contemporaneous transgression of the sea in North America. In South America the Cenomanian sea transgressed widely over the valley of the Amazon, its sediments extending as far as Bahia. Farther south, Cretaceous rocks underlie the Pampas of Argentine, and middle Cretaceous fossils are found up to Lat. 35°30' S. in Patagonia.

Corresponding deposits are also found on the west

coast of Africa, from near the equator to Mossamedes (Angola, Lat. 15° S.). On the east coast, however, the Cretaceous is of a different type, being closely allied to that of India. This fauna accompanies the transgression on the coast of S. E. Africa, at Natal (Lat. 30° S.), and again on the coast of south-east India at Trichinopoli, (10°50' S., 78°44' E.) and finally in the Shillong Plateau, Assam (25°33' N., 91°55' E.), all of which regions were opened on the separation of India and Africa. West Australia also is covered by transgressive early Cretaceous beds, but of these little is known.

"This wide transgression, however, excluded Greenland, Spitzbergen, perhaps the north of Scandinavia, northern Russia, Siberia, and the whole of northern China. Throughout this region, Tertiary leaf-bearing beds rest directly on the Jurassic or Volga stage, even in east Greenland and Spitzbergen.

"The great extension of the Cretaceous seas was followed after the Senonian by an extraordinary retreat (comparable to that which followed the Portlandian)." [3]

The Cretaceous Pulsation Period shows the following marine succession in Peninsular India; all represented by highly fossiliferous strata :

DECCAN (LARAMIDE) INTERPULSATION PERIOD	
Deccan volcanic flows	6000-10,000 ft

Hiatus

CRETACEOUS PULSATION SYSTEM	3000 ft +
3. *Aryalour Stage, (Upper Cretaceous)*	1000 ft
Bedded sands and argillaceous strata, becoming calcareous upward, all highly fossiliferous.	
2. *Trichinopoly Stage*	1000 ft
Sands and grits often cross-bedded and pebbly. Fossils abound.	
1. *Utatur Stage*	1000 ft
Argillaceous beds, but including coral limestone in the lower part. Littoral marine fauna, but also an abundance of ammonites. Age	

3) SUESS: *Face of the Earth*, II, pp. 290-292.

Cenomanian-Turonian.   Sometimes overlapped by Trichinopoli
beds.

Hiatus and Unconformity

ARCHÆAN GNEISSES

The higher Cretaceous beds are covered by the great
Deccan Trap sheet which reaches a thickness from 6,000
to nearly 10,000 ft, but is usually much thinner.   It
represents the igneous activity during the post-Cretaceous
Interpulsation Period, and is covered by Eocene marine
beds.   These lavas are still mostly horizontal sheets, and
they have greatly influenced the erosion topography and
scenery of this part of India.

5

Along the Pacific coast of South America, from the
folded ranges of Mexico southward, we search in vain for
beds of Cenomanian or Turonian age, even Albian may be
absent because of erosion.   This seems to be a characteristic
feature of the whole Andean geosyncline, down to south-east
Graham Land, where the last of Comanchean (Neocomian)
is represented, on Snow Hill Island, by red limestone with
Zoantharian corals, which lies disconformably below richly
fossiliferous Upper Senonian beds.   Two species of ammonites were found in the red limestone; these are
closely related to *Desmoceras latidorsatum* and indicate an
Upper Albian age.

The overlying Cretaceous beds are glauconitic marls,
with sandy nodules which contain ammonites (*Phylloceras*
2 sp., *Gaudryceras* 2 sp., *Anisoceras notabile*, *Kossmaticeras* 7 species) and corals, crustacea, lamellibranchs and
gastropods.   All the species, with 2 or 3 exceptions, are
well known in the later Cretaceous of India and in the
Senonian of British Columbia.   According to Kilian they
belong in the horizon of the Upper Trichinopoli beds of
India.   They are succeeded, on Snow Hill, by sandy
marls which are also well developed in the neighbouring
Seymour Island (now in 64°20′ South Latitude, 56°45′
W. Long.) inside of Weddel Sea, where ammonites

(*Gaudryceras varagurense, Pseudophyllites indra, Kossmaticeras* 5 sp., and *Parapachydiscus* sp. near a Mæstrichtien species of Europe) are found. With these ammonites occur pelecypods (*Lucina, Lahilia, Astarte,* etc.).

It is of great significance that the affinities of this fauna are with that of the corresponding beds of India (Aryalour and Valoudayour), but also with the beds of Quiriquina on the Pacific border (Lat. 36°30′) of Chili and the Senonian of South Patagonia. Thus, according to Haug, the genus *Phylloceras* and *Lytoceras* are found in Japan, in British Columbia, in California, in Chili, in Patagonia and in Graham Land, in other words circum-Pacific. This shows that, as yet, none of these regions were subject to polar influence, since under such conditions as exist in the south polar region, these ammonites would be excluded. Hence we conclude that although the South Atlantic rift had probably progressed considerably by Senonian time, yet it was not extensive enough for Antarctica to have again entered the realm of south polar glaciation. Meanwhile, on the opposite side, the north polar regions did not influence the North Pacific.

In the Turgai Province, of Central Asia (S.E. end of Urals) beds of Mæstrichtien age rest unconformably and horizontally on folded and truncated Palæozoics, and contain remains of *Pycnodonta vesicularis, Ostræa* and *Belemnitella vesicularis,* the Russian Mæstrichtien fauna. Farther west occur Cretaceous beds with *Inoceramus, Hemiaster humphriesianus* and with many pelecypods (*Avicula, Nucula, Lucina, Solemya, Dentalium, Entalis, Vanikoro*) of Fort Pierre and Fox Hill age of North America.

6

The beginning of the North Atlantic rift is marked by the marine Cretaceous beds of the border geosyncline, in which the non-marine Comanchic sediments were deposited. These fossiliferous Cretaceous sands and clays are now preserved on the Atlantic coast, in the strata of the Coastal Plain. In western Alabama these beds are about 2,400 ft thick and include the following distinct formations :

SUPERFORMATION: EOCENE.

Hiatus and Disconformity

CRETACEOUS SERIES		Approximate European equivalent
Selma chalk	900 feet	Senonian
Eutaw formation [4]	400-500 feet	Emscherian
Tuscaloosan formation [4]	1,000 feet	Turonian and Cenomanian

Hiatus and Unconformity

SUBFORMATION: PALÆOZOIC.

It will be observed that the highest Cretaceous deposits (Danian) are absent here; they were either never deposited, or, if formed have again been eroded. The Comanchean is absent too, the Cretaceous resting by overlap directly upon the eroded Palæozoic of the old Appalachian Mountains. This overlap continues northward and eastward, so that in Tennessee and in eastern Alabama the Eutaw rests directly upon the older rocks, as shown in the following diagram.

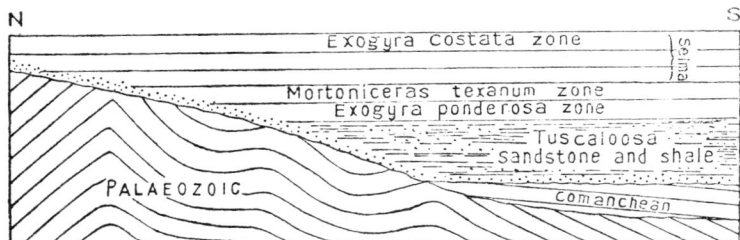

Fig. 107. Ideal section of the Gulf Coast region of the southern United States showing the relationship and overlap of the Cretaceous formations. The inclination of the Comanchean beds is exaggerated; they are essentially horizontal.

Further north in Maryland and New Jersey the Atlantic rift apparently began later, for here the first marine formation is the Magothy clay and sands of the age of the CONIACIAN (Emsherian) of Europe. Thus the Turonian and

4) The upper part of the Eutaw ranges into the Senonian, and the upper part of the Tuscaloosan into the Emscherian.

the underlying Cenomanian, the recognized base of the Cretaceous series, are absent here, for the underlying Raritan clays are separated by a pronounced disconformity and carry a Dakota flora. This suggests that the Raritan, which is from 300-500 feet thick and rests unconformably either on the Newark, or on the crystallines, represents the Interpulsation continental formation, which was not covered by the transgression in this region until post-Turonian time.

The relationships of the Raritan (more particularly the Amboy clays) to the Dakota is, however, fairly close. Out of 156 described species of the Amboy clays at least 40, or one fourth of the whole, are also found in the Dakota flora. This latter has 460, or almost 3 times as many described species as the Raritan, and when the flora of the latter is better known, the number of species in common may be vastly increased. Fifty, or about one third of the Amboy species of plants, occur in the Cretaceous rocks of Greenland, described by Heer, while Newberry finds a species of the Amboy also occuring in the Cretaceous flora of the Aachen plant beds, although this is a much younger (Senonian) flora.

In Maryland and Virginia, the Raritan is underlain by an older plant-bearing series, the POTOMAC Group. This is referable to the Comanche and is divisible into 3 groups as follows:

SUPERFORMATION: CRETACIC PULSATION SYSTEM:

(Magothy—basal member)

Hiatus and Disconformity

DAKOTA INTERPULSATION PERIOD OR RARITAN
CLAYS AND SANDS   300-400 ft

Hiatus and Disconformity

COMANCHE PULSATION SERIES: POTOMAC GROUP

Patapsco Formation	200 ft
Arundel Formation	125 ft
Patuxent Formation	350 ft

Hiatus and Unconformity

SUBFORMATION: NEWARK SERIES OR CRYSTALLINES

According to Fontaine and Newberry the Potomac flora is of Neocomian or Comanchean age (Lower Cretaceous in the older classification, if not actually of Wealden Interpulsation type), and Newberry concludes: ". . . probably a somewhat long interval of time separated the epoch of the Potomac Group from that of the Amboy Clays. This is indicated by the almost entire distinctness of the floras of the two formations, which shows that a great change took place during that interval in the character of the vegetation which clothed the eastern shore of North America. From the Potomac group of Virginia and Maryland, Fontaine has described 365 species of plants of which not one is certainly found in the Amboy Clays; and the difference in the character of the vegetation is shown by the fact that in the long list furnished by Professor Fontaine there are but 75 angiosperms (about one-fifth of all), whereas in the New Jersey clays, throwing out fragmentary and doubtful remains, of 156 described species all but 10 are dicotyledonous plants."[5]

However, the Potomac flora has only a few species in common with the Wealden of Europe.

There are a few fresh or brackish water-shells found locally in the Raritan, and Hollick reports finding drift blocks on Staten Island in which marine mollusca are associated with the characteristic Amboy flora[6]. Weller[7] also lists a number of species.

"All of these localities are in the lower portion of the Raritan formation, and, although the generic relations of all the species are more or less in doubt, all seem to be of brackish-water types, just such forms as might be expected to occur in beds having the estuarine origin of these Raritan clays and sands."[8]

Weller also mentions a slab of concretionary sandstone from near the base of the section in the clay banks at Sayreville, N. J., with species of *Turritella* cf. *jerseyensis* and "an imperfect impression of a small pelecypod shell, which has the general form and proportions of *Cymbophora lintea*, a species particularly abundant in the Cliffwood

---

5) NEWBERRY: *Flora of the Amboy Clays*, p. 23.
6) HOLLICK: *Trans. N. Y. Acad. Sci.* Vol. XI, pp. 96-104.
7) WELLER: *Cret. Pal. N. Jersey*, p. 28.
8) IBID: *Op. cit.* p. 28.

clay, and also occurring in several of the higher
Cretaceous formations in the State." Since it is not
positively known that the slab was *in situ* and not a
fragment slid from the overlying Cliffwood formation, it
cannot be accepted as evidence of near approach of the
sea during the deposition of the Raritan clays, if that was
during the Interpulsation Period, as seems to be indicated
by the flora.

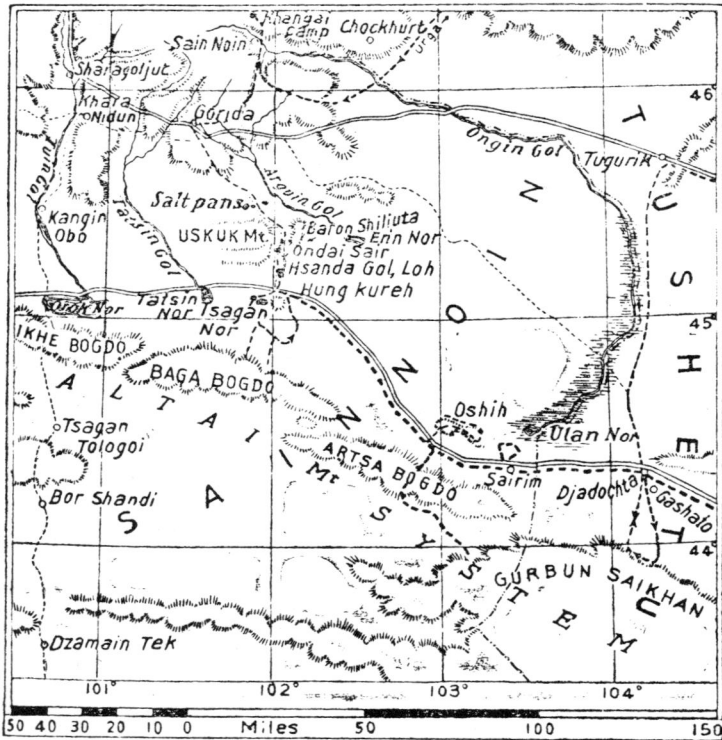

Fig. 108. Sketch map of the eastern Altai region, showing location of
stations mentioned in the text. (After Berkey and Morris, Bull. Am. Museum
of Natural History, Vol. 51, 1924).

7

When the Atlantic rift finally reached Maryland and
New Jersey during the Cretaceous transgression, only the
thin wedges of the coastal plain strata were formed in the

regions now open to study. These show the following succession in descending order.[9]

SUPERFORMATION: PALÆOCENE (SHARK RIVER FORMATION).

CRETACEOUS:		APPROXIMATE EUROPEAN EQUIVALENTS
Manasquan formation	50 ft	Danian
Rancocas formation	125 ft	Danian
Monmouth group	150 ft	Mæstrichtian
Matawan group	275-400 ft	Campanian / Santonian
Magothy formation	100 ft	Emscherian

Hiatus and Disconformity

SUBFORMATION: RARITAN FORMATION.

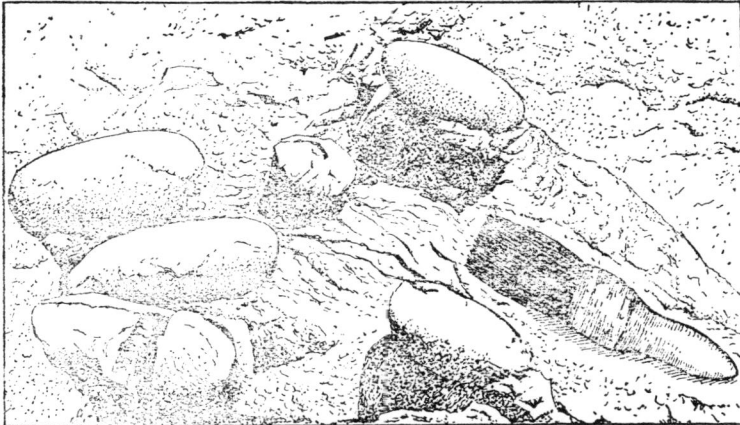

Fig. 109. *Protoceratops andrewsi.* A number of eggs after removal of the rock cover which has again disintegrated into sand. (After photograph in Natural History of Mongolia Vol. I). The hand-brush is for comparison of size.

Here then the series begins with higher members than it does farther south (Emscherian as compared with Cenomanian), showing that the older beds are overlapped. But in the New Jersey section the continental Raritan

9. For the progress of marine sedimentation and overlaps during the Cretaceous, including Europe, and for a summary of the life of this and the Mesozoic as a whole, the reader is referred to my Text-Book of Geology, Vol. II, pp. 715-778 and the illustrations there given.

(Dakota), and Comanchean beds are present. Here too
we find the highest Cretaceous division, the Danian, which
is followed disconformably by the Palæocene Shark River
beds. The Monmouth is the equivalent of the upper part
of the Selma, being characterized by *Belemnitella americana*

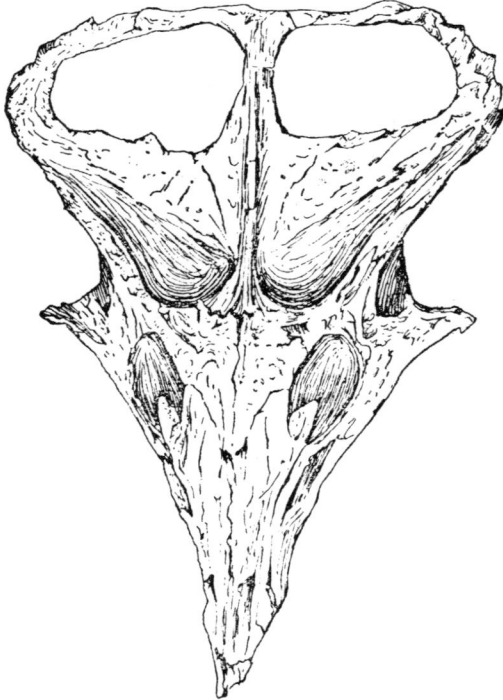

Fig. 110. *Protoceratops andrewsi.* Small, young adult skull, possibly a
female. A. M. No. 6408. Viewed from above. About two-sevenths natural
size.
   The occipital frill is composed exclusively of the expanded parietals.
There is no good evidence of separate interparietal or fused dermo-supra-
occipitals. The squamosals are limited to the anteroexternal border of the
frill. There being no horns on the postorbitals, there is no secondary skull
roof above the frontals and consequently no "pseudopineal" opening. (After
Gregory and Mook. Am. Museum Novitates, No. 156, 1925).

and *Exogyra costata.* The Matawan carries the *Mor-
toniceras* zone in its lower part and therefore corresponds
to the top of the Eutaw and to the European Santonian,
near the base of which this fossil occurs. Thus the

Cretaceous beds of New Jersey represent only the Senonian
(comprehensive sense, inclusive of Emscherian) and the
Danian; that is, only the upper part of the European
Upper Cretaceous. The fauna too, is closely allied to that
of the northwest European Senonian.

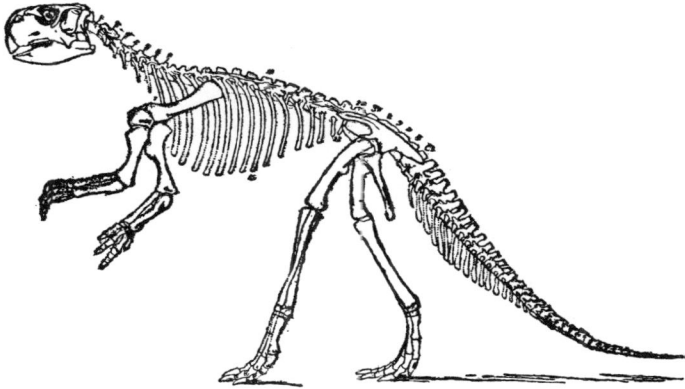

Fig. 111. Reconstruction of the *Psittacosaurus mongoliensis*; type skeleton
(Amer. Mus. 6254) in its lateral aspect. 1/16 natural size.
The skull in this reconstruction combines sclerotic ring and dental char-
acters observed in the type (Amer. Mus. 6254), also in the referred specimen
(Amer. Mus. 6261) from the same geologic formation. Eight maxillary teeth
are restored from the referred skull (Amer. Mus. 6261); the corresponding
dentary teeth are conjectural. These restored parts are indicated by dotting.
(After Am. Museum Novitates, No. 127, 1924).

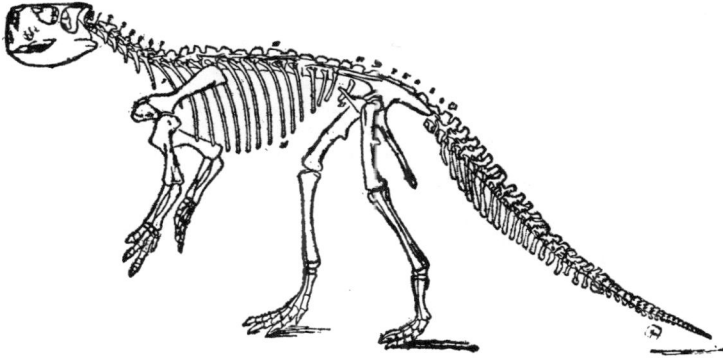

Fig. 112. Reconstruction of the *Protiguanodon mongoliense*; type skeleton
(Amer. Mus. 6253) in its lateral aspect. 1/16 natural size. (After American
Museum Novitates, No 127, 1924).

8

In Asia the best known continental Comanche and
Cretaceous strata are those of the Mongolian basins or
GOBIS of which four have been determined as sites of Cre-
taceous deposits. (Map Fig. 108 and Sections Figs. 113-114).

These beds are all continental and are characterized
by the remains of a succession of *Dinosaurs* [10] (Figs. 109-
112), while in one of them, the Djadochta formation,
the famous dinosaurs eggs have been found. This formation consists of red friable sandstones and clayey sands
300 ft thick, of which more than 200 feet are exposed
in cliffs of the Bad-land type.

The eggs (Fig. 109), which are found in these beds
and believed to be those of *Protoceratops* from their close
association with the skeletons of this dinosaur, are of elongate
oval form compressed, and on the average, of fairly uniform dimensions. They have been found in nests of many
individuals, as well as isolated. The shell is commonly
cracked but present, the interior being filled with the
fine mud of the formation, which, from oxidation, has
assumed a red or dark brown color. Frequently the partly
developed skeleton of the embryonic dinosaur is still
preserved within this matrix. The microscopic structure
of the shell shows the following detail. [11]

"The parts of the shell which have been fossilized are the mamillar
zone, *i. e.*, the zone in contact with the chorion, and the prismatic zone.
The shell, formed of calcite mixed with phosphate of lime, is about 1 mm
thick. Its color is red-brown, due to infiltrated iron oxide. High, subparallel,
and meandriform hillocks, separated by very large and deep valleys, give
a vermiculate aspect to the outer face. On the contrary, the inner face is
nearly smooth, the mamillæ being very small and closely crowded together.
The fibro-radious structure is therefore finely packed and the prisms of the
prismatic zone are small. The pores are extremely reduced both in number
and size, and consequently the aeriferous canals maintain their diameter and
are not ramified. The prisms of calcite are crossed by thin black layers of
organic matter.

"Secondary mineralization of the shell took place during the fossiliza-
tion, so that the pores and canals are entirely choked with calcite."

In conclusion van Strælen remarks:

"The eggs of *Protoceratops andrewsi* are of the utmost interest. From
the rugosities of the outer surface together with the rare and extremely
small pores, it is right to infer that the eggs had no outer cuticle. This

10) For details see the reports of the 3d Asiatic Mongolian Expedition
    and the Summary in my Stratigraphy of China, Vol. II, p. 680, *et seq.*
11) VAN STRÆLEN: *Micro-structure of Dinosaur Eggs;* p. 4. Quoted by
    GRABAU: *Stratigraphy of China, Vol. II,* p. 702-703.

is a character shown to-day by birds and turtles which lay their eggs in very dry regions. We may find herein a confirmation of the desert conditions prevailing in Mongolia during the formation of the Djadochta beds."

9

From the Cretaceous on, the Atlantic rift constantly widened throughout the Cretaceous, Tertiary and Quaternary and not until modern time did it attains its full width. But for a time at least in the Lower Eocene the northern lands remained in contact. The view here advanced implies that migration of Tertiary animals from America to Asia was via Europe and the Greenland route and not via Alaska and the Siberian extension. This is opposed to the current view, but harmonizes with the concept of a pangæa in process of fragmentation and does no violence to the concept of inter-continental migrations. For this the Alaska-Bering route might have constituted a modern post-Polycene pathway, but was certainly not able to function as such in pre-Quaternary time, for only by constant widening of the Atlantic rift were the Pacific margins brought within reach of each other, and not until the beginning of the modern period when the lands had their present polar relation and the north polar ice acquired its present position, was the Bering land-bridge available for primitive man and his contemporaries to enter North America. But by that time he had become fully established in the old world of Asia.

CHAPTER XXXVII

THE OLDER TERTIARY SEAS

1

THE Tertiary or Cenozoic Period of the earth's history is principally the history of the separate continents which had come into existence by the opening of the Atlantic rift (See Map; Plate XV). Eurasia was still a unit, though the Indian Ocean had probably begun to open on the south. North America, however, of which Greenland was a part, was still attached to Fennoscandia, though the rift in the Iceland region was becoming imminent, and the building up of the Icelandic volcanoes over the fissure had begun. North and South America were throughout connected by the Isthmian land mass, and South America retained its hold on Antarctica, while Graham Land continued to function as a part of the extension of the Patagonian end of the Andean geosyncline.

The classification of the Cenozoic elements of the geological time scale in general is as follows, in descending order:

NEOGENE PULSATION SYSTEM

Hiatus

ALPINO-HIMALAYAN INTERPULSATION SYSTEM

Hiatus

EOGENE (PALÆOGENE) PULSATION SYSTEM
  III. OLIGOCENE OR TONGRIAN (Narian)
    C. Upper or Chattian
    B. Middle or Rupelian (Stampian, non-marine)
    A. Lower or Lattorfian (Sannoisian, non-marine)
  II. EOCENE OR PARISIAN (Sindian)
    C. Upper or Priabonian

2.   Ludian (Ligurian)
1.   Bartonian
   B.   Middle or Auversian (Sables Moyans)
   A.   Lower or Lutetian (Calcaire grossier)

I.   PALÆOCENE SUESSONIAN (Libyan)
   C.   Upper or Londinian (Cuisian estuarine, Spar-
      nacian fresh water, Ypresian in part)
   B.   Middle or Thanetian (Hessian, non-marine;
      Ypresian lower part; Cernaysian)
   A.   Lower or Montian (Vitrollian, non-marine)

Hiatus and Disconformity (or Unconformity)

LARAMIAN INTERPULSATION SYSTEM or
CRETACEOUS PULSATION SYSTEM

2

The Cretaceous ended with the complete withdrawal of the waters from all the geosynclines and their concentration in the deepened ocean basins, of which the incipient Atlantic was one.

The Interpulsation Period is, however, marked by continental deposits (Laramie) and by a pronounced orogeny in western North America, where continued westward drift and the resultant compression, produced the long chains of pre-Tertiary folds which can be traced from Alaska to Terra del Fuego.

The Cretaceous and older strata of the Central American Isthmus, which, it will be remembered, formed a broad, much elongated belt along the Pacific border between North and South America, and connected the Cordilleran with the Andean geosyncline, were also folded at this time, though the breaking up of this Isthmus into the West Indian Islands did not take place until the end of the Pleistocene. The virgations and directions of the mountain ranges, and those of the later but similar folds of the Tertiary strata, are the result of the bending of the Isthmus into the hair-pin-loop and the splitting of the parallel ranges into divergent and re-uniting strands, as discussed on a subsequent page. During the westward drift, the strata of the Andean geosyncline of western

South America also suffered their first deformation, and during much of the remainder of the Tertiary these mountains underwent erosion, ending with the formation of a peneplane. This in late Tertiary time was bodily elevated, warped, and cut by erosion into the present ridges and peaks of the Andes.

A new geosyncline appeared to the west of these newly formed Andes, and in this the marine Tertiary strata (chiefly later Tertiary) of western South America were deposited.

During the Laramian Interpulsation Period, the main folds of the Rocky Mountains were formed, together with a parallel series, extending more or less continuouly from western Wyoming to Mexico, where it joined the extensions of the Antillean ranges already referred to. The old-land area between the Coloradoan and the western geosyncline was apparently arched into a broad anticline and broken into a series of north-south blocks which, by tilting, became the block-mountains of the Great Basin region. The western border of this faulted region was marked by the eastward-facing fault-scarp of the Sierra-Nevadas and the eastern, by the westward-facing scarp of the Wasatch Mountains. Movements continued in this region more or less throughout Tertiary time, and in some parts appear to be still in progress. Within this great interior region, as well as in the intermontane valleys, and in the remnant of the Coloradoan geosyncline east of the Rockies, only continental strata were deposited during Tertiary time.

The Appalachian Mountain region was lifted bodily withont deformation, and the great erosion cycle began, which carved the present mountains and valleys, including those of the Great Lakes, the Hudson, the Susquehanna, and the other transverse and longitudinal valleys of the Appalachians.

The New Jersey section of the Cretaceous given on p. 429, is more complete in its later half than any other of the Atlantic coast, because either the series represents a more complete deposit, or is less eroded. It is the only one in which the Danian is represented, for here the Upper Senonian (Monmouth) is followed by 125 feet of RANCOCAS and this by 50 feet of MANASQUAN. These *Jerseyan Beds*

(=Danian), as they are collectively called, carrry a fauna
almost wholly distinct from that of the underlying Ripley-
an series. The lower Rancocas or Hornerstown marl
fauna comprises only a few species, only one of
which, *Gryphæostrea vomer*, is found in the older beds.
Most of the species are, however, derived from the preceding fauna, except some conspicuous brachiopods like
Terebratula harlani, which suggests Tertiary age. [1] The
next fauna, that of the Vincentown, is unique and has little
in common even with the preceding, but the higher
(Manasquan) again shows a relationship to the earlier
Danian (Jerseyan), but not to the older Cretaceous. Thus,
these retreatal beds show a distinctive fauna, one derived
from the preceding, but modified under stress of narrowing environment.

The Jerseyan (Danian) series is succeeded by 12 feet
of Shark River green-sands, the only representative of the
Palæocene on the Atlantic coast of North America. The
Shark River fauna is a wholly new one, with no relationship to the preceding Cretaceous fauna, as might be
expected, when we consider the long Interpulsation Period
which separates the two. A few of the weathered out
fossils of the older beds are secondarily included with the
Eogene species, and this has led to the mistaken belief,
that the succession is a continuous one. There is, however,
a great disconformity between the Cretaceous and the
Eocene, though reworking has masked it. A slight representation of the Danian is preserved in Maryland
(about 20 ft), resting on the Monmouth, but this is directly
succeeded by Eocene beds, the Palæocene being entirely
absent by overlap.

4

Only on the Gulf coast of Louisiana, Arkansas and
Texas do we meet again with transgressing Palæocene

---

1) This is also doubtfully reported from the "Eocene" beds of the Maryland
section, and it would seem probable that its occurrence in the Danian beds
is due to reworking of the older beds during the Eocene transgression.
Since this section has always been thought a conformable and continuous
one from Cretaceous into the Tertiary, any evidence for such reworking
along a disconformity would naturally be overlooked. This section
deserves careful re-analysis.

beds. These are the Midway limestones and marls from
20-260 feet in the former region and 200-400 feet in
the Texas region. But here they rest on the beds of the
Ripley group, which is Senonian, or they may lie on
older beds. The Danian is wholly unrepresented except
by a hiatus.

The Danian, the retreatal uppermost Cretaceous, is
typically developed on the opposite side of the Atlantic,
but is confined chiefly to the Baltic Region. It rests upon
the highest Senonian and begins with a coral or bryozoan
limestone of Malmö (Faxo limestone), with its peculiar
fauna of sea urchins (*Cidaroida holastis*, etc.), bryozoa and
peculiar corals, and the probably pelagic *Nautilus*, (*Herco-
glossa*) *danicus*. This is followed disconformably by the
lower Palæocene, or Montian, which at Mons is a brackish
water limestone, but elsewhere is marine. In France it
overlaps and rests disconformably on the Mæstrichtian,
and is followed by the Thanet sands (Thanetian), which
by overlap rest on the Cretaceous in the British Isles.
The highest Palæocene beds (Sparnacian) often contain
brown coal and plant remains, the marine invasion being
as yet oscillatory. These formations also occur in the
Paris Basin. Into the Russian region the Palæocene transgressed, and the later beds rest disconformably by overlap
on Pre-Danian Cretaceous. Similar conditions are found
in South Europe, North Africa and South Asia. None of
these transgressions are from the Boreal sea, all represent
the West European transgression of the Atlantic and
especially the North Sea portion. The extension was
eastward to the north shore of the Caspian sea (Tzaritzin),
but the Boreal sea was still kept out by the Greenlando-
Scandian land-bridge, which formed the main line of
migration for the Tertiary vertebrate fauna from the
plains of North America to those of Europe and Asia.
Not until the transgression had reached its maximum in
late Eocene time or early Oligocene time, was the connection broken by the Turgai Straits in the Ural region,
and after that, and throughout the Oligocene, Europe,
though perhaps still accessible to migrants from America,
was separated from Asia.

Meanwhile the Atlantic rift was widening, though
for a long time, probably up to Pliocene time, the Green-

lando-Scandic route of migration remained intact, the Iceland volcanoes healing the breach with volcanic plaster until it had become too wide.

The Eocene which follows the Palæocene conformably,

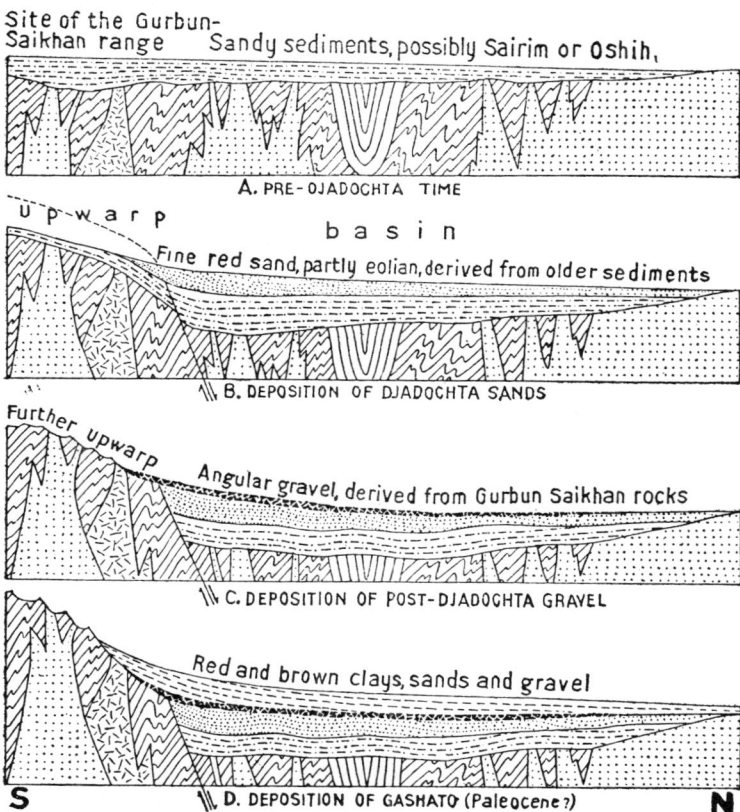

Site of the Gurbun-
Saikhan range   Sandy sediments, possibly Sairim or Oshih,

A. PRE-OJADOCHTA TIME

Upwarp   basin

Fine red sand, partly eolian, derived from older sediments

B. DEPOSITION OF DJADOCHTA SANDS

Further upwarp

Angular gravel, derived from Gurbun Saikhan rocks

C. DEPOSITION OF POST-DJADOCHTA GRAVEL

Red and brown clays, sands and gravel

S   D. DEPOSITION OF GASHATO (Paleocene?)   N

Fig. 113. Four stages of warping indicated in the Djadochta basin. A, shows an old basin of sandy sediments, deposited upon a floor of complex old rocks. B, shows initial upwarp, that caused the Djadochta sands to be washed down into the newly formed basin, without, however, exposing crystalline rocks of the present Gurbun Saikhan range. C, shows the deposition of a gravel of angular pebbles, derived from the Gurbun Saikhan rocks, and hence implies that those rocks have been laid bare. D, shows the deposition of the Eocene Gashato beds, the peneplaned remnant of which is at least 300 feet thick, and may be much more. The structure of the rock floor of the basin is inferred from the granite, exposed to the north of the basin, and the rocks observed in the Gurbun Saikhan range. The successive faultings of relatively small throw suggest the method by which this wing of the Altai was made.

is typically seen all around the borders of the Atlantic breach from the Gulf coast of North America northward and on the east of the Atlantic, southward to the Mediterranean. The former region was approximately in the latitude of the Anglo-Baltic embayment, while the Mediterranean embayment, or Alpino-Himalayan geosyncline, were tropical water bodies, in the neighborhood of the equator of the

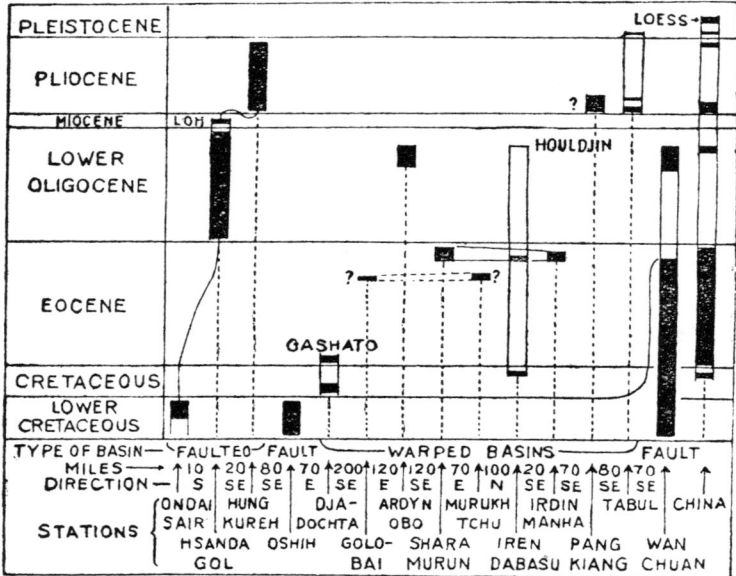

Fig. 114. Columnar diagram of the formations thus far observed in Mongolia.

Thicknesses are plotted to scale. It will be seen that the faulted basins contain far thicker deposits than do the warped basins. The diagram brings out clearly the shifting of the locus of deposition from place to place. The distances indicated at the bottom of the diagram are measured between stations or camps, not between the limits of the basins. The section for northern China is compiled from J. G. Andersson's "Essays on the Cenozoic of Northern China." (After Berkey and Morris, Bull. Am. Museum of Natural History, Vol. 51, 1924).

time (See Map Pl XV). This included the borders of North Africa too, that is the Atlas geosyncline, for which the Moroccan land-mass and its probable former continuation in Turkey and Irania formed the old-land. For it must be understood that neither the Mediterranean as such, nor the Black or Caspian Seas, were in existence in early Tertiary time. They are rift-valleys, formed after the folding of the strata of the Alpine and Himalayan geosynclines, by the reverse movement of the land-masses after that orogeny. This will be discussed later.

5

The migration of faunas and the limiting causes are well illustrated by the distribution of the *Baluchitherium*

Fig. 115. Preliminary restoration of *Embolotherium andrewsi*, based upon the structure of the cranium only. The other portions of the neck and body are restored with the characters of *Brontotherium platyceras*. About one-twentieth natural size (After Am. Mus. Novitates, No. 353, 1929).

(or *Indricotherium*), the giant hornless rhinoceras, that has recently risen to fame as the result of the discovery of its bones in China and Mongolia by the Third Asiatic Expedition of the American Museum of Natural History, and the restorations which have been made by Dr. Walter Granger and his associates (See Fig. 116).

Remains of this animal were found in Baluchistan, south of the Himalayan Range, which was not in existence at that time. Instead the region was an extensive low river plain, not unlike the broad Indo-Gangetic lowland of

Fig. 116. *Baluchitherium grangeri*, the giant hornless rhinoceras of the Oligocene plains of eastern Asia (Baluchistan to Mongolia). A restoration (Central Asiatic Expedition Am. Mus. Nat. History.)

today, over which these animals could roam far and wide. Since the climate was essentially a subtropical one, the region being a broad, well watered and forested plain, they could extend their migration widely, eventually reaching northern

Siberia, which was then in latitudes 20° to 30° north of the equator.   However, they were prevented from entering Europe by the Turgai Straits, which had come into existence in later Oligocene time, completely separating Asia from Europe by animpassable marine barrier. Thus these animals were confined to Asia, where their remains are found widely scattered in the sands of that period.   When, shortly after the end of the Oligocene period, during the succeeding interpulsation time of low sea-level, the Himalayas began to rise, cutting off India on the south, their habitat was confined to the plains of Asia. There they eventually succumbed as the climate became more and more arid, since this resulted in the death of the forest trees, which gave them sustenance. Thus was brought about the final extinction of these browsing animals.

The arboreal flora of the Oligocene period included the giant redwood (*Sequoia langsdorfii*), which still survives in California at the present time.   It also included the poplar (*Populus latior*) the *Glyptostrobus europæus*, the Iron-wood or Hornbeam (*Carpinus grandis*), the hazel (*Corylus macquarri*) and the alder (*Alnus keferstcini*), and of course, a host of herbaceous plants.   Scattered over the area were marshy basins in which peat beds were forming, and these have been preserved as the Oligocene coals, of which the Fushan coal of Fengtien, and many other plant-bearing beds of the period [2] are witnesses.

The mammal fauna, of course, included many other species, some 30 or more having so far been recorded from Mongolia and western China. These include 11 species of rodents, 10 of carnivores, 4 of insectivores, several other perissodactyla (odd-toed ungulates) besides *Baluchitherium*, several small deer and even a mastodon.

All these became extinct when, during the next Inter-pulsation Period, the great Himalayan chain began to rise by the compression of the early Tertiary and older strata, which had accumulated in the Himalayan geosyncline of Mesozoic and early Tertiary time. This great barrier between India and Tibet was the main cause of the gradual growing aridity, which finally embraced most of Asia and turned large tracts of country into desert basins.

---

2) For further details see GRABAU, *Summary of Cenozoic and Psychozoic* 1927, pp. 191 *et seq.*

6

Turning now to the Pacific coast of North America, we find that the strata deposited in the coastal geosyncline, which extended from Alaska to Patagonia, as well as those formed in the Colorado or Rocky Mountain geosyncline, in the region approximately some distance east and west of the 110th meridian of today, were primarily under the control of the westward drifting American landmass. These important Cretaceous and Tertiary geosynclines had developed in obedience to the drift pressure on the one hand and the plastic resistance of the Pacific sima on the other. Finally this same force was responsible for the folding of the strata in these geosynclines into north-south ranges, which came into existence wherever and whenever the force exceeded the strength of the strata.

The Eocene Arago group of the coast ranges of Oregon consists of sandstones and shales with an average dip of 70° N.E., and a thickness of over 3,000 ft. According to Dall[12] these strata contain *Cardita planicosta*, *Ampullina* sp., and other middle Eocene forms which suggest their correlation with the Claibornian of the Gulf-coast column.

According to Arnold[13] a wide-spread unconformity exists between the Eocene and the Cretaceous on the Pacific coast of North America, this being recognized in the states of Washington, Oregon, and certain parts of California. Over considerable areas in California, however, and at one locality in Oregon, it is replaced by a disconformity which is indicated by a more or less marked hiatus in the faunas.

Arnold stresses the "noteworthy fact" that, wherever the line between the marine Eocene formations (Martinez, Arago, Tejon, etc.) and the Cretaceous beds is marked by an angular unconformity, the underlying beds are either of Paskenta or Horsetown (Shasta) age, i.e. Comanche or Neocomican[14] and that, wherever the Eocene rests on the Chico, or Upper Cretaceous,[15] (with the one exception at San Diego) the unconformity is not angular and, as far as

12) DALL W.H. *North American Tertiary Horizons*. 18th Ann. Rept. U.S. Geol. Survey, pt. 2, 1898, p. 343.
13) ARNOLD: *Environment of Tertiary faunas*, pp. 512-513
14) That is Lower Cretaceous in the older Classification.
15) See *ante:* p. 410

the stratigraphic evidence goes, the two formations represent an apparent uninterrupted period of sedimentation.

This is entirely in accord with the principle repeatedly stressed in this book : namely, that an unconformity (involving an orogeny) also implies the elimination from that section of the strata of one or more entire pulsation systems, the time period of these pulsation system being occupied, in the folded region, with erosion and truncation of the folds into a peneplane, before further sediments can be deposited over them. In the present case, the entire Cretaceous is omitted, that being the period of erosion of the Comanche (Shasta) beds, which were folded during the Dakota Interpulsation Period.

The concordant superposition of the Eocene on the Cretaceous, together with the superficial similarity of their faunas, led Gabb and Whitney, of the early California Survey, to class the Martinez and Tejon formation as Cretaceous; but White, Stanton, and Merriam have demonstrated the Eocene age of these formations.

According to Merriam "The Martinez group, comprising in the typical locality between 1,000 and 2,000 feet of sandstones, shales, and glauconitic sands, forms the lower part of a presumably conformable series, the upper portion of which is formed by the Tejon. It contains a known fauna of over sixty species, of which the greater portion is peculiar to itself. A number of its species range up into the Tejon, and a very few long-lived forms are known to occur also in the Chico. Since the Martinez and Chico are faunally only distantly related, it is probable that an unconformity [disconformity] exists between them." [16]

7

In the Eocene beds of the Roseburg region Oregon, occur oysters which are so similar in appearance to the characteristic Cretaceous fossil *Gryphœa*, that they might readily be mistaken for Cretaceous forms. This may be a case of re-interment of weathered-out Cretaceous fossils in Eocene strata.

Marine Eocene rocks are found at many localities in

16) MERRIAM : *Section through John Day Basin*, p. 71.

Washington and Oregon, west of the Cascade Range, as well as in many areas of the Coast Ranges in central and southern California. "Although Eocene rocks probably once fringed the greater part of the western base of the Sierra Nevada, they are now all removed by erosion or covered by later formations except at one locality near Merced Falls." (Arnold)

Most of the Eocene rocks of the Pacific coast are either sandstones or shales, but conglomerates are found at the base of the formation throughout southeastern Oregon. They occur north of San Diego, and at a few localities along the northeastern flanks of the Coast Range. At Port Crescent, Wash., Eocene fossils are associated with tuff, but such occurrences are exceptional. Diatomaceous shales also occur at the top of the Eocene series in the vicinity of Coalinga, Cal., where they are believed to be the source of important deposits of petroleum. Coal, and other indications of more or less continental conditions, are found over much of Washington, Oregon and California, usually overlying marine Eocene beds.

Maximum thicknesses of the Eocene sediments range from 8,500 feet east of the Cascades, to 10,000 and 12,000 feet in western Oregon, and to 9,000 $\pm$ feet in southern California.

The classification of the Eogene beds in the standard west coast section and in the Santa Clara Valley is shown in the following table :

General Table of Cretaceous and Tertiary Strata
in the Palæocordilleran Geosyncline

NEOGENE PULSATION SYSTEM
UPPER MIOCENE or Monterey formation
LOWER MIOCENE or Vaqueros formation

Eo-Neogene Interpulsation Hiatus and Disconformity

EOGENE PULSATION SYSTEM
OLIGOCENE SERIES
San Lorenzo Series (may be cut out by Eo-Neogene disconformity)
EOCENE Series (8,500-12,000 ft)
Tejon formation

Martinez formation	
(Not folded Regions)	*(Folded Region)*

LARAMIDE INTERPULSATION SYSTEM	
Hiatus and Erosion	Erosion continued
CRETACEOUS PULSATION SYSTEM	Cretaceous absent
(Cenomanian - Senonian)	
CHICO SERIES	Profound Erosion
5,700-2,600 ft of	of folded strata
fossiliferous	down to Horsetown
clastic sediment	or Knoxville
DAKOTA INTERPULSATION	DAKOTA HIATUS
PERIOD-	
Normal Interpulsation	Folding of Shasta
Erosion Hiatus and	and Knoxville (and
Disconformity	older) formations
COMANCHE PULSATION	COMANCHE PULSATION
PERIOD-	SYSTEM
SHASTA SERIES	SHASTA SERIES
Horsetown group	Horsetown ⎫ (may
(7,600-12,500 ft)	beds ⎪ be absent
Paskenta group	Paskenta ⎬ by erosion
(5,300-11,000 ft)	beds ⎭
WEALDEN INTERPULSATION	WEALDEN[17] HIATUS
PERIOD-	

Interpulsation Erosion Hiatus

JURASSIC PULSATION SYSTEM-	JURASSIC PULSATION SYSTEM
KNOXVILLE SERIES	KNOXVILLE SERIES

In the Santa Ynez Range a thick series of marine
sediments represents both Upper Eocene (Tejon formation)
and Lower Miocene (Vaqueros formation), though it appears
as a continuous succession of marine sediments of detrital
origin. The disconformity is marked at the base of the
Vaqueros by a coarse conglomerate, which represents an
interpulsation hiatus and disconformity, that cuts out much
or all of the Oligocene San Lorenzo Series.

---

17) In some sections the Knoxville beds were folded during the Wealden
Interpulsation Period and eroded during Lower Shasta time.

It is significant that there is a pronounced unconformity between the Comanche and Eocene on the West
Coast, the former having been folded and truncated during
Cretaceous time.   This implies a westward movement of
the continent in Dakota time.   On the other hand, the
Oligocene-Miocene contact is a disconformable one, implying lack of folding, because of stationary conditions, or the
reverse movement.

We must, however, note that Arnold and Anderson,[18]
report finding a violent unconformity between the Vaqueros
formation (Lower Miocene) and beds which they tentatively referred to the Tejon formation (Upper Eocene)
in the Coalinga oil district, west of Tulare Lake in southern
California (Lat. 36° N., Long. 119°40′ W.).   If the identi-
fication of the fomations is substantiated, this would indicate
that locally at least, the Eogene-Neogene Interpulsation
Period was marked by orogenic disturbances in western
America as a mild echo of the great disturbances which
characterized Europe and Asia during that interval.

This can be explained by the probable positions of
the North Pole during the respective periods and the
shift of the pangæal cap during the interpulsation period.
I would locate the North Pole for Eogene time (in Lat.
150° W. and Long. 75° N. i. e. about 5 degrees north
of the northern border of Alaska, while for Miocene time
(Neogene period) I would place it approximalely in Long.
160° W. and Lat. 45° N., or in the Pacific ocean about
10 degrees south of the Alaska Peninsula.

This is the reverse of the positions given by Köppen
and Wegener[19] for the North Pole in these two periods,
they placing the Eocene Pole in the Pacific, where I place
the Miocene Pole, and the Miocene in the Arctic Sea,
where I place the Eocene Pole.   This reversal in position
is demanded by the evidence we have for crustal movement, which would be inexplicable on the Köppen and
Wegener location.

The older tillites of Bristol Bay Alaska contain mol-
uscan shells, which were regarded by Dall as indicating

18) ARNOLD AND ANDERSON: *Geology and oil resources of the Coalinga
district, California:* p. 48.
19) *Climates of Geological Ages.*

Miocene age. [20] These older Moraines are frequently disturbed, which is quite consistent with the post-Miocene westward movement of the sial cap, to bring Baffin Bay in polar location for the early Polycene (Baffin [21] or Günzian) glaciation.

It is true that Dall concluded that the Miocene marine fauna of Alaska indicated a much colder climate than that of Eocene time, while in the Pliocene the climate was truly boreal down to the state of Washington. This may, perhaps, be explained by lack of drainage from the Eocene ice sheet into Pacific waters, whereas the Miocene ice-cap dominated the coast from the Seward peninsula south to the Miocene embayments in the north western United States.

9

The change from the location of the Eocene Pole to the Miocene involved a shifting of the crust of about 30° of latitude, almost due north of the polar localities on the modern map [22] and, of course, an equal southward shifting, in modern terms, of points on the opposite side of the polar site of today.

If this shifting is regarded as rotation on a pivot at the apex of an isosceles triangle, we might conceive of the location of such a pivot in central Idaho (Boardman Mt. —nearly 11,000 feet, or Hyndman Peak, over 12,000 ft) or in the region between the Yellowstone and Wind River Mts. of western Wyoming, where there is an abundance of Tertiary igneous activity to form an anchoring plug.

Such a rotary movement would also be effective in

---

20) STEPHEN RICHARZ *Tertiary Glaciation of Alaska and the Polar Migration.* Zeitsch. d. deutsch geol. Gesell. Mon Ber. 1922; 74, 180-190.

21) I prefer this term Baffin for this first of the Polycene glaciations, since the center (pole) was located in Baffin Bay. The Günz was a local Alpine phase.

22) It must be borne in mind, that new polar location is not brought about by polar wandering, but by the shifting of the sial cap. Thus the change from the Eocene Pole north of Alaska to the Miocene south-west of Alaska appears like a southward (p. d.) polar wandering to the extent of 30 degrees, whereas it is a shifting of the entire sial crust 30 degrees in the opposite direction. so as to bring the more southerly point into polar location.

producing the folded Eocene strata of the Coalinga district
of southern California, for, since that is south of the
pivot-point, the movement would be in the opposite direction, *i.e.* westward, with the necessary pressure to produce
the folding of the Eocene beds of southern California.

The Eocene-Miocene sial shift of 30 degrees to produce
the new Miocene polar center had, however, other and
much more far-reaching effects. A comparison of our
maps (Plates XV and XVI) shows the greater extent in
Eogene time of both southern Europe, and southern Asia.
The Mediterranean is replaced by a shallow geosyncline,
in which the bordering countries are mostly submerged.
The northern border of this geosyncline lies near latitude
40° to 45° in Europe and 30° in Asia on the modern
map, but in Eogene time this was 25° to 35° and 15
degrees respectively. In this geosyncline accumulated the
Palæocene, Eocene and Oligocene strata, which were
partly marine, and partly huangho or continental river
deposits, and which now form the Alps of South Europe
the Atlas of Africa, and the Apennines etc., and the
Himalayas and some other ranges of Asia. The old
land which supplied the clastic material for the Himalaya
sediments, was situated in what is now North India, in
approximately 25° N. latitude (10° N. for Eocene time),
and North Africa and Algeria for the South European
areas, which were situated then somewhat farther south
than now.

10

During the Oligocene-Miocene Interpulsation Period
the Miocene center moved into polar location—a distance
of 30 degrees, and this resulted in the intense folding of
the strata of the geosynclines, and the production of both
the Alps and the Himalayas. The folding was clearly
the result of the pressure on these geosynclinal strata
caused by the moving northern land masses, which forced
them against the old massive fronts of Africa and India,
which were essentially stationary and immersed in the
resistant sima. Thus the Alps formed by the folding of
the strata of the Mediterranean geosyncline, occupied
essentially the position of the present day Mediterranean,

while the Himalayas probably lay further south than they
do today. (Pl. XVI)

The final adjustment came in Pliocene or early
Polycene time, when Baffin Bay moved into polar location.
This was a westward movement of the crust, for more
than 50 degrees on a great circle, and it inaugurated the
Baffin or Günzian polar glaciation.

It was this powerful wrench which tore the Alps
and Europe from Africa, leaving the Mediterranean gap,
erected Italy, formerly a fragment of Africa, and the
Greecian peninsula, into their modern positions (these having
likewise suffered some folding during the previous epoch),
and produced the fault-rifts, which opened to form the
modern Mediterranean and Black Seas. Many other faults
resulted from this wrenching assunder of the land, and
this was accompanied by an intense period of volcanicity.
(Plate XVII). There can be little question that most of
the volcanoes of the Mediterranean were born at about
this time, developing over the main rift fissures.

If the line of the Miocene-Baffin course of movement,
which brought Baffin Bay into polar location is prolonged,
it will strike the Gulf of Sidra on the north coast of
Liberia, which is the most likely region from which Italy
was torn during the post-Miocene cataclysm, and after the
folding of the Apennines (most probably a former continuation of the Atlas ranges of Algeria). The Mediterranean-
Black Sea lines of faulting were thus essentially at right
angles to the polar movement, and this probably explains
the disruptive result of the wrench in this part of the
earth's crust. On the other hand, North India and the
Himalayan region — while in a more direct line during
the earlier movement which produced the folding — now
lay in a line more nearly at an angle of 30° to the direct
line of pull, and as a result only minor riftings occurred,
while no great dismemberment of the land took place.
Instead, India appears to have been pulled bodily poleward
to some extent, so that in the later adjustments of the
poles it was left essentially in its modern position.

This poleward pull at this time seems also to have
affected the south coast of Malaysia, for the volcanoes of
the southern border of Sumatra and Java, both of which
were then a part of the main-land, may have begun their

activities at this time, arising over submarine rifts, produced
by the wrenching force. These regions had also been
affected by the earlier post Oligocene polar movement, for
then the Eocene and older beds of the coast regions suffered their main deformation. The formations of the
great rifts, which separate these islands to-day from the
main land did, however, not begin until late Polycene or
early Pleistocene time (Pls. XX, and XXI).

CHAPTER XXXVIII

THE MID-TERTIARY INTERPULSATION AND NEOGENE PULSATION PERIODS

1

THE complete withdrawal of the sea at the end of Oligocene time brought the older Tertiary to a close and inaugurated the Mid-Tertiary Interpulsation Period and its orogeny. (Plate XVI)

The Americas had their orogenic disturbance during the Laramie (pre-Palæocene) Interpulsation Period, when only erosion prevailed in the Old World area. But by Mid-Tertiary time Europe and Asia had reached the point where they were ready for the greatest mountain-making event that the world had seen since the beginning of Palæozoic time.

The most important of these great mountain chains in Europe were the Alps, Carpathians and Caucasus, and in Asia, the Himalayas and the Yung-ling Range or Eastern Alps of Tibet. In each case, the folding of the strata of the older geosynclines resulted in the migration of the geosyncline into the old-land region, and in these new geosynclines the later strata come to rest unconformably upon the very much older rocks. The folding of the Alps produced the MOLLASSE CHANNEL; that of the Carpathians the MIODOBARIAN BASIN, and that of the Caucasus, the CRIMO-CAUCASIAN BASIN. With the first folding of the strata of the Himalayan geosyncline, the formation of the SIVALIK GEOSYNCLINE in the former old-land may have commenced, but apparently no deposition was going on in this new geosyncline until Tortonian or Upper Miocene time, this being the age of the oldest member of the Sivalik series. The absence of older deposits in this region indicates that perhaps the geosyncline as such was not developed until the beginning of the Upper Miocene, the earlier portion of this period being occupied by the deformation unaccompanied by much erosion. Or it may be due to the fact that early Miocene deposits formed in the young Sivalik geosyncline were removed again by erosion during a renewed period of uplift, and before the final development of the Sivalik geosyncline had begun.

"Folding also characterized the geosynclines which bordered the Pacific; in Korea and Japan the older Tertiary is folded, and is unconformably succeeded by the Miocene. On Sakhalien, too, the older Tertiary strata are strongly folded and much dislocated, while the Miocene beds rest unconformably upon their eroded ends and are themselves less strongly folded.

"This folding seems to have been more extensive in eastern Asia, for we find that in the Sichota-Alin Tertiary Beds have an east-north-east strike and a dip of 18° to 20°. These beds contain *Sequoia langsdorfi*, *Alnus kefersteini*, *Carpinus grandis*, *Corylus macquarii*, and *Taxodium distichum miocenum*. Though these beds are generally considered as Miocene, the flora is one equally at home in the Oligocene deposits of Eastern Asia, and it may be that these beds are to be regarded as of Oligocene age".[1] Such folding is not inharmonious with the movements of the sial crust between the Eogene and Neogene polar locations especially if this movement was a rotary one, as a glance at the maps will show (Plates XV-XVI).

"That post-Oligocene folding affected this eastern region of Asia is shown by the fact that in the coal basins of the lower Bureja in the Amur country, the Miocene rests unconformably on the strongly folded older Tertiary, which has a strike of north 10° east. The Miocene itself is also folded, but less strongly than the older Tertiary, and it strikes in a direction of N. 70° E."[2]. The former is the product of crustal shift, which effected the Eocene to Miocene polar translation, the latter of the counter-clockwise rotation in early Polycene time or of the similar movement in the Mindel-Wisconsin polar shift.

"In the Anadyr Basin the folding of the older Tertiary is also pronounced, the strikes being meridional or N. N. W. These beds are unconformably succeeded by the Miocene, which is less strongly folded. Here too, the flora of the

1) GRABAU: *Summary of Cenozoic and Psychozoic* 1927. p. 197.
2) IBID.

older beds includes *Sequoia langsdorfi*, *Glyptostrobus ungeri*, *Alnus kefersteini*, and *Taxodium distichum miocenum*. This flora is definitely regarded in this region as older Tertiary, either Oligocene or even late Eocene, and this clearly shows that it is quite possible that many of the strata of eastern Asia referred to the Miocene may actually belong to the older Tertiary.

"Folding of the older Tertiary is also indicated in the Arctic region where, in the new Siberian islands, Tertiary sands and clays are strongly tilted and eroded, and unconformably succeeded by beds of Quaternary age. The highest member of this Tertiary series encloses remains of *Populus arctica* and *Sequoia langsdorfi*, while lower members have yielded *Taxodium distichum miocenum*. Here too, the deposits may represent Oligocene, though they have been regarded as Miocene. If the former classification is correct we have here also evidence of folding at the close of the Lower Tertiary.

"Thus, it would appear that on the south, east and north, orogenic disturbances were pronounced. . . [in post-Oligocene time, that is during the mid-Tertiary Interpulsaton Period,] and this suggests that in the interior of the Asiatic continent (China, Mongolia, and Siberia) other orogenic disturbances occurred at this time." [3] Whether these are waves of the Alpine-Himalayan equator-ward thrust, or the crustal movement on the periphery of the lands opposite, that resulted in the polar shift, or whether all of these disturbances were manifestations of the same force, must remain for future investigation to determine.

"No disturbances are recorded in western Siberia or in the region to the south, where in the Lake Aral district and northward, marine or semi-marine Miocene strata follow disconformably upon the Oligocene, while farther north, in the region of the old Turgai Straits, plant-bearing beds of early Miocene age follow disconformably upon the Oligocene *Indricotherium* Beds." [4]

Since the Alpine and Himalayan folds were the most spectacular achievements of the tectonic forces during this Interpulsation Period, we may conveniently name it, the ALPINO-HIMALAYAN INTERPULSATION PERIOD and orogeny.

---

3) GRABAU: *Ibid* p. 198.
4) *Ibid.*

2

The subdivisions of the Neogene System here adopted are as follows:

TABLE OF FORMATIONS OF THE NEOGENE PULSATION:

SUPERFORMATION: *Polycene Norwich Crag*

Interpulsation Hiatus and Disconformity

NEOGENE PULSATION SYSTEM
   B. PLIOCENE SERIES
      III. ASTIAN OR RED CRAG (140 ft)
         3. Rutleyan, or zone of *Cardium groenlandicum* (Amstelian);
         2. Newbournian, or zone of *Mactra constricta;*
         1. Waltonian
            b. Pœderlian, or zone of *Mactra obtruncata;*
            a. Scaldisian, or zone of *Neptunea contraria.*
      II. PLAISANTIAN
         2. Gedgravian, or Casterlian (Coralline Crag 40-60 ft) or zone of *Mactra triangula*
         1. Lenhamian, or Diestian (Lenham beds) or zone of *Arca diluvia.*
      I. ANVERSIAN (Redonian, Sahelian or Pontian).
   A. MIOCENE OR TRANSGRESSIVE SERIES
      III. UPPER MIOCENE OR VINDOBONIAN
         3. Sarmatian
         2. Tortonian } Second Mediterranean Stage.
         1. Helvetian }
      II. MIDDLE MIOCENE, OR }
         BURDIGALIAN   } First Mediterranean
      I. LOWER MIOCENE, OR }   Stage
         AQUITANIAN   }

Hiatus and Unconformity or Locally Disconformity

ALPINO - HIMALAYAN INTERPULSATION PERIOD AND OROGENY
SUBFORMATION :
   EOGENE OR MORE GENERALLY MESOZOIC-PALÆOZOIC

The lowest Miocene beds are naturally to be looked for where the Eogene strata are not folded. In the Himalayas, Alps, and other folded mountains of this period the whole of the marine Miocene and Pliocene beds are probably absent, since it would take more time than the Neogene to accomplish peneplanation. Continental strata of these periods may, however, be looked for, especially in the intermontane valleys, for here both lake and river deposits were forming.

In the attempt to reconstruct the geography of the Tertiary, the first step must be the location of the poles, and after that the amount of the "Atlantic overlap", that is, the amount by which the width of the Atlantic was diminished over its present extent. Taking the latter at 70 degrees[5] in round numbers, and 20° for Eogene and Neogene time we have a difference of 50° which is the amount of overlap. Obviously this influenced the location of the South Pole on the modern map, though on the map of the period it was always opposite to the North Pole.

3

In our reconstruction we place the Eogene North Pole at a point which is today in Long. 150° W. and Lat. 75° N., or 15 degrees south of its modern location. This point at the time was 50° nearer the 0° meridian (Greenwich) because of the Atlantic overlap; thus it lay on what is today the meridian of 100° W. That would bring the South Pole on meridian 80° East, of today.

Because of the Atlantic overlap, Greenland was wedged between Baffin Land and Fenno-Scandia, making a continuous land-bridge between Europe and America, which served for intermigration of Eogene land faunas and floras.

---

5) A convenient line on which to measure the change in width of the Atlantic rift is on the modern parallel of 30° between Agadir Bay, Morocco, on the African coast and St. Augustine Florida. These points are 72°+apart. In Eocene and Miocene time the distance apart, that is the Atlantic rift, was only about 20° as represented on our map, and this is expressed as a 50° (really 52°) overlap. This brings the Eocene South Pole into East Antarctica between Victoria and Enderby Quadrangles, whereas longitude 30° without overlap, would bring it into western Enderby Quadrangle. The Miocene South Pole was located a few degrees north of Kergulen Islands which by their strong glacial scouring testify to their former nearness to the center of ice accumulation.

The Boreal Sea, however, was for the most part frozen; the Siberian shore lay in latitude 60° to 65° N. or within 15 to 10° of the Ice front. North Greenland was more or less ice bound and so was probably the arctic end of the Uralian geosyncline. The southern margin of Europe was close to North Africa, the shallow Alpino-Mediterranean geosyncline covering both this and the North African border. The boot of Italy knelt in the Gulf of Sidra, and the expanded South Europe, now crumpled into the Alps and other mountains ranges, occupied the rest of the Mediterranean territory, and was partly covered by the shallow waters of the Alpine geosyncline. This was extended south-eastward across the Nile and Red Sea country, and eastward to the Himalayan geosyncline. From the North Atlantic, which the Alpine geosyncline joined on the west, and which was then only 20 degrees wide, a narrow North Sea channel extended to southern Scandinavia and South Greenland, but as yet no connection was established between it and the Boreal Greenland Sea, the land bridge keeping out the cold waters of the ice bound region. Not until Pliocene or early Polycene time were the cold waters of the north able to spill over this land barrier, thus allowing the animals of the Boreal Sea to mingle with the warmer water types of the Atlantic, which were typical of what is now the North Sea.

During the Alpino-Himalayan Interpulsation period, when the waters drained away into the sunken ocean beds, the folding of the strata into the Alpine ranges produced a great fore-shortening of southern Europe, because Africa, the mass against which it was crushed, remained essentially stationary. The north Moroccan littoral, a part of the Eogene geosyncline, was also folded, producing the Atlas mountain ranges, of which the Apennine ranges were a part, for Italy continued to kneel in its North African niche.

4

At the opening of the Miocene period (Plate XVI) the North Sea, which was probably entirely drained during the Interpulsation Period, was again invaded by the marine waters of the rising Atlantic, these spilling over and encroaching on the low coastal areas. In the Cologne region

of Germany however, non-marine brown-coal beds ranging up to 104 meters in thickness, and with trunks of
the bald cypress (*Taxodium distichum*), rest directly upon
the Oligocene.   These are immediately followed by marine
lower Pliocene (Anversian) beds.   In Belgium the sea
transgressed somewhat earlier, for here marine Upper
Miocene (Vindobonian) is the first product of the invasion.
This is followed here also by the marine Lower Pliocene.
Miocene is not found in Great Britain, which was land,
and subject to continuous denudation, after which the
south-east counties of England were submerged by the
rising sea-level, and sand-banks with shell rubble were
deposited.   Such is the "Coralline crag" of Suffolk, and
similar deposits have been preserved in Cornwall.

Though there was, as yet, apparently no connection
with the Arctic ocean of the time (for Greenland had
not yet become wholly separated from the Scandic region
and Barents shelf), still the waters were becoming cooler.
The plants, though no longer subtropical as were the
plants of the older Tertiary, nevertheless pointed "to a
climate rather warmer than that of southern England at
the present time".[6]   There was apparently an older
Pliocene sea invasion, the deposits of which have, however,
been largely removed again by erosion, but its former
presence is indicated by fossiliferous fragments in a basal
conglomerate along with fragments of older formations.

The coralline crag (40-60 ft), includes the so-called
Lenham beds found in hollows or pipes dissolved out of
the chalk and indicating a subsequent relative lowering
of the sea-level of 860 ft since then.[7]   These and the
equivalent remnants in Cornwall (St. Erth beds), are
locally followed by the Red Crag, about 140 ft thick,
which registers the final junction of the North Sea with
the Arctic (though not yet the completion of the Atlantic
rift), by the invasion of the Arctic cold water fauna.
The deposits of the Red Crag range from the warm water
type in Essex on the south, to the Rudly cold water
Crag in the north.   The sequence from south to north
includes:

---

6) GEIKIE: *Text-Book* p. 1281.
7) This implies that the depth at which these shells, etc., accumulated, was
   not less than 40 fathoms, or 240 ft (Geikie).

(a) WALTON CRAG which has a Molluscan fauna of 520 species mostly southern forms, these being chiefly extinct, while northern forms are rare or absent. (b) The *Oakly Crag* farther inland from the Walton Crag into which it merges; it contains upward of 350 species and varieties, many still of the older, even Coralline Crag, types. There is here the first indication of the northern types (*Trophon islandicus, Scala grœnlandica* and others). This marks the beginning of the overflow of the waters of the north, across the former effective barrier between Greenland and Scandinavia.

The next, or RED CRAG of Newborn, (c) shows still further extinction of the southern and increase of northern forms. *Mactra (Spisula) constricta* is typical, as well as species of *Tellina*, and *Scala grœnlandica, Purpura lapillus* and others, many of them still living in the North Atlantic. Finally the RUTLEY CRAG, (d) or zone of *Cardium grœnlandicum* lies farthest north and is marked by further diminution in the southern and a corresponding increase in the northern types (Geikie), while the *Tellinas*, as well as *Mactra constricta* and *Cardium angustatum* constitute a large part of the assemblage. The essentially northern types: *Tritonofusus altus, Buccinum grœnlandicum, Natica grœnlandica,* and *Cardium grœnlandicum* are more abundant than in the other divisions.

Dr. R. von Königswald [8] has convincingly presented the lithological, faunal and floral arguments for the gradual aridification of the climate of West Europe in Pliocene time, which was primarily a period of regression of the sea, as the Miocene was one of marine transgression. His contention is that the cold northern currents which influenced the fauna of the British Crag, *i. e.* the result of the periodic overflow across the Greenland Scandinavian land-bridge, were primarily responsible for the reduction in precipitation, while the general temperature of the lands continued to be higher than now. He cites the modern example of the south-west African coast, where the water temperature is lowered, by cold currents, below that of the land. Although the mean annual temperature of the water is only $0.9°$ C lower than that of the air over dry

---

8) RALPH VON KÖNIGSWALD: *Klimaänderung im Jung Tärtiär etc.,* pp. 11-20.

land, this is sufficient to prevent evaporation over the water and condensation over the land, and because of this the annual precipitation is only 18.5 mm.

5

Above the Red Crag of the British coast follows the Interpulsation hiatus and disconformity, and here we meet with a distinctly new feature, that is the first record of ancient man in this region.

In 1919 Reid Moir discovered what he regarded as a definite occupation layer in the Red Crag, near Foxhall, Ipswich in Suffolk. From 2 to 3 ft below the top of the Red Crag occurs a thin layer with bones, shells, and implements, while from 2 to 5 ft lower is a second layer which also contains worked flints. Here not only the flint implements but cores and flakes were found, and what he regards as evidence of the action of fire. These flints are unlike those of the Chellean or pre-Chellean of France, being chiefly fashioned from flakes and not from the cores. They include hafted specimens, side scrapers, and a number of arrow-head-like points, also borers and scrapers of the ordinary type. A specimen of a rostro-carinate implement was also found here. The human origin of these flints has been accepted by Abbé Breuil, by Osborn, and by many other authorities, though Boule thinks that the intrinsic character of the new finds, convincing though they appear, may be not incapable of interpretation as other than that of intentional dressing.

Of course the dating of this occupational level as Pliocene is scarcely substantiated by the facts in the case. It must be remembered that we have here a great dividing line between geological eras, a change from the end of the Tertiary to the beginning of the Quaternary, with a long Interpulsation Period between.

Primitive man lived here during the Post-Pliocene Interpulsation land period, when the North Sea was completely drained, and active migration of late Tertiary and post-Tertiary animals from America to Europe and the reverse took place. That his implements and even the evidence of fire (the making of which was an accomplishment acquired long before this — See Chapter XL) should be found embedded in these sands, is not surprising, for the cross-bedded character of the Red Crag sands shows repeated reworking by streams, if not by wind, for the arid conditions of the Pliocene probably continued into the Interpulsation Period.

The Red Crag consists of a series of local accumulations of highly oxidized dark-red or brown ferruginous sand, which is evidence of a long exposure after its formation as a shallow water marine deposit. The gradation from the warmer water phase in the south to the colder water phases northward, due to the influence of the cold currents from the north, has already been given. These sands were evidently much worked over by streams and winds during the succeeding Interpulsation Period, and this reworking and reinterment would affect the distribution of the human artifacts as well.

The record of the migration across this land area between Europe and America is seen in the deposits of the large number of animal bones, chiefly of the extinct species of elephants *(Elephas antiquus,* and *Mastodon avernensis),* horses *(Equus stenonus),* deer, gazelle, etc. which make up the bone beds (MAMMIFEROUS CRAG) of the Norwich Crag (Icenian), which directly overlies the sands of the Red Crag and is mingled with them at the contact. These are terrestrial deposits of the Interpulsation Period.

Analogous conditions are indicated by the deposits on the Belgian and Holland coast and to some extent on the north coast of France. In the Paris basin, however, the Miocene is represented by fresh water deposits only, and the Pliocene by continental river deposits, enclosing the remains of a temperate climate flora and fauna, and also many early types of mammals.

6

We return to the marine Miocene. The main Atlantic embayment was in the south-west of France (Aquitania), where are located the type regions of the Aquitanian and Burdigalian.

The Atlantic had also by now gained access to the Mediterranean which had come into existence after the rising of the Alps and the tensional rifting which followed

as the result of the polar shift to the Baffin Bay (Günzian) center.   Where at first only small areas around its
margin were transgressed by the Lower Miocene (Aquitanian) sea, a far more extensive transgression occurred in
the Middle or Burdigalian period.

As a result of the folding of the strata of the Alpino-
Carpathian, and other geosynclines in the Interpulsation
Period, new geosynclines came into existence in what was
previously the old-land.

The Molasse Channel formed by the Alpine folding
at first received only continental sediments (Aquitanian),
but in Burdigalian time it was flooded from the Mediterranean.

In Sarmatian time[9] it was again the site of continental
sedimentation, including those of the famous Steinheim
and Oennigen (Switzerland) basins.   The former is known
for its evolutional series of *Planorbis* and the latter is noted
for its great variety of well preserved insects.

The eastern end of the Molasse Channel later widened
out into the Vienna basin, and this became confluent with
the Pannonian basin.   Transgression began in Burdigalian
time with overlap of marine strata of shallow water type
(Schlier) followed by purer marine beds in Tortonian
time, and then by brackish water beds in Sarmatian time.
Finally in Pliocene time these basins became great fresh
water lakes.

The folding of the Carpathian geosynclinal strata into
the young Carpathian mountains produced the Miodobarian
new or migrated geosyncline.   Here beds of Helvetian age
rest unconformably on the older Palæozoic or Mesozoic,
as seen at Kilce.   "These Miocene beds are rich in marine
shells of the type found in the Vienna basin and they
are succeeded by *Lithothamnium* limestone, the so-called
"Leithakalk", which is of Tortonian age.

"This basin was shut off from the main marine
basins in later Miocene time, whereupon these enclosed
waters became subject to evaporation, with the result that
at first gypsum, and later rock salt was deposited, these
including the famous Wieliczka salt beds near Krakow.
These salt beds are probably of late Tortonian age, for
in the succeeding Sarmatian time, the basin was again

9) GRABAU: *Summary* 206 map fig. 5.

flooded with marine waters, and extensive *Bryozoa* reefs were developed in this closing stage of the Miocene.

"The new geosyncline formed in the Caucasian old-land upon the folding of the strata of the Caucasian geosyncline, also became a nearly enclosed water body, resembling in some respects the Black Sea of today, but probably not of so great a depth.

"Here in Burdigalian and Helvetian time were formed the Meletta shales which are so-called because of the small sardine-like fish *Meletta* which abounds in them. With these occur Radiolaria and Diatoms, while a small spiral shell, *Spirialis*, probably the protoconch of a gastropod, occurs in vast numbers. On this account this old water-body has also been called the *Spirialis* sea.

"These shells, and the immature pelecypods which occur with them, evidently represent planktonic organisms, which floated into this water-body from the outer sea, and, sinking to the bottom, died without further development. In this respect the conditions are essentially similar to those of the Black Sea of today. The absence of scavengers on the sea floor permitted the accumulation of the organic matter, which by destructive distillation has produced the petroleum and asphalt of this region.

"This basin was again flooded by normal marine waters in early Tortonian time, this flooding proceeding through the Kertch Straits in the Crimea. The inlet was once more closed in middle Tortonian time, when the stagnant Spaniodon sea was formed. In Sarmatian time, however, the sea again entered this basin as well as the Miodobarian, Pontian (Black Sea) and Aralo-Caspian basins, the marine deposits of this period widely overlapping those of the earlier Miocene.

"Marine Miocene beds.....cover large areas of the Mediterranean region (Italy, Sicily, Corsica, Sardinia, the western Balkans, Greece and the Aegean districts). They show overlapping relationships and extend over the lowlands of Syria into the Euphrates Valley, as far as Urfa and to Killiz in Syria.

"In the closing stages of the Miocene, these marine beds were replaced by the great Persian gypsum and salt formation of upper Sarmatian age. These extend into Asia Minor where they form the basis of the great salt

deserts of Armenia, etc., and where they and the underlying marine Miocene beds constitute the Supra-Num-
mulitic Series.   Similar deposits extend into Egypt and
the Gulf of Suez." [10]

Throughout the Mediterranean region the early
Pliocene is still marine (Sahelian-Plaisantian), though in
some regions the retreatal beds are followed or replaced
by non-marine sediments (Pontian).

The basal marine Pliocene or Anversian is known
as the Redonian in the broad sea-channel that severed
Brittany and the Cotentin from the rest of France and
formed one of the main connections between the Atlantic
and the North Sea.   The deposits in this region consist
"of calcareous sands, which in the region of the lower
Loire (Nantes, etc.), rest directly on the old crystallines,
but in the more northerly districts (Rennes, etc.) they
rest on the eroded Vindobonian or on the Tongrian.
Although there are still many species which pass upward
from the underlying Vindobonian (Upper Miocene) (about
65% of the fauna) [11] the fauna of the Redonian is a
distinctive one, consisting of corals, echinoids, brachiopods,
pelecypods, scaphopods, and gastropods, as well as teeth
of salachians.   About 50% of the fauna consists of species
still living.   While this fauna has a number of species
in common with the Anversian of Belgium, such as
*Lucina borealis, Venus multilamellata, Saxicava arctica,
Calyptræa sinensis, Turritella subangulata, Natica mille-
punctata*, etc. it is nevertheless sufficiently distinct to indicate
a separate center of origin.

7

One of the most remarkable facts in the stratigraphy
of Asia is the almost complete non-development of the
deposits of Miocene age, except in the marginal regions,
i. e., the Hongkong-Nippon geosyncline, the islands and
coastal regions of the southern Pacific and Indian Oceans,
and their extensions, and in the Sivalik geosyncline, where
beds of Upper Miocene or Tortonian age form the lower

---

10) *Ibid:* p. 201-202.
11) Some of these may be secondarily included.

division of the Sivalik series, though these are fossiliferous only in the Salt Range region.

No marine Miocene deposits are positively known from China unless the LUTZUKOU FORMATION of northwest Shansi proves to be such. This underlies the Pontian of that region and consists of from 20 to 25 meters of marl, marly limestone, cross-bedded sands, etc., with fresh-water mollusks, fish-remains and mammal bones and teeth, including those of the horse, rhinoceros, rodents, etc. The Pontian *Hipparion clays*, which were formerly considered in part Miocene, are here placed in the base of the Pliocene. This leaves a complete gap in China so far as the Miocene is concerned, and even the Oligocene is only sparingly represented.

The only Miocene at present known from Mongolia is the LOH FORMATION, a comparatively local deposit north of the Altai mountains, with a thickness of less than 100 ft. Since exploration has been rather extensive in recent years in this region, the fact that no other Miocene deposits have been positively recognized in Mongolia, is strongly indicative of the general non-deposition of rocks of that age in this part of Asia. This again would argue that the Mongolian region, too, was one of elevation and erosion throughout practically the whole of the Miocene time. The fact that both on the south and the north of this vast continental region, Miocene beds are also largely lacking or only locally developed, raises the question of the disposition of the products of erosion during the Miocene, if these were at all voluminous.

Perhaps the region was one of excessively dry climate and no very pronounced elevations, with a minimum of rock disintegration and the removal of the products of erosion by the winds to far distant regions. Against this, however, is the evidence of the Loh fauna which, according to Osborn, seems to indicate forested and swampy habitats. The rather scanty occurrence of this fauna, however, might be taken as indicative of only temporary existence, either in early or late Miocene time, of such moister conditions.

Even in Siberia, Miocene deposits appear to be very scanty, and, as already suggested, some of those which have been classified as Miocene, may actually belong to the older Tertiary. Miocene plant-beds have been recorded from southwestern Siberia, from the old Saisan depression, in the region of the Black Irtysch River, approximately west of Long. 85°. These plant beds, however, do not exceed 60 or 70 metres in thickness. At Tomsk, (Long. 85°30′ East) a plant-bed, 4 meters thick, is regarded as representing the Miocene. This is followed by 8 metres of concretionary sands of unknown age which are in turn succeeded by post-Pliocene deposits. Miocene beds are said to be widespread in the Kirghiz Steppe, but seldom exceed 15 m in thickness.

What is probably the thickest deposit of Miocene beds, if the age is correctly determined, is found to the south of Lake Baikal, where it is said to reach a thickness of 1,500 m. It is, however, distinctly stated, by those who have studied this formation, that its Miocene age is not certain. If the Miocene age could be established we would have here a region of deposition of the products of erosion from the vast area over which Miocene beds are absent.

The Miocene beds of the Amur basin range only from 40-70 m in thickness, while those reported from the Lena River basin probably do not exceed 75 m. The Tertiary beds of the new Siberian Islands are probably not over 80 m in thickness, and these, as we have seen, contain a flora suggestive of Oligocene rather than Miocene age.

It is important to bear in mind the fact that the Indo-Pacific province had been distinct from the west European since Oligocene time, and that hence differentiation of species had been going on in these basins independently for a considerable period of time, long enough, that is, to develop a distinctive Indo-Pacific fauna. Hence correlation with European formations on the basis of identity of species with European types, is not possible in this region, though the general state of development of the faunas may be taken to some extent as indicative of approximate parallelism with European horizons which contain species in the same stage of development.

Plant remains are just as unsatisfactory as indicators of correspondence in age of Asiatic with Europan horizons, since we do not yet know to what extent older plants continued in this region without undergoing modification. Intermigration of land forms between Europe and Asia was possible only after the opening of the Miocene period, and it may be that species existing in Europe in Oligocene time migrated to Asia at the opening of the Miocene, arriving in the east without having undergone any pronounced modification. The reverse may, of couse, also have been true, since Oligocene species of Asia, retaining their essential characteristics into Miocene time, may have migrated westward at the opening of the Miocene, and reached European fields without having undergone any essential modification.

It is thus evident that a separate standard, both for the marine and the plant-bearing terrestrial formations must be established for the Neogene in Eastern Asia, and this must be in the first place, based on stratigraphic superposition. For correlation with the European succession we may have to rely on palæogeographic development. It may, of course, be possible to establish certain standard reference horizons for the Asiatic Neogene on the basis of the vertebrate distribution, and such appears to have been effected for the Pontian of the Lower Pliocene.

8

"Perhaps the oldest Miocene or Malaysian beds of the Indo-Pacific province are those of the island of Nias off the Sumatran coast and in the Philippines. Here the basal beds are characterized by large foraminiferal shells *Nummulites*, (*Operculina*, and *Lepidocyclina*) types which have generally been considered as indicative of the older Tertiary. It may indeed be, that these Nummulitic beds represent the later Oligocene in this region, but it is also possible, that, as Dickerson has suggested, the *Nummulites* and their congeners persisted in the Indo-Pacific province into early Miocene time, though in the Mediterranean province they became extinct at the end of the Oligocene."[12] With these foraminifera occurs a molluscan fauna which has a more distinctive Miocene aspect.

The molluscan fauna of the Nias Miocene contains 26 per cent of species still living in the neighbouring

12) GRABAU: *Ibid* p. 205.

waters, although all are distinct from European types. Only 10 per cent of the species, however, are known from the higher Miocene beds of Java.

A formation of essentially the same if not somewhat older age is found in southern Sumatra in the Bencoolen district though this has been correlated with the European Burdigalian or Middle Miocene. Out of 32 species of Mollusca obtained from the Eburna marls, 6, or less than 19 per cent, are living at the present time in Indo-Pacific waters, and even these are chiefly represented by varieties.

"A somewhat younger fauna has been obtained from the KROE MARLS of southern Bencoolen, for although this contains only 24 per cent of living species as compared with the 26 in the Nias deposits, 35.7 per cent of the total number are identical with species occurring in the higher Miocene of Java, whereas the fauna of the Nias beds has only 10 per cent in common with these Javan deposits. Taken as a whole the Nias group, the Bencoolen or Eburna marls and the Kroe marls appear to constitute a unit, which represents the older division of the Miocene in the Indo-Pacific province. Because of its extensive development on Sumatra and the bordering islands, it may be designated the SUMATRAN SERIES, and tentatively correlated with the Aquitanian of the Mediterranean Province.

"On the Island of Java, the oldest bed referable to the Miocene is an *Orbitoides* limestone, which Verbeek classed as the latest member of the older Tertiary. This may be the equivalent of the basal Foraminiferan bed of Nias. On Java, this basal bed is succeeded by from 1,000 to 3,000 meters of unfossiliferous sandstone which apparently represents the Sumatran Series. This is overlain by the richly fossiliferous Java marls and sands i. e. the JAVAN GROUP. Of the extremely rich molluscan fauna very few are as yet known elsewhere except in the Miocene of the Philippines. In these islands the entire Miocene is named the VIGO GROUP, the thickness of which is estimated at from about 3,000 to about 10,000 ft (approximately 900 to 3,000 meters). The basal portion is characterized by the *Numulites subniasi* fauna and corresponds to the basal bed of Nias Island. The lower portion of the succeeding series is also principally characterized by Foraminifera,

(*Lepidocyclina, Operculina,* etc.) and apparently represents the Sumatran series. The upper part of the Vigo group, however, abounds in mulluscs, the species being essentially those of the Java series, this being expecially true of the index fossils of this series, all of which are now extinct. According to Dickerson, this fauna has 75 per cent of its species still living in the neighbouring waters, which, when compared with the European standard, is extremely high. Dickerson, however, considers that this is due to the greater persistence of specific types in tropical waters, according to which a part of the living faunas would still consist of Miocene species. In spite of the high percentage of living species, this UPPER VIGO or JAVAN series is generally correlated with the European Burdigalian, rather than with the higher Miocene, which is believed to be unrepresented by marine strata in this region, though this is by no means an established fact.

"The Persian embayment of Miocene time extended to the Teheran region and Lake Urmia and even into Armenia. Despite the fact that it approached the region submerged by the Mediterranean waters, there is no evidence that junction with these waters was effected. Some of the most extensive deposits of this gulf are today found near Bidjar where the series consists of 200 meters of limestones, charactized by molluscs and echinoids."[13]

Following upon these marine beds comes the great IRANIAN SALT AND GYPSUM FORMATION of Upper Miocene age, in many respects the most striking deposit of this period. It covers practically the whole of Persia and extends into the adjoining regions of Mesopotamia, Syria, Asia Minor, and Armenia. The only fossils found in it are the fish *Clupea lanceolata* and *C. humilis* which apparently represent desert-lake types, the formation itself representing to a large extent the drying up of extended bodies of residual sea-water. The series reaches a thickness of 1,000 meters, and consists of an alternation of red, brown, and gray sandstones, gray marls, and marly limestones with gypsum and rock-salt. To the south-east this formation merges into a river plain deposit, the UPPER MOKRAN FORMATION of Baluchistan. This desert

13) GRABAU: *Ibid* p. 204, 205 (For the position of some of the localities see the maps there given).

undoubtedly formed one of the most effective barriers to the migration of land animals between India and the Africo-Arabian region. Beds of Pliocene age are much less in evidence in Eastern Asia, where the Pliocene sea transgressed only along the Pacific border.

In the Philippines, the Pliocene rests with a hiatus upon the Vigo group of Miocene age. The oldest division is known as the MAHIMBANG FORMATION. The molluscan element is much less prominent than it is in the underlying Miocene. A very large percentage of the mollusks which do occur, however, exists in these waters today.

A younger formation of these islands, the BANISILAN FORMATION has all its specifically identified mollusks still living, and may be of Polycene age.

9

Post-Pontian continental beds form a part of the Sivalik series of northern India, the lowest beds of which belong to the Upper Miocene. The subdivisions of this series recognized are as follows:

3. Upper Sivalik. Later Pliocene.
2. Middle Sivalik. Pontian. (Lower Pliocene)
1. Lower Sivalik. Tortonian (Iranian) (Upper Miocene).

The middle Sivalik consists of clays and shales distinguished from the lower division by the absence of bright colors. It contains the Pontian vertebrate fauna. The upper Sivalik, which probably represents the remainder of the Pliocene, consists of thick boulder conglomerates and grits and of thick earthy clays, with an abundance of bones of vertebrates. The fauna includes few extinct species, but a great abundance of living ones.

Sedimentation in the Sivalik geosyncline began in Upper Miocene time, after the second elevation of the Himalayas, which occurred in Mid-Miocene time, probably at the end of the Burdigalian or Javan. This would make the older sedimentation of the Sivalik region contemporaneous with the period of the Iranian salt deposit, formed in the great Persian desert of that time and acting as an effective barrier to westward migration.[14] The third

---

14) For further detail see GRABAU: *Summary of Cenozoic and Psychozoic* etc.

elevation, which also disturbed the Sivaliks, lifted the Himalayas to something like their present height. It probably occurred near the close of the Pliocene.

10

Returning now to North America we find both Miocene and Pliocene strata represented by marine beds on the Atlantic Coastal Plain south of 40° north Latitude and all the way to Mexico.

The Miocene beds (CHESAPEAKE FORMATION) consist mainly of sands and clays in Maryland and Virginia, but they represent only Upper Miocene. Since these rest on beds of Lower Eocene age (Pamumkey-Clayborne), it is apparent that there is a great hiatus and unrepresented interval between, which extends over the whole of the Oligocene period as well as parts of both Eocene and of Miocene. The later Miocene beds also locally overlap the older members showing normal transgressive character.

No such marked hiatus exists in the Gulf coast formations, for here Lower Miocene (ALUMN BLUFF BEDS) rest upon Oligocene beds with only the normal Interpulsation hiatus. Nevertheless there appears to be an oscillation indicated here also, unless the age determination is erroneous, for beds of Middle Miocene and early Upper Miocene age appear to be absent, while the late Upper Miocene rests on the early Middle Miocene.

The Antillean Group of Islands was in Miocene time still a portion of the long connecting Isthmus, and was folded with the strata of the Cordilleran and Andes by the westward striving of America. Since at this time the Atlantic had come into existence, and since this Isthmus faced both oceans, it may well have acquired deposits with either or both faunas whenever any part of it became submerged.

Marine Pliocene beds are found only from North Carolina southward and westward to Florida, beyond which the known Pliocene beds are non-marine, though marine beds may occur under cover of the younger deposits. They are found, however, in Mexico, especially in Tabasco, Chiapas, and Yucatan. In the Carolinas they are represented by thin fossiliferous marls (Waccamaw beds), and in Florida by similar marls of no great thickness (Nashua and Caloosahatchee beds). Above them lie non-marine later Pliocene beds. Marine conditions thus continued for a time in the Pliocene, after which the final retreat of the sea from the present land-surfaces took place".[15]

11

The standard Coast Range section of western North America comprises the following subdivision in the Neogene:

SUPERFORMATION :   PLEISTOCENE   *San Pedro*
            (or Polycene ?)

Interpulsation Hiatus and Disconformity

NEOGENE PULSATION SYSTEM
    PLIOCENE BEDS
        b. MERCED SERIES   5,830 ft
            Upper marine sandstone and shales;
            Lower marine clays, sandy shales;
               Sandstone, fine pebble conglomerate
        a. PURISOMA SERIES
               ("a" and "b" elsewhere combined
                   into Fernando Series)
    MIOCENE BEDS
        c. SAN PABLO
           (Disconformity?)
        b. MONTEREY SANDSTONES AND SHALES
        a. VAQUEROS FORMATION
INTERPULSATION *Hiatus and Disconformity*

EOGENE PULSATION SYSTEM
    OLIGOCENE San Lorenzo beds etc. (see p. 446)

Extensive continental Miocene and Pliocene deposits, as well as similar deposits of Eocene age, are preserved in the western Interior Basins of the United States and Canada, mainly west of the 100th Meridian of today. These are not only famous for the scenery carved on a

---

15) GRABAU: *Text Book of Geology*, II, p. 788.

gigantic scale by streams and wind in later geological time, but also because they contain the remains of the vertebrate life of the Tertiary in bewildering form and astonishing variety, and have been fully described not only in scientific monographs but in popular books as well.[16]

_____

16) See among others, H. F. Osborn. *The Age of Mammals.*

CHAPTER XXXIX

THE PSYCHOZOIC ERA
AND THE QUATERNARY GLACIATIONS

1

THE latest era which the earth has reached so far in its physical evolution is commonly called the Psychozoic Era, which signifies the supposed development of that mentality which, by now, claims to have assumed control of the evolutionary forces.

Although we may regard man as the dominant type of organisms in all periods of the Psychozoic, we must recognize the fact that he arose earlier, probably during the Neogene period, as will be outlined in the final chapter.

The Pyschozoic Era naturally falls into three divisions with one to several glacial periods in each. These are in descending order : [1]

PSYCHOZOIC OR QUATERNARY ERA

III. HOLOCENE PERIOD or modern period with north polar Glaciation, at 90°.

II. PLEISTOCENE PERIOD: or Period of younger glaciations mainly in Europe

I. POLYCENE PERIOD: or period of older Quaternary glaciations, primarily in North America

Hiatus and Disconformity

CENOZOIC OR TERTIARY ERA
NEOGENE PULSATION SYSTEM
    Pliocene Series
    Miocene Series

Interpulsation Hiatus and Disconformity

EOGENE PULSLTION SYSTEM

---

1) These correlations are necessarily provisional, but they serve to illustrate glaciation under polar control. The polar locations are given in terms of the modern map.

In the following table the succession of glacial and interglacial periods is shown :

Table D. Quaternary Glacial Periods

III. HOLOCENE PERIOD

XII. MODERN POLAR GLACIATION (90°)
Neolithic and Mesolithic (Azelian) Culture Stages

II. PLEISTOCENE PERIOD

XI. POST-WÜRM GLACIAL (N.P. 85° N., 5° E. 20° overlap).
G=Forestian in non-glaciated areas, Upper Palæolithic, Magdalenian, Solutrian, Aurignacian Culture Stages, Upper Cave man of Choukoutien, China.

VIII. WÜRM GLACIAL (N. P. 67° N., 25° W. 20° overlap) (Mousterian culture period) *Homo neanderthalensis,* Ordos tooth and industry.
F=*Riss-Würm Interglacial*
Lavalloisian Culture Periods 3 units (60,000 years)
Early Neanderthal, including Krapina, Ehringsdorf, etc.

Europe	*North America*
VII. RISS (SAALE) GLACIATION (Greenland : 40° W., 77° N.) 20° overlap	E. Post-Glacial Interpulsation
D=*Mindel-Riss Interglacial* (12 units — 240,000 years) Acheulian *(sens. strict.)* 1-3 ; Clactonian Culture Stage 1-3.	VI. WISCONSIN GLACIAL (Baffin Land 70° W., 65° N. 20° overlap)
V. MINDEL (ELSTER) GLACIAL (Bothnia — 23° E., 65° N., 20° overlap).	C. *Toronto Interglacial* (75,000 — ? × 3) (Includes Peorian)

I.　POLYCENE PERIOD

B. *Abbevillian* or *Chellean*)
　Interglacial Period
　(Long Post-Günzian
　Hiatus in Europe)
　Homo heidelbergensis
　(Also *Sinanthropus* may
　have continued through
　this period)

IV. LABRADORIAN GLACIAL
　　30° overlap
　IV-3. IOWAN GLACIAL
　　　(76° W., 56° N.)
　　B3 *Sangamon　Inter-
　　　glacial*
　IV-2. ILLINOIAN GLACIAL
　　　(75° W., 54° N.)
　　B2. *Yarmouth　Inter-
　　　glacial*
　IV-1. KANSAN (JERSEYAN)
　　　GLACIAL (75° W.,
　　　52° N.)
　　B1. *Aftonian　Inter-
　　　glacial*
III. KEEWATIN ICE SHEET
　　(90° W., 59° N. 30° over-
　　lap) (Includes *Nebrascan
　　Glacial*)
II. CORDILLERAN ICE SHEET
　　(130° W., 60° N. 40 over-
　　lap)

I. BAFFINIAN, GÜNZIAN OR SCANIAN ICE SHEET (60° W., 70° N.) 40° overlap
NORTH SEA OR SCANDINAVIAN GLACIATION
Contorted drift ⎰⎱ Cromer Moraine, Norwich Brick Clay
Cromer till ⎰⎱ High Plateau drift of Oxford[2]
A = *Pre-Günzian (Polycene)* Delta Deposits
　*Arctic Plant Bed (Reduction in Temperature of
　20° F). Leda (Yoldia) myalis Bed*　1-20 ft.

Cromer Forest Bed	3. Upper Fresh-water Bed	with marine
	2. Forest Bed	intercalation
	1. Lower Fresh-water Bed	0-10-60 ft.

　Weybourn crag (marine)　1-22 ft.
　Chillesford or Norfolk crag (marine)　5-15 ft.
　Norwich crag (marine) up to　150 ft.

---

2) The Oxford till shows especially well the result of the leaching during the long Günz-Mindel interglacial period, practically the greater part of the Polycene. It is virtually reduced to a gumbotil.

Inter-Pulsation Hiatus and Norwich Bone Bed.

Occupation period of Piltdown man *Eoanthropus dawsoni* (Foxhall and Cromer man). Also probably *Pithecanthropus* (Java) and *Sinanthropus* in (China.)

PLIOCENE PULSATION PERIOD

Astian or Red Crag (Oxidized and reworked)
(Foxhall Flints, etc., secondarily enclosed)

Table of Changes in the Character of the Molluscan fauna in the British "Crags" (Only characteristic and abundant species included).

Zone	Local Name	Not known living %	Living only in distant seas %	Southern %	Northern %	Types in both N. and S. Waters %
Newer Pliocene or Polycene						
Z. of *Elephas meridionalis*, Z. of *Tellina baltica*	Cromer Forest Beds Weybourn Crag	11	—	—	33	56
Z. of *Leda oblongoides* (Estuarine)	Chillesford Clay and Sand	—	—	—	—	—
Z. of *Astarte borealis* *Mactra subtruncata*	Norwich Crag	11	—	7	32	50
Interpulsation Hiatus and Bone Bed						
Z. of *Cardium groenlandicum*	Crag of Butley	13	4	13	23	47
Z. of *Mactra constricta*	Crag of Newbourne	32	5	16	11	36
Z. of *Mactra obtruncata*	Crag of Oakley Crag of Walton	36	4	20	5	35
Z. of *Neptunea contraria* Z. of *Macta triangularis*	Coralline Crag	38	4	26	1	31
Older Pliocene sens. strict.						
Z. of *Aca diluvii*	Lenham Beds Boxtone Fauna	— —	— —	— —	— —	— —

(Red Crag spans the zones from Crag of Butley through Coralline Crag.)

2

The gradually growing influence of the cold waters from the north on the North Sea Upper Pliocene fauna, continued as the Scandian-Greenland rift opened, or the dividing land bridge was submerged by the rising waters. This is shown in the change of faunas in the late Tertiary and early Quaternary deposits of the British succession as given in the table on the opposite page.[3]

The POLYCENE of the North Sea basin begins with the NORWICH CRAG of the English coast, which ranges in thickness up to 150 ft, thinning southward and eastward. It consists of sands, clays and gravels, which rest directly on the London clay (Upper Palæocene) and southward abuts against the Coralline Crag (Pliocene). The known deposits represent merely the overlapping ends of the formation on a small portion of the British coast, where they are now found between Yarmouth and Norwich. They rest disconformably upon the Norwich Interpulsation bone breccia, which includes the remains of horses (*Equus stenonis*), mastodon (*Mastodon avernensis*), and elphant, (*Elephas meridionalis*), etc, and is the only representative of the Interpulsation period. The beds overlying the bone bed represent the advent of the Polycene sea, and include impersistent shelly beds, with a fauna which is quite distinct from that of the Pliocene RED CRAG. It contains fewer extinct forms, and a larger number of Boreal types are present. Altogether about 150 species have been obtained from it. Nearly two thirds of the molluscs live in the North Sea today, while others are found in the modern Arctic waters, the most important of these being *Astarte borealis*.

The formation is apparently a delta-like deposit, and includes about 30 species of land and fresh-water shells, which were washed into it from the land.

The next succeeding formation, the CHILLESFORDIAN or Norfolk formation (Crag), rests upon the Norwich Crag, or by overlap on older beds. It consists of 5 to 15 feet of laminated micaceous clays and sands, with a marine fauna of still more northern types than the preceding.

---

3) Taken with slight rearrangement from Wright: Quaternary Ice Age, Chapter VII, p. 103-105.

The upper beds have an estuarine character.

The third division is the Weybourn Crag (WEY-BOURNIAN). It covers still less of the English coast than the preceding formation, occurring chiefly in the Cromer region on the north-east coast of Norfolk and attaining only slight thickness. The chief index fossil is *Tellina baltica*. Of 52 species of shells obtained from it 5, or 9.6 per cent are extinct, while 9 are arctic forms.

The next higher division of the Polycene preserved in the North Sea Basin is the famous CROMER FOREST BED or the CROMERIAN. In and about Cromer cliff the following members are exposed in decending order.

5. Arctic plant-bed.
4. *Yoldia (Leda) myalis* bed, marine, partly replacing the beds below.
3. Upper fresh-water bed ⎫
2. Forest bed   ⎬ Cromer Forest Bed
1. Lower fresh-water bed ⎭

This bed rests on the Weybourn marine beds and partly replaces them and is replaced by them.

The forest bed is so called because of the numerous rafted tree-trunks. Of the 56 known species of plants preserved in it, all but 2 still live in the region (Norfolk). The forest trees are mostly of types indicating a mild and moist climate. They include the maple, sloe, hawthorne, cornel, elm, birch, alder, hornbeam, hazel, oak, beech, willow, yew, pine and spruce.

Land and fresh-water shells are most characteristic of the Lower and Upper fresh-water beds; of 58 species obtained from the entire formation, 3 appear to be extinct, and 5 no longer live in Britain.

The Forest bed itself has yielded 19 species of marine shells, all of them, however, common Weybourn Crag species. This is understandable when it is recognized that these rocks represent a marginal delta, of the Huangho type, on the edge of a large continent, with its outer portion repeatedly subjected to brief inundations by the sea (marinings).

This formation has also yielded 14 species of fish, 4 amphibians, 2 reptiles, 5 birds, and 59 species of mammals, including among them Carnivores (*Machærodus, Canis*

lupis, C. vulpus, Hyæna crocuta, Ursus spelæus etc.); Artio-
dactyls (*Bison bonanus, Ovibos moschatus, Alces latifrons,*
many deer; *Hyppopotamus major, Sus scrofa*); Perrisso-
dactyls; (*Equus caballus, E. stenonis, Rhinoceros etruscus*);
Proboscidia (*Elephas antiquus, E. meridionalis, E. trogon-
therii*); rodents, insectivores, etc. "The most abundant and
conspicuous forms are the three elephants, while the
hippopotamus and rhinoceros are of common occurrence.
Of the two horses one is extinct, the bison and wild boar
have survived elsewhere, while the whole of the remark-
ably numerous species of deer have disappeared, with the
exception of the red deer, ... The carnivores embraced also living and extinct forms, for the long vanished
Machærodus haunted the same region with our still
surviving fox, otter and marten, and with other animals
which, like the hyena, wolf and glutton, though no longer
found in Britain, continued to live elsewhere. The total
species of land mammals (exclusive of bats) found in the
Forest bed is 45, while the corresponding series of the
living British fauna numbers only 29 species. Of the 30
large land mammals found in this deposit only three are
now found living in Britain, or have died out there within
the historic period, and only 6 species have survived in
any part of the world."[4]

5

That primitive man continued to live in a region so
rich in game to hunt, and within reach of the sea, which
abounded in fish and molluscan food, and withal in a
region well forested and for a long time characterized by
a mild climate,[5] is to be expected. It was a veritable
"Garden of Eden," and only the advent of the polar ice
drove him out. This is shown by the stone implements.

4) REID, C.: *Pliocene deposits of England* p. 182. Quoted by *Geikie; II.* p. 1288.
5) According to Professor Boswell, the so-called Weybourn Crag glaciation and Cromer Forest Bed warm periods are myths. He thinks the
   British climate and sea temperatures were becoming steadily colder from
   Middle Pliocene times onward. (P. G. H. BOSWELL : *Summary* Proc.
   Geol. Assoc. XLII, 2, 1931, pp. 87-111; also K. S. Sandford: *Pleistocene
   Succession in England,* Geol. Mag. LXIX, No. 811, 1932, pp. 1-18). A distinction between 'warm' and 'mild' climate seems, however, permissible,
   the latter prevailing over the land, though the influx of the cold currents
   from the north was changing the temperature of the water.

From a bed immediately underlying the forest beds the so-called "rostro-carinate stone implements similar to types found secondarily included in the upper beds of the Red Crag (Pliocene, see p. 456) have been obtained by Reid Moir along with other stone flakes, and these he regarded as having been used by primitive man, probably the Piltdown man *Eoanthropus dawsoni*.

The location of the early Polycene polar center, which we have identified with the older or Günzian (Baffin Bay) ice-sheet of Europe, is given by Köppen and Wegener as 60° W., 70° N. on the modern map. This would place it in Baffin Bay west of Greenland. With the Atlantic still only 30° wide between Morocco and Florida, and Greenland only just beginning to break away from Scandinavia, a polar ice cap of 20° radiation would cover practically the whole of Great Britain except the southern counties, and cover Scandinavia as well (See map Plate XVII). This may be therefore accepted as the location of the ice-cap which is responsible for the North Sea glaciation or the Cromer moraine and Oxford Plateau drift. It may be identified with the Günzian of the Alps or with the Baffin-Bay center of North America.

But we must remember that previously, in Neogene (Miocene and Pliocene) time, the ice-center was west of Alaska (160° W., 45° N. p. d.), and that the change from Pliocene to Polycene (Baffin or Günzian) location involves a movement of about 50° on a great circle, *i.e.* a clock-wise rotation of the sial over that distance to bring Baffin Bay over the polar site. This must have occupied a long time-period, as long as that of the Eogene to Neogene shift, unless we are prepared to consider these changes sudden and catastrophic, for which there is no evidence.

On our Middle Polycene (Cordilleran) map the South British region lay approximately 10° farther from the pole than it does today, and this implies a correspondingly higher temperature, especially as the connection between the North Sea and the Boreal Sea was not yet open. It was then that the Red Crag was formed. By Cromer-Forest time (Keewatin), the climate was essentially as today, hence we may assume that the region had approached 10° nearer to the pole, while the Greeno-Scandian rift had begun to open.

When the bed characterized by *Yoldia myalis,* and
the Arctic Plant Bed with remains of the dwarf birch
(*Betula nana*) and dwarf willow (*Salix polaris*) which overlie this were formed, the definite approach of the polar
glaciation was apparent. The indications are that at this
stage the temperature of the region was lower by 20° F.
than now, a difference as great as between the South of
England and the North Cape at the present day, and
sufficient to allow the seas to be blocked with ice during
the winter and to allow glaciers to form in the hilly
districts.[6]

The clockwise rotation of the sial, which brought
about the polar translation from the Miocene to the
Baffin Bay (Günzian) site, was translated into pressure on
the Pacific coast of America and into strain on the Atlantic
side. No doubt this contributed to the further extension
of the Atlantic rift, though it did not yet effect the junction of the North Sea with the Boreal Sea. But it also
caused the profound catastrophe in the history of south-
western Asia, which severed it from the African land-
mass, and, together with the events which succeeded and
were in part conditioned by it, gave to this part of the
earth essentially its modern physiognomy. The changes
which then took place resulted in the formation of the
Erithræan, Aden and Persian rift-valleys as part of the
stupendous system of canyon-faulting which affected the
east African and west Asiatic regions. Long blocks of
the earth's crust sank between parallel fault-lines or, what
is perhaps more probable, the sides of the rifts were
merely pulled apart. This resulted in the formation of
the Red Sea basin, and its extension to the Jordan Valley,
in the Gulf of Aden, and the Persian Gulf, and in the great
rift-valleys on the eastern past of the African land-mass.

Into these rift-valleys the waters of the Indian Ocean
entered, penetrating far into the old-land, approaching,
and even for a time joining the Mediterranean. This
was essentially along the line occupied by the earlier

---

6) REID: *Op. cit.* In Middle Polycene time during the existence of the
Keewatin Ice Cap, the south of England was about 25° from its front
(Map Pl. XIX). During the existence of the Labradorian ice-cap
(Plate XX) the distance was reduced to 15°. In Mindel time the ice
covered most of the British Isles (Pl. XXI).

Tertiary sea, but through a deeper and more restricted channel, which was virtually the basin of the Red Sea of today. But the waters along the borders rose to greater relative heights than their present level, for the region was depressed below its modern altitude. This permitted the formation of the extensive coral reefs and shell-beds which are now seen exposed on the borders and in the region of the Gulf of Suez, because of subsequent crustal movements or sinking of the sea-level.

The shifting of the crust which brought the Cordilleran region of western North America (130° W., 60° N. today) on the pole, was an eastward shifting of the sial, in modern terms. The resulting thrust transmitted to the opposite margin of the sial-crust, which was still sufficiently united to act as a unit, was essentially at right angles to the south-east Asiatic coast. It appears to have been the force that produced the strong folding of the younger Tertiary rocks in the center of the Kendeng ranges of Java, which are unconformably overlain by the Polycene and Pleistocene (Djetis, Trinil and Ngandong) beds. The moderate arching of these latter on an east-west axis may (if these beds are Polycene) be referred to the Mindel-Riss interglacial period, *i. e.* to the Wisconsin glacial of America, at the beginning of which there was another clock-wise crustal rotation which, however, brought a less pronounced pressure on Java, but one more directly on Borneo and the Phillippines, etc.[8]

8) The distance between the south Baffin Land station of the Wisconsin ice-center and the Bothnian (Baltic) center of the Mindel station was about 30° on a great circle. At the end of the Mindel period the North Pole moved relatively southward for about 30°, *i. e.* the sial crust shifted northward (eastward p. d.) to that amount, perhaps on a pivot (igneous plug (?)) located at the apex of an isosceles triangle somewhere in northeast Siberia (Anadyr), until southern Baffin Land 30° beyond the pole on the opposite side had been brought into polar location. On the Asiatic side the movement would be southward (eastward p. d.) and would press Java, Borneo and the Phillippines squarely against the sima, and produce the second folding, on an east-west axis. These two movements, at the beginning of Polycene time and during the Mindel-Riss interglacial or Wisconsin glacial, about mid-Pleistocene, were the only periods of pressure on the south coast of Asia, all the others since the Palæocene having been movements which produced a drag and rifting on the south Asiatic, but compression on the West American coast. This is in harmony with the 'mid-Pleistocene' folding of Java, which is more pronounced and complex in the west, and simpler in the east. According to *J. Duyfjes* (*Geol. und Strat. Kendenggebiets* p. 137), one after another different anticlines split from the complex bundle in the western part of the Kendeng range and disappear eastward, until finally at Sœrabaja only a single symmetrical anticline remains.

4

The Polycene glaciation of North America began with the North Pole in Baffin Bay, and this we have identified with the Günzian glacial stage of the Alps, or the North-Sea glaciation of Europe, which included the Oxford plateau drift and the Cromer till of the east coast. But the next change of location, which moved the Cordilleran site over the pole, though it moved through 70 degrees of longitude, covered only a distance of 30° on a great circle. It was this poleward movement of the Cordilleran region that, as we have noted, resulted in the later folding of the Tertiary rocks.

The Cordilleran ice-center, which was also the Cordilleran North Pole, was situated in the mountains on the borders of Yukon in latitude 60° and longitude 130° of today. A 20° ice radiation from this center covered northwest North America east to the 90th degree of longitude, but only the north-western corner of the United States, as far as latitude 40° on the coast. Then the pole advanced relatively eastward, by the westward drift of the crust, until in Keewatin time the pole emerged west of Hudson Bay, and America's Arctic winter was well under way. Europe meanwhile still enjoyed equable climates and did not suspect the coming invasion, though the pole approached nearer. By the time Quebec was ice covered (Labradorean ice-sheet) Europe had some chilling experiences. England was 10 degrees nearer the pole than now, yet Greenland was probably unglaciated, and migration of Arctic types was probably still possible along the ice-margin.

In the Alpine foreland, as in northern Europe, there can be detected everywhere two belts of moraines, an outer or older belt and an inner or younger belt. They show striking contrasts in surface features. The older moraine shows a mature surface drainage and is everywhere covered by loess or loam, which is completely wanting from the surface of the younger moraine, which moreover retains its original character of form without change.

5

The more detailed classification of Penck and Brückner recognized four out-wash gravels of which the oldest is the highest, though only partly preserved. Each of the four can be traced into a moraine, so that four distinct glaciations can be recognized, as follows:—(the numbering is in the order of age).

Glacial stage   Out-wash gravels
4. Würm Glaciation which
   formed the.....   4th. Lower Terrace gravels
                      3. Interglacial Loess deposit
3. Riss Glaciation which
   formed the.....   3rd. Higher Terrace Loess
                      2. Interglacial Loess deposit
2. Mindel Glaciation which
   formed the.....   2nd. Younger or Munich
                                  blanket gravel;
                      1. Interglacial Loess deposit (?)
1. Günz[8] Glaciation which
   formed the.....   1st. Older Blanket gravel[9]

Zones of weathering also mark the surfaces of the older outwash gravels, while the loess, lies upon the weathered surface. The Würm moraines and gravels are free from erosion or weathering.

The Munich gravel (No. 2) is far more intensely weathered than is gravel No. 3, from which fact Penck deduced that the interglacial period between the Mindel and Riss must have been much longer than, even 4 times as long as, that between the Riss and the Würm, while on a like basis he concludes that the Riss-Würm interglacial periods must have been considerably longer than the whole of post-glacial time since the Würm Ice Age.

The proportional lengths of the interglacial periods as deduced by Penck are:
4. Post-Würm — Post glacial period 1 unit
3. Riss-Würm Interglacial period 3 units
2. Mindel-Riss   ,,   ,,   12 units
1. Günz-Mindel   ,,   ,,   3 units (more probably n
                                          $\times$ 3 where n is
                                          4 or more).

I would suggest further that the Mindel-Riss interglacial period corresponds to the Wisconsin glacial

8) This is here identified as belonging nearer the base of the Polycene, which would make the first interglacial period much longer.
9) German term "Deckenschotter".

period in America, while the preceding Toronto interglacial period was represented in Europe by the Mindel
glacial period. On the other hand, there is a possibility
that the Günz glaciation of the Alps and perhaps an early
Skanian glaciation may belong to the base of the Polycene
and to an ice-sheet which Köppen and Wegener Place
between Greenland and Baffin Land. That would mean,
of course, that the Günz-Mindel Interpulsation Period is
much longer, and corresponds to the greater part of the
Polycene.

If I am correct in regarding the Toronto interglacial
as essentially equivalent to the Mindel glacial of Europe [10],
and the Mindel-Riss interglacial is represented in America
by the Wisconsin glaciations, then the Pleistocene must be
made to begin with the Toronto interglacial in America
and the Mindel glacial period in Europe. After the Wisconsin came to an end, glacial conditions in America were
confined to the northern regions of Canada.

6

The insignificant thickness of the fossiliferous later
Tertiary and Polycene beds of the North Sea Basin, which
had caused the periods they represent to be considered of
such slight value, is primarily due to non-deposition,
because of land conditions, as is also the absence of the

---

10) In discussing the question of the significance of the first European glacial
deposits, the Günz needs to be considered. Was it the first of the
Pleistocene (sens. strict.) glaciation of Europe or did it belong to the
Polycene? If, as I believe, it belongs to the latter, it must be regarded
as earliest Polycene, for only at the beginning of that period was the
pole favorably situated for European glaciation. This was the Baffin Bay
center, the first post-Pliocene location, given by Köppen and Wegener (Long.
60° W., Lat. 70°). This, with the Atlantic narrowed to 30 degrees between
Florida and Africa (40 degrees overlap) and the North Atlantic rift un-
opened as in Neogene time, would, with 20° radiation, include Scandinavia,
the British Isles (except the southern counties), parts of Newfoundland,
Labrador and, of course, Arctic America and Greenland. A few degrees
extension and it would cover Germany, while the influence of the cold
would probably cause the Günz glaciation of the Alps.

But this interpretation would require that the Günz-Mindel interglacial period should be greatly lengthened, to correspond, in fact, to the
whole of the Polycene with the early American glaciations. Perhaps a
re-examination of the evidence for the length of this interglacial period
in the light of this reasoning will lead to a conclusion which favours this
extension.

Miocene beds in Britain. Another factor towards the end, was the passage of this region into the control of the Polycene North Pole. Though it passed out of this control and remained free during the remainder of the Polycene (measured by the whole time of the Cordilleran, Keewatin and Labradoran ice ages) Britain shows no record of this, for the succeeding periods of glaciation in the Pleistocene, all of which involved the North Sea Basin, obliterated all these older records. An incursion of the cold waters from the Arctic sea across the slightly submerged Barents Shelf occurred, as we have noted above, at the beginning of Polycene time before the advent of the Günz Glaciation, but the actual rift probably did not take place until some time later in Polycene time. Barents Shelf today has an average submergence of about 200 meters, but there are local depressions up to 400 meters, which probably represent submerged river valleys. West of a line connecting west Spitzbergen and Bear Island with the Scandinavian coast (approximately Long. 20° E. of today), the shelf drops off abruptly, — within a short distance to 3,200 meters —, and this marks the line of final rift between the American continent (of which Greenland is a part) and the European coast.

The folds in the Tertiary and older rocks of Spitzbergen are probably an echo of the pre-Miocene or Alpine folding, but the many north-south faults which characterize the whole island of Spitzbergen and Bear Island as well, lie parallel to the main rift and most probably represent the continued vibrations of the great paroxism which finally tore the continents apart after the beginning of the Pleistocene period, when the Scandinavian countries moved into polar location, and Greenland and America were pushed south on the opposite side of the pole. This was also marked by the great volcanism which built up the still active creaters of Iceland over the rift.

7

To summarize: The thesis which I have previously advanced[11] implies that Quaternary glaciation was not a unit, that, in other words, European glaciation (Mindel

11) GRABAU : Ice Ages or Polar Glaciation.

and later) did not begin until the Labrador ice-sheet vanished. This means that at the end of Polycene or Lower Quaternary time the pole or center of ice radiation migrated from east of Hudson Bay to the Bothnian Gulf, while the Atlantic was widened until the overlap was 20°. European glaciation began after the main early glaciation of America had come to an end, *i.e.* in Mid-Quaternary or Pleistocene time. While the Scandinavian region was the center of glaciation, America passed out of the ice embrace and we had a return of spring in the Ontario interglacial period, when a climate comparable to that of Cincinnati of today favoured Toronto with the mildest days it had probably known since Eocene time.

But our trials were not ended. Further arctic experiences lay in store for eastern North America, when the ice-center returned to Baffinland, and the Wisconsin glaciers held the land in their freezing grip. This meant another poleward shift of the sial crust, to the extent of 30° on a large circle, while on the opposite side the equator was moved 30° south, and the edge of the sial, submerged in the sima for seven-eights of its thickness, suffered compression and folding, as previously noted.

Europe, however, had a respite, for during the Wisconsin glaciation of America, Europe was in the main ice-free; its long mid-Pleistocene interglacial period was upon the land.

But, though the world of early Pleistocene animals may have rejoiced at the retreat of the invader, and though the Peking Man *Sinanthropus pekinensis* had learned by this time to stand erect and justify his destiny as the ancestor of *Homo*, they and he had more to learn.

For it appeared that the rejoicing was premature. Destiny in the form of the North Pole readvanced (relatively). The Riss glaciation and, later, the Würm came upon the land, when the pole migrated relatively to Greenland, which was still 20 degrees closer to Europe than it is today. From there it could send its destroying glacier over Europe, which once more experienced the baneful climate of a long Arctic period. Even distant Asia felt its influence, for though the equator reached the coast at the Yangtze Mouths, while indeed, as Black has pointed out, all parts of the east coast of Asia from

Kamtchatka south experienced milder climates, this was not an unmixed good. Anadyr Peninsula, the farthest north on the coast of Siberia, lay some 20° farther south. So did Peking. But the strong contrast which this made with the cold of the Würm glaciers accentuated the force of the cold winds that commenced to blow with great velocity from the area of high pressure in the west across the plains of the interior, where rock disintegration had been proceeding probably at a rapid rate. Passing across Sinkiang and the Tarim basin, these winds picked up the finer dust particles and carried them to East China, where, in the moister climate, they settled as loess, while the pure quartz sand remained behind in central China to form the desert dunes. It was then that Peking Man took to the caves, for living in the open was not agreeable. Not until the disappearance of the Greenland icesheet from North Europe, with the final relative retreat of the pole to its present location, did Europe again offer him a suitable habitat such as had favoured the migration to western Europe and America of the Piltdown man and the other older stone age men in Polycene time. But by then all chances for easy migration to America had come to an end.

The final change to its present location was not without effect on America. We have seen that the Atlantic progressively widened, and this rift extended north until it separated Barents Shelf with Spitzbergen and Scandinavia, from Greenland. As the rift deepened and widened, the Iceland volcanoes developed in and over it, and for a while formed a fiery link that attempted to heal the breach, but only succeeded in forming a temporary bridge, on which the ice could cross the Gulf. Today it lies 20 degrees distant from Scandinavia; but a submarine ridge, not falling as a rule below 500 fathoms, still connects it with Great Britain on the one hand and Greenland on the other.

9

Throughout these periods South America remained the faithful adherent of Antarctica, as is shown by the structure and stratigraphy of the west coast of Argen-

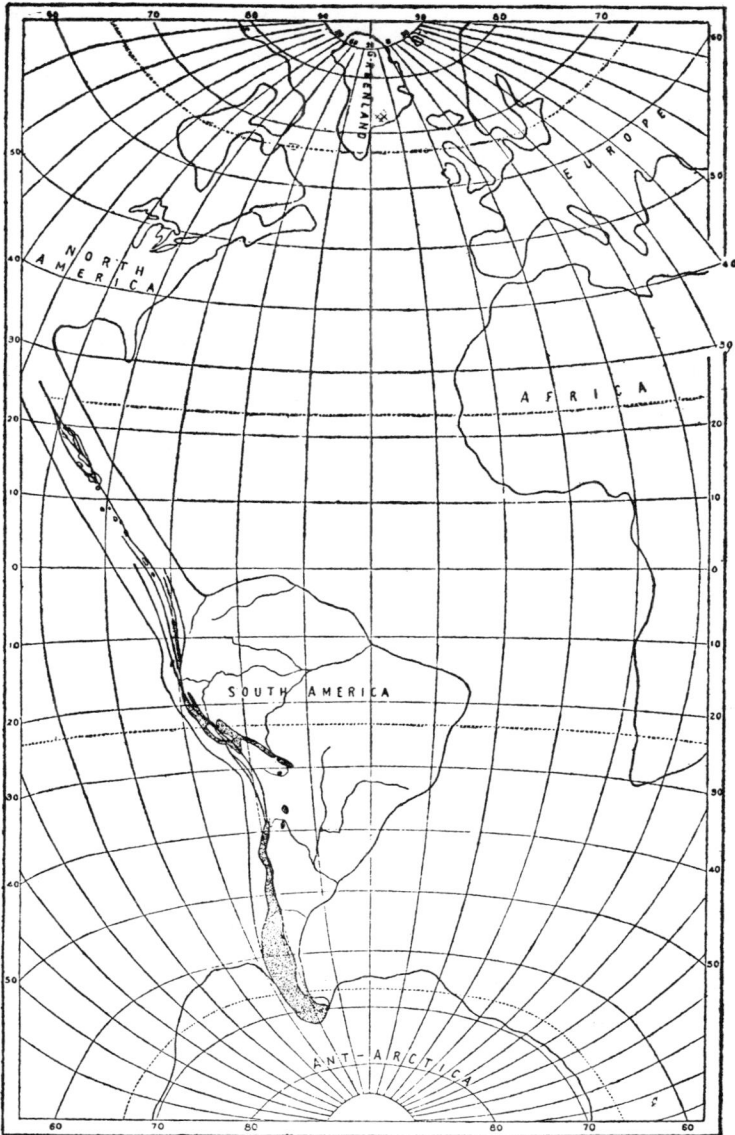

Fig. 117. Map of early Holocene South America in contact with Antarctica showing area of glaciation (stippled). Central America connects South and North America as a continuous, though folded, land-mass.

tine and of Graham land. Though the South Pole was
near, it was not close enough to account for the late
Quaternary glaciation of Patagonia, until the North Pole
assumed a position in Latitude 85° N., Long. 5° East,
which meant 165° East Long. for the South Pole. Even
then it required a 30° ice radiation to cover the area
involved (See Fig. 117).

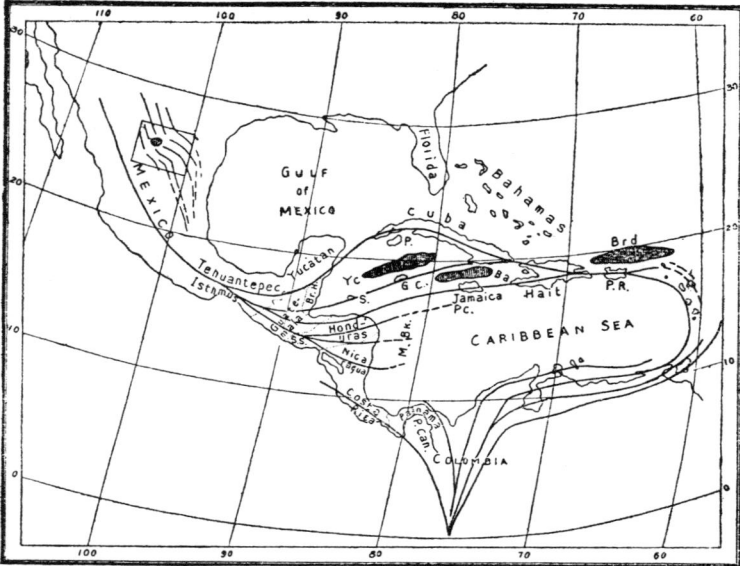

Fig. 118. The hair pin loop formed by the Antillian or West Indian
Islands today as the result of collapse of the orignial Central America. The
small rectangle in Mexico indicates location of map Fig. 119.

But the most significant part is the Isthmus, which
connected the two Americas since the earliest time (Fig.
117), when it was a broad convexly westward bending
old-land of crystalline rocks covered by marine and continental Mesozoic and Tertiary strata, with Palæozoic
marine beds in the region of the Caribbean epi-seas.

During the westward drift South America remained
joined to Antarctica, but its free drifting margin was folded
into the Andean Cordilleras. Because of the greater
drifting freedom of North America, South America lagged
behind, until there was 30° or more difference in longitude, which was expressed by the oblique direction of

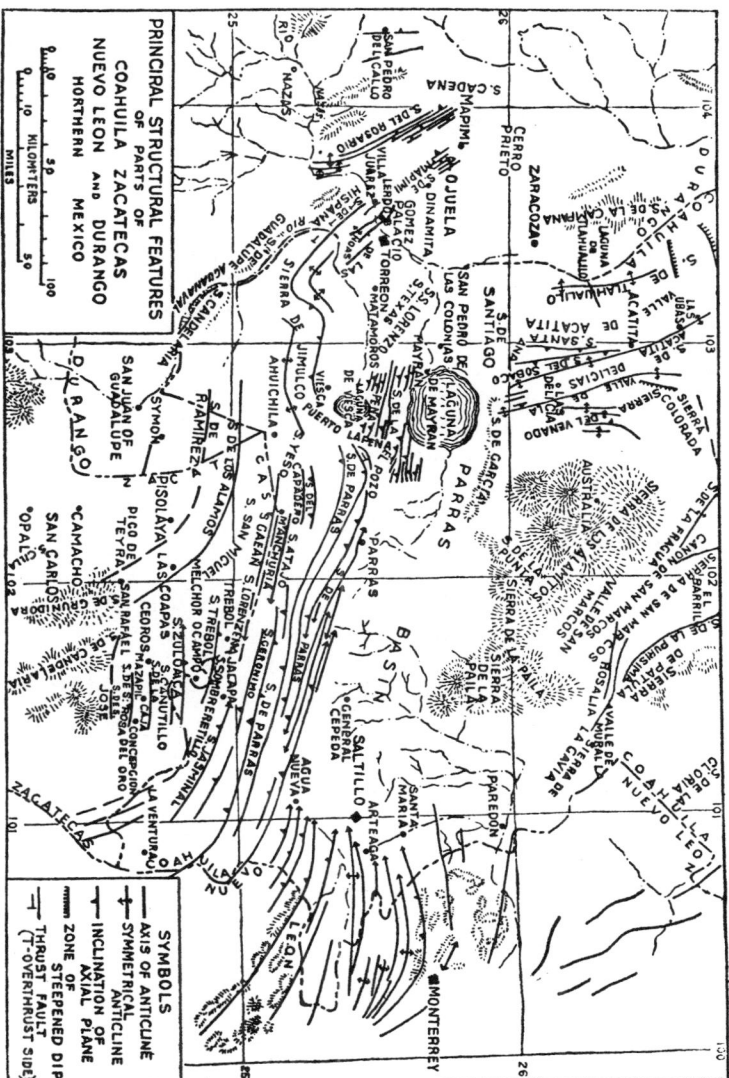

Fig. 119. Details of the folded Ranges of a part of northern Mexico formed during the Antillean collapse. (After Imlay; G. S. A. Bull. 49).

the Isthmus (by then folded into a series of parallel folds),
and its forward curving, as opposed to its original outline
and trend when a part of the Pangæa.

Finally at the beginning of Holocene time, South
America was torn violently from the grip of Antarctica,
leaving the gap of Weddell Sea and the scattered fragments of its mass which are seen in South Georgia, the
South Orkneys and the Falklands.   But divorced from its
steadying companion-continent Antarctica, South America
now became the prey of the centrifugal forces of the
rotating globe, and "Pol-flucht" escape from polar captivity
resulted.   This caused the crumpling of the Isthmus into
the remarkable 'hair-pin' or 'tuning-fork loop' which it
now presents, with the virgations at either end recalling
those of a split bamboo rod or a bundle of wire strands,
which separate at the point of most profound bending.
(Fig. 118).   The remarkable eastwest directions of the
Mexican ranges between the parallels of 24° and 26°,
which is so anomalous when compared with the general
north-west, southeast direction (Fig. 119) must also be
regarded as a displacement of the ranges during this collapse
of the Isthmus.

The northward pressure of these moving masses may
have been responsible for the remarkable circular lagoonas
of the Parras Basin, as it certainly was for the east-west
depressions or deeps of the Caribbean which were formed
in front of the moving fragments of the loop.   This mass
movement was thus responsible for all the deformations
known in the Antillean region, and the gashes torn in the
original anchorage permitted the volcanic activities for which
the region is famous, and the earthquake shocks, which
show that the vibrations of the prongs of the tuning fork
have not yet wholly ceased.   For this is the latest major
event in the geological history of America.

CHAPTER XL

THE COMING OF MAN [1]

1

WE must now go back a little way, to the Tertiary
period, in order that we may be able to trace a
connected outline of the origin and early history of man.
For although his development belongs essentially to the
Quaternary or Psychozoic era of the earth's history, his
roots lie deeply buried in the Tertiary rocks, where we
must look for his beginning.

We have in an earlier chapter described the major
physical events that separated the older from the newer
Tertiary. This was the rising of the Himalayas and the
Alps, events which followed the Oligocene and preceded
the opening of Miocene time, and which, therefore, occupied
a mid-Tertiary Interpulsation Period of long duration. This
was a time of supreme orogeny in the old world, though
in America it passed almost unnoticed.

Before the Himalaya Mountains arose, the region where
they now stand was one of subsidence and sedimentation.
In other words, it was a geosyncline, in which were
deposited the sediments which were derived from an old
upland that lay to the south, that is, in the region where
today lies the great Indo-Gangetic plain of North India.
While this geosyncline was occupied by an arm of the sea
at frequent intervals during the early periods, it had become
completely drained of its waters by the end of the Eocene
period. At that time the region formed a broad level
plain, not unlike the great China plain of today. Let
us try to visualize such a plain, broad, level, flat, with
rivers lazily meandering through it towards their distant
outlets to the sea. On the south, the low upland rises, more
or less extensively wooded and swept by moisture-bearing

1) The greater part of this chapter appeared in 1930 in the China Journal,
   Vol. XII under the title *Asia and the Evolution of Man.*

winds, which came from the oceans on the south. Though we have no direct evidence to indicate the character of the plain which at that time preceded the present mountain systems, we can, nevertheless, infer that it probably was more or less strongly wooded. The same thing was probably true of the low plain, which was the Tibet of that period, a marginal plain on the north of the Himalayan geosyncline. That none of these regions had characteristics which would form a barrier to the animals living at that period, is shown by the fact that the largest of these, the hornless rhinoceros, *Baluchitherium* (Fig. 116 p. 442), wandered freely across the whole of Asia in Oligocene time, for its bones have been found in Baluchistan, a region now shut off from other parts of Asia by the existence of the Himalayan Mountain barrier. Had those mountains existed when *Baluchitherium* was alive, that animal would have been confined to southern Asia, for it was totally unfitted to cross mountain barriers of even moderate elevation. But *Baluchitherium* was not confined to southern Asia. It roamed far and wide across the whole of Pal-Asia, which is the name given to the predecessor of modern Asia in the Tertiary and Mesozoic eras. Its bones have been found on the Tobal River in western Siberia, and in the sandstones of Mongolia, and even in some parts of western China. It was the giant beast of the Oligocene forests and plains, and as it browsed on leaves and shoots of trees, the region over which it wandered must, to some extent at least, have been forest-covered. That this animal did not enter Europe was due to the fact that Pal-Asia was separated from Pal-Europe by an impassable marine barrier, the ancient straits of Turgai.

2

Among the contemporaries of the *Baluchitherium* were the early Tertiary anthropid apes, those highly specialized sub-human types which were destined to become the ancestors of man. Like their modern representatives, the giant anthropoids, members of a collateral evolutionary line, they, too, were forest dwellers and more at home in trees than on the ground. And these ancient forests served not only to shelter them and provide them with dwelling places, but also gave them food and provided them with every necessity of life. It needs little imagination to picture these animals as contentedly living in the forests of that day, satisfied to find their food and rear their families with little or no interference from others of their kind, since food no doubt was plentiful, and with little danger from other forest dwellers, since their dwellings were above the ground among the branches of the trees. These branches also served them in their wanderings, for, while the forest trees grew close together, they could easily swing from branch to branch by the use of their powerful arms, and with the aid of their grasping feet they rarely found it necessary to descend to the ground, where progress of necessity was slow for most of them, for they were poor pedestrians.

Into the midst of this wide-flung garden of a happy Eden, entered the disharmonious forces which produced the great Himalayan range by the folding of the strata that underlay the great plains which formed the surface of the Himalayan geosyncline. Slowly they rose under the irresistible forces of compression, which brought the north and south into closer juxtaposition, bent and crumpled the strata as they rose, broke and forced them to slide across one another, and in this process sent shock after shock through the trembling Earth, to the terror and dismay of its inhabitants, who had not before experienced the effects of such great earth disturbances. And when the movement at last slowed down, there stood the newborn Himalayan Range, transsecting the former garden that was Pal-Asia and dividing the happy hunting ground of the baluchithere and the anthropoids into a northern portion that comprised Tibet and China proper, with Mongolia and Siberia on the north, while to the south, the Indian peninsula remained, not much the worse for the great change that separated it from the rest of Asia. With this exception, that now the ancient upland, which had furnished the sediments which were folded into the Himalayan Mountains, had sunk, to become in turn the site of the deposition of sediments brought by the new rivers from the recently formed Himalayan Mountains on the north. It had become, in other words, the Sivalik geosyncline of the later Tertiaries, and the Indo-Gangetic geosyncline of the present day.

Baluchitherium did not survive this change, but the anthropoids which remained in India, found little reason to alter their mode of life. Indeed, there was every incentive to continue their arboreal existence, since the conditions for the growth of forests became more favourable over this region than they had ever been before. This was due to the fact that the moisture-bearing winds, sweeping in from the warm southern seas, met the new-formed barrier of the Himalayas, which forced them to rise, and in consequence to deposit their moisture. Thus the southern border of the Himalayas became a region of intensive precipitation, and this in turn produced a luxuriant forest growth on the plains that served as a home for the descendants of the survivors of the great partition, and confirmed them in their ancient mode of life.

But it was different on the north of the great mountain barrier, and the forest dwellers there soon found reason to bewail the fate that cut them off from communication with their more happily placed brethren on the south. The same winds that brought the moisture to the southern slopes, so that forests grew in luxuriance, became, on crossing the Himalayas, winds greedy for moisture, with drawing it from the lands over which they blew. This is the effect which mountain barriers everywhere have upon winds. The windward side is always the side of heavy precipitation. As the winds are forced to rise into the upper cooler regions, they become condensed and cannot hold the moisture which they held before. But on crossing the barrrier these winds descend, they expand and become warmed and thereby their capacity for moisture is increased. They are like the winter winds that blow across North China, drawing up the moisture from the ground and setting the dust devils dancing.

Tibet, a former region of equable climate, began to dry up when, in Miocene time, it found itself north of the new born Himalayas. It did not then stand as high as it does today, for this final elevation is of later date. But it suffered then, as now, from the effect of the drying winds. The first result of the desiccation was the sinking of the water table until most trees found it difficult to gain their requisite supply of ground water, because their roots did not reach deeply enough. Then the trees began to die and the forests dwindled, and the forest dwellers witnessed the destruction of their forest-homes and the diminution of their food supply. As the sheltering trees disappeared, these animals were forced more and more into the open, where many strange and new conditions confronted them. Instead of an arboreal life, where hands and feet alike were used in swinging from branch to branch, they were compelled to walk, and their feet, formerly adapted for grasping, underwent slow modifications. Walking like other quadrupeds was not for them, for their arms had long become accustomed to other uses, and so an upright bi-pedal locomotion was forced on these creatures, the difficulty of which they mitigated by supporting themselves by the use of broken branches which were plentifully strewn about on the edge of the dying forest. Thus came into use the first walking-sticks.

No doubt all the weaker individuals, those with less ability to withstand the difficulties of their new environment and to adapt themselves to it, died. They were weeded out because they were the less responsive ones to the call for the development of latent powers. This was natural selection, and, after the weaklings had been destroyed, the survivors were able to go forward and make the most of the new conditions which confronted them.

3

It is a well-known fact that dry climates are also regions of great temperature changes, for in the rarefied air, free from much water vapour, the accumulated heat of the day will radiate and the nights will be intensely cold.

We can picture these creatures, descendents of ancestors which were accustomed to an equable climate, protected by the forests, now huddling together for warmth, and suffering intensely from the temperature changes. Once again the weaker were eliminated, the stronger seeking shelter under over-hanging rock ledges or in caverns, if such were to be found, until the great discovery which made living in the open a possibility.

In the happy days when these pre-humans still lived in the forests, their greatest terror was the forest-fire, for forest-fires not infrequently resulted when trees were struck by lightning. But when the forest ceased to serve as their home, such fires were no longer of immediate concern to them. Indeed, the blazing dead trees formed a spectacle which these creatures could contemplate with equanimity from a distance, and which might even tempt the bolder to investigate at closer hand. And, inevitably, on approaching the burning tree the grateful warmth of the fire would at once strike the shivering creature, and this discovery would lead the shrewder to view the phenomenon of fire from a new angle. Unable to induce his shivering but more timid mate and off-spring to approach, the boldest of the group would seize a burning stick and carry it to where the huddled family sat terrified. At once they, too, would feel the grateful warmth, and note the relative harmlessness of the individual brand, and they would hasten to throw more sticks together to keep the blaze from dying down. In some such manner came into existence the first camp-fire. Prometheus was unbound. Primitive man had made a servant of a former enemy. Round the camp-fire, now jealously kept burning, gathered the clan. Outsiders were unwelcome, as the radius of its influence was limited, and, for further protection against outsiders, the clan was organized and leaders selected, and this was the beginning of primitive communities.

In their daily forage for their ever-decreasing food supply, the males of this pre-human clan roamed far and wide across the desolate plains and plateaux. Each bore a stout walking stick, which on occasion could be used as a war club, when males of other clans contended for the food. For necessity compelled these creatures to become enemies and rivals for the scanty food supply, where formerly, when food was plentiful, their ancestors lived in amicable relationship. Since that day, rivalry for the goods that the earth supplies has remained the chief cause of war among clans, communities and nations.

4

Growing aridity had other effects upon the land than the destruction of the vegetation. The soil, no longer

bound together by the roots of plants, became the sport
of the winds, which season by season increased in violence.
From every exposed place the soil was blown away and
the underlying rock was uncovered.   Diurnal changes of
temperature, especially in desert regions, effect the shattering of the rock, and gradually the solid ledges became
surfaces of broken fragments.   Across such surfaces plodded
the weary seekers after food, leaning heavily upon their
primitive walking sticks.   These from much use, had
begun to split, and it happened that occasionally a broken
rock fragment became wedged into the split end of the
stick.   Now enemies hove in sight, and the wary plodders
prepared to give them battle.   Vigorously they used their
sticks to strike and thrust, and behold, again and again,
such thrusts drew blood, to the amazement of both at-
tackers and attacked.   And when at last the enemy was
routed, examination of the blood-stained weapon disclosed
the chip of flint wedged into the bottom of the split
stick.   A new weapon had been discovered and *æoliths*,
the primitive natural stone chips, came into use.   And
soon the primitive warrior learned to modify these fragments by chipping, that they might better be secured to
the sticks which now had gained new values.

By this time, the increasing rigour of the climate
had, no doubt, caused these proto-humans to wander
northward across the inhospitable Tibetan plateau, where
a broad fertile valley had come into existence during the
period of stress which formed the Himalayan barrier.
This was the modern Tarim Basin in Sinkiang Province,
a desert today, but in late Tertiary and early Quaternary time a region of abundant streams, as is shown by
the deposits of stratified clay which underlie the modern
desert sands.   And here late Tertiary man found a new
and more congenial home, and when these strata are
explored, no doubt his bones will be found preserved in
many an ancient layer of clay.

It was from this new found home in Western China
that early man set out upon his wanderings.   Eastward
he came along the route which before him many a Pliocene
animal had followed, and which in much later days
formed the caravan route of man in the commercial age.
And along with many animals he reached East China

early in Polycene time, as we know from the fact that
his skull and other skeletal remains were preserved in
the Choukoutien cave, twenty-five miles south-west of
Peking. (Fig. 120 B)

Westward, too, he migrated, across the Eurasiatic
continent, and, as the North Sea did not then exist, he

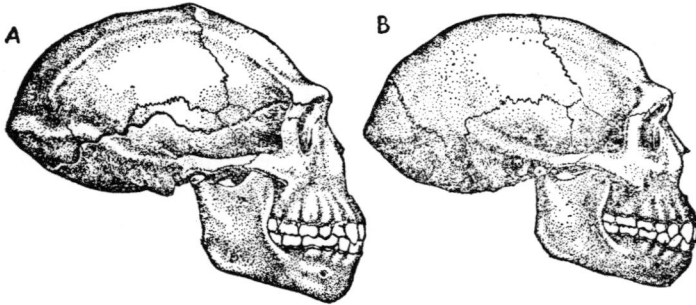

Fig 120. A. *Pithecanthropus erectus* du Bois. The skull of the 'Java
Ape Man' (restored) B. *Sinanthropus pekinensis* Black and Zdansky. The skull
of the 'Peking Man'; side view (restored) (From Romer)

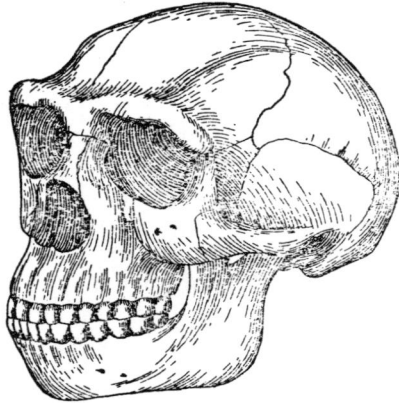

Fig 121. Reconstruction of entire skull (adult Female) of *Sinanthropus
pekinensis* Bl. & Zd. by Prof. Franz Weidenreich in collaboration with Lucile
Swan.

came to what today is England, and there his remains
have been found in Piltdown in Sussex. *Eoanthropus
dawsoni*, the Piltdown man of England, was a contemporary and close relative of *Sinanthropus pekinensis*

(Figs. 120-122), the Peking man of North China.   Both
were descendants of more primitive ancestors, who, it
appears, first settled in the Tarim Basin.   Perhaps the Java
ape-man, *Pithecanthropus erectus* (Fig. 120 A), was also
a descendant of these ancestors of Peking man and Piltdown man, but, if so, he early wandered east and south,
and in southern Cathaysia found physical conditions so
easy, and requiring so little exertion to maintain a live-
lihood, that there was no inducement to endeavour, and
hence no stimulant to further development.   So *Pithe-*

Fig 122. Restoration of the head of *Sinanthropus pekinensis* Bl. & Zd.
by Lucile Swan under the direction of Prof. Franz Weidenreich.

canthropus, though the contemporary of the Piltdown and
the Peking men, remained more primitive than they,
because their environment had the stimulating effect
which results in progressive evolution at an accelerated rate.

Of course, it is also possible that *Pithecanthropus* represents an independent evolution from simian ancestors
that remained on the south of the great Himalayan
barrier, and that evolution toward a human type was
slow because the stimulus of the environment was slight.
In any event *Pithecanthropus* can probably not be regarded
as in the direct line of ancestry of the modern human race.

But early man was not destined to continue his evolution unhindered on the Asiatic continent, for great changes were in progress in the physical evolution of the earth, and these made themselves felt at the opening of the Middle Quaternary or Pleistocene period.

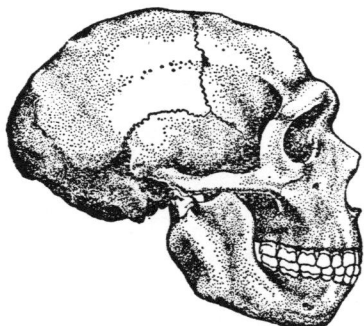

Fig. 123. *Homo neanderthalensis.* The skull of the extinct Neanderthal man restored (From Romer)

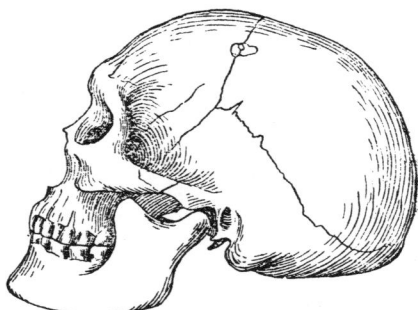

Fig. 124. *Homo sapiens.* Skull of an old male, Pre-mongoloid type, or type comparable with the late Palæolithic races of Western Europe. Upper Cave, Choukoutien. Associated with skull shown in Figs. 125 and 126. (After Weidenreich).

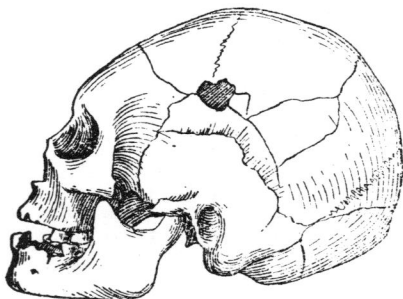

Fig. 125. *Homo sapiens?* Skull of relatively young female, Melanesoid type. Upper cave, Choukoutien. (Associated with skulls shown in Figs. 124 and 126). (After Weidenreich).

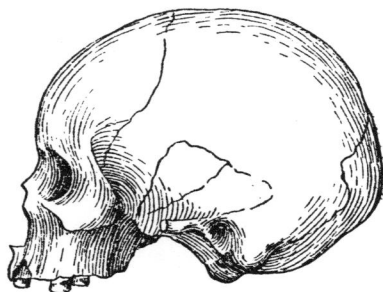

Fig. 126. *Homo sapiens.* Skull of relatively young female. Eskimoid type; Upper cave Choukoutien. (Associated with skulls shown in Figs. 124 and 125). (After Weidenreich).

This was the great crustal shifting that brought the Skandinavian (Bothnian) region over the North Pole of the earth and allowed the formation of the Bothnian or Mindel glacier which radiated over most of Europe and drove Pleistocene man into the ice-free regions of the south.

5

Pleistocene man is known to us as the man of the "Old Stone Age," that is, Palæolithic man. He had learned to make more suitable stone implements by chipping the natural stone, or what is technically called dressing them, and slowly he developed the art of making highly effective implements of various kinds. He may have been driven to this by finding, when he had wandered northward into the Tarim depression, that the stones which the rivers brought from the Kwenlun Mountains were all well rounded, as such stream pebbles usually are. An accidental fracture, when such a pebble was thrown against a larger boulder, would have given him the idea of chipping out the desired implements, and after much trial he would probably learn to pick out those stones that by chipping and by flaking made the best tools.

The cultural stages of Palæolithic man are classified, on the basis of the implements he made, into a number of divisions, the names of which are taken from localities in southern Europe where these stages were first found. (Chapter XXXIX Table p. 476-477). Of the men themselves we know almost as much as we know of their cultures, since their skeletons have been found in great numbers in many localities of southern Europe. There are two outstanding types, Neanderthal man, who lived in the early Palæolithic, and Cro-magnon man, who lived in the later Palæolithic along with other types (Grimaldi, Brünn, Chancelade and Furfoss-Grenelle). Neanderthal man was distinct enough from modern man to be designated *Homo neanderthalensis*, but Cro-magnon man was, like ourselves, a *Homo sapiens*.

The abundance of actual remains of Neanderthal man in the caves and ancient burial grounds of Europe is in striking contrast with the relative scarcity of his remains in eastern Asia.[2] It is not that Palæolithic man ceased

2) I am indebted to my colleage and former student, Dr. W. C. Pei for the details regarding Palæolithic man in China.

to come to the east, but that here he was compelled to struggle in a dusty, arid and cold climate. It is now recognized that migration from central Asia into China continued in Pleistocene time, as it did to Europe, though on a considerably reduced scale.

6

The presence of Pleistocene man in China was discovered by Pères Licent and Teilhard de Chardin, who investigated middle Palæolithic sites in the Ordos. Besides a rich fauna, contemporary with the Loess of North China, these Ordos sites have yielded a great quantity of stone implements. The artifacts, chiefly of quartzite pebbles, were studied by Abbé H. Breuil, who thinks that the Ordos industries correspond to the European Mousterian and Aurignacian epochs so far as the types of implements are concerned. However, a correlation of Palæolithic stages in Europe and in Asia can hardly be made on the sole basis of culture, since there is a wide difference in the material available for tool-making in the two distinct areas which were occupied by man, though he came from the same center of origin.

Unfortunately the Ordos sites did not furnish any other human remains than a single isolated tooth, from which naturally only a limited conclusion could be drawn regarding the degree of relationship of Pleistocene man in Asia to either Neanderthal or Cro-magnon man of Europe. But that the latter were descendants of Asiatic ancestors is shown by the history of development of ancient man.

Licent and Teilhard and, later Teilhard and Young, have also located a considerable number of Palæolithic sites at the base of the loess or in its lower part, along the Yellow River, but so far these have not been thoroughly studied. Some questionable occurrences have also been found by Nelson in Mongolia.

All these represent the Mousterian culture stage of Europe, though this does not necessarily imply close relationship of these human types. So far as our knowledge goes there is no evidence that the culture stages, which have been adduced purely from the artifacts, and the development of which chiefly depended upon the local supply of material available, are contemporaneous in these two continents, though their geological ages may be the same.

Recent discoveries in China have, however, thrown much light on the presence of man here during later periods. These discoveries were made in the 'Upper Cave' of the Choukoutien Hill by Dr. W. C. Pei, and excavated systematically under his supervision.

The old solution caverns in the Choukoutien limestone hill not only provided shelter for Polycene hominids but also for later Pleistocene man, who did not essentially differ from existing *Homo sapiens*. Of course these two types of ancient man, though found in the same caverns, were not contemporaries; when the man of the upper cave took up his residence there the remains of *Sinanthropus* had already become mineralised.

The remains of the Upper Cave man, according to the investigations of Prof. Weidenreich, prove to be a mixture of human races. This eminent anthropologist determined that the Upper Choukoutien Cave was at one time the home of a family of seven members: four adults, one adolescent and two children. Associated with them was an abundance of animal remains.

Interesting observations were made by Weidenreich on the physical appearance of the members of this family. Only three skulls of fossil man in the Upper Cave are available for such a study. One of them was an old male, resembling some of the European skulls of the late Palæolithic time, or essentially a primitive Mongoloid. Another skull was that of a female, and younger than the first one. It bears a close resemblance to members of the modern Melanesian race, as well as to the pre-historic (Lower Neolithic and Mesolithic) races of Indo-China. The third skull of another young female, shows many peculiarities that are frequently seen in the modern Eskimo skulls.

How did these representatives of three different types of the human race come together in the Upper Cave? Weidenreich deprecates the opinion that they were the victims of hostility between different tribes, buried together in the same spot.

In Europe such mixtures of human races have also been found. In the well-known late Palæolithic site of Oberkassel in Germany, two distinct races of humans were found buried at the same place. Certainly the assumption that a mingling of distinct human races took place only in modern times is incorrect. Inter-marriage among races occurred probably at least as early as Pleistocene time.

The characteristics of the human types, found mingled in the 'Upper Cave', are likewise found among the North American Indians. Migration of the human races from Asia to North America, via Bering Strait must, therefore, have occurred during this or the immediately succeeding period,[3] since migration via Europe was no longer possible (See the map, Pl. XXIV).

With respect to the industry of the Upper Cave, it may be said that it compares favorably with the late Palæolithic culture of Europe as a whole, according to Dr. Pei. However, he has so far failed to establish closer relationship to any one of the sub-divisions of this period known in the West. This, once more, proves that the development of culture is not necessarily contemporaneous in two distinct continents, though it might be parallel. In other words, a certain cultural stage found in Asia might reappear at a later date in Europe.

So far the Upper Cave is the only late Palæolithic cave known in China. On the contrary, in Europe, especially in the south of France, numerous caves of the same cultural period are known, located often close together. In spite of the fact that the Chinese territory has been extensively explored, the infrequency of finds indicates that Pleistocene man did not migrate eastward as extensively as he journeyed to Europe. [4]

7

In attempting to answer the question, "Why was Palæolithic man so rare in China when he was so abund-

5) Post-Würmian or later, See table Chapter XXXIX, p. 476.
4) WEIDENREICH F:—*Earliest Representatives of modern mankind in East Asia*, 1939; pp. 161-174, 4 plates.
   PEI: W.C: *On the Upper Cave Industry:* 1939, pp. 175-179, 3 plates.

ant in Europe, and when the centre of distribution was nearer the former than the latter region?" we must first, I think, take cognizance of the loess. Loess is a characteristic Chinese deposit. It is a peculiar yellow earth composed of very fine particles, deposited so as to leave no evidence of bedding, and compact enough so that it will form vertical cliffs wherever it is transected by stream channels, or by roadways, which centuries of travel have lowered into narrow canyons.

Most authorities are agreed that the loess is composed of windborne material, and that its development in China began and ended with the glacial occupancy of West Europe. Where did this material come from? Representing as it does the final products of rock decay, it constitutes only a fraction of the bulk of the original rock from which it was derived. We have learned that rocks, which by decomposition produce materials like the loess, also produce large quantities of quartz sand, and that at the time of such production the clays, which form the loess, and the sand are intimately commingled. To separate them requires the activity of certain natural agents, most potent of which is the wind, for wind will carry away the fine material and leave the sand behind.

The loess of China seems to be a deposit of such fine material, the product of rock disintegration separated from the sand and carried to China by the winds. If that is so, then somewhere we should find pure sands which have been freed from the finer material by wind, and this sand should show characteristics impressed upon it by its origin as a wind-blown deposit.

Such a sand deposit we find in Western China, in Sinkiang Province and elsewhere. The Tarim Basin, which lies between the Kwenlun Mountains on the south and the Tianshan ranges on the north, forms a forbidding desert, which is bounded along its northern border by the Tarim River (Fig 34 p. 123). This river, fed by intermittent streams from the two mountain ranges, terminates at the eastern end of the basin in the salt lake of Lop Nor (Fig 34 p. 123). Here evaporation disposes of the water, leaving deposits of salt, which the streams have brought from the mountains, where these salts were liberated by the decomposition of ancient marine rocks which carried fossil, or connate, sea-water imprisoned in their pores. This desert, known as the Takla-Maken desert is throughout a sand dune desert, with dunes reaching heights of a hundred meters or more. These dunes, according to Sven Hedin, rest on stratified clays, which represent deposit of a pluvial period which preceded the present desert period. It is quite within the range of probability that the sands which now form the dunes were originally deposited as river sediments over much of the basin, and, if so, the material was not pure sand but a mixture of sand and fine rock-flour and clay, such as is characteristic of normal river deposits. It is also apparent that at the time when these deposits were formed, the region was not a desert, but possessed a more pluvial climate.

To change such a region into a desert, requires the long-continued activity of drying winds. As residents of North China well know, the cold monsoon winds blowing during the winter from the Mongolian region to the warmer coastal region of China, are typical examples of such drying winds. Blowing outward from a region of high pressure, where, because of their greater density, their capacity for moisture is moderate, they expand as they reach the warmer coastal region, and as a result their capacity for moisture is increased. They thus deprive the country over which they blow of all the moisture in the soil, and this soil becomes a loose shifting mass easily lifted by the wind, which blows it about in clouds of dust. It is a popular fallacy that Peking dust comes from the distant Gobi Desert. Peking dust is nothing more than the fine soil of the plain, picked up by the winds that come, dust-free from the Gobi, after these winds have first sucked up the moisture and so rendered the soil defenceless against its onslaught. Peking dust, then, is the local loess material of the plain stirred up by passing winds, and, though these dust-laden Peking winds are often strong enough to irritate man almost beyond endurance, they are as gentle zephyrs compared to the winds which brought the loess to China far back in Pleistocene time. For that the loess was deposited in China in Pleistocene time, when Europe was buried under ice, is now a well established fact, and that the loess was formed of wind-borne dust is equally well established.

8

Now let us try to correlate the facts so far discovered. The loess, which covers so much of China, is a thick deposit of wind-blown dust. This dust is the fine material sorted out from an older deposit, which was a mixture of sand and dust. In the Takla-Maken of Sinkiang Provinces, and to some extent in the Gobi of Mongolia, we have great areas covered with sand that has been freed from its dust deposit by an agent, which, taking all the facts into consideration, could only have been the wind. The loess thus appears to be that dust which was taken by the winds from the Tarim and the Gobi Basins. But the winds of today are far too weak to carry dust from the Gobi and from Sinkiang to China. They can only play about with the dust which they pick up locally. Hence the winds of Pleistocene time, the period during which loess was deposited, must have been many times as powerful as the strongest monsoons of today. The modern monsoons are the product of the great cold of Mongolia, and their drying effect is due to their flow to warmer regions. The Pleistocene winds, which carried the dust from the Tarim and the Gobi to China, must have been the product of a much greater and more continuous cold than that which prevails in the interior of Asia today. If now we recall that this was the period when Europe was covered with ice, we realize that we have found the source of this greater cold, and if glaciation was due to the shifting of the pole 20°-25° south of its present location, then the equator in the coastal region of China must have been 12°-15° farther north than it is today. Thus Peking would have enjoyed such a climate as is characteristic of Formosa and the South China coast today. (See Pl. XXI)

But even if North China was bounded by the latitudes which bound it today, the difference in the temperature between the glaciated region of Europe and the coastal region of China would have been sufficient to produce a vast difference in atmospheric pressure, and this would lead to the production of the powerful winds which brought the loess. It was these winds which changed the Tarim Basin to a desert, because it was there that their drying effect would first be felt, and it was these winds

which carried the dust and did not drop it until it
reached the Chinese plains. We find life burdensome
when the dust winds blow today, but the dust-winds
of Pleistocene time must have made existence in
China nigh to impossible. It was bad enough in the
Tarim Basin when the cold wind from the ice fields
blew across it, but these winds only picked up the dust
and carried it away, and so existence was still possible.
Nevertheless, these changing conditions forced early man
to seek a region where the climate was more genial or
where at any rate protection from its rigours was available. Such protection he found in the caverns of South
Europe, and there, too, he found a more genial climate
during the warmer interglacial periods when the cold winds
ceased to blow. But in the east, little or no protection
was found from the cold blasts of the glacial winds, and,
even if man had been able to endure the cold, he would
have been defenceless against the devastating effects of
the dust storms. No wonder, then, that Pleistocene man
rarely came to China, and that most of the larger animals,
which went with him to southern Europe, likewise kept
out of China. There are, to be sure, occasionally found
remains of animals whose presence seems incongruous with
the climate which we have pictured as the prevalent
one. Such is the giant ostrich, or *Struthiolithus*, whose
eggs are found in Chinese loess. But it is probable that
these creatures wandered into the loess country during
the milder interglacial periods, when the dust-bearing
winds were in temporary abeyance, and that they came
from the coastal belt, which then was much farther east
than it is today. For at that time old Cathaysia had
probably not yet become wholly dismembered, and neither
these animals nor early man could have been permanent
inhabitants of North China during the period of loess
deposition. And so it is in Europe, rather than in Asia,
that we find the records of his evolution in mid-Quaternary time.

When the Pleistocene era came to a close, the
glaciers waned, the North Pole wandered to its present
location, and as it receded the glaciated country again
became free from ice. And with the disappearance of
the ice-caps, the strong winds ceased to blow across Asia,

and the loess deposition came to an end. Then Neolithic man could spread eastward once more, and occupy the territory from which his predecessors were excluded. The evidence of the presence of the later Neolithic man in North China and Siberia and Japan is as striking as is the scarcity of the record of the presence of Palæolithic man. Not only are the artifacts of Neolithic man found everywhere in great abundance, but the remains of man himself have been obtained from many an ancient Neolithic burial ground. And the striking fact, first revealed by the researches of Dr. J. G. Andersson, is that the culture of this man was essentially uniform, from the Mediterranean to the islands of Japan. Not content to occupy the Asiatic continent, man wandered across the Bering Bridge to North America, and there established himself, and some of his descendants, the last of the Amer-Indians, still linger among the later population, which migration in historic time has brought to America from the shores of Europe.

Correct the following misprints:

Page 285 — Chapter Heading
    For Missippian read Mississippian
Page 302 — Chapter Heading
    For Palæozoic Period read Palæozoic Periods
Page 379 — 7 lines from bottom
    For Ostrea rastellaris read Ostrea rostellaris
Page 435 — 3 lines from top
    For (Sables moyans) read (Sables moyens)

LITERATURE REFERENCES

AGASSIZ, A.—1889 *The coral reefs of the Hawaiian Islands.* Mus. Comp. Zool. Bull. Vol. 17, pp. 121-170, plates 1-13.

ANDERSON, F. M.—1938 *Lower Cretaceous in California and Oregon.* Special Paper 16, Geol. Soc. Am. 339 pages, 84 plates.

ANDREWS, ROY, CHAPMAN—1932 The New Conquest of Central Asia. *Natural History of Central Asia* Vol. I, Am. Mus. Nat. His.

ANTEVS, ERNST (a)—1928 *The Last Glaciation.* American Geographical Society Research Series, No. 17, pp. 292, Chart Pl. XI.

(b)—1929 Maps of The Pleistocene Glaciation, Bull Geol. Soc. Am. Vol. 40, pp. 631-720.

ARNOLD, RALPH—1910 *Environment of the Tertiary Faunas of the Pacific Coast of the United States.* In "Outlines of Geological History," (Willis and Salisbury), pp. 226-250.

ARNOLD, RALPH, and ANDERSON, ROBERT—1910 *Geology and Oil Resources of the Coalinga District California.* Bull. U. S. Geol. Survey, No. 398.

BENSON, W. N.—1925 *Stratigraphy and Structure of the Northern and Eastern Margin of Australia.* Gedenkboek Verbeek, July, 1925.

BERKEY, CHARLES, P. and MORRIS, F. K.—(a) 1926 *The Geological Background of Fossil Hunting in Mongolia.* Natural History, Vol. 26, No. 5, pp. 527-530, September-October (New York).

(b)—1927 *The Geology of Mongolia.* Natural History of Central Asia, Vol. II, pp. xxxi 507, 44 plates and 161 illus. in text; 6 maps in pocket at end.

BISAT, W. S.—1936 *The Faunal Stratigraphy and Goniatite Phylogeny of the Carboniferous of Western Europe, with notes on the connecting links with North America.* XVI Intern. Geol. Congr. U. S. A. Rept. Vol. I, p. 529-537, 2 plates.

BLACK, DAVIDSON.—1931 *Palæogeography and Polar Shift.* Bull. Geol. Soc. China Vol. X, pp. 105-131.

BOSWELL, P. G. H.—1931 *Summary Proceedings* Geol. Association XLII, 2, pp. 87-111.

CASTER, KENNETH C.—1938 *A Restudy of the Tracks of Paramphibius* Bull. American Pal., Vol. XII, No. 1, pp. 3-60, 13 plates.

CLARK, T. H.—1934 *Structure and Stratigraphy of Southern Quebec.* Bull. Geol. Soc. Am. Vol. 45, pp. 1-20.

CLARKE, J. M. and RUEDEMANN, R.—1912 *Eurypterida of New York with Atlas.* Mem. N. Y. State Mus. 14, Text & plates.

COLEMAN, A. P.— 1926 *Ice Ages Recent and Ancient* (New-York).

COOKSON, I. C.—1937 *Fossil Plants from Warrentima Tasmania.*

CUSHING, H. P. AND OTHERS—1910 *Geology of the Thousand Islands Region.* Bull. N. Y. State Mus. 145.

DALL, W. H.—1898 *North American Tertiary Horizons.* 18 Ann. Rept. U. S. G. S., pt. 2, p. 345.

DALY, R. A.—1934 *The Changing World of the Ice Age.* New Haven, Yale University Press.

DANA, JAMES D.—1847 *Geological Results of Earth's contraction in Consequence of Cooling.* Am. Journ. of Science, 2nd Ser., Vol. 3, pp. 176-188; Vol. 4, pp. 88-92.

DARWIN, CHARLES.—1845, The *Voyage of the Beagle* (2nd Ed.).

DAVID, T. W. E. and SÜSSMILCH, C. A.—1936 *The Carboniferous and Permian Periods of Australia.* XVI Int. Geol. Congr. pp. 629-644.

DAVID, SIR T. W. EDGEWOOD and TILLYARD, R. J.—1936 *Memoir on Fossils of* the *Late Pre-Cambrian from the Adelaide Series South Australia.* pp. 122, 13 plates; Sydney;—Angus and Robertson.

DEISS, CHARLES—1938 *Cambrian Stratigraphy and Trilobites of Northwestern Montana.* G. S. A. Special Papers, No. 18.

DU TOIT, A. L.—1926 *The Geology of South Africa.* pp. 445, 64 text-figs. 39 plates, and geological map. London, Oliver and Boyd.

DU TOIT, A. L.—1937 *Our Wandering Continents.* Edinburg and London; Oliver and Boyd.

DUYFJES, J.—1936 Zur Geologie und Stratigraphie des Kendenggebietes zwischen Trinil und Sverabaja (Java), De Ingenjeur in Nederl.—Indie. Jrg. III No. 8 ; IV, pp. 136-149, 5 figs. (incl. Geol. Sk. maps).

EASTMANN, C. N.—1907 *Devonic Fishes of New York Formations.* Memoir X, New York State Museum.

HAYASAKA, ICHIRO.—1936 *The Japanese Carboniferous.* XVI International Geological Congress. Reported 1933 Session U. S. A. pp. 585-592, 1 fig.

FETTKE, CHARLES R.—1933 *Subsurface Devonian and Silurian Sections across Northern Pennsylvania and Southern New York.* Bull. G. S. A. 44-3. pp. 601-660.

GEIKIE, A.—1903 *Text-Book of Geology, Vol. II.* 4th Ed. London Macmillan & Co.

GRABAU, A. W.—(1) 1894 *The Pre-Glacial Channel of the Genesee River;* Proc. Bost. Soc. Nat. Hist., Vol. XXVI, pp. 259-369.

(2)—1897 *The Sand-Plains of Truro, Wellfleet and Eastham (Mass.);* Science, New Ser. Vol. 3, pp. 334-335, 361, 1897.

(3)—1895-99 *Geology and Palæontology of Eighteen Mile Creek and the Lake Shore Sections of Erie County, New York;* Buffalo Soc. Nat. Hist. Bull., Vol. VI, pp. i-xxiv and 1-403, 263 Text-Figs; pls. I-XXVII.

(4)—1900 *Siluro-Devonic Contact in Erie County.* N. Y. Bull. Geol. Soc. Amer., Vol. XI, pp. 347-376, pls. 21-22, figs. 1-8,

(5)—1900 *Lake Bouvé, an Extinct Glacial Lake in the Southern Part of the Boston Basin;* Boston Soc. Nat. Hist., Occasional Papers, IV, pt. III, pp. 564-600, Map.

GRABAU, A. W. (6)—1900 *Palæontology of the Cambrian Terranes of the Boston Basin*; Boston Soc. Nat. Hist. Occasional Papers IV, pt. III, pp. 601-694, pls. XXXI-XXXIX.

(7)—1901 *Guide to the Geology and Palæontology of Niagara Falls and Vicinity*; Buffalo Soc. Nat. Sci. Bull., Vol. 7, pp. 1-284, 18 pls., 190 figs. and Geologic Map (Also: New York State Museum Bull. 45).

(8)—1902 *Stratigraphy of the Traverse Group of Michigan;* Mich. Geol. Surv. Ann. Rept. for 1901, pp. 163-210, 2 pls., 2 figs.

(9)—1902 *Hamilton Group of Thedford, Ontario* (with H. W. Shimer). Geol. Soc., Am. Bull. Vol. 13, pp. 149-186, 5 figs.

(10)—1903 *Stratigraphy of Becraft Mountain Columbia County, N. Y.* N. Y. State Mus. Bull. 69, pp. 1030-1074, 13 figs.

(11)—1903 *Palæozoic Coral Reefs.* Bull. Geol. Soc. Am. Vol. 14, pp. 337-352, pls. 2.

(12)—1904 *On the Classification of Sedimentary Rocks.* Am. Geol., Vol. 33, pp. 228-247.

(13)—1906 *Guide to the Geology and Palæontology of the Schoharie Valley in Eastern New York;* N. Y. State Mus. Bull. 92. pp. 77-386, 24 pls. 216 figs.

(14)—1907 *The Medina Sandstone Problem; Abstract:* Science, N. S., Vol. 25, pp. 771-772.

(15)—1907 *Types of Sedimentary Overlap.* Bull. Geol. Soc. Vol. 17, pp. 567-636, 17 figs.

(16)—1908 *New Upper Siluric Fauna from Southern Michigan:* Bull, G. S. A. Vol. 19, pp. 540-553, 1 fig. (Sherzer and Grabau).

(17)—1909 *A Revised Classification of the North American Lower Palæozoic:* Science, N. S. Vol. 29, pp. 351-356, Feb. 26.

(18)—1909 *Physical and Faunal Evolution of North American during Ordovicic, Siluric and Early Devonic Time:* Journ. Geol. Vol. 17, No. 3, pp. 209-252, 11 figs.

(19)—1910 *The Monroe Formation of Southern Michigan and adjoining Regions* (with *H. W. Sherzer*): Mich. Geol. & Biol. Survey; Pub. 2.

(20)—1911 *On the Classification of Sand Grains:* Science, N. Ser. Vol. 33, pp. 1005-1007.

(21)—*Über die Einteilung des Nordamerikanischen Silurs:* Intern. Geol. Congr. XI; Stockholm, 1910. Compt. Rend. pp. 979-995, 6 figs.

(22)—1913 *Continental Formations in the North American Palæozoic:* Intern. Geol. Congr. XI; Stockholm, 1910. Compt. Rend. pp. 997-1003.

GRABAU, A. W. (23)—1913 *Principles of Stratigraphy*. N. Y., A. G. Seiler,
1185 pages, 263 figs. Second Edition in 1924. Third
Edition 1932.

(24)—1913 *The Origin of Salt Deposits, with Special Reference
to the Siluric Salt Deposits of North America:* Min. &
Met. Soc. America, Bull. No. 57 (Vol. 6, No. 2) pp.
33-44, Feb. 28.

(25)—1913 *Early Palæozoic Delta Deposits of North America.*
Geol. Soc. Am. Bull., Vol. 24, No. 3, pp. 399-528, 14
figs., 1 pl.

(26)—1913 *A Classification of Marine Deposits; Abstract:* Geol.
Soc. America, Bull. Vol. 24, pp. 711-714.

(27)—1913 *Preliminary Report on the Faunas of the Dundee
Limestone of Southern Michigan:* Mich. Geol. and Biol.
Surv. Publ. 12 (Geol. Ser. 9), pp. 327-378.

(28)—1916 *Comparison of American and European Lower Ordovicic Formations:* Bull. Geol. Soc. Am., Vol. 27, pp.
555-622, 10 text-figs.

(29)—1917 *Stratigraphic Relationships of the Tully Limestone
and the Genesee Shale in Eastern North America:* Bull.
Geol. Soc. Am. Vol. 28, pp. 945-958, 3 figs.

(30)—1917 *Were the Graptolite Shales as a Rule Deep, or Shallow Water Deposits? (with Dr. Marjorie O'Connell):* Bull.
Geol. Soc. Am. Vol. 28, pp. 959-964, 1 fig.

(31)—1917 *Age and Stratigraphic Relations of the Olentangy
Shale of Central Ohio, with Remarks on the Prout Limestone and so-called Olentangy Shales of Northern Ohio:*
Journal of Geology, Vol. XXV, No. 4, May-June, pp.
337-343, 1 text fig.

(32)—1919 *Significance of the Sherburne Sandstone in Upper
Devonic Stratigraphy:* Bull. Geol. Soc. Am. Vol. 30,
pp. 423-470, 2 text-fig.

(33)—1920 *Geology of the Non-Metallic Mineral Deposits other
than Silicates;* Vol. 1, *Principles of Salt Deposition:* pp.
xvi and 435, text-figs. 125, 1 plate, Mac Graw-Hill
Book Co., N. Y.

(34)—1920-21 *Text-book of Geology.* Vol. I. pp. viii to xviii,
and 1-864, 734 text-figs. and frontispiece; Vol. II, 976
pages, 1245 figs. D. C. Heath & Co.

(35)—1920 a) *A New Species of Eurypterus from the Permian
of China;* b) *A Lower Permian Fauna from the Kaiping
Coal Basin.* Bull. Geol. Survey China, No. 2, pp. 61-
79, pl. IX.

(36)—1922 *Ordovician Fossils from North China:* Palæntologia
Sinica, Ser. B, Vol. 1, pp. 1-127, 9 pls., 20 text-figs.,
April 28.

GRABAU, A. W. (37)—1922 *The Sinian System:* Bull. Geol. Soc. China, Vol. 1, pp. 48-88.

(38)—1923 *Cretaceous Fossils.* Bull. Geol. Survey China, No. 5, pt. 2, pp. 143-218.

(a) *Cretaceous Fossils from Shantung:* 19 pages.

(b) *Cretaceous* Mollusca from *North China:* 15 pages.

(c) *Lower Cretaceous Ammonite from Hongkong*: 10 pages.

(d) *Contribution to the Fauna of the Kweichow Formation of Central China:* 9 pages.

(39)—1923 *What is a Shantung?* Bull. Geol. Soc. of China. Vol. 2, No. 1-2.

(40)—1923 *Carboniferous Formations of China* (with W. H. Wong). Congress Geologique International XIIIeme Sess; 1922 Belgique. Compte Rendu Liege, 1923, pp. 657-689, 4 text-figs.

(41)—1924 *Geological Conditions Bearing upon Potash Prospecting in China.* Bull. III. The China Institute of Mining and Metallurgy, pp. 1-28, 2 figs.

(42)—1924-28 *Stratigraphy of China.* Pt. 1, pp. 1-528, 306 text-figs, 6 pls, 1924; Pt. 2, 774 p. 449 figs, 4 pls. 1928.

(43)—1924 *Migration of Geosynclines:* Bulletin of the Geological Society of China, Vol. III, No. 3-4, pp. 207-349, 12 text-figs.

(44)—1925 *Summary of the Faunas of the Sintan Shale:* Bull. Geol. Survey of China, No. 7, pp. 77-85, 3 plates.

(45)—1926 *Silurian Faunas of Eastern Yunnan:* Palæontologia Sinica, Ser. B, Vol. 3, Fasc. 2, pp. 1-100, plates I-IV, 2 text-figs., March 25.

(46)—1927 *Summary of the Cenozoic and Psychozoic Deposits, with Special Reference to Asia;* Bull. of the Geological Society of China; Vol. VI, No. 2,-3, pp. 151-264.

(47)—1924-1928 *Stratigraphy of China,* Pt. I, pp. 1-528; 306 text-figs, 6 pls; Pt. II, pp. 1-774, 449 figs. 4 pls.

(48)—1929 *Origin, Distribution and Mode of Preservation of the Graptolites:* Mem. Research Inst. China; August, pp. 1-52.

(49)—1930 *Asia and the Evolution of Man.* The China Journal, Vol. XII, No. 3, March, pp. 152-163.

(50)—1930 *Corals of the Upper Silurian Spirifer tingi Beds of Kweichow:* Bull. Geol. Soc. China, Vol. 9, No. 3, pp. 223-240, pls. I-III.

(51)—1931 *The Permian of Mongolia; A report on the Permian Fauna of the Jisu Honguer Limestone of Mongolia and its Relations to the Permian of other Parts of the World;* Nat. Hist. of Central Asia, Vol. IV, pp. i-xlii and pp. 1-665, text-figs. 1-72, pls. 1-35.

GRABAU, A. W. (52)—1931 *Devonian Brachiopoda of China*: Part I, Devonian Brachiopoda from Yunnan & other Districts in South China; Palaeontologia Sinica; Ser. B, Vol. III, Fascicle 3, pp. 1-538, text-figs. 67. *Ibid:* 1933, plates I-LIV.

(53)—1933 *Early Permian Fossils of China*, Pt. I ; Brachiopods, Pelecypods and Gastropods of the Lower Permian Beds of Kweichow: Palæontologia Sinica, Ser. B, Vol. VIII, Fascicle 3, pp. 214, 11 plates. (Pt. II, issued in 1936).

(54)—1933 *Oscillation or Pulsation*, Report of the XVIth Int. Geol. Cong. Washington July-Aug. 1933. Vol. I, pp. 539-553; Washington 1936; (Preprint issued August 1934).

(55)—1933 *The Carboniferous of China and its Bearing on the Classification of the Mississippian and Pennsylvanian;* (with V. K. Ting); Report XVI Intern. Geol. Cong. Washington, 1936, Vol. I, pp. 555-571, 1 plate (Preprint issued Aug. 1934).

(56)—1933 *The Permian of China and its Bearing on Permian Classification* (with V. K. Ting); pp. 663-677. (Preprint issued Aug. 1934).

(57)—1933 *Palæozoic Formations in the Light of the Pulsation Theory*: Vol. I, Taconian and Cambrian Pulsation Systems, 1st Ed. 1933, 2nd Ed. 1936, 680 pages, 17 figs, 5 plates. *Ibid*, Vol. II, 1936 Cambrovician Pulsation Systems, Pt. I, 751 pages, 42 figs, 1 map. *Ibid*, Vol. III, 1937 Cambrovician Pulsation System, Pt. II, 850 pages, 58 figs, 9 maps and Charts. (Includes Skiddavian). (Vols. I-III; Pei-ta; University of Peking.) *Ibid*, Vol. IV, 1938. Ordovician Pulsation System, Pt. I, 941 pages, 67 figs, Henri Vetch, Peking.

(58)—1934 *The Beginnings of the Human Race*. Lecture, Shanghai, Sept. 29, 1933. Journal of the North China Branch of the Royal Asiatic Society. Vol. LXV, 1934, pp. 1-20.

(59)—1935 *Did Man Originate in Asia?* Asia Magazine. pp. 24-27.

(60)—1935 *Tibet and the Origin of Man:* Sven Hedin 70th Anniversary Publ., Geografiska Analer 1935, pp. 317-325.

(61)—1936 *Early Permian Fossils of China*, Pt. II; Fauna of the Maping Limestone of Kwangsi and Kweichow; Pal. Sinica; Series B, Vol. 8, Fasc. 4; 31 pls. 1 text-fig., pp. 1-411.

(62)—1936 *Revised Classification of the Palæozoic System in the Light of the Pulsation Theory;* Reprinted from the Bull. of the Geol. Soc. of China; Vol. XV, No. 1, 1936; pp. 23-51.

(63)—1936 *The Great Huangho Plain of China:* Reprinted from the Journal of the Association of Chinese and American Engineers; Vol. XVII, No. 5, pp. 247-266.

(64)—1937 *Fundamental Concepts in Geology and their Bearing on Chinese Stratigraphy:* Reprint, Bull. Geol. Soc. China, Vol. 16, pp. 127-176, 1 map, 2 figs.

(65)—1938 *Sea-Salts and Desert Salts:* Pt. I, Reprint, Journal Assoc. Chinese and American Engineers; Vol. XIX, No. 3, May-June Part II. *Ibid:* No. 4, July-August.

(66)—1938 *The Significance of the Interpulsation Periods in Chinese Stratigraphy;* Reprint, Bull. Geol. Society of China; Vol. XVIII, pp. 115-120, 2 tables, Changsha.

(67)—1938 *Ice Ages or Polar Glaciation.* (Abstract.) Bull. G. S. A, 49: pp. 1883-4.

(68)—1939 *Present Status of the Polar Control Theory of Earth Development.* Bull. Geol. Soc. China, Vol. XIX, No. 2, pp. 189-205, 1 fig.

Gracht, W. A. Van der—1928 *The Theory of Continental Drift.* Am. Ass. Petroleum Geologists.

Granger, Walter—1926 to-date *Am. Museum Novitates,* many papers.

Gutenberg, B.—1936 *Structure of the Earth's Crust and the Spreading of the Continents.* Bull. G. S. A., Vol. 47, No. 10, pp. 1587-1610.

Hall, J.—1847 *Summary Report 4th Geol. Dist.*

Hall, J.—1859-1895 *Pal. N. Y.* Vols I-VIII.

Hartnagel, C.—1907 *Geol. maps of Rochester and Ontario Beach Quadrangles.* Bull. 114 N. Y. State Mus.

Haug, E.—*Traité de Géologie.* Paris Armand Colin, 1786 pages.

Heim, Arnold—1936 *Energy Sources of the Earth's Crustal Movements.* XVIth Int. Geol. Cong. U. S. A. 1933, pp. 1587-1610.

Hebert, E.—1857 *Les mers anciennes et leur rivages dans le bassin de Paris.*

Hollick, Arthur:—1892 *The Palæontology of the Cretaceous Formations of Staten Island,* Transact. N. Y. Acad. Science Vol. XI, pp. 96-102.

Huntington Elsworth:—1907 *The Pulse of Asia.* Boston, Houghton Mifflin & Co.

Imlay, R. W.—1938 *Studies of Mexican Geosyncline.* Bull. Geol. Soc. Am. Vol. 49, pp. 1651-1694.

Joly, John.—1924 *Radioactivity and the Surface History of the Earth.* Halley Lecture, May 28, 1924, Oxford, Clarendon Press.

Kayser, E.—1913 *Lehrbuch der Formationskunde,* 5th Ed. Ferd. Enke, Stuttgart.

Kelus, Alexander von—*Devonische Brahciopoden von Petzca.*

Kindle, E. M.—1938 *A Pteropod Record of Current Direction.* Journ. of Pal. Vol. XII, No. 5.

Königswald, Ralph von (1)—1930 Klima-änderungen im Jung-Tärtiär Mitteleuropas und ihre Ursachen. Zeitschr. für Geschiebeforschung Bd. VI, Heft 1, pp. 11-16.

(2)—1936 *Der Gegenwärtige Stand des Pithecanthropus Problems.* Handelingen 7th Ned-Ind. naturwetenshappelik Congress, pp. 724-732, 1936.

(3)—1937 Erste Mitteilung über einen fossilen Homini-
den aus dem Altpliocän Ostjavas. K. Akad. Wet-
ens, Amsterdam, Sec. Sci. Pr. Vol. 39 No. 8, pp.
1000-1009, 1 map 1 pl.

(4)—1939 *Hipparion und die Grenze zwischen Miocän und
Pliocän.* Zentralbl. für Miner. etc. Jahrg. 1939, B.
6 pp. 236-245.

(5)—1939 *Das Pleistocän Javas.* "Quatär" Bd-2, pp.
27-53, pls. IX-XI.

Köppen W. and Wegener, A.—1924 *The Climate of the Geological Ages.* (Translated from the German). Borntræger Bros.

Leith, C. A. and Mead, W. J.—1915 *Metamorphic Geology,* New York, Henry
Holt & Co.

Logan, W.—1863 *Geology of Canada,* Summary Report.

Mather, K. F.—1922 *Front Ranges of the Andes between Santa Cruz Bolivia and
Embarcation Argentina.* Bull. Geol. Soc. Am. Vol. 33, pp.
703-764.

Merriam, John C.—1901 *A Geological Section through John Day Basin Oregon.*
Bull Geol Soc. Am. Vol. 12, pp. 496-497; Journ. Geol. Vol.
9, pp. 71-77.

Morris, F. K.—(See Berkey and Morris).

Moore, R. C.—1936 *"Carboniferous" Rocks of North America.* XVI Int. Geol.
Congr. Comptes Rendus Vol. I, pp. 593-617, 3 pls.

Nelson, W. A.—1922 *Volcanic Ash in Ordovician of Tennessee.* Bull. Geol. Soc.
America, Vol. 33, pp. 605-616.

O'Connel, Marjorie.—1916 *The Habitat of the Eurypterida.* Bull. Buff Soc.
Natural Science. Vol. XI, No. 3, pp. 1-278, pl. (See also
Grabau and O'Connel, 30).

Osborn, H. F.—1910 *Age of Mammals in Europe, Asia and North America.*
MacMillan Co., N. Y.

Pækelmann, W. and Schindewolf, O. H.—1937 *Devon-Carbon Grenze.* Heer-
len Congress du 2me Congr. pour l'advancement des études
de Stratigraphie. Heelen 1935, pp. 703-714.

Park, W. A.—1931 *Geology of Gaspé Peninsula.* Bull. Geol. Soc. Am. Vol.
42, pp. 785-800.

Peach, B. and Horn, J.—1930 *Chapters on the Geology of Scotland* (Posthumous)
232 pages, 27 text-figs, 28 plates (including Geological map),
Oxford, University Press.

Pei, W. C.—1939 *A Priliminary Study on a new Palæolithic Station known as
Loc. 15 within the Choukoutien Region.* Bull. Geol. Soc. China,
Vol. XIX, No. 2, pp. 147-188, 25 figs.

Penck, A. and Brückner E.—1901-1909 *Die Alpen im Eiszeitalter.* Vols. 1-3,
Leipzig.

Raash, Gilbert O.—1939 *Cambrian Merostomata.* Geol. Soc. Am., Special
Papers, No. 19, pp. 1-146, 21 plates.

Richarz, Stevens.—1922 *Tertiary Glaciation of Alaska and Polar Migration.*
Zeitschr. d. d. Geol. Gesellsch. Mon. Ber. 74, pp. 180-190.

Ruedemann, B.—1919 *On Some Fundamentals of Pre-Cambrian Palæogeography.* Proceeding·Nat. Acad. Science Vol. 5, Washington. (Map of ancient trend lines).

Ruedemann, R.—1904-8 *Graptolites of New York.* Memoir N. Y. State Mus. Mem. 7, 1904, Mem. 11, 1098 (also numerous articles on Palæozoic of North America).

Schindewolf, O. H.—1936 *Probleme der Devon-Karbon Grenze.* XVI International Geol. Congr. Washington Vol. I, 1933, pp. 505-514.

Schuchert, C. a.—1928 *The Hypothesis of Continental Displacement.* In Theory of Continental Drift-Van der Gracht Symposium, pp. 104-144.

Schuchert, C. b.—1930 *Orogenic Times in the Northern Appalachian.* Bull. Geol. Soc. Am. Vol. 41, pp. 711. etseq.

Schuchert, C. and Dunbar, Carl O.—1934 *Stratigraphy of Western Newfoundland.* Geol. Soc. Am. Memoir, p. 123, pl. 8 figs.

Simpson, G. C.—1930 *Past Climates,* Mem. and Proceed. Manchester Literary and Philosophical Soc. Vol. 74, pp. 1-34. (See also *Ibid*—1927 Quart. Journ. Roy. Met. Soc. Vol. LIII No. 223, pp. 213-232)

Staub, Rudolf.—1928 *Der Bewegungsmechanismus der Erde.* Berlin, Borntraeger pp. 270; text-figs. 44, 1 plate.

Stauffer, C. B.—1912 *Oriskany Sandstone of Ontario,* pp. 371-376.

Stille, Hans—1924 *Grundfragen der Vergleichenden Tektonik.* Borntræger Bros.

Stoyanow, A. A.—1936 *Correlation of Arizona Palæozoic.* G. S. A. Bull. 47, No. 4, 1936, p. 484.

Stubblefield, C. J. and Bulman, O. N. B.—1927 *The Skincton Shales of the Wreking District etc.* Q. J. G. S. Vol. 83, pp. 96-144, plates 3-5.

Suess, E.—1906 *The Face of the Earth.* (Das Antlitz der Erde) translated by Sollas and Sollas, 4 Vol, Bd II.

Swartz, Frank McKim.—1934 *Silurian Section near Mount Union Central Pennsylvana.* Bull. G. S. A. Vol. 45, pp. 81-134.

Teilhard de Chardin, P. and Stirton, R. A.—1934 *A Correlation of some Miocene and Pliocene Mammalian Assemblages in North America and Asia with a Discussion of the Mio-Pliocene Boundary.* Bull. Dept. Geol. Univ. Calif. Vol. 23, No. 8, pp. 277-290.

Ting, V. K. and Grabau, A. W.—1936 *The Carboniferous of China.* XVI Int. Geol. Congr. C. R. 1933, pp. 555-571, 1 plate.

Twenhofel, W. H.—1926-1932 *A Treatise on Sedimentation.* Baltimore, The William and Wilkins Company.

Trowbridge, A. C. and Atwater, G. I.—1934 *Stratigraphic Problems of the Upper Mississippi Valley.* Bull. Geol. Soc. Am. Vol. 45, pp. 21-80.

Ulrich, E. O.—1911 *Revision of the Palæozoic System.* Bull. G. S. A. Vol. xxii, pp. 281-680.

Umbgrove, J. H. F.—1939 *On Rhythms in the History of the Earth.* Geol. Mag. Vol. LXXVI, No. 897, pp. 116-129.

Vaughan Arthur.—1904 *Palæontological Sequence in the Bristol Area.* Q. J. G. S. Vol. 61, pp. 181-266.

Van Hise Charles R.—1892 *Correlation Papers, Archaean and Algonkian.* U. S. Geol. Survey, Washington Bull. 86.

Van Strælen, Victor—1925 *The Microstructure of Dinosaurian Eggs from the Cretaceous Beds of Mongolia.* Am. Mus. Novitates No. 173, pp. 1-4, 2 figs.

Walther, Johannes (a)—1924 *Das Gesetz der Wüstenbildung.* Quelle & Meyer Leipzig, pp. 1-421, 203 figs.

Walther, Johannes (b)—1893-94 *Einleitung in die Geologie als Historische Wissenschaft,* Gustav Fischer Jena, pp. 1055.

Wegener, Afred—*The Origin of Continents and Oceans.* 3rd Ed. London Methuen & Co. Ltd. 312 pages, 44 text-figs.

Weidenreich, Franz (a)—1937 *The Fore-runners of Sinanthropus pekinensis.* Bull. Geol. Soc. China Vol. XVII, No. 2, pp. 138-144.

Weidenreich, Franz (b)—1939 *Six Lectures on Sinanthropus pekinensis.* Bull. Geol. Soc. China Vol. XIX, No. 1, pp. 1-92, pls. I-IX.

Weller, S.—1907 *Cretaceous Pal. New-Jersey.* N. J. Geol. Survey, Pal. Ser. Vol. 4, Text 871 pages, Plates: pp. 875-1106, 111 plates.

Willard, Bradford—1936 *Continental Upper Devonian of Northeastern Pennsylvania.* G. S. A. 47, No. 4, pp. 565-608.

Willis, Bailey—1912 *Index to the Stratigraphy of North America with Geological map.* U. S. G. S. Professional Paper 71, 894 pages.

Willis, B. and Salisbury R. D.—1910 *Outlines of Geological History with especial Reference to North America.* Univ. Chicago Press.

Wright, W. B.—1937 *The Quaternary Ice Age.* MacMillan & Co. 478 pages.

Zittel, Karl A.—1900 *Text Book of Palæontology.* Vol. I (Trans. Eastman) MacMillan & Co., New York.

Zittel, Karl A. and Rohon, J. V.—1886. Über Conodonten. Sitzungsber. Bayr. Akad. d. Wissenschaften, Bd-XVI.

Chester 315; (—age) 268
Chetang Cliff 85
Cheviot Hills 294
Cheyenne sandstone 404
Chiapas 264, 472
Chicago area 172
Chicamauga Limestone 142
Chico 409, 416, 444, 445; (—overlap)
410; (—series) 410, 413, 414, 447
Chihli 113
Chili 406, 409, 424
Chillesford 477; (—Clay) 478; (—Sand)
478
Chillesfordian 479
China 50, 69, 71, 87, 145, 161, 248,
279, 310, 393, 441, 455, 466, 476,
478, 506, 507, 509, 510, 511, 512;
(—, central) 490; (—coast) 511; (—,
East) 71, 113, 161, 311, 490, 501;
(—, Great Plain of) 36; (—, North)
130, 140, 158, 307, 310, 311, 498,
503, 506, 510, 512, 513; (—region,
North) 312; (—, northern) 422, 440;
(—plain) 495; (—proper) 497; (—,
South) 130, 297, 306, 307, 311, 312,
364; (—, South-West) 266;(—, south-
ern) 307; (—, western) 443, 496, 501,
506
Chinese deposit 509, 512; (—territory)
508; (—Turkestan) 191
Chisholm shale 102
Chitina Valley 414
Chiussuan 27, 259, 280, 292, 306
Choukoutien 476, 504, 507; (—cave)
502; (— —, Upper) 507; (—Lime-
stone hill) 507
Christiania (Oslo) 91
Cincinnati 239, 250, 489; (—beds)
143; (—group) 144, 251; (—section)
144
Clactonian Culture Stage 476
Claibornian 444
Clark, T. H. 126
Clarke, J. M. 202, 204, 224
Cleveland black shale 263; (—shale)
253
Cliffwood formation 428
Clinch 176
Clinton 47, 176, 180; (—Limestone,
Upper) 174; (—pelecypod) 180
Clisiophylium zone 271
Clunian 197
Coal Measures 346; (— —flora) 347
Coalinga 446; (—district) 450; (—oil
district) 448
Coast Range 446; (— —section) 473
Coastal Plain 424; (— —strata) 342
Coblenz 212
Coblenzian 261, 305; (—fauna) 212
Cobleskill 193, 224, 285; (—Lime-
stone) 211
Coeymans 222; (—Limestone) 211
Coldbrookian shale 150

Coleman, A. P. 139
Colle di Muntijella 395
Colombia 406
Cologne region 458
Colorado 265, 267, 401, 402, 403, 420;
(—geosyncline) 419; (—River) 266,
401; (—, southern) 404
Coloradoan 420, 436; (—geosyncline)
436
Columbus 213, 236, 251; (—Limestone)
228
Comanche 283, 398, 399, 401, 410,
416, 426, 432, 444, 448; (— (Shasta)
beds) 445; (—flora) 416; (—, Lower),
398; (—Peak) 403, 404; (— —Limestone) 402; (—Pulsation Period) 447;
(— —Series) 426; (— —System) 398;
(—series) 398, 402, 404, 405; (—
strata) 418; (—subdivision) 401; (—
system) 409, 447; (—time) 419; (—,
Upper) 405
Comanchean 412, 423; (—age) 427;
(—beds) 389, 425, 430; (—Interpulsation System) 390; (Pulsation System)
390, 416; (—sea) 406; (—system) 400
Comanchic 418; (— -Cretacic) 401;
(—sediments) 424; (— (Shasta)time)
413
Comley 81
Conasauga 104; (—beds) 100, 101;
(—Formation) 99; (—sea) 99, 100
Coniacian 399, 425
Coniston Lake 131
Connecticut Valley 330
Continental beds 99, 157; (—clastics)
14; (—deposition) 155; (—deposits)
45, 73, 132, 154, 195; (—drift) 387;
(—geosynclinal deposits) 34; (—In-
terpulsation deposits) 58; (—sands)
94, 113; (—sedimentation) 22, 115;
(—sediments) 34, 214; (—series)
156; (—strata) 105, 198, 199
Contorta beds 376
Cookson, I. C. 199
Coral oölite 379
Corallian 379, 382; (—Coral rag) 382
Coralline Crag 455, 459, 460, 478, 479
Cordilleran 295, 435, 472, 488; (—
Epi-Sea) 397; (—ice-center) 485; (—
—sheet), 477; (—North Pole) 485;
(—region) 419, 484, 485; (—site) 485
Corcy Limestone 126
Cornbrash 380, 385
Cornwall 216, 217, 459
Corona beds 131
Corsica 373, 464
Corvara 359
Cotentin 465
Cottonwood beds 410; (—district) 415
Cow Creek Beds 403
Cox 48
Crag of Butley 478; (— —Oakley)

437, 438, 449, 450, 457, 489; (—, Upper) 447, 448
Eogene 456, 457, 482; (—beds) 446; (—geosyncline) 458; (—land faunas) 457; (—North Pole) 457; (—polar location) 454; (—Pulsation System) 434, 446, 473, 475; (—series) 446; (—species) 437; (—strata) 457; (—time) 448, 450
Eo-Neogene disconformity 446
Eozoon 75
Epi-Sea 22, 59, 60, 112, 129, 143
Erath county 404
Erian 283; (—orogenies) 294
Eriboll (Eireboll) quartzite 67
Erithræan 483
Erth beds, St. 459
Escabrosa Formation 268; (—Limestone) 266, 268
Eskimo skulls, modern 507
Esmeralda county 103
Esopus 220; (—grit) 211
Essex 459
Esthonia 84, 97, 127, 128
Etcheminian Limestone 149
Etroeungtian 271, 272, 273, 292; (—beds) 272; (— —, Russian) 272; (—division) 271; (—, European) 272; (—, false) 273; (—type) 272; (—, West European) 272
Euphrates Valley 464
Eur-Africa 342
Eurassia 277, 434
Eurasian 85, 86, 88; (—mass) 154
Eurasiatic continent 502
Eureka 263; (—shale) 263
Europe 84, 86, 93, 134, 146, 155, 195, 198, 199, 213, 223, 265, 272, 280, 291, 295, 297, 298, 310, 314, 347, 349, 354, 365, 366, 367, 373, 374, 382, 390, 397, 400, 407, 409, 421, 424, 425, 427, 429, 433, 438, 443, 448, 450, 451, 453, 457, 458, 461, 462, 468, 475, 482, 485, 487, 489, 490, 496; (—, central) 84, 310, 340, 340, 342; (—, North) 311, 335, 490; (—coal beds) 347; (—northern) 485; (—north-western) 213; (—, South) 310, 450, 458, 512; (—, southern) 94, 450, 458, 505, 512; (—, West) 320, 331, 387, 460; (—, western) 155, 167, 272, 299, 381, 384, 490, 504
European 146, 305; (—affinities) 287; (—areas, South) 450; (—brachiopod elements) 306; (—Burdiagalian) 469, 470; (—coast) 488; (—continuation) 168; (—deposits) 195; (—development) 147; (—equivalent) 425; (—examples) 113; (—extension) 278; (—fields) 468; (—Formations) 467; (—geologists) 378; (—glacial de-

posits) 487; (—glaciation) 487, 488, 489; (—horizons) 467; (—lands) 87; (—literature) 279; (—mass) 291; (—Mousterian) 506; (—Namurian) 312; (—Neocomian species) 409; (—part) 154; (—regions) 89, 218, 385; (—western regions) 272; (—rocks) 213; (—Santonian) 430; (—sections) 104; (—skulls) 507; (—, southern) 212; (—standard) 470; (—subdivisions) 213; (—succession) 195, 468; (—Taconian faunas) 85; (—transgression, West) 438; (—types) 467, 469
Eurydesma beds 346
Eurypterid beds 289
Eutaw 425, 430; (—Formation) 425
Extensus zone 111

F

Falkland 305, 494; (—Islands) 126, 304, 305
Falls of the Ohio 235
Faxo Limestone 438
Fenestellids 273
Feng-oric 280, 283, 292, 293; (— — orogeny) 288
Fengninian 27, 256, 259, 263, 264, 268, 280, 283, 292, 294, 303, 304, 306, 314, 344; (—, Chinese) 259; (—, Lower) 253, 306; (—Mountain series) 303; (—Pulsation Period) 253, 292; (— —system) 259, 261, 263; 329; (—sea) 252, 254; (—time) 303
Fengshan 113; (—Limestone) 113; (—series, Upper) 113
Fengtien 443
Fennoscandia 85, 98, 407, 434, 457
Fennoscandian mass 151; (—old-land) 151
Fernando series 473
Ffestiniog 109
Fifeshire 323
Finistere basin 291; (—, Department of) 291
Finland 72
Finnish coal-bed 73
Flat-rock 204
Florida 457, 472, 473, 482, 487; (—coast 174
Floyd group 260
Foraminifera 469; (—bed) 469
Foreland 51, 55
Forest Bed 477, 480
Forestian 476
Formosa 314
Fort Payne Chert 261
Fort Pierre 424
Fort Worth Limestone 401
Fossil Trees 136, 137
Fox Hills 405; (— —age) 424

N

Shaly limestone 211; (—Siberian Islands) 97, 173; (—South Wales) 350 (—world Island) 288

New-York 77, 130, 142, 143, 187, 193, 194, 201, 212, 220, 223, 224, 227, 232, 239, 242, 244, 245; 247, 253, 285, 288; (— —, eastern) 39, 205, 210, 282; (— —reefs, western) 245; (— —region) 209; (— —, western, section) 175; (— —, southwestern) 254; (— —State) 82, 145, 186, 203; (— —, western) 204, 245.

New-Zealand 364

Newark 426; (—beds) 374; (—geosyncline) 335, 342; (—series) 426; (—systems) 337

Newberry, J. S. 253, 427

Newborn 460

Newbournian 456

Newcastle stage 350

Newfoundland 81, 83, 84, 135, 140, 144, 288, 487; (—, central) 84; (—, North-West) 142, 216; (—, northern) 286; (—, South-eastern) 83

Ngandong bed 484

Niagara 47, 174, 175, 195; (—Falls) 174; (— —region) 181; (—Gorge) 174; (—region) 223; (—River) 221, 227

Niagaran 181, 188, 229, 231; (—age) 180, 250; (—coral reefs) 240; (— — —period) 176; (—fauna) 212, 289; (—limestone) 184, 187, 190, 193, 228, 234; (—marine) 182; (—sea) 194; (—shale) 209; (—time) 227

Nias 469; (—beds) 469; (—deposits) 469; (—group) 469; (—Island) 469; (—island of) 468; (—Miocene) 468

Nife-sphere 6

Niger country 331; (—Valley) 331

Nile 458; (—delta) 276

Niobrara 405; (—limestone series) 420

Nishni-Novgorod 384

Noel shale 262, 263

Nolichucky 100; (—shale) 100

Norfolk 480; (—crag) 477; (—formation (Crag)) 479

Norian 372

Noric 338, 355, 356, 360, 363, 368, 371, 372, 377, 378; (—, Lower) 368; (— —Rhætic, Upper) 368; (—stage) 376; (—time) 360; (—, Upper) 369

Normal Interpulsation 447

Normanskill 143; (—divisions) 130; (—graptolite shales) 212; (—shales) 284

North Africa 276, 438

North America 100, 169, 174, 178, 207, 278, 386, 407, 409, 416, 418, 419, 421, 424, 427, 433, 434, 435, 437, 438, 444, 472, 475, 482, 485,

489, 491, 508, 513; (— —, Arctic) 387; (— —, eastern) 205; (— —, Gulf coast of), 440; (— —, western) 473, 484

North American 278; (— —region) 198; (—Cape) 483

North China 205; (—England) 130, 160; (—India) 248; (—India geosyncline) 33, 96; (—Pole) 85, 334, 407, 448, 484, 485, 489; (—Scotland) 99, 247; (—Sea) 134, 458, 459, 461, 465, 477, 479, 482, 483, 502; (— —basin) 479, 480; 487, 488; (— —glaciation) 485; (—Wales) 104, 108, 112

Norway 134; (—sandstone) 219; (—south-eastern) 81

Norwegian coast 421

Norwich 479; (—Bone Bed) 478; (—Brick Clay) 477; (—crag) 462, 477, 478, 479; (—Interpulsation) 479

Notre Dame 288; (— —Bay), 288; (— —mountains) 54, 125

Nottingham 373

Nova Scotia 299

Novo-Sibirsk region 347

Numismalis marls 375

Nummulitic beds 468

Nyassa lobe, 304; (—location) 290

O

Oakly Crag 460

Oban 215

Obercassel 508

Oberkarbon 279

Ocean 397; (—, southern) 408

Oceania 369

Ochsenius, C. 183

Oenningen (Switzerland) basin 463

Ogygopsis shale 102

Ohio 116, 204, 228, 245, 250, 251, 252, 260, 262, 263; (—black shale) 263; (—, central) 252; (—, Falls of the) 227; (—, northern) 252; (—shale) 253, 262; (—, southern) 234

Okhotskian Epi-Sea 312

Oklahoma 115, 116, 118, 404; (—, southern) 406

Olcostephanids 415

Old Field 248

Old-land 30, 31, 32, 33, 34, 45, 51, 55, 68, 95, 96, 106, 107, 146, 150, 156,158, 161, 174, 176, 182, 283; (— —margin) 42; (– – —marginal) 14

Old Red Devonian 214; (— —Interpulsation Period) 218; (— —, Late) 283; (— —, Lower) 215, 216; (— —, Middle) 215; (— —, Upper) 214, 216, 218, 219, 270; (— —rivers) 216; (— —sandstone) 213, 215, 248, 249, 269, 271, 294; (— — —beds) 214; (— — —, Lower) 213; (— — —,

R

S

Saalian 283, 298; (—folding) 299
Sables Moyens 435
Sahara 205; 337, 421; (—region) 103; (—, western) 135, 150, 276
Saharan, North, 331
Sahelian 456; (—Plaisantian) 455
Santa Clara Valley 445
Saint Lawrence-Champlain fault 235; (see also St. Lawrence); (— — geosyncline) 235
Saisan depression 467
Sakhalin 312, 454
Salamanca quartz pebble conglomerate 255
Salina 181, 229, 240; (—Interpulsation) 288; (— —deopsits) 196; (— — Period) 149;167, 287; (—salt basin) 195
Salinan 167, 180, 194, 198, 207, 218, 229; (—age) 198, 218; (—beds) 184; (—deposits) 182, 195, 193, 208; (—desert period) 210; (—eurypterids) 181; (— (Gansvickian) Interpulsation) 196; (—Interpulsation beds) 183; (— —Period) 27, 178, 181, 194, 196, 202, 211; (—interval) 198; (—loess) 182; (—period) 195; (—Red beds) 182; (—salts) 184, 187; (— —basin) 194; (— —deposits) 191, 209; (— — series) 194; (—time) 182; (—type) 195
Salopian 161, 165, 281
Salt district 186; (—Range region) 466
Salzburg facies 373
Salzkammergut 356
Samara 421; (—beds) 321, 325; (—series) 320
Samfrau geosyncline 69, 77, 232, 293, 304, 306, 307, 314, 364
Sarmatian 456; (—age) 464; (—time) 463, 464
San Diego 444; (—Lorenzo series) 446
Sands, Lower 403; (—rock) 211; (—, Upper) 403
Sangamon Interglacial 477
Sannoisian 434
Santa Rosa formation 264; (—Ynez Range) 447
Santonian 399
Saone 392
Sao Paulo 328
Saxony 283, 382, 387
Scaldician 456
Scandian-Greenland rift 479
Scandic region 459
Scandinavia 68, 81, 84, 87, 97, 216, 422, 458, 460, 482, 487, 490; (—, southern) 107
Scandinavian 87, 107; (—basin) 106;

(—ceast) 488; (— —, southern) 91; (—countries) 488; (—form) 83; (—glaciation) 477; (—region) 107, 439
Scania 130, 373, 421
Scanian ice-sheet 477
Scaur 269
Schaumburg-Lippe syncline 388
Schindewolf, O. H. 273
Schlern district 368
Schoharie 220; (—shale) 211
Schuchert, C. 53, 212, 235, 286, 237, 288, 304, 323, 331, 336
Schuylkill river 179
Scoresby Sound 335
Scotland 54, 73, 143, 158, 161, 163, 198, 213, 216, 259, 381, 421; (—, central) 84; (—, Highlands of) 50; (—, North) 59, 66, 67, 81, 82, 214, 215, 232, 283, 373; (—, northern) 219; (—, South) 214, 215, 323; (—, South-East) 232, 234; (—, southern) 215; (—, West, Highland of) 66
Scottish Highlands 67, 81, 84, 216; (—Scandinavian axis) 82
Scythian 354, 358;(—transgression)355
Scythic 355; (—stage) 354, 355
Sdansky, O. 502
Seefeld 377; (—asphalt shales) 377
Seiser beds 356
Selma 430; (—chalk) 425
Semicrystalline limestones 402
Seneca Co., N. Y. 221
Senonian 399, 410, 421, 422; 423, 424, 425, 431, 438; (—beds, Upper) 423; (—flora) 426; (— (Monmouth), Upper) 436 (—time) 424;
Sepai 268
Serpulite 393
Seward peninsula 449
Seymour Island 423
Shakopee dolomite 117
Shall-Shall cliff 362
Shaly series, Upper 220
Shangssuan 306
Shansi, northwest 466
Shantung 35, 50, 160, 161; (—Peninsula) 160; (—province) 50
Shantungs 50
Shark river 437; (— —fauna) 437; (— —formation) 429
Shasta 412, 416; (— -Chico Interpulsation disconformity) 412; (— — — hiatus) 412; (—(Comanchic) fauna) 407; (—county) 410; (—, Lower) 410; (—series) 409, 410, 413, 414, 447; (—time, Lower) 447; (—, Upper) 410
Shawangunk 179, 181, 208; (—conglomerate) 178, 179, 180, 182, 239; (—grit) 181, 211
Shelf 488
Shelve district 160; (—region) 164
Sherwood 213

INDEX OF GENERA AND SPECIES

A

Acervularia davidsoni 266
Acrioceras 410
Acrodus 371
Acrolepis lotzi 346
Acrothele sagittalis 98
Aegoceras capricornis 375; (—jamesoni) 375; (—pettos) 375
Agnostus 92
Agraulos 92
Albertella 101
Alces latifrons 481
Alethopteris 347
Allorisma 321
Alnus kefersteini 443, 454, 455
Amaltheus spinatus 375; (—margaritatus) 375
Ammonites affinis 380; (—psilonotus) 376; (—turneri) 376
Ampullina sp. 444
Ancyloceras 410
Annularia 347
Anisoceras notabile 423
Anomozamites 385
Anthracoceras 321
Anthrapalæmon 346
Araucarites alpinus 370
Arca diluvia 456
Arcestes gigantogaleatus 372; (—intuslabiatus) 372
Archæopteryx 383
Archæoscyphia 142
Arietes 376; (—obtusus) 376
Arthrophycus 180; (—harlani) 178; (——alleghoniensis) 178
Asaphus bröggeri 128; (—lepidurus) 128; (—raniceps) 128
Asperdites 359
Aspidoceras perarmatum 379
Asplenium 385
Astarte 408, 424; (—borealis) 478, 479
Athyris trigonalis 307
Aucella 386, 394, 397, 410, 411, 414, 415; (—crassicollis) 386; (—volgensis) 336
Aulacostephanus eudoxus 379; (—pseudomutabilis) 379
Avicula contorta 361, 372, 373, 376; (—speciosa) 373

B

Baculites 420, 421
Baiera schencki 365
Bakevellia antiqua 323; (—ceratophaga) 323, 325; (—tumida) 323
Baluchitherium 63, 441, 443, 496; (—

grangeri) 441
Bathyuriscus rotundatus 102
Beaumontia eckersleyi 70
Belemnitella americana 430; (—vesicularis) 424
Belemnites 408, 411; (—paxillosus)375
Beltina 70; (—danai) 69
Betula nana 433
Bison bonanus 481
Bothryolepis 257
Brachyphyllum 409
Brontosaurus excelsus 389
Brontotherium platyceras 441
Burgessia bella Walcott 103
Buthotrephis 204;
Buxenum grœnlandicum 460

C

Calamites 323, 347
Callavia 81, 84, 85, 86, 95; (—bröggeri (Walc.)) 80, 85; (—crosbyi (Walcott) Grabau) 81
Callipteris 325, 347
Callognathus serratus 253
Calyptræa sinensis 465
Camarocrinus 207
Camarophoria humbletonensis 323; (—schlotheimi) 323
Camarotœchia 209
Caninia 270; (—cylindrica) 271
Cardinia listeri 376
Canis lupis 430; (—vulpus) 481
Capitosaurus 349
Cardiopteris polymorpha 291
Cardita planicosta 444
Cardium angustatum 460; (—grœnlandicum) 456, 460, 478
Carpinus grandis 443, 454
Centropleura 97; (—loveni) 98; (—vermontana) 98
Cephalaspis 216
Ceratiocaris 204
Ceratites 357; (—compressus 353; (—dorsoplanus) 358, (—evolutus) 358; (—intermedius) 358; (—nodosus) 336, 353; (—schmidti) 358;(—semipartitus) 353; (—spinosus) 353; (—subrobustus) 363; (—trinodosus) 355, 363
Ceratodus 359
Chondrites 204
Chonetes papilionaceous 306
Choristites haueri 376
Choristoceras marshi 373
Cidaris florigemma 279; (—olifex) 376
Cidaroida holastis 438
Cistecephalus 348

EXPLANATION OF PLATE II

Palæozoic Centers of Glaciation (South Pole)

(The modern latitudes and longitudes are indicated in each country)

No.	Pulsation Period:	Location on modern map:

I. *Sinian:*   *Egypt* (Lat. 32° N, Long. 31° E)
     12. Kuruk-Tag; 13. Geology Fjord, Greenland;
     14. Veranger Fjord, Finland.

II. *Tacorian or Georgian and*   *Touareg* (Lat. 24° N, Long. 6° E)
  Cambrian or Acadian Period
     21. Petra, Syrabia (10); 22. East Sinai (11).

III. *Cambrovician and Skiddavian*   *Taodini* (Lat 22°31' N, Long. 4° W)
  Periods
     (*Ordovician* represented by movememt from III to IV)
     30. Straits of Belle Isle, between Newfoundland and Labrador;
     30-a. Corswell Point, Wigtownshire Scotland; 30-b. Carrick, Ayr-
     shire, Scotland; 31. Boston, Roxbury and Squantum; 32. Rideau,
     Thousand Islands; 33. Quebec City; 33-a. North-West Vermont
     (Swanton, etc.); 34. End of Gaspé Peninsula; 35. Melrose N. Y.;
     35-a. Levis. 36. Johnson City, Tenn.; 37. Rock mart, Georgia; 38.
     Scaumenac Bay; 39. St. John, New Brunswick. (All were closer to
     the ice-front before the Taconian folding.)

IV. *Silurian Period*   *Levant* (35° N, 38° E)

V. *Siluronian Period*   N. Rhodesia *Nyassa Ld.* (10° S, 33° E)
     41. Cape Town; 42. Pretoria; 43. Waterberg; 44. Matsap; 45. Ma-
     galiesberg; 46. Griquatown; 47. Orange River Valley (all in South
     Africa).

VI. *Devonian Period*   *Mozambique:* (15° S, 40° E)
     50. Irwin River; 51. Arthur River; 52. Mt. Marmion.
     53. Port Keats (50-53 all in West Australia).
     54. South Madagascar; 55. Talchir of Orissa, India; 56. Nagpur,
     India; 56-a. Junction of Penganga and Wardha Rivers; 59. Older
     Tillite of Adelaide, Australia.

VII. *Fengninian and*   *Porto Alegre, Brazil:* (30° S, 50° W)
  Visèmurian Periods
     61 & 62. Parana Basin, Brazil; 63. Fraile Muerto;
     64. Sierra de la Ventana; 65. Leoncita Encima;
     66. Jachal; 67. Sierra Chica de Zonda; 68. Campinas,
     Brazil; 69. Ponte Grossa, Parana, Brazil.

VIII. *Moscovian* ⎫   *E. Basuto Land* (29°30' S, 28°30' E)
    (Dwyka) ⎬ *Donbassian Period*
IX. *Donetzian* ⎭   *E. Antarctica* (80° S, 110° E)
     73. Cape District; 74. Bulowayo, Rhodesia; 75.
     Lubilache, Lualaba Valley; 76. Katanga, Belgian Congo.
     77. Falkland Islands; 79. Inman Valley, near Adelaide, Upper Till.

X. *Uralinskian Period*   *Musgrave Range, Australia* (26° S, 132° E)
     81. Bowen; 82. Yatton; 83. Rockhampton; 84. Springsure; 85. Kemp-
     sey; 86. Hunter River; 86-a. Rydal and Bathurst; 87. Tasmania;
     88. Bacchus Marsh, near Melbourne, Victoria.

XI. *Permian Period*   *Burma* (29° N, 95° E)
     91. Salt Range; 92. Blaini Boulder Bed, Simla, Himalayas (most
     probably pre-Cambrian); 93. Orissa District; 94 Rajputana, Aravalli
     Hills; 95. Rajmahal; 97. Satpura Hills; 98. Ranigang (all in India).

XII. *Post Permian*   *Aravalli Hills, India* (25° N, 74° E)
  Metazoic Interval)

XIII. *Triassic*   *Antarctica* (90° S, Long. 0°)

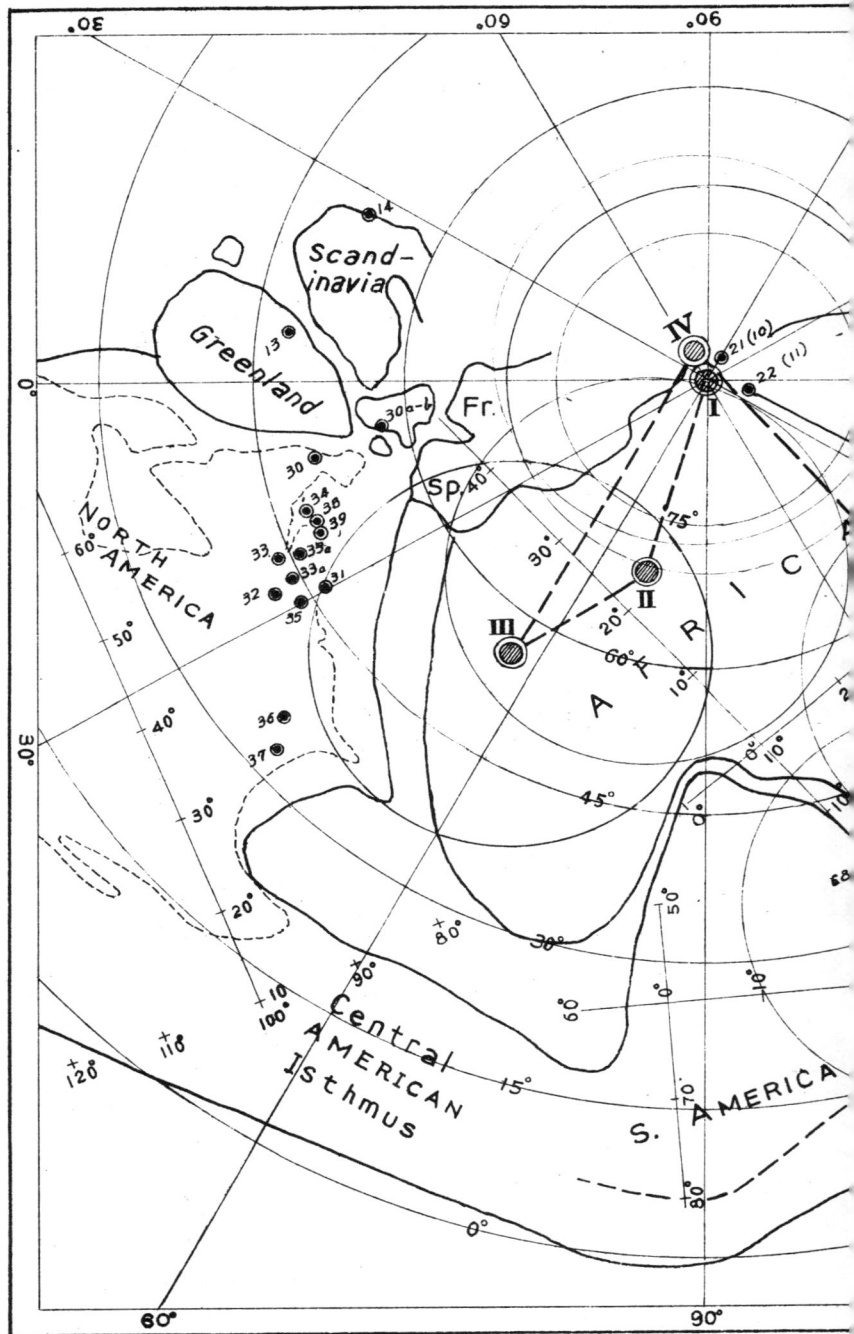

14

Scand-
inavia

13

Greenland

IV 21(10)
22(11)
I

30a-b Fr.

North 0°

Sp. 40°

30

34
38
39
33 35
32 33 31
35

30°

II

III 60°L

NORTH
60° AMERICA

50°

75°

30°

20°

10°

A R I C

40°

36
37

45°

30°

30°

20°

80°

30°

50°

60°

0°

+10
100°

+90°

Central
AMERICA
Isthmus

+118

+120°

15°

S. AMERICA

70°

80°

0°

30°

60° 90°

Pl. II

TIBET

Himalayan G.

Cathaysian Geosyncl.

INDIA

N.Z.

N.Z.

AUSTRALIA

Antarctica

Samfrau G.

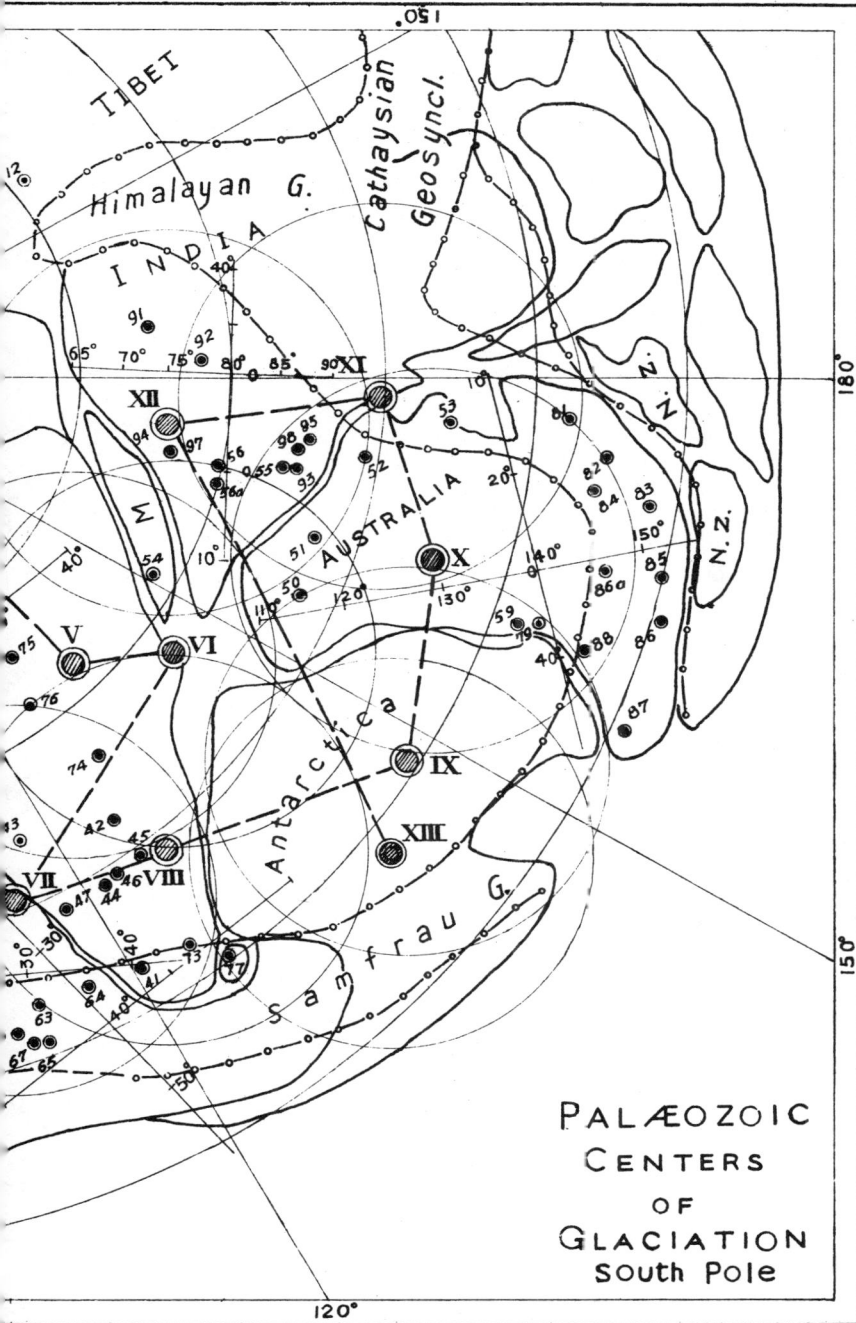

PALÆOZOIC
CENTERS
OF
GLACIATION
South Pole

12 f 40° 65° 70° 75° 80° 85° 90° XI 10
91 92 XII 94 97 56 96 95 55 93 52 53
54 51 AUSTRALIA 20° 82 84 83
40° 10° 50 120 130° X 140° 150° 86a 85
V 75 VI 110 59 119 88 86
76 40 87
74 IX
43 42 45 XIII
VII 46 VIII
47 44
50 40 73 77
30 41
63 64
67 65

180°

180°

150°

150°

120°

Plate III

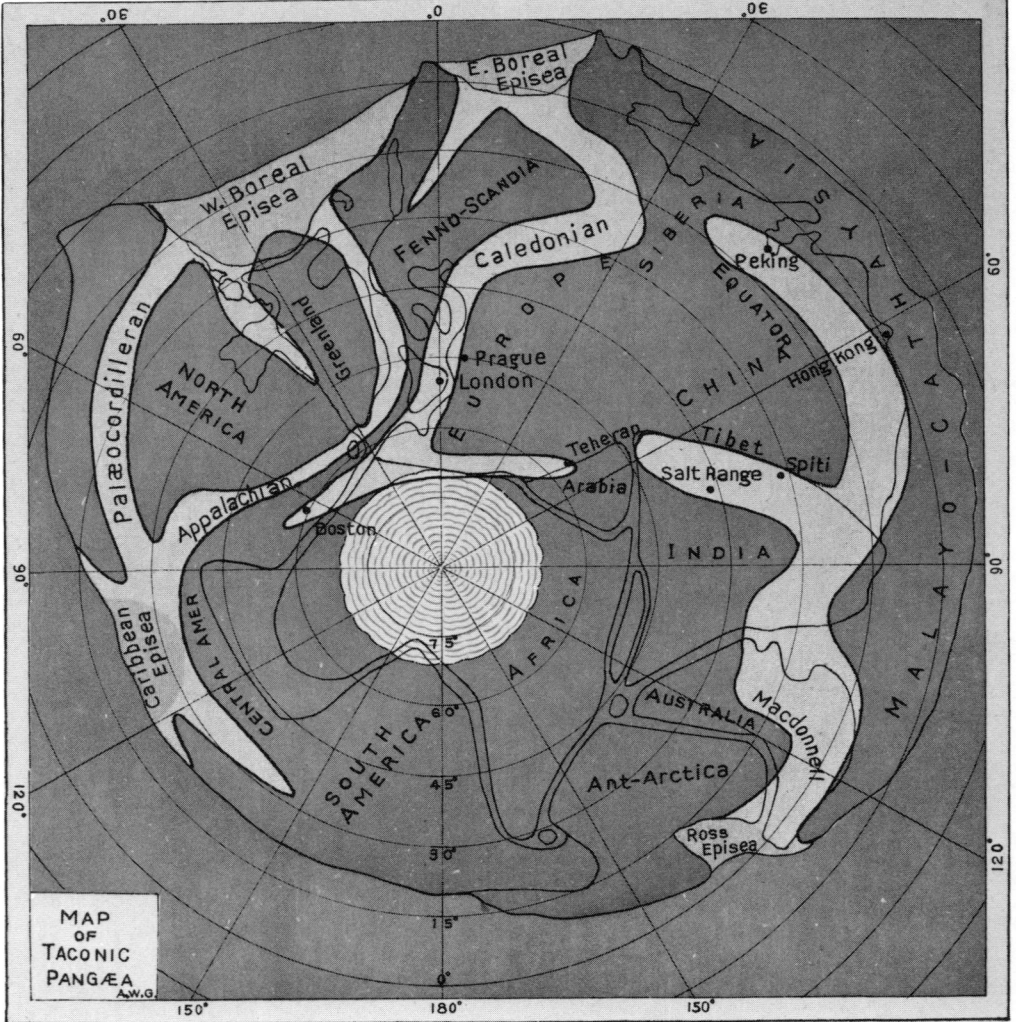

MAP
OF
TACONIC
PANGÆA
A.W.G.

Plate IV

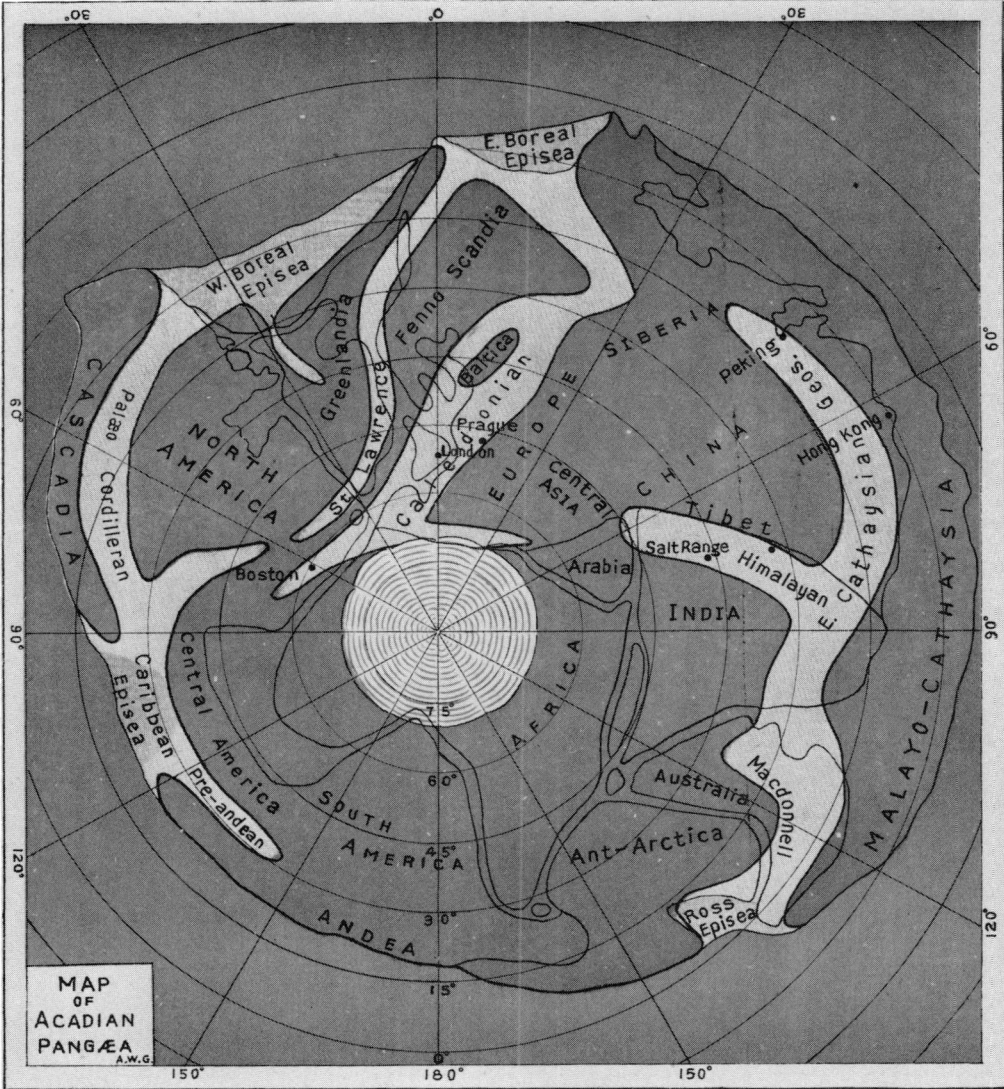

MAP
OF
ACADIAN
PANGÆA
A.W.G.

Plate V

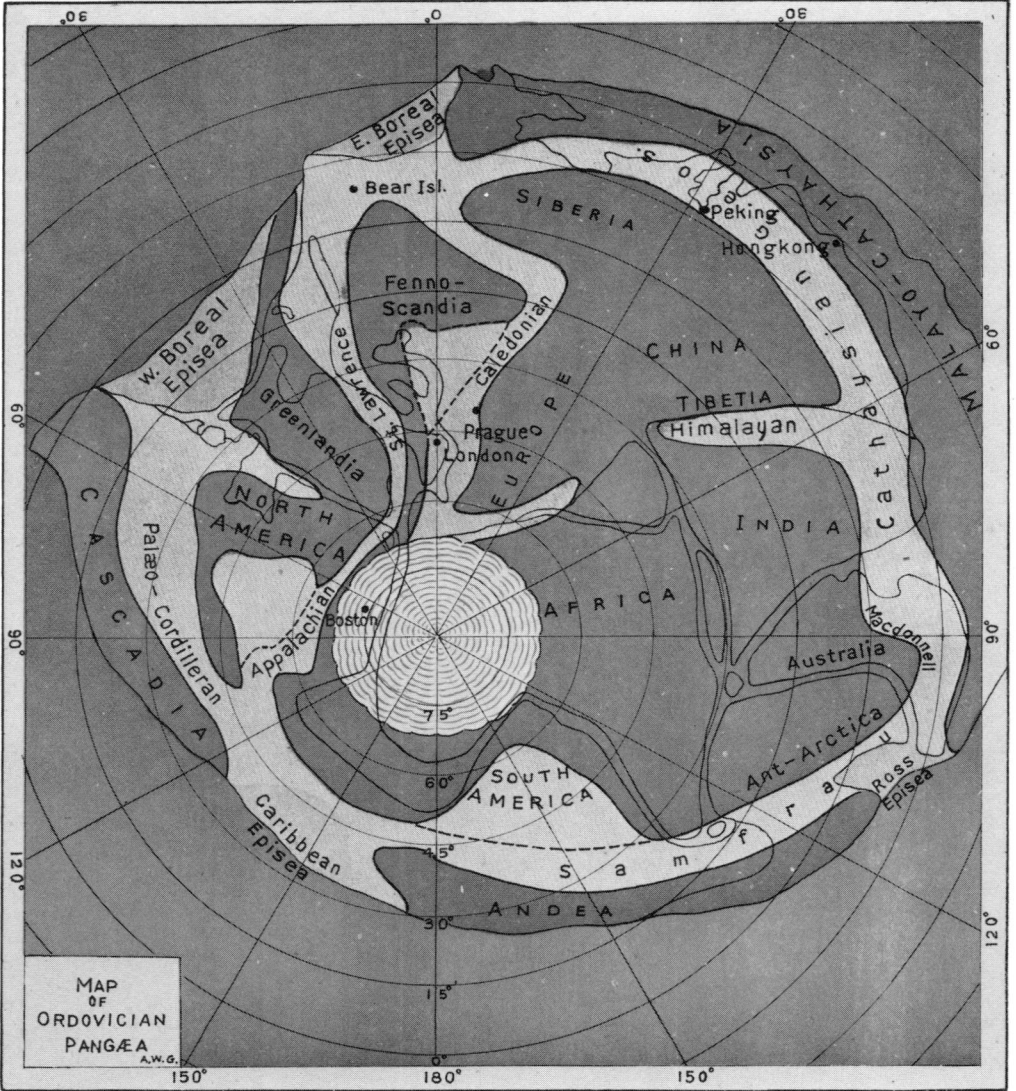

MAP
OF
ORDOVICIAN
PANGÆA
A.W.G.

Plate VI

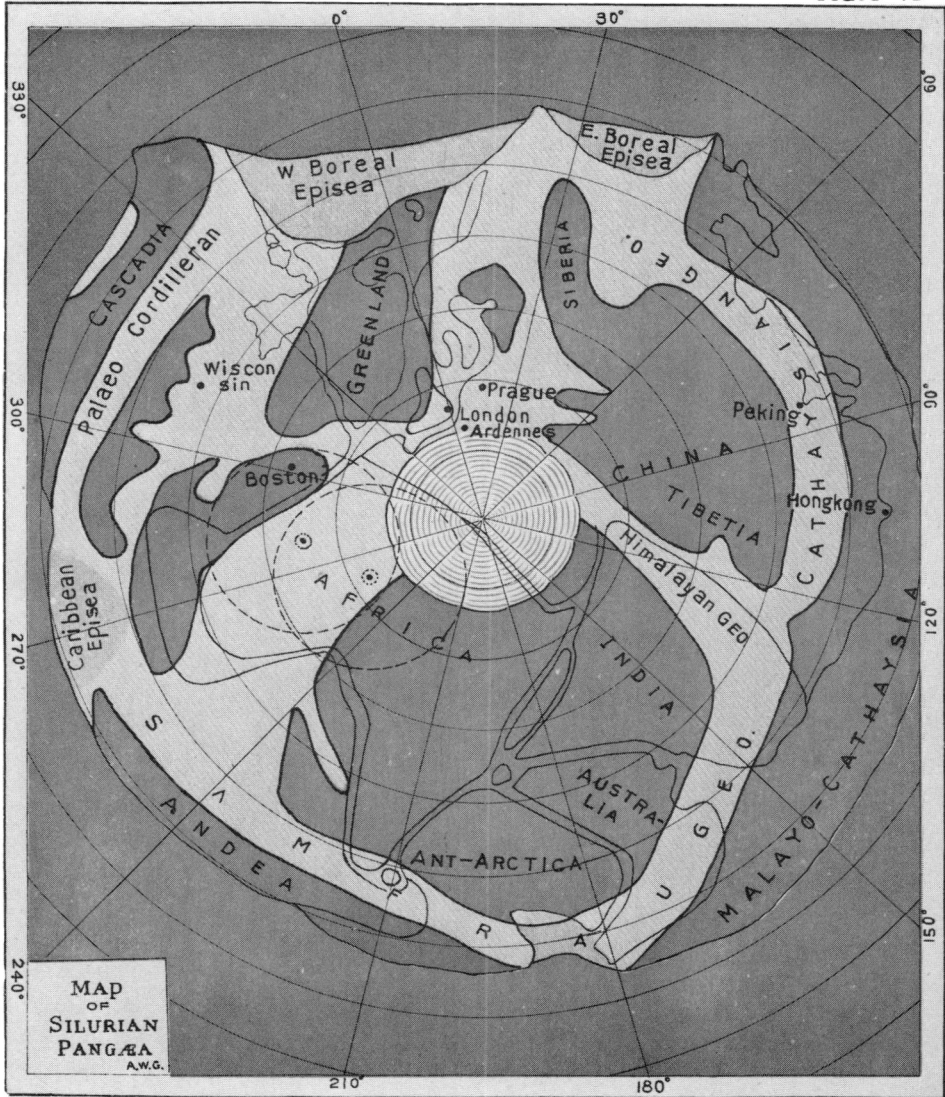

MAP
OF
SILURIAN
PANGÆA
A.W.G.

Plate VII

Map of Siluronian Pangæa

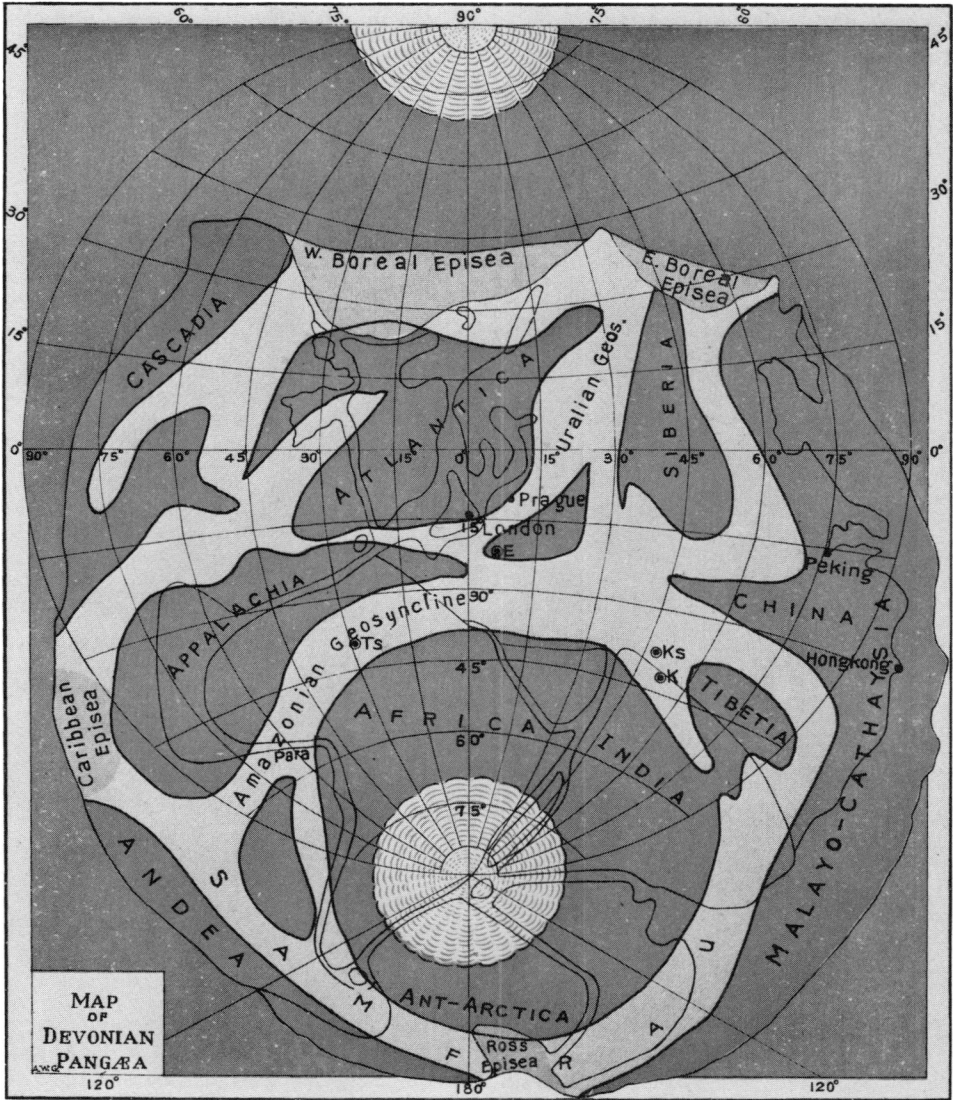

Plate VIII

MAP
OF
DEVONIAN
PANGÆA

Plate IX

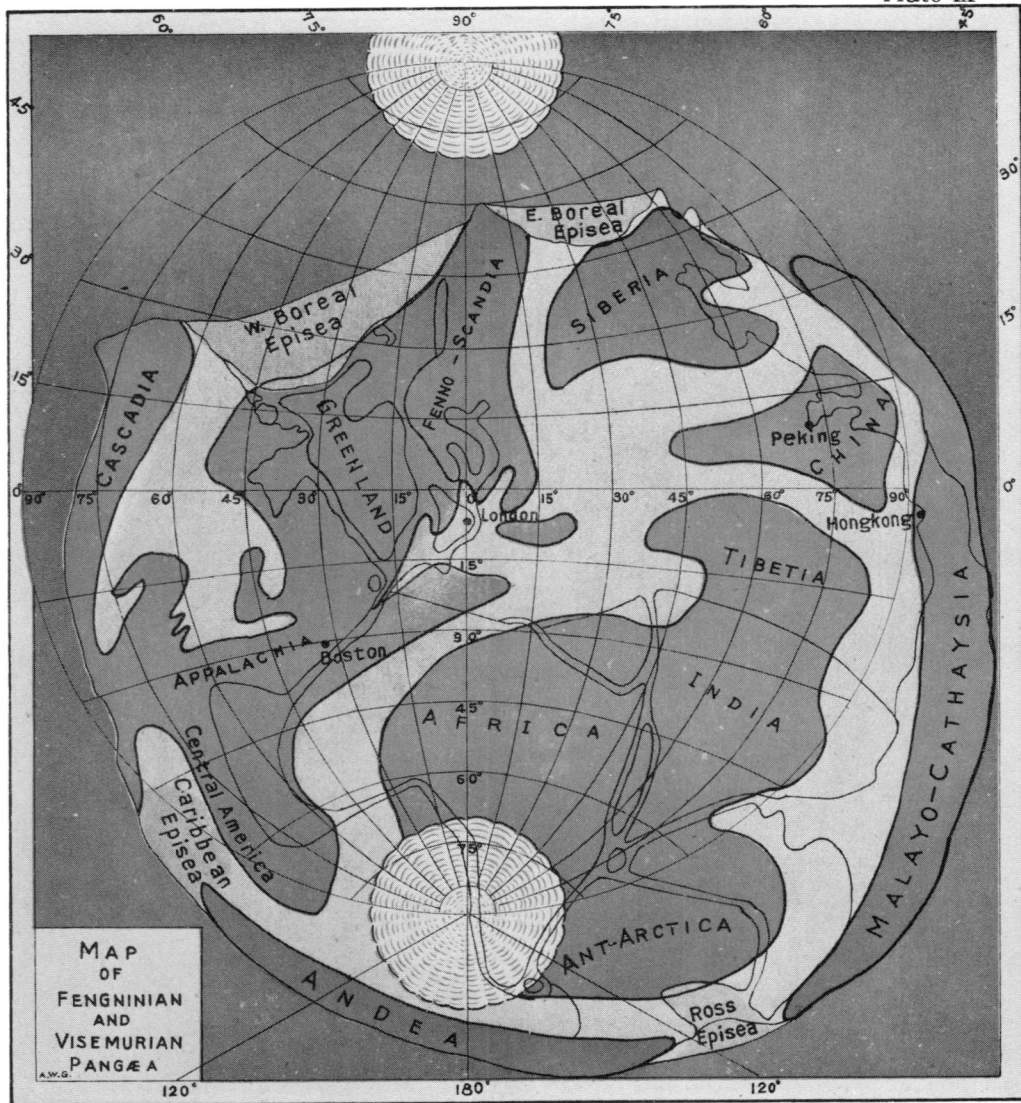

MAP
OF
FENGNINIAN
AND
VISEMURIAN
PANGÆA

Plate X

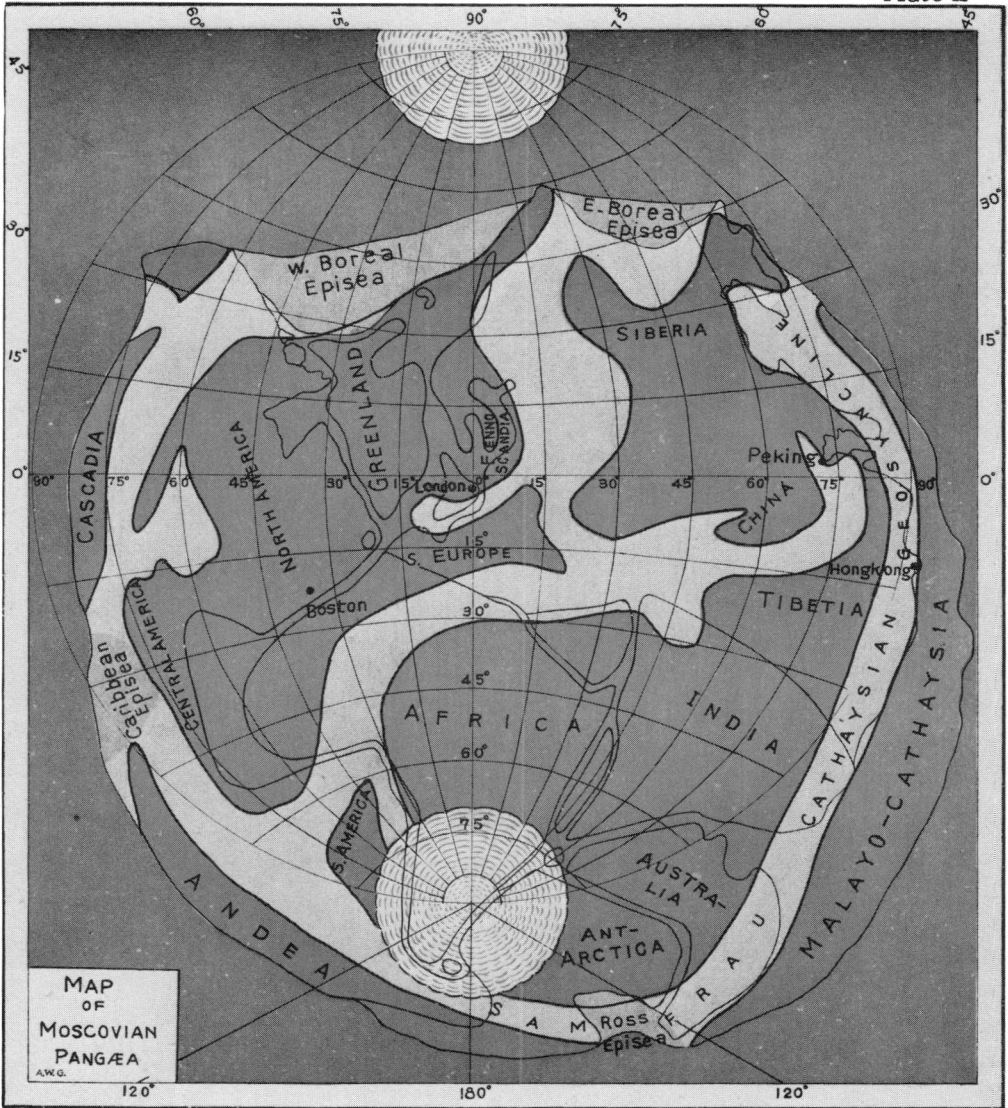

MAP
OF
MOSCOVIAN
PANGÆA

A.W.G.

Plate XI

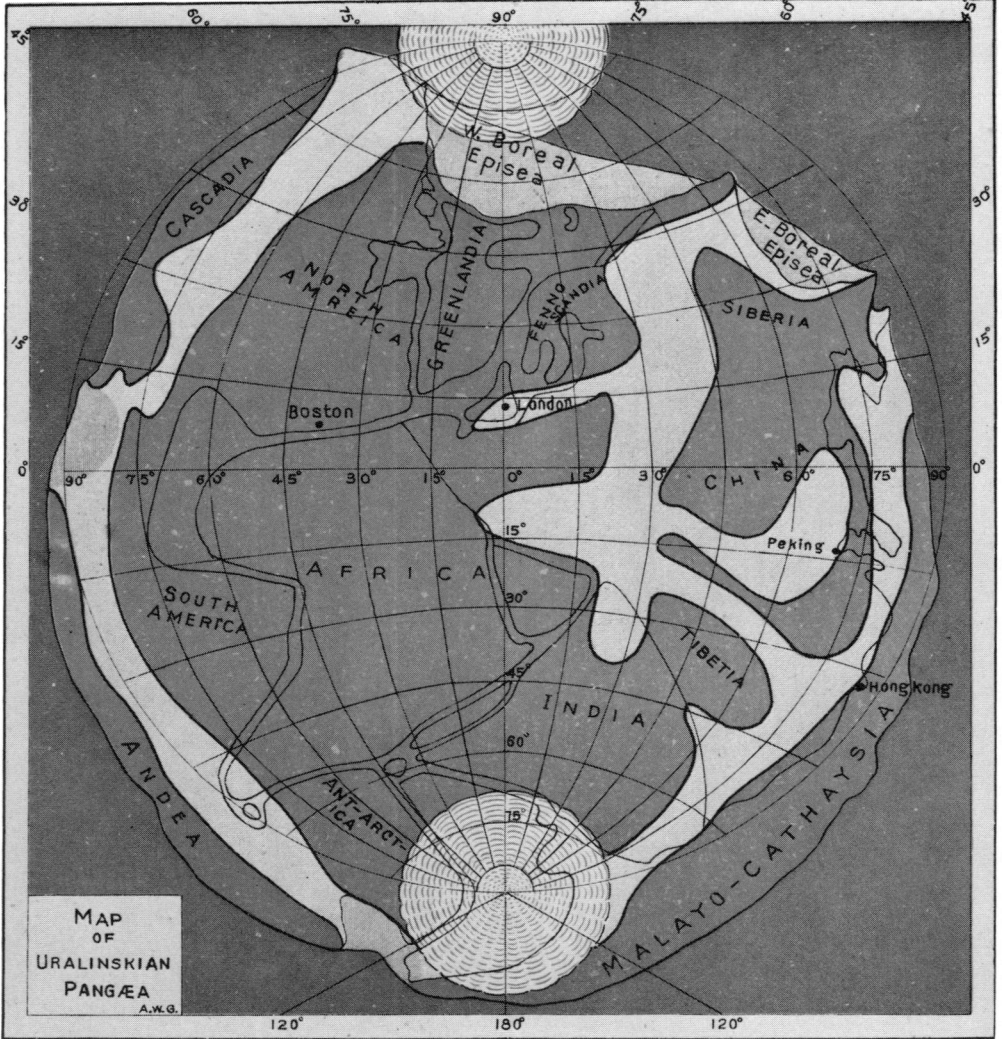

MAP
OF
URALINSKIAN
PANGÆA
A.W.G.

Plate XII

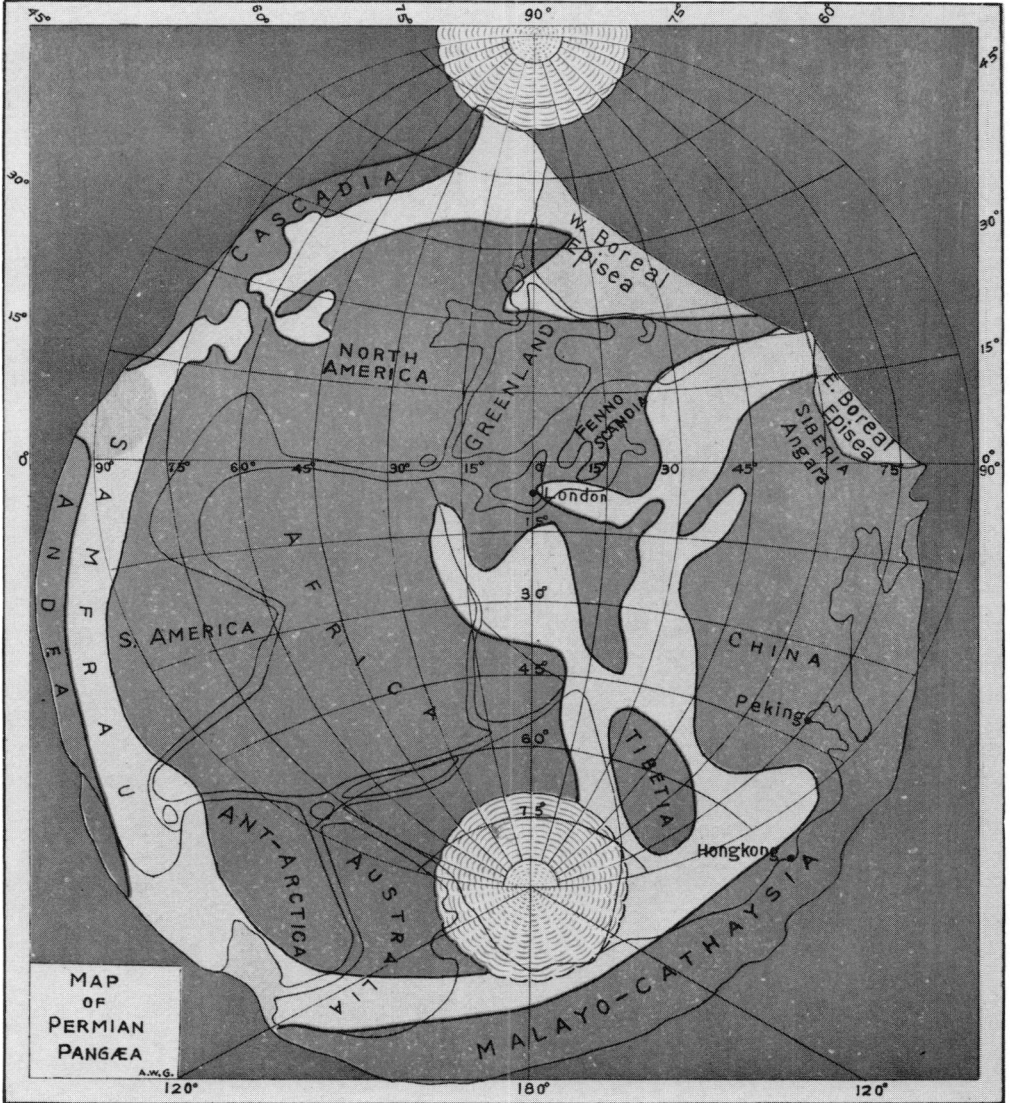

MAP
OF
PERMIAN
PANGÆA

Plate XIII

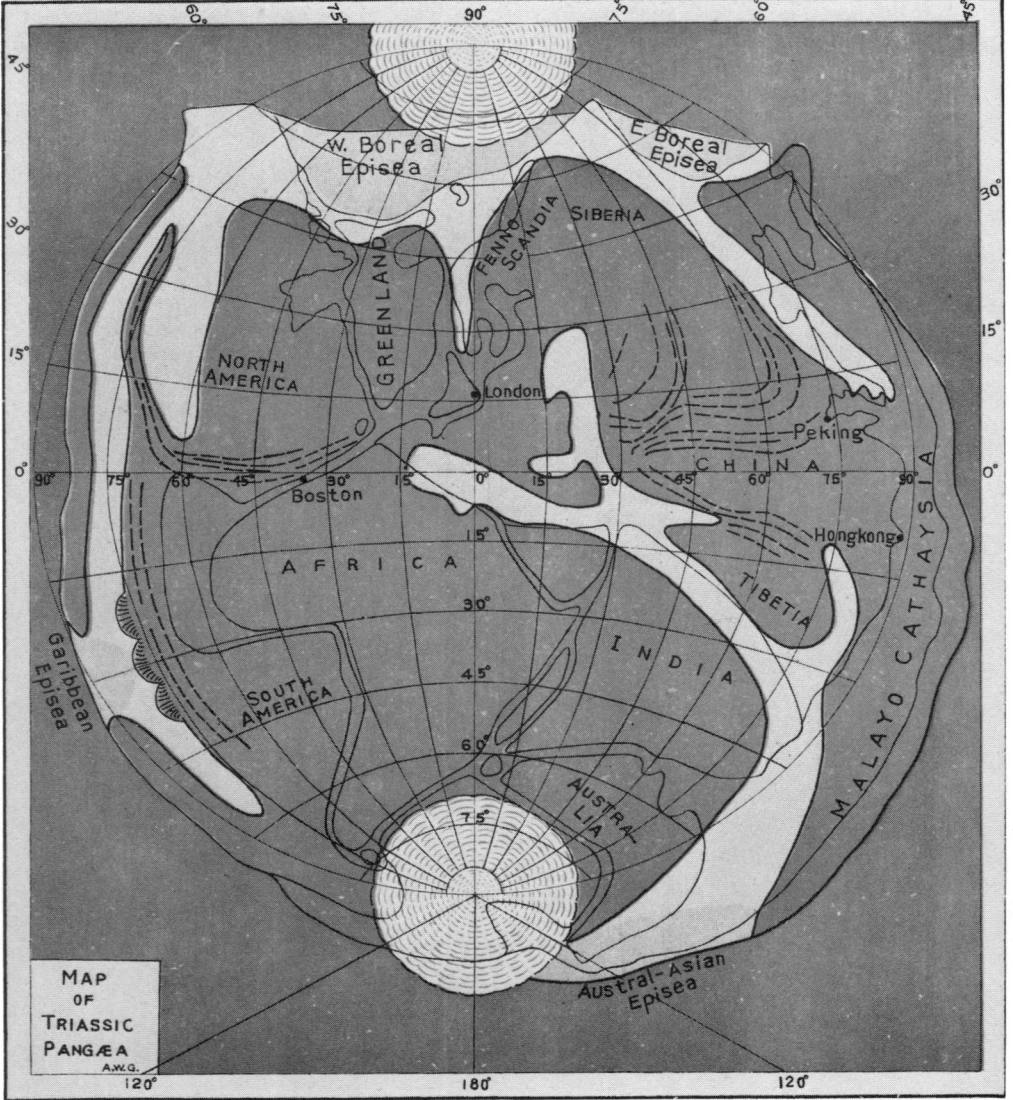

MAP
OF
TRIASSIC
PANGÆA
A.W.G.

Plate XIV

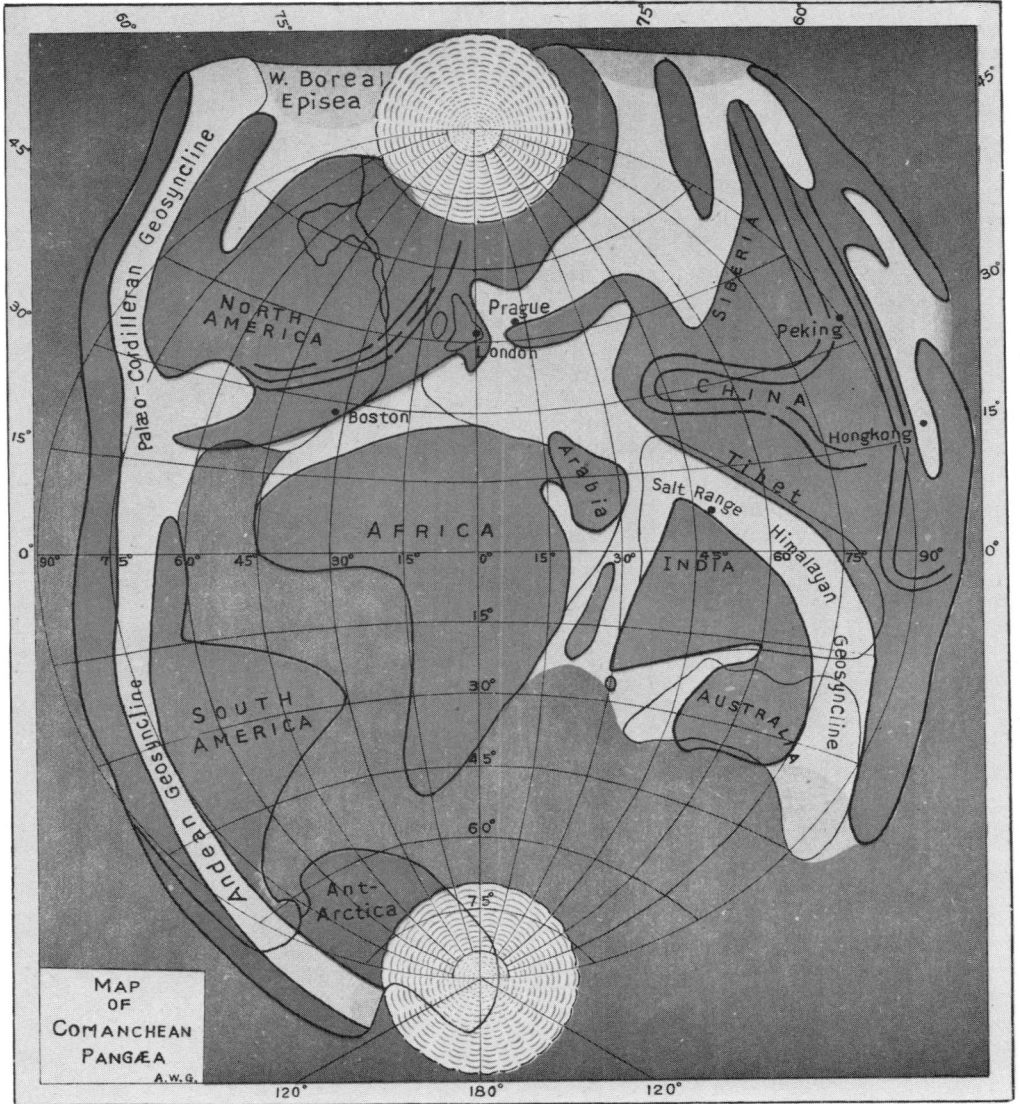

MAP
OF
COMANCHEAN
PANGÆA

A.W.G.

W. Boreal Episea

NORTH AMERICA

SIBERIA

Prague
London
Boston

Peking

CHINA

Hongkong

Tibet

Arabia

Salt Range

AFRICA

INDIA

Himalayan Geosyncline

Palæo-Cordilleran Geosyncline

Andean Geosyncline

SOUTH AMERICA

AUSTRALIA

Ant-Arctica

PLATE XV

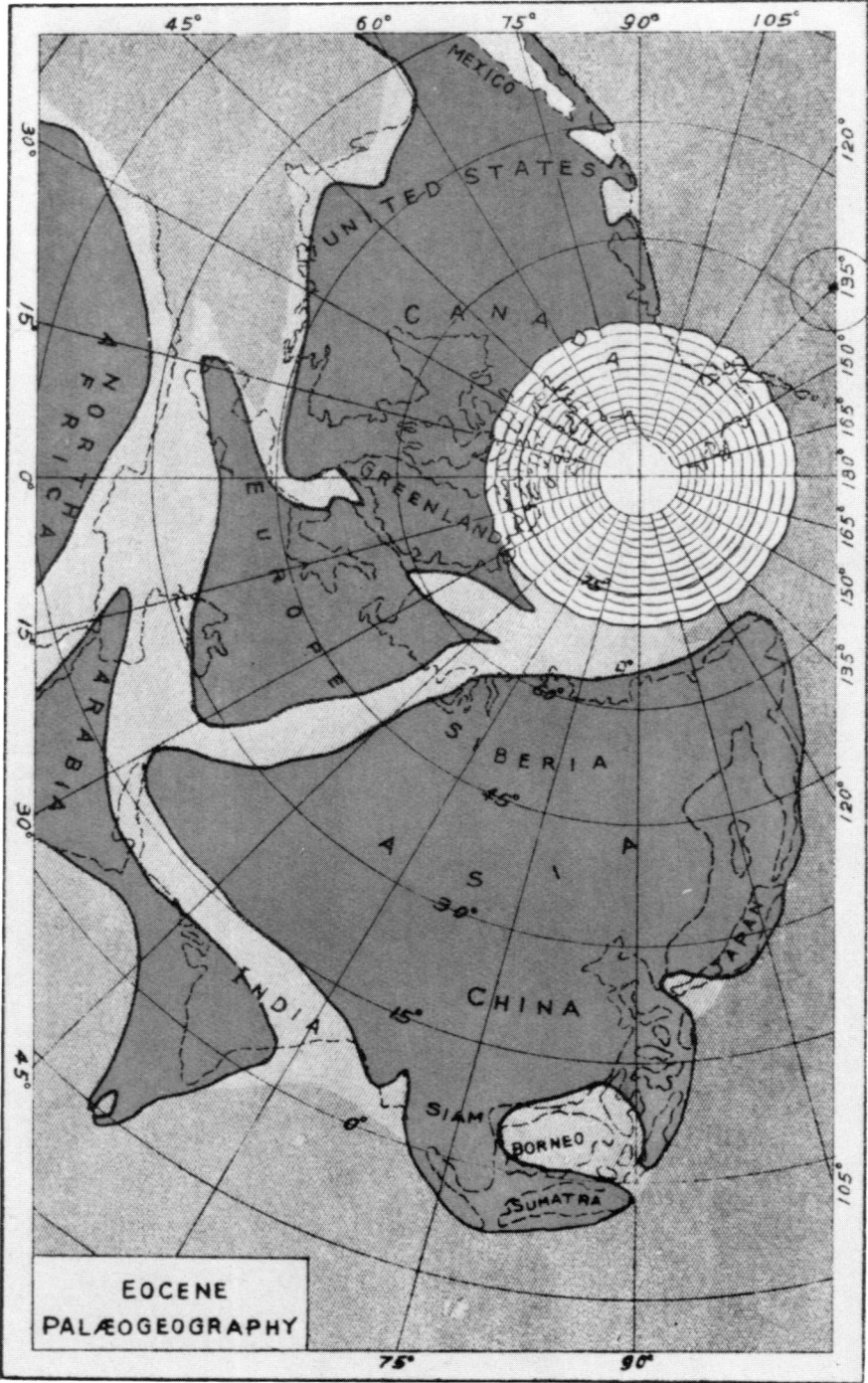

EOCENE
PALÆOGEOGRAPHY

PLATE XVI

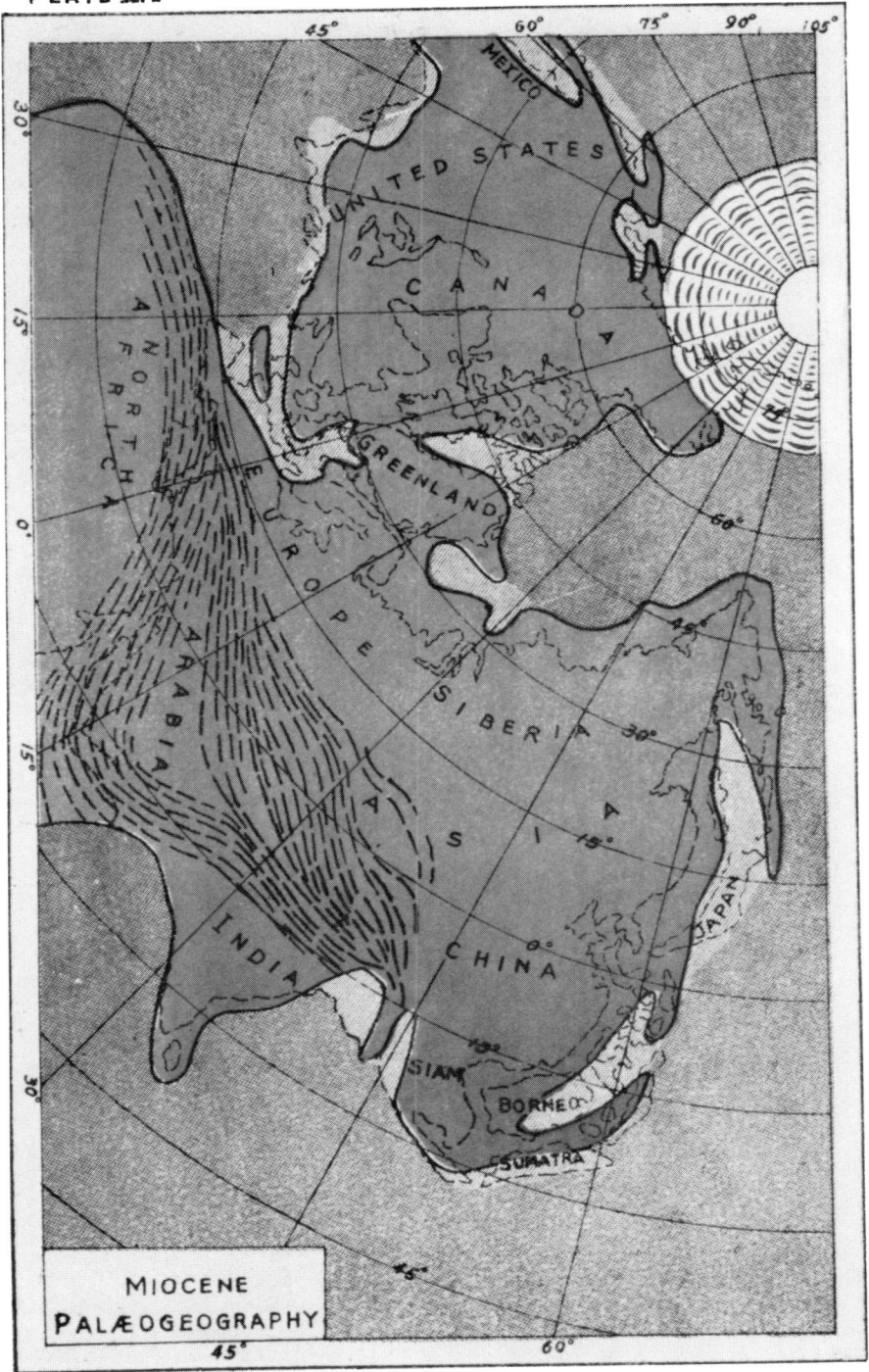

MIOCENE
PALÆOGEOGRAPHY

PLATE XVII

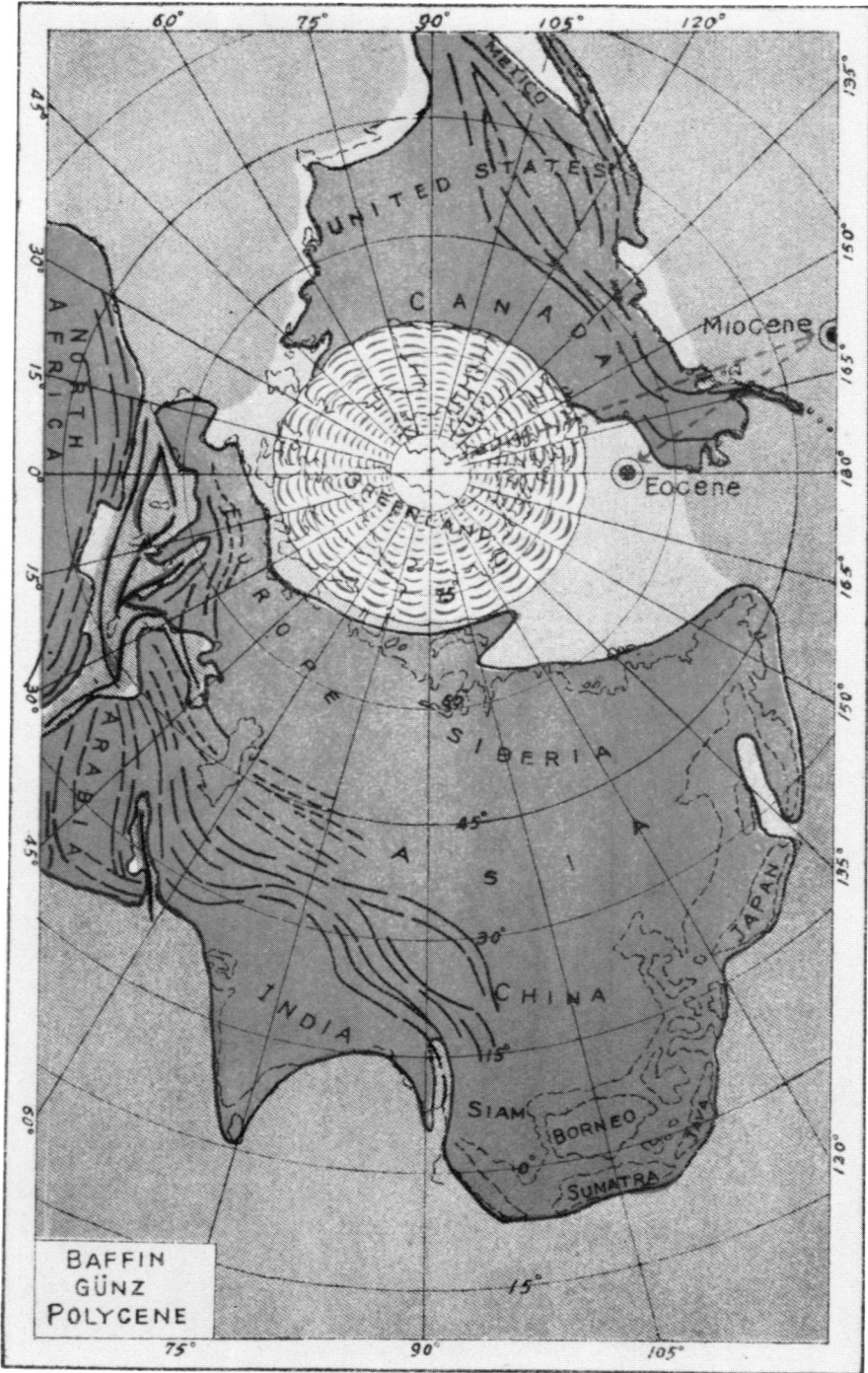

PLATE XVIII

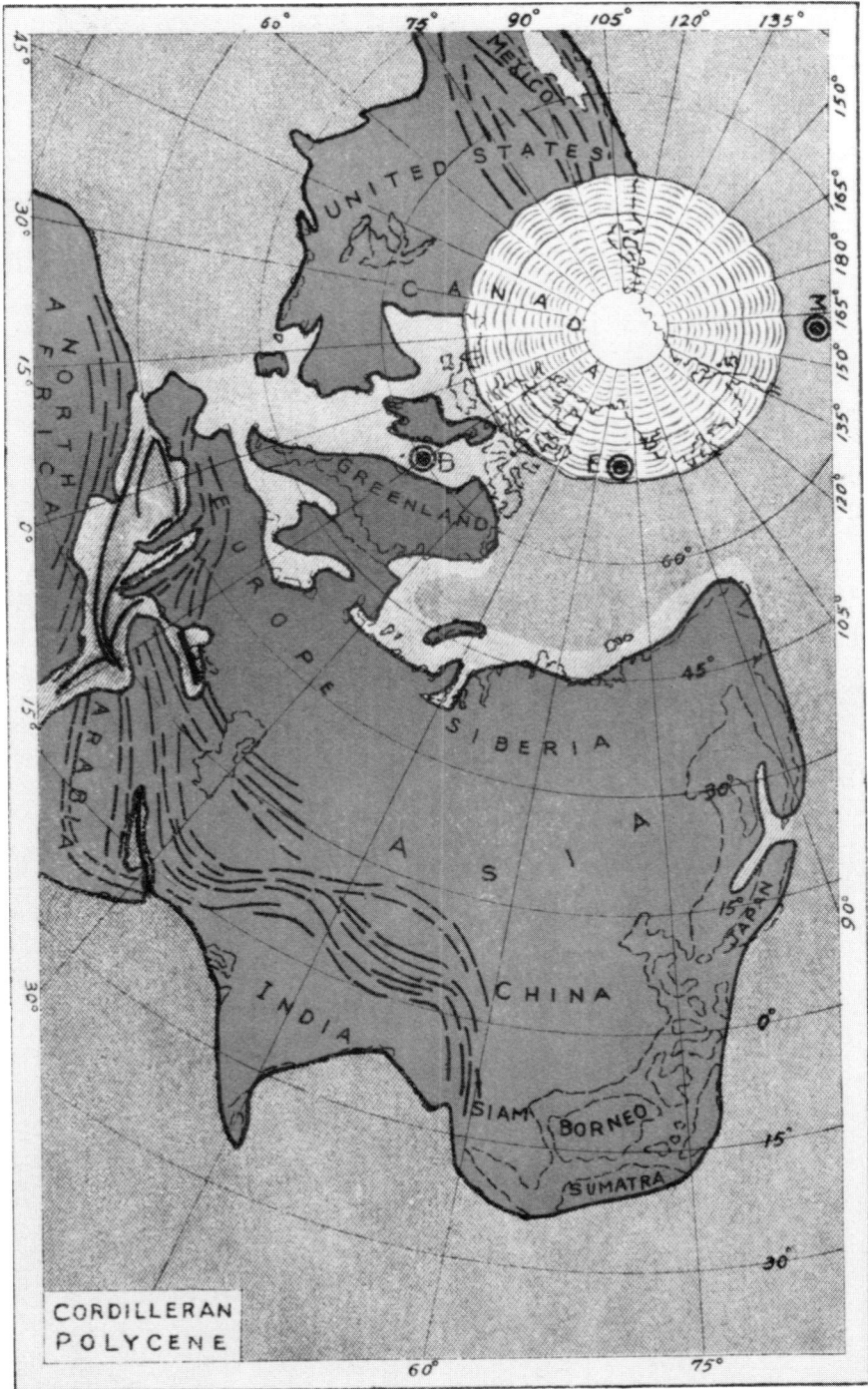

CORDILLERAN
POLYCENE

PLATE XIX

KEEWATIN
POLYGENE

PLATE XX

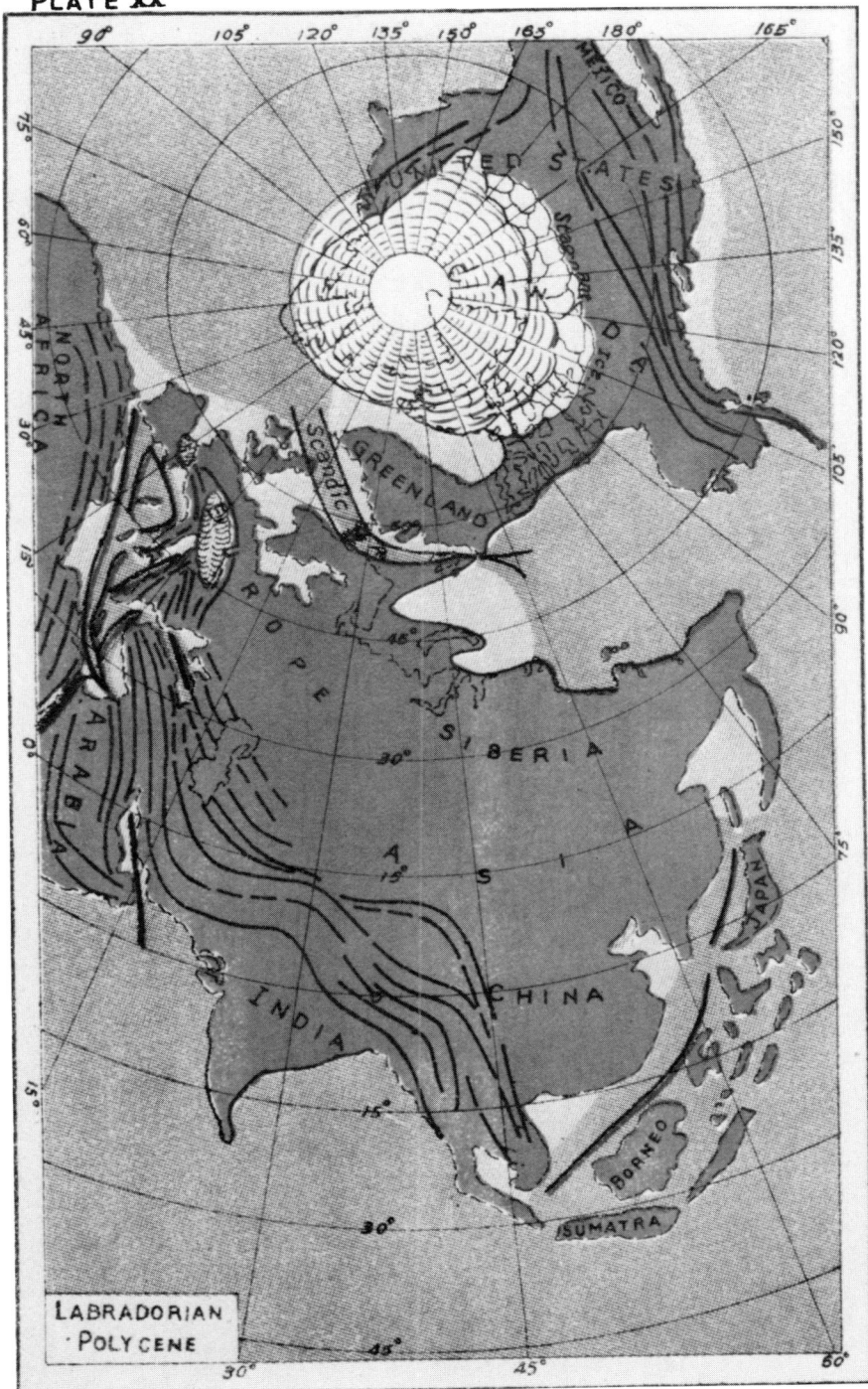

LABRADORIAN
POLYGENE

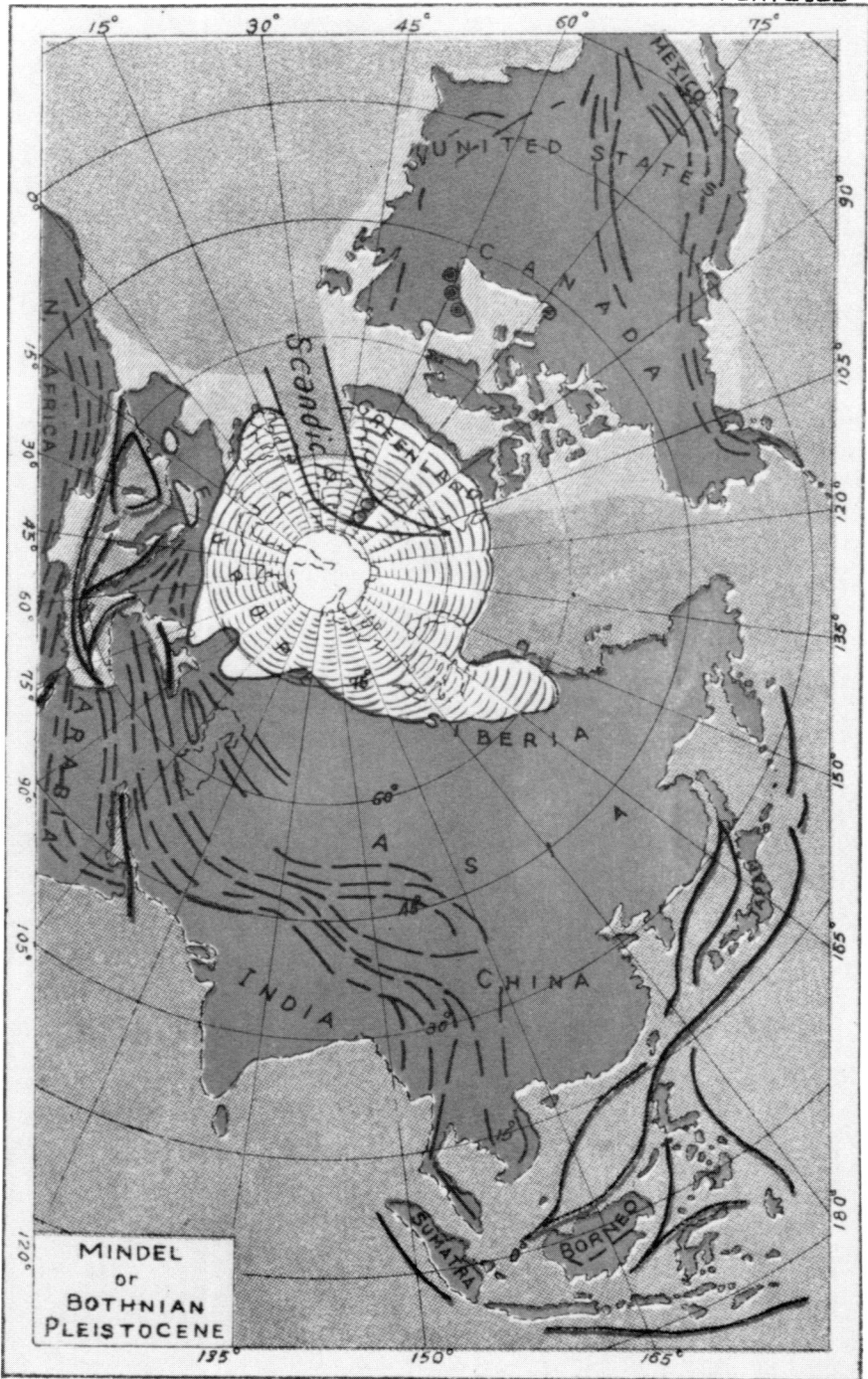

PLATE XXI

MINDEL
or
BOTHNIAN
PLEISTOCENE

PLATE XXII

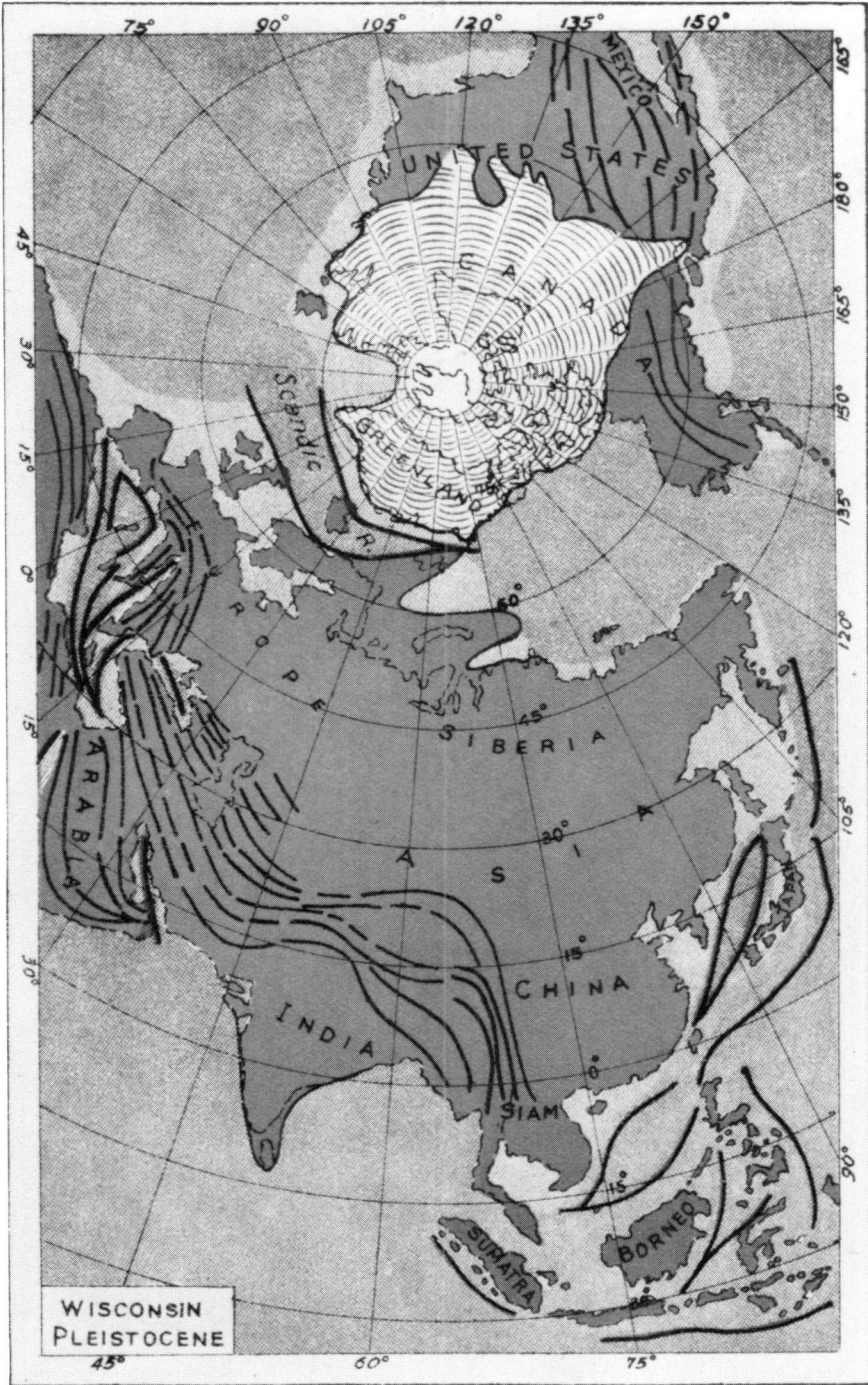

WISCONSIN
PLEISTOCENE

PLATE XXIII

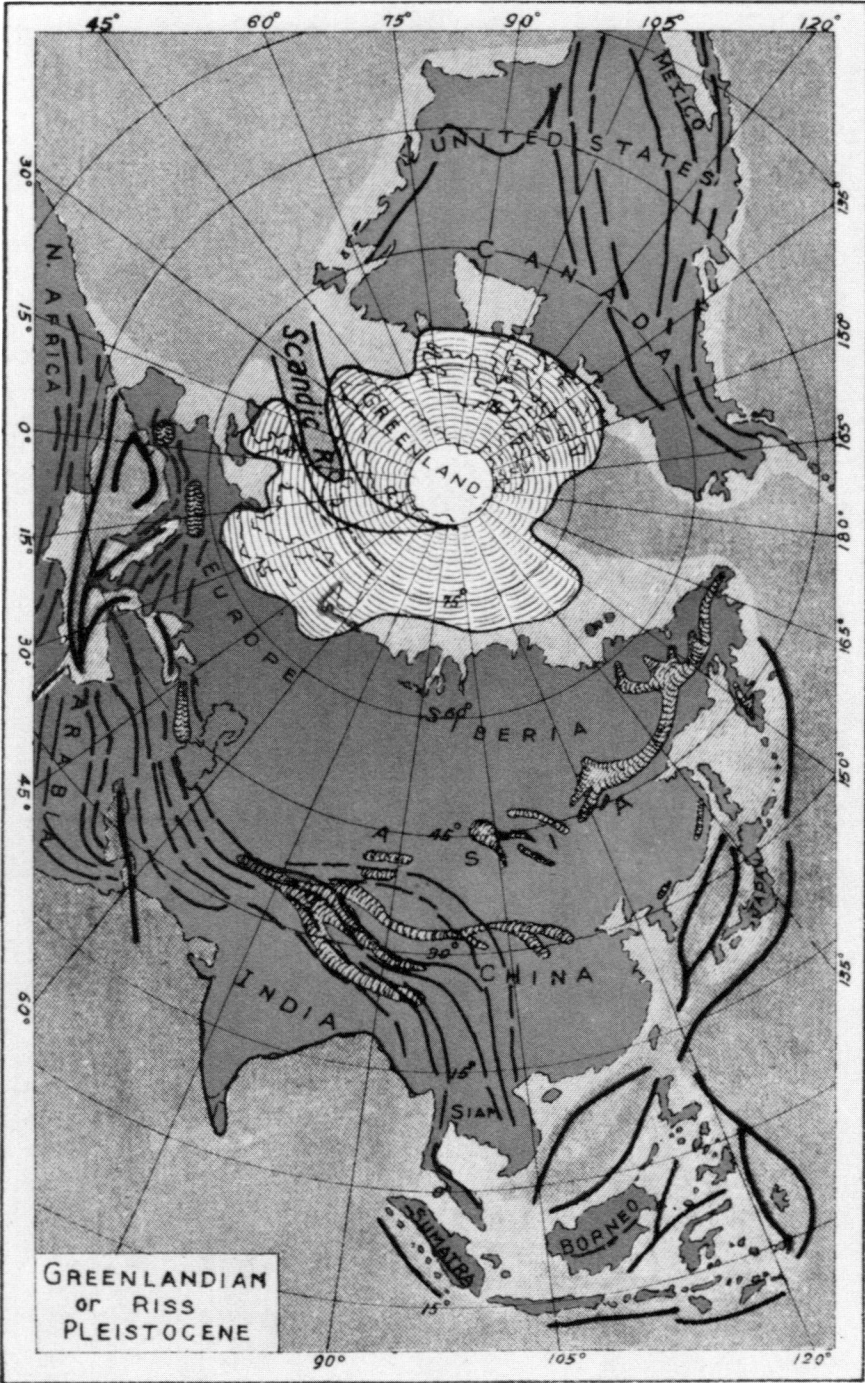

GREENLANDIAN
or RISS
PLEISTOCENE

PLATE XXIV

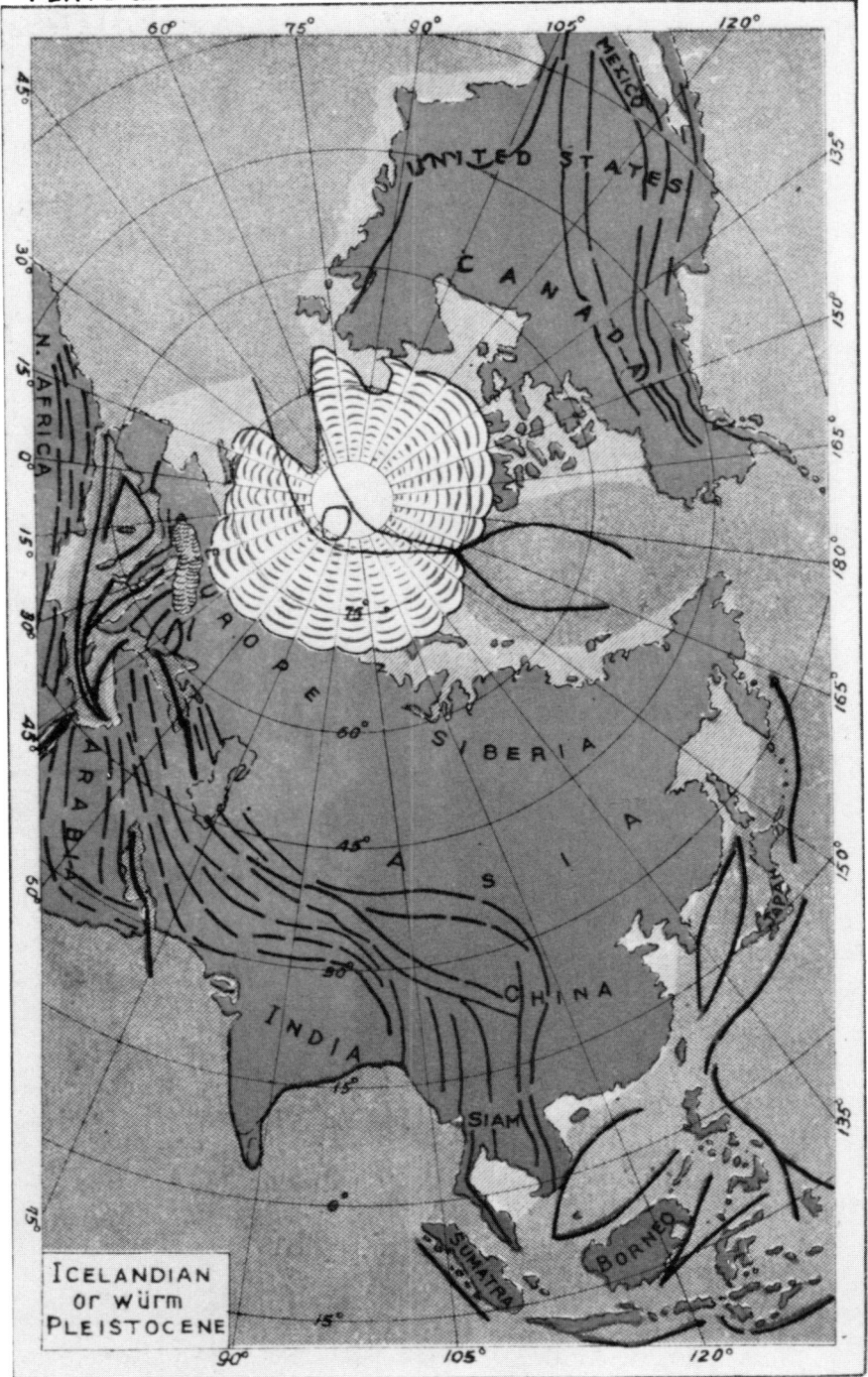

ICELANDIAN
or WÜRM
PLEISTOCENE

PLATE XXV

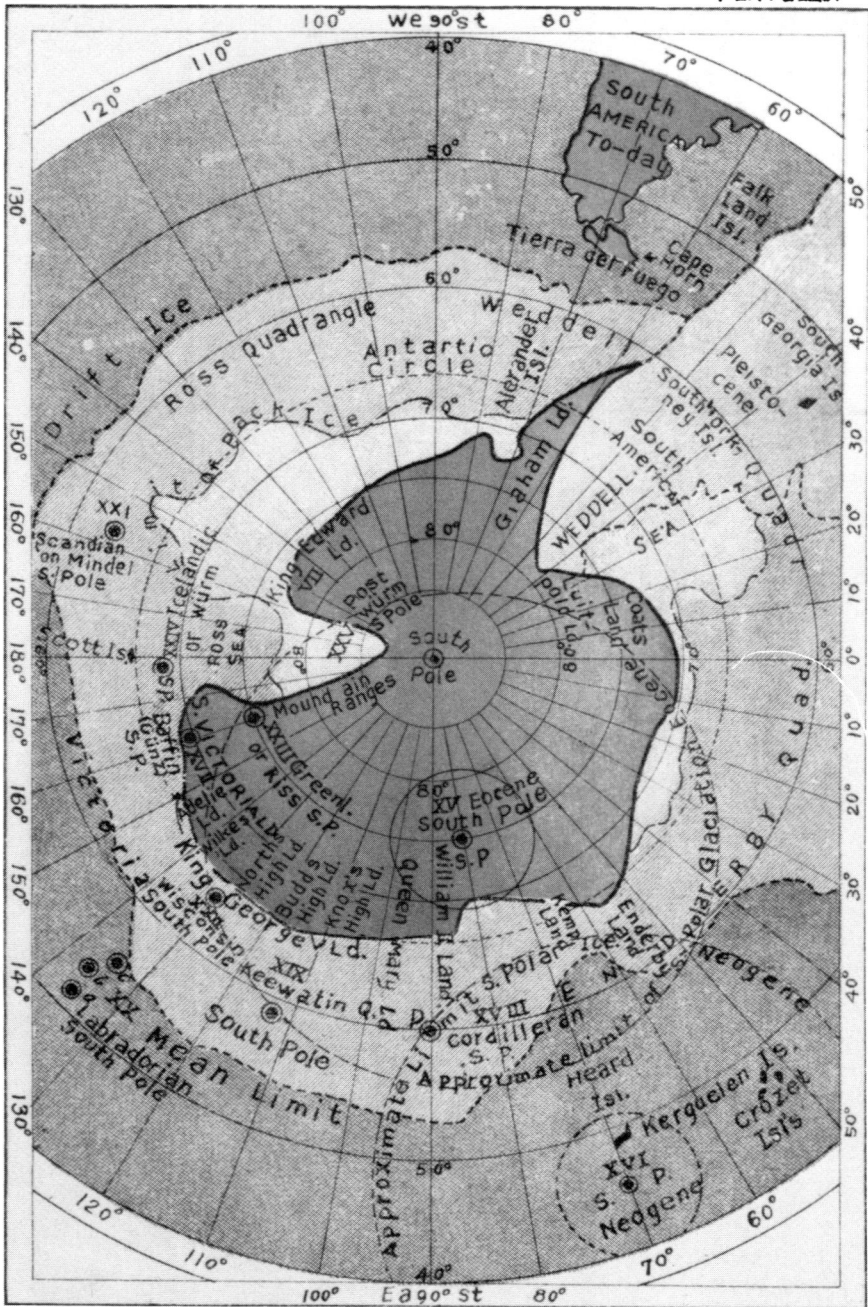

MAP OF SOUTH POLAR MIGRATION FROM EOCENE TO MODERN TIME